Antimicrobial Resistance and Food Safety

Antimicrobial Resistance and Food Safety

Methods and Techniques

Editors

Chin-Yi Chen
US Department of Agriculture, Agricultural Research Service, Wyndmoor, PA, USA

Xianghe Yan
US Department of Agriculture, Agricultural Research Service, Wyndmoor, PA, USA

Charlene R. Jackson
US Department of Agriculture, Agricultural Research Service, Athens, GA, USA

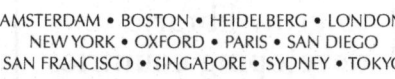

AMSTERDAM • BOSTON • HEIDELBERG • LONDON
NEW YORK • OXFORD • PARIS • SAN DIEGO
SAN FRANCISCO • SINGAPORE • SYDNEY • TOKYO

Academic Press is an imprint of Elsevier

ELSEVIER

Notices
Knowledge and best practice in this field are constantly changing. As new research and
experience broaden our understanding, changes in research methods, professional practices, or
medical treatment may become necessary.

Practitioners and researchers must always rely on their own experience and knowledge in
evaluating and using any information, methods, compounds, or experiments described herein.
In using such information or methods they should be mindful of their own safety and the safety
of others, including parties for whom they have a professional responsibility.

To the fullest extent of the law, neither the Publisher nor the authors, contributors, or editors,
assume any liability for any injury and/or damage to persons or property as a matter of
products liability, negligence or otherwise, or from any use or operation of any methods,
products, instructions, or ideas contained in the material herein.

ISBN: 978-0-12-801214-7

British Library Cataloguing-in-Publication Data
A catalogue record for this book is available from the British Library

Library of Congress Cataloging-in-Publication Data
A catalog record for this book is available from the Library of Congress

For information on all Academic Press publications
visit our website at http://store.elsevier.com/

Typeset by MPS Limited, Chennai, India
www.adi-mps.com

Printed and bound in the USA

Working together
to grow libraries in
developing countries

www.elsevier.com • www.bookaid.org

Contents

11. Methods for the Detection of Antimicrobial Resistance
and the Characterization of *Staphylococcus aureus*
Isolates from Food-Producing Animals and Food of
Animal Origin

Kristina Kadlec, Sarah Wendlandt, Andrea T. Feßler and Stefan Schwarz

12. Non-Phenotypic Tests to Detect and Characterize
Antibiotic Resistance Mechanisms in *Enterobacteriaceae*

*Agnese Lupo, Krisztina M. Papp-Wallace, Robert A. Bonomo and
Andrea Endimiani*

16. Antibiotic-Resistant Bacteria and Resistance Genes in the Water–Food Nexus of the Agricultural Environment

Pei-Ying Hong

17. Development and Application of Novel Antimicrobials in Food and Food Processing

Yangjin Jung and Karl R. Matthews

List of Contributors

María Ángeles Argudín Laboratoire de Référence MRSA-Staphylocoques, Department of Microbiology, Hôpital Erasme, Brussels, Belgium

Craig Baker-Austin Centre for Environment Fisheries and Aquaculture Science, Weymouth, Dorset, UK

Clara Ballesté-Delpierre ISGlobal, Barcelona Ctr. Int. Health Res. (CRESIB), Hospital Clínic – Universitat de Barcelona, Barcelona, Spain

Robert A. Bonomo Department of Pharmacology, Molecular Biology and Microbiology, Case Western Reserve University, Cleveland, OH, USA; Research Service, Louis Stokes Cleveland Department of Veteran Affairs Medical Center, Cleveland, OH, USA; Department of Medicine, Case Western Reserve University, Cleveland, OH, USA

Patrick Butaye Department of Biomedical Sciences, Ross University School of Veterinary Medicine, Basseterre, St Kitts and Nevis, West Indies; Department of Pathology, Bacteriology and Poultry diseases, Ghent University, Salisburlylaan, Merelbeke, Belgium

Juliany Rivera Calo Center for Food Safety and Department of Food Science, University of Arkansas, Fayetteville, AR, USA

Chin-Yi Chen US Department of Agriculture, Agricultural Research Service, Eastern Regional Research Center, Wyndmoor, PA, USA

Jinru Chen Department of Food Science and Technology, The University of Georgia, Griffin, GA, USA

H. Gregg Claycamp Center for Veterinary Medicine, US Food and Drug Administration, Rockville, MD, USA

Louis Anthony (Tony) Cox NextHealth Technologies, Cox Associates and University of Colorado, Denver, CO, USA

Philip G. Crandall Center for Food Safety, Food Science Department, University of Arkansas, Fayetteville, AR, USA

Emily Crarey Food and Drug Administration, Center for Veterinary Medicine, Laurel, MD, USA

Andrea Endimiani Institute for Infectious Diseases, University of Bern, Bern, Switzerland

Anna Fàbrega ISGlobal, Barcelona Ctr. Int. Health Res. (CRESIB), Hospital Clínic – Universitat de Barcelona, Barcelona, Spain

Andrea T. Feßler Institute of Farm Animal Genetics, Friedrich-Loeffler-Institut (FLI), Neustadt-Mariensee, Germany

Anuradha Ghosh Department of Diagnostic Medicine and Pathobiology, Kansas State University, Manhattan, KS, USA

Marja-Liisa Hänninen Department of Food Hygiene and Environmental Health, University of Helsinki, Finland

Lee H. Harrison Department of Medicine, Division of Infectious Diseases, University of Pittsburgh, Pittsburgh, PA, USA

Pei-Ying Hong Water Desalination and Reuse Center, Division of Biological and Environmental Sciences and Engineering, King Abdullah University of Science and Technology (KAUST), Thuwal, Saudi Arabia

Charlene R. Jackson US Department of Agriculture, Agricultural Research Service, Russell Research Center, Athens, GA, USA

Nathan A. Jarvis Center for Food Safety, Food Science Department, University of Arkansas, Fayetteville, AR, USA

Yangjin Jung Department of Food Science, School of Environmental and Biological Sciences, Rutgers, The State University of New Jersey, New Brunswick, NJ, USA

Claudine Kabera Food and Drug Administration, Center for Veterinary Medicine, Laurel, MD, USA

Kristina Kadlec Institute of Farm Animal Genetics, Friedrich-Loeffler-Institut (FLI), Neustadt-Mariensee, Germany

Vinayak Kapatral Igenbio, Inc., Chicago, IL, USA

Rauni Kivistö Department of Food Hygiene and Environmental Health, University of Helsinki, Finland

Keith A. Lampel Food and Drug Administration, Laurel, MD, USA

Agnese Lupo Institute for Infectious Diseases, University of Bern, Bern, Switzerland

Jane W. Marsh Department of Medicine, Division of Infectious Diseases, University of Pittsburgh, Pittsburgh, PA, USA

Karl R. Matthews Department of Food Science, School of Environmental and Biological Sciences, Rutgers, The State University of New Jersey, New Brunswick, NJ, USA

Corliss A. O'Bryan Center for Food Safety, Food Science Department, University of Arkansas, Fayetteville, AR, USA

Satu Olkkola Department of Food Hygiene and Environmental Health, University of Helsinki, Finland

Krisztina M. Papp-Wallace Department of Medicine, Case Western Reserve, University, Cleveland, OH, USA; Research Service, Louis Stokes Cleveland Department of Veteran Affairs Medical Center, Cleveland, OH, USA

Steven C. Ricke Center for Food Safety, Food Science Department, University of Arkansas, Fayetteville, AR, USA

Mati Roasto Institute of Veterinary Medicine and Animal Sciences, Estonian University of Life Sciences, Kreutzwaldi, Tartu, Estonia

Mirko Rossi Department of Food Hygiene and Environmental Health, University of Helsinki, Finland

Stefan Schwarz Institute of Farm Animal Genetics, Friedrich-Loeffler-Institut (FLI), Neustadt-Mariensee, Germany

Heather Tate Food and Drug Administration, Center for Veterinary Medicine, Laurel, MD, USA

John Threlfall European Food Safety Agency (EFSA) Biological Hazards (BIOHAZ) Panel, Parma, Italy

Jordi Vila ISGlobal, Barcelona Ctr. Int. Health Res. (CRESIB), Hospital Clínic – Universitat de Barcelona, Barcelona, Spain; Department of Clinical Microbiology, Hospital Clínic, School of Medicine, University of Barcelona, Barcelona, Spain

Guangshun Wang Department of Pathology and Microbiology, University of Nebraska Medical Center, Omaha, NE, USA

Siyun Wang Food, Nutrition and Health, Faculty of Land and Food Systems, The University of British Columbia, Vancouver, BC, Canada

Sarah Wendlandt Institute of Farm Animal Genetics, Friedrich-Loeffler-Institut (FLI), Neustadt-Mariensee, Germany

Xianghe Yan US Department of Agriculture, Agricultural Research Service, Eastern Regional Research Center, Wyndmoor, PA, USA

Ludek Zurek Department of Diagnostic Medicine and Pathobiology, Kansas State University, Manhattan, KS, USA; Department of Entomology, Kansas State University, Manhattan, KS, USA

Chapter 1

Introduction to Antimicrobial-Resistant Foodborne Pathogens

Patrick Butaye[1], María Ángeles Argudín[2] and
John Threlfall[3]

[1]Department of Biomedical Sciences, Ross University School of Veterinary Medicine,
Basseterre, St Kitts and Nevis, West Indies; Department of Pathology, Bacteriology and Poultry
diseases, Ghent University, Salisburlylaan, Merelbeke, Belgium, [2]Laboratoire de Référence
MRSA-Staphylocoques, Department of Microbiology, Hôpital Erasme, Brussels, Belgium,
[3]European Food Safety Agency (EFSA) Biological Hazards (BIOHAZ) Panel, Parma, Italy

Chapter Outline

Antimicrobial resistance is no longer just a potential threat, it is a serious health problem that is rapidly increasing across the world. Since the discovery of penicillin, resistance has been described. With the advent of the massive use of antibiotics, appropriate or not, resistances have been continuously selected, both in commensal bacteria, zoonotic bacteria, and pathogenic bacteria. According to the report of the European Centre for Disease Prevention and Control (ECDC) and the European Medicines Agency (EMA), in Europe, each year 400,000 patients suffer from infections caused by multidrug-resistant bacteria, and 25,000 die (Anonymous, 2009a). The ECDC, as well as the World Health Organization, considers antimicrobial drug resistance to be one of the major health threats in Europe in the twenty-first century (Anonymous, 2011a, 2013a,b). In addition to direct healthcare costs, infectious diseases caused by drug-resistant bacteria result in indirect costs such as days away from work and lost output. The report by ECDC/EMA (Anonymous, 2009a) estimates the overall cost to society at €1.5 billion each year.

Antimicrobial resistance is an ever-growing problem, but **what is an antimicrobial agent?** The term "antimicrobial agent" includes all compounds that kill microorganisms or inhibit their growth. The antibiotics are included within these agents. Antibiotics are natural substances and are produced by fungi or bacteria. Next to these antimicrobial agents, there are purely chemically derived products that are named synthetic antibacterial drugs or chemotherapeutics. At

Antimicrobial Resistance and Food Safety. DOI: http://dx.doi.org/10.1016/B978-0-12-801214-7.00001-6

the microbiological level, they exert the same activity. Against both chemotherapeutics and antibiotics, resistances have been selected. Currently, the term antibiotic is so extended that it is often used as synonym of the general term "antimicrobial agent".

In this chapter we first deal with the different criteria of determining how antimicrobial resistance is defined. The aim is to provide guidance for understanding how people discuss resistance, whilst they are using different definitions. We will then discuss how resistance is spreading between ecosystems and determine the importance of each for foodborne pathogens.

HOW ANTIMICROBIAL RESISTANCE IS DEFINED?

Antimicrobial resistance is a complex item in which several ways of measuring are applied. Resistance in bacteria can be determined according to several different criteria. First, there is the microbiological or epidemiological criterion. This is subdivided into a phenotypic determination or genotypic determination. The latter can also be seen as a separate criterion, since the phenotype does not always accord with the genotype. Secondly there is the pharmacological criterion; and finally, there is the clinical criterion.

The **phenotypic criterion** for the determination of antimicrobial resistance relates to the characteristics of the bacterium itself and relies only on *in vitro* testing. It deals with the minimal inhibitory concentrations (MICs) or inhibition zones of antibiotics for bacteria of one species and looks at how the bacteria are distributed over the MICs/inhibition zones, which are normally doubling dilutions of the respective antibiotics. As such, when in a specific bacterial population, the strains are distributed as a normal Gaussian distribution over the doubling dilutions, this should be regarded as the normal, susceptible, or wild-type population (Figure 1.1).

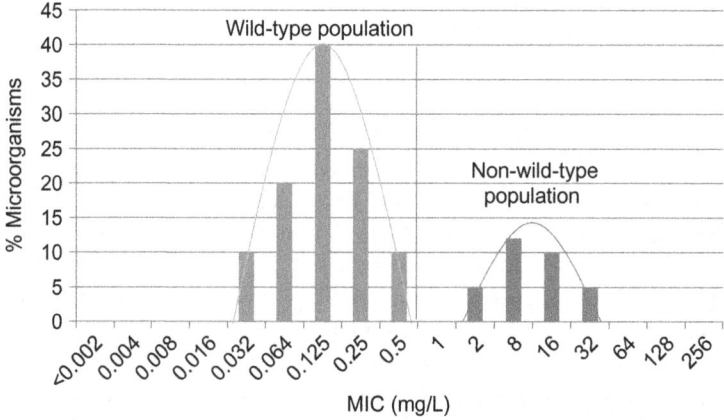

FIGURE 1.1 Example of a hypothetical MIC distribution, in which the epidemiological cutoff is set in 0.5 mg/L. The wild-type (from 0.032 to 0.5 mg/L), as well as the non-wild-type (from 2 to 32 mg/L), populations follow normal Gaussian distributions.

The *in vitro* tests generally rely on two different kinds of tests. One is a "dilution test", which may be in broth or agar, and the other is the diffusion test. Both tests are widely used as antimicrobial susceptibility testing methods in clinical laboratories. They are suitable for testing the majority of bacterial pathogens, including the more common fastidious bacteria, are versatile in the range of antimicrobial agents that can be tested, and require no special equipment. In the dilution tests, agar plates or tubes or microtiter trays with twofold dilutions of antibiotics in agar or broth, respectively, are inoculated with a standardized quantity of bacteria and incubated. After 24 h the MIC is recorded as the lowest concentration of the antimicrobial agent with no visible growth.

In contrast, diffusion tests are primarily qualitative methods, in which a known quantity of bacteria is grown on an appropriate culture plate (such as Mueller–Hinton agar) in the presence of antibiotic-impregnated filter paper disks or tablets. During incubation the antimicrobial agent diffuses into the agar and inhibits growth of the bacteria if sensitive. The presence or absence of growth around the disks or tablets is an indirect measure of the ability of that compound to inhibit that organism. There exists also a quantitative diffusion test, the E-test. This test also allows determination of the MIC because a concentration gradient of the antibiotic is made in the medium.

Apart from the above classical methods, different approaches have been published regarding the use of matrix-assisted laser desorption ionization–time of flight mass spectrometry (MALDI-TOF MS) as a tool for resistance detection (Kostrzewa et al., 2013). MALDI-TOF MS was applied as a fast monitoring tool for the different effects of an antibiotic to resistant or susceptible strains, with the advantage, in contrast to established standard methods, of a reduction of time results. This technology has been suitable for yeast profiling, as well as antimicrobial tests based on enzymatic activities including the beta (β)-lactamase and aminoglycoside-modifying enzyme tests. This technique is still evolving and new developments are underway.

The phenotypic criterion is thus focused on one specific bacterial species, and in most cases may not be extrapolated to other species, since different bacterial species may have different susceptibilities to a particular antibiotic. Typically, within a species, the normal susceptibility, as measured by phenotypic means, is distributed over a specific concentration range. This range can be larger or smaller, depending on the bacterial species and the chemical characteristics of the antimicrobial agent. The normal susceptibility is dependent on the test methods used and for some antibiotics, even small differences may have an effect on the normal susceptibility of the bacteria (Butaye et al., 1998, 1999, 2000, 2003). Therefore standardized methods have been developed for phenotypic susceptibility testing. Different standardized methods are in use (e.g., CLSI, EUCAST, BSAC). Because of this the results (in the case of disk diffusion, the mm, and in the case of dilution tests, the MIC) may differ and as such require specific breakpoints/cutoff values. These differences have led to discussions in the interpretation of what is regarded as "sensitive" or "resistant", particularly in relation to fluoroquinolone antibiotics (see later).

The breakpoints for the microbiological criterion indicate the differences between the normal susceptible population (wild-type population) and the resistant population (non-wild-type population) (Figure 1.1). The wild-type cutoff, used for the microbiological criterion, is commonly named the epidemiological cutoff value (ECOFF) and is determined for both disk diffusion and broth dilution tests by the European Committee on Antimicrobial Susceptibility Testing (EUCAST). This breakpoint may differ from the clinical breakpoints set by EUCAST and most other methods.

For some bacteria–antimicrobial agent combinations the wild-type and non-wild-type populations may overlap, making it very hard to define an accurate breakpoint and, if one is defined, the sensitivity will not be 100% for the determination of a resistant strain. This phenomenon is known as "tailing", and is shown in Figure 1.2. In fact, this tailing is caused by an overlap of the susceptible and resistant populations. This has also been found for disinfectants and makes it difficult to determine resistance against disinfectants. Here, till now, only genetic detection of known resistance genes is possible. It should be clear that this method of determination of the susceptibility may not always have a clinical implication. It may even be the case that a strain defined as having no acquired resistance may not be treatable by the antibiotic and a strain defined as being "resistant" may still be treatable when using this microbiological breakpoint (Aarestrup et al., 2003).

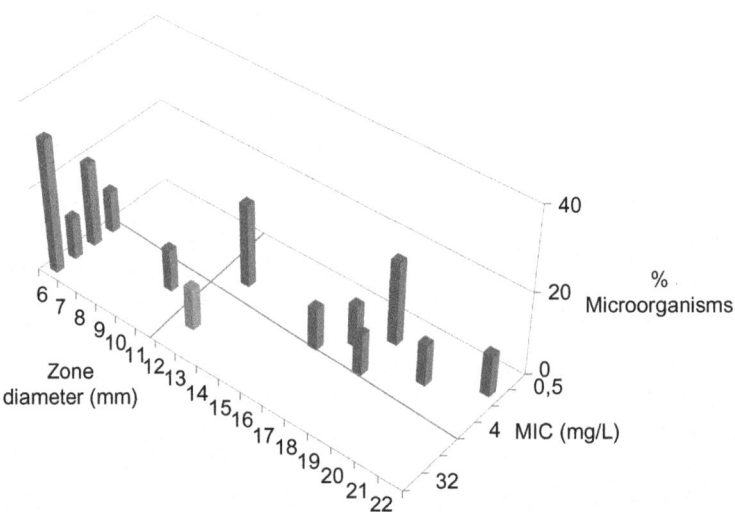

FIGURE 1.2 Hypothetical example of overlapping of wild-type and non-wild-type populations. In this example the MIC epidemiological cutoff is set at 8 mg/L, and the resistance breakpoint diameter is set at 11 mm. The non-wild-type or resistant population (with zone diameter from 6, to 11 mm and MIC of 32 mg/L), as well as the wild-type or susceptible population (with zone diameter from 11 to 22 mm and MIC of 0.5-4 mg/L), follows both criteria. However, there is a part of the population (with zone diameter of 12 mm but MIC of 32 mg/L) that could not be classified as either non-wild type or wild type.

In some cases after defining the wild type, two or more different resistant populations may be identified. Indeed, different resistance genes/mechanisms may be present in a population and these different genes/mechanisms may have a different distribution over the MIC range tested (Figure 1.3). This is a well-known phenomenon for some antibiotics, and for some, the multitude of resistance genes in the population results in the resistant population representing a wave-formatted distribution beyond the MICs of the non-wild type.

As stated above, the cutoff, separating resistant and susceptible bacteria, is only to be seen at species level. There are some exceptions in which this is not even valid for a single species, although this is not a frequent finding. One example is the susceptibility of *Salmonella enterica* for colistin, a polypeptide antibiotic. *Salmonella* taxonomy divides the species *S. enterica* in subspecies and serovars. As for the *S. enterica* subspecies *enterica* serovar Enteriditis and Dublin, the normal susceptibility for colistin is slightly lower than for the other serovars (higher MICs for colistin than the other species) (Agersø et al., 2012). As such, specific breakpoints are necessary for these two serovars to separate the wild-type from the resistant population. In reality the apparent "resistance" to colistin in these serovars is not clinically significant (Anonymous, 2013c), and

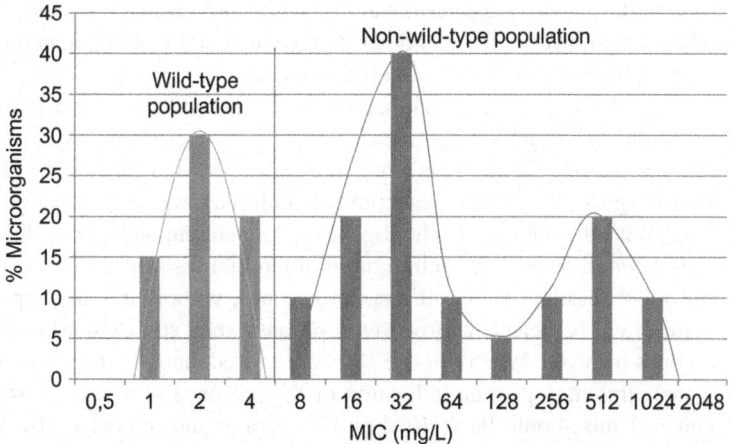

FIGURE 1.3 Hypothetical example of a MIC distribution variation depending of the resistance mechanism: MIC range of mupirocin of a hypothetical population of *Staphylococcus aureus*. For this antibiotic, depending on the MIC, different resistance mechanisms have been discovered (Patel et al., 2009; Seah et al., 2012). MICs ≤4 mg/L corresponded to the susceptible wild-type population (blue). MICs from 8–64 mg/mL corresponded to low-level mupirocin resistance via base changes in the native isoleucyl-tRNA synthetase gene (*ileS*). Isolates with MICs of 128 or 256 mg/mL are uncommon, but also considered as presenting low-level mupirocin resistance. Isolates with MICs ≥512 mg/mL showed high-level mupirocin resistance associated with the plasmid-mediated *mupA*, which encodes a novel isoleucyl-tRNA synthetase. Isolates with MICs ≥1024 mg/L carried the novel isoleucyl-tRNA synthetase encoded by *mupB* (Seah et al., 2012).

in the clinical laboratory "resistance" to colistin in *S. enterica* isolates is almost inevitably an indication of contamination of the specimen with *Proteus* spp.

Closely related to the phenotypic criterion is the **genetic criterion**, which also deals only with the bacterium itself and is therefore frequently not also classified as a separate criterion, but is considered as another testing method for defining microbiological resistance. The results of this determination are not always in accordance with the phenotypic result. This may be due to a poor or absence of expression of the resistance gene, or eventually non-functionality in the genetic background of the bacterium. In fact, the expression should be monitored but that is a more labor-intensive method. Eventually the protein can be looked for, but in most cases such tests are even more labor-intensive. Therefore, they are not done routinely but merely for describing a new resistance mechanism. Expression of the resistance gene may also be variable due to the genetic background in which it is included. The simplest example is the position of the resistance gene in an integron and the differences in strength of the integron promoters. Gene cassettes (that are without their own promoter) are integrated into integrons at the site closest to the promoter, and are thus best expressed. The further from the promoter the poorer the expression, and when, for example, an enzyme responsible for a resistance phenotype is not produced enough due to the poor expression of the gene, MICs of the antibiotic concerned will be lower and eventually will not fall above the cutoff for resistance. Nevertheless, the resistance gene still belongs to the mobilizable resistance gene pool and may be transferred into another bacterium, thereby giving it an importance in the epidemiology of antimicrobial resistance.

Detection of the resistance genes, if done by polymerase chain reaction (PCR) (eventually combined with sequencing of the resistance genes as is necessary for the β-lactamases) is time-consuming and tedious. Then microarray technologies allowed the detection of multiple resistance genes at the same time. With the advent of whole-genome sequencing and prices that are still going down, molecular detection of antimicrobial resistance may become more and more feasible. Here, all resistance genes, without any assumptions made before, will be detected. However, a problem may still exist when there are new genes involved. When suitable software is used that also detects homologous genes, this may give an indication of the presence of a new resistance gene, however this should be verified phenotypically and of course also with deletion and complementation studies. Either way, whole-genome sequencing offers lots of opportunities for fast resistance gene detection.

The **pharmacological criterion** is based on the bacterium and the behaviors (pharmacokinetics) of the concerned antibiotic in the (healthy) body. To be able to determine this, pharmacokinetics of the antibiotic should be determined and it should be known what concentrations of the antibiotic can be reached in the blood or in the tissues. In general the blood levels are taken, since tissue concentrations are hard to determine and may be highly variable. It is then assumed that when blood levels of the antibiotic are reached that are higher than the *in vitro* MIC, the bacterium is susceptible to the antibiotic. The antibiotic is of course applied at a non-toxic dose.

Although the pharmacological criterion comes closer to the clinical situation, there are some drawbacks. The assumption that an *in vitro* determined concentration in a totally different environment (oxygen, proteins, fatty acids, solid/liquid medium, etc.) reflects what happens in the body is not always accurate. As for when MICs *in vitro* are much higher than what can be reached in the blood, one has however a high chance of having no clinical effect. Likewise this criterion is very functional in the development of new antimicrobials, where this criterion will determine whether it is useful to continue with the development of the antimicrobial and go to clinical trials, which are extremely expensive. In general, the pharmacological criterion gives valuable results for systemic infections and results correlate well with the clinical findings. This criterion also forms the basis for the clinical breakpoints but sometimes may be adjusted after it has become clear that for some infections the breakpoint is too high and should be lowered. This has, for example, been the case for fluoroquinolones and *Salmonella* infections. After years of use, it has been noted that there is a "low level" of resistance for which there is possible treatment failure (see above). This exemplifies also that when different levels of resistance are possible (like in a stepwise resistance development) there may be difficulties in determining the pharmacological breakpoint.

Another drawback of the pharmacological criterion is when the bacterium is contained in an abscess, hiding intracellularly or forming biofilms. In an abscess or biofilm, the bacterium may be protected from the antibiotic because it cannot penetrate the abscess (as is the case for mycobacterial infections) or it may be that the bacterium is in a state of low metabolic activity, and antibiotics need growing bacteria to be active. Another drawback is the local application of an antimicrobial. This definitely does not correlate with the blood levels and may be more variable. Local application is included in the treatment of intestinal infections. The concentrations that can be reached in the intestine for the local application (non-absorbed antibiotic) may moreover vary according to the intestinal compartment as has been shown for tylosin (Burch, 2011). In the ileum, very high concentrations can be reached, which in this case is even toxic for Gram-negative bacteria (tylosin has a mainly Gram-positive spectrum, but also affects certain Gram-negative bacteria). The Gram-negative enteric bacteria, *Escherichia coli* and *Salmonella* spp., are intrinsically resistant to tylosin (Shryock et al., 1998), although at high, non-clinical concentrations, this antibiotic has some activity. The effect of macrolides is also exemplified by the presence of acquired resistance genes in *E. coli* (Nakamura et al., 2000). More difficult to determine are the levels of antibiotic that can be reached, for example, in cases of mastitis in cows where intramammary or topical applications are frequently applied. As such, this criterion implies also that for different infections with the same bacterium (e.g., *E. coli* in urinary tract infections, intestinal infections, and systemic infections) different breakpoints should be applied related to the site of infection.

The third and final criterion, and also the most important, is the **clinical criterion**. This criterion is of importance for the clinician since it will provide information as to whether administration of a particular antibiotic might result in cure of the infection. This criterion takes into account the bacterium and the sick body, but it may be difficult to determine what is a sick body since it is dependent on many factors as there are the immune status, progression of the disease, underlying conditions, combined infections (virus–bacterial and eventually two or more bacterial species). With this criterion, it is the aim to correlate the result of an *in vitro* test with a clinical outcome. The dose applied for the treatment is in this case predefined. In general these breakpoints are determined in the first instance empirically and, as shown above, can be adjusted if necessary. In veterinary medicine, frequently, experimental trials are performed. In such trials, animal models are used, but even these are not always accurate since in this case, only healthy animals are used, infected with an in general massive infection dose to make the infection model reproducible. Also, results from clinical trials/field studies can be taken into account, though these should be interpreted with care, since many underlying factors may play a role in the outcome of the trial. Viral infection together with the bacterial infection, for example, may complicate the disease and give a different outcome. Of course, an accurate diagnosis is necessary in these cases.

It is clear that each of the criteria has its use. It makes no sense to talk to a clinician about a microbiological criterion, since a primary requirement of a clinician is whether or not application of an antibiotic at a certain concentration and dosage regimen will result in the cure of an infection. Similarly, a pharmaceutical company wants to know whether an antibiotic makes a chance of success in clinical trials, and as such it wants to eliminate, as much as possible, a potential clinical failure, therefore the pharmacological criterion is of great interest. And, finally, for scientists who study the evolution and dynamics of antimicrobial resistance, the microbiological criterion (ECOFFs) is of primary concern.

HOW DOES RESISTANCE SPREAD BETWEEN ECOSYSTEMS?

The study of the dynamics of the spread of antimicrobial resistance is very complicated since many factors are involved, not all of which have been elucidated. It involves many factors of which many are still unknown and compromise our understanding. Certainly antimicrobial resistance is much older than the period in which humans have used antimicrobials, though we cannot be sure how long we have been using antibiotics. For sure Alexander Fleming only rediscovered antibiotics since it is known that in ancient times certain antibiotics and antiseptics were used. A Sumerian table from 2150 BC showed that the medical doctors of ancient Mesopotamia used beer to treat wounds (García Rodríguez et al., 2006). In ancient Egypt tetracyclines, natural antimicrobial compounds (such as onion, garlic, *Aloe vera*, beer yeast) and metal compounds (copper, antimony) were used (García Rodríguez et al., 2006; Aminov, 2010). It is also well known, that medical doctors from ancient China knew of antibiosis treatment in the third millennium BC, and, for example, they used soya for skin infections and chaulmoogra oil for the treatment of leprosy (García Rodríguez et al., 2006). In ancient Israel the use of hyssop was common; a plant in which *Penicillium notatum* (a source of penicillin compounds) was discovered at the beginning of the twentieth century. Moreover, pre-Columbian cultures used diverse plants (such as root beer, guaiacwood, cinchona bark, cress) to combat infectious diseases (García Rodríguez et al., 2006) in the Americas. During the Middle and Modern Ages, "medical staff" used as reference the encyclopedia *De Materia Medica* of the Greek botanist Dioscorides. In this encyclopedia a variety of plants are described for the treatment of infectious diseases, including plants with antimicrobial properties, such as garlic, lichen, or *A. vera.*

Resistance genes are most probably as old as the bacteria are since antibiotics are natural products, used by bacteria/fungi in their communication with one another, or in their competition with one another. To defend themselves, the bacteria needed to protect themselves against these molecules, excreted by their competitors. The re-discovery of antimicrobial agents in the mid-twentieth century revolutionized the management and treatment of bacterial infections. Infections that would normally have been fatal became curable. Ever since then, antimicrobial agents have saved the lives of millions of people. These gains are now seriously jeopardized, because our recent massive use of antibiotics has somehow largely disturbed a certain bacteriological equilibrium and bacteria have defended themselves against our actions, and they have massively acquired resistance genes.

Resistance can spread between ecosystems in two different ways. The first way is a **direct** way by which the resistant bacterium itself establishes in another ecosystem. Transfer of resistant bacteria from food-producing animals to humans can be by direct contact, by the ingestion of contaminated food, or via intermediate ecosystems, such as sludge that may contaminate vegetables. In all these cases, the bacterium has to colonize or infect humans. In the case of zoonotic agents, the situation is clear, in that the bacterium causes a disease that is more difficult to treat. On many occasions this has been shown for different pathogens including *Salmonella* serovars, *Campylobacter jejuni*, and enterohemorrhagic *E. coli*.

S. enterica subspecies *enterica* is widely distributed in nature, several serotypes colonize a range of different animal hosts (unrestricted serovars) while others are more host-adapted and others are host-restricted (typhoid *Salmonella*) (Singh, 2013). The unrestricted serovars may colonize or infect food animals that are the main reservoir for human infections, with the majority of illness associated with contaminated meat products and eggs (Butaye et al., 2006). Despite the different antimicrobial uses in different countries, global and local outbreaks of multidrug-resistant *Salmonella* have been reported (Butaye et al., 2006). The best example is the emergence and dissemination of *S. enterica* serotype Typhimurium definitive phage type (DT) 104, of which only a few highly related clonal subtypes have been documented. This strain has a penta-resistance phenotype due to the chromosomally located *Salmonella* genomic island I (SGI I), which confers resistance to ampicillin, chloramphenicol/florfenicol, streptomycin, sulfonamides and tetracyclines. *S.* Typhimurium DT 104 was first identified in exotic birds in the United Kingdom in the early 1980s, but later spread to other animal species on a global scale (Threlfall, 2000; Butaye et al., 2006).

The exact epidemiology of *C. jejuni* is less clear, though it is also a zoonotic pathogen widespread in the environment. It is generally accepted, however, that chickens are a natural host for *C. jejuni*, and that colonized broiler chicks are the primary vector for transmitting this pathogen to humans (Hermans et al., 2012). Several potential sources and vectors for transmitting *C. jejuni* to broiler flocks have been identified (Hermans et al., 2012). Initially, one or a few broilers can become colonized, after which the infection will rapidly spread throughout the entire flock. Such a flock is generally colonized until slaughter, and infected birds carry a very high *C. jejuni* load in their gastrointestinal tract. This eventually results in contaminated carcasses during processing, which can transmit this pathogen to humans. Recent genetic typing studies showed that chicken isolates can frequently be linked to human clinical cases of *Campylobacter* enteritis (Hermans et al., 2012). Though this is the most frequently reported foodborne pathogen (Anonymous, 2009b; 2011a; 2013b), the resistance problem is less of an item in this species since the necessity for treatment is even less than for *Salmonella*.

Enterohemorrhagic *E. coli* (EHEC, also named STEC (Shiga-like toxin-producing *E. coli*) or VTEC (verocytotoxin-producing *E. coli*)) compromises a diverse group of *E. coli* serotypes, though the serotype O157 is best known (Johnson et al., 2006). The reservoir of the enterohemorrhagic *E. coli* includes different mammals and birds, thus its transmission to humans is primarily through consumption of contaminated foods, such as raw or undercooked ground meat products and raw milk (Anonymous, 2011b). Fecal contamination of water and other foods, as well as cross contamination during food preparation will also lead to infection. Moreover, an increased number of outbreaks of enterohemorrhagic *E. coli* are associated with the consumption of fruits and vegetables, whereby contamination may be due to contact with feces from domestic or wild animals at some stage during cultivation or handling (Anonymous, 2011b). Furthermore, person-to-person contact is also an important mode of transmission through the oral-fecal route, since an asymptomatic carrier state has been reported (Anonymous, 2011b). The definition of enterohemorrhagic *E. coli* has recently been fine-tuned. This was primarily because of the recent large outbreak of EHEC O104:H4 in Germany in May 2011. This atypical strain also carried, next to the classical virulence genes, virulence genes associated with enteroaggregative *E. coli*, while the intimin-encoding gene (*eae*) was not present. This strain was moreover highly multi-resistant (Mellmann et al., 2011). This contrasts with most of the EHEC strains, which are in general rarely resistant (Threlfall et al., 2000; Threlfall, 2009).

While it is better not to treat infected humans since it may lead to greater toxin production and consequently a higher pathogenicity (Davis et al., 2013), resistance in these bacteria may be transferred to other bacteria of the intestinal flora, even without selective pressure, as has been shown for commensal *E. coli* in an *in vitro* model mimicking the human intestinal flora (Smet et al., 2011).

Source attribution is not always that evident. Indeed, a good surveillance system is necessary to eventually find the source, when solely relying on microbiological data. Including an epidemiological investigation can determine the true source of the infection. Moreover, the colonization of commensal bacteria from animals, transferred via the food chain, is not evident. For many bacteria, it seems that they are quite well adapted to the specific host and do merely pass through the intestinal tract upon ingestion, while for others, it may be more evident. Even within a species, there seem to be differences depending upon the clone, whether they can remain for a more-or-less longer period in the new host. A good example is the transmission of livestock-associated meticillin-resistant *S. aureus* (LA-MRSA) of the clonal complex (CC) 398. Research performed so far indicates that it has a high transmission potential and also a high probability of persisting in the intensively farmed pig population (Crombé et al., 2013). Moreover, persons working or living on pig farms have an increased risk of being colonized or infected with this LA-MRSA (Pletinckx et al., 2013). However, human infections and outbreaks are infrequently reported, and it has been shown that MRSA CC398 has lower nosocomial transmissibility and virulence than other hospital-associated MRSA clones more adapted to the humans (Crombé et al., 2013). These MRSA are highly multi-resistant and, moreover, very receptive for foreign DNA, which poses the risk of increasing pathogenicity and pan-resistance (Vanderhaeghen et al., 2010).

To determine where a bacterium originates from, several molecular techniques are available. It is not the aim here to go into detail for these methods since their discriminative power may differ according to the bacterial species under investigation. With the advent of whole-genome sequencing (WGS) by next-generation sequencing, a method readily applicable to whatever bacterial species became available. Indeed, this method allows the most detailed description of the genomic content of the bacterium and allows to determine the relationships between different isolates. It omits the need for identification, serotyping as in the case for *Salmonella*, and allows also to determine the resistance potential. On the latter, care should be taken since it may miss yet unknown resistances and it cannot determine whether the resistance is effectively expressed. The cost of next-generation sequencing is diminishing continuously and it is becoming more and more accessible for laboratories. The evolution of this technology is also continuously improving. A potential bottleneck is data storage and subsequent analysis. A large amount of information is gathered on the whole bacterium and needs to be stored appropriately. The software available for analyzing these data is also evolving rapidly, as are different methods for analyzing sequence data.

The second way to spread resistances is the **indirect** spread of resistance genes. This exchange of new genetic material can happen by transformation, conjugation, or transduction (van Hoek et al., 2011).

In the process of *transformation*, naked DNA is taken up by the recipient bacteria (termed "competent" because they are capable of taking up DNA from the environment) and either incorporated into the host genome by homologous recombination or transposition. This process occurs in both Gram-positive and Gram-negative bacteria. There are some microorganisms that are competent at a specific stage in their growth (such as streptococci).

Conjugation is the most common way of exchanging resistance genes between bacteria. Elements that conjugate or mobilize include plasmids, transposons, or genomic islands. These can be self-conjugating or may be dependent on the conjugation of transfer genes located on other mobile genetic elements (MGEs).

Bacteriophages also play a role in the spread of DNA between bacteria, they do this by a process called *transduction* in which bacterial DNA (any segment of DNA adjacent to the phage insertion site), rather than phage DNA, is packaged into the phage head and injected into the recipient bacteria. There are also elements which are capable of translocation to new sites in the genome but are not themselves capable of transfer to a new host (transposons and mobile introns); however, they can be moved into a new host if they transpose to a conjugative element.

The simplest mobile genetic unit is the *gene cassette*, which only contains a gene (usually conferring antimicrobial resistance) and a recombination site (called *attC* site). The gene cassettes may exist freely as circular DNA or incorporated into a grade more complex genetic element, the *integron*. Integrons are genetic elements that include components of a site-specific recombination system enabling them to capture and mobilize genes. The basic components of an integron include a site-specific integrase gene (*intI*), the primary recombination site (*attI*), and a promoter or promoters. The integrase carries out the recombination between the *attC* and the *attI* sites, allowing the integration of gene cassettes (Figure 1.4). The horizontal transfer of integrons is mostly mediated by plasmids or transposons.

Apart from these genetic structures there are MGEs, which are acquired by transformation or conjugation (van Hoek et al., 2011). The simplest MGEs are *insertion sequences (IS)*. These elements consist only of the gene required for element mobility and the inverted repeat ends of the element. When these elements contain accessory genes not involved in element translocation, they are called *transposons*. A simple transposon contains an accessory gene (often encoding antimicrobial resistance) and the transposase gene. This simplex element is not capable of conjugal transfer, but there are more complex elements called integrative conjugative elements (ICE) or conjugative transposons that contain an origin of transfer and the genes required to make the conjugation apparatus (van Hoek et al., 2011).

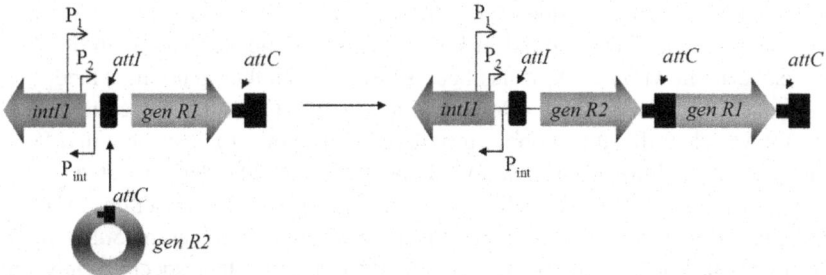

FIGURE 1.4 Example of gene cassette insertion in an integron. The integron carries initially a resistant gene (gen R1), but after insertion of the free gene cassette it carries an additional resistant gene (gen R2).

Another important MGE in bacteria is *plasmids*. Typically plasmids are linear or circular extra-chromosomal elements that contain their own origin of replication, and commonly also contain an origin of transfer and genes encoding functions that allow them to transfer to new hosts via conjugation. In addition to replication and transfer functions, plasmids commonly encoded antimicrobial resistance genes integrated (or not) into transposons or integron elements. Plasmids can have a broad host range and transfer between different species, or have a narrow host range and be confined in one genus or species. Moreover, there are plasmids that can be transferred to a particular host, but they cannot replicate or do not replicate well in this new host. In these circumstances, the plasmid may be lost, but if it contains antimicrobial resistance genes on a transposon, it can translocate to the bacterial chromosome and be maintained in the absence of the plasmid, and thus contribute to the spread of resistances (van Hoek et al., 2011).

The epidemiology of the spread of resistances is complicated, since it has to take into account two components. First there is the chance of the bacteria getting into contact. Secondly, there is the speed and possibility of exchanging DNA. To what extent zoonotic disease or the passage of commensal bacteria contribute to this spread of resistance is not known. A good example is the mobility of SCC*mec* in *S. aureus* and the spread of the *Salmonella* genomic island SGI1 in different *Salmonella* serovars and other bacterial species. The mobility of SCC*mec* has never been shown under laboratory conditions, however it is clear from the epidemiological data that it is capable of spreading since identical SCC*mec* have been found in different staphylococcal species. On the other hand, when looking at its possibility of spreading within one ecosystem, then it is not always obvious. Very recently it has been shown that meticillin-resistant *Staphylococcus sciuri* is highly prevalent in poultry (Nemeghaire et al., 2014a,b). In that same sample, meticillin-resistant *S. aureus* could be isolated only rarely, and moreover, the livestock-associated MRSA did carry different SCC*mec* (Nemeghaire et al., 2013). This indicates that they do not share completely the same resistance pool (however it should be noted that there are similar resistance genes found in both species). While SCC*mec* mobility has never been shown *in vitro*, it has been shown for another genomic island, the *Salmonella* genomic island (Douard et al., 2010). From the epidemiological data it was clear that it is mobile since it has been shown to be present in different serotypes. However, it took a while before it could be demonstrated in the laboratory since it needed a helper plasmid, offering mobility functions in *trans* to be mobile. Striking is that, while it is quite widespread in certain *Salmonella* serotypes it is not in others, and it is rarely found in other species. Only on a few occasions, was it found in a *Proteus vulgaris* (Bi et al., 2011; Siebor and Neuwirth, 2011), although it is becoming more frequent at hospital level, where recently it was shown that 9% of the strains tested carried this genomic island (Siebor and Neuwirth, 2013). Since this genetic element integrates at a specific site, potential integration sites have been identified in *Shigella* spp., *Vibrio* spp., *Pseudomonas* spp., *Brucella* spp., *Legionella pneumophila*, and *Klebsiella pneumoniae* (Doublet et al., 2007). To our knowledge, the transfer of the *Salmonella* genomic island SGI1 has not yet been observed in these bacteria. Nevertheless, we should remain vigilant since this element is capable of collecting and subsequently transferring multiple resistance genes.

The horizontal transfer of resistance is difficult to assess since genes are frequently the same and may even be located on the same mobile genetic element. This does not allow determination of the origin of the element, when encountered in different ecosystems. It is also not possible to quantify the event, since once introduced in an ecosystem, it may amplify in this system, thus making it impossible to estimate what come from which ecosystem. Moreover resistant bacteria are plentiful in the environment. This in turn largely complicates our possibilities of risk analysis for the resistance on MGEs. It is even so for non-self-transferable elements, in one ecosystem the transfer may be very efficient while it is deficient in another.

CONCLUSION

Infections with extreme-resistant or pan-resistant bacteria are increasingly being reported in Europe. Moreover, the emergence of antimicrobial resistance is accompanied with a decline in the discovery of new antimicrobial agents. As highlighted by ECDC and EMA, the current pipeline of new antimicrobials is running dry, and at the same time resistance to antibiotics is escalating, creating an ever-increasing gap. We are facing the possibility of a future without effective antibiotics for some infections and a scenario where infections that hitherto were considered harmless are now a serious health problem and a major cause of morbidity, mortality, together with major financial and social repercussions. It has been estimated that most of the antibiotics used presently for common infections will be useless within 5–10 years, turning back the clock to the pre-antibiotic era. Without access to effective antibiotics, medical practice procedures, such as organ transplantations, prosthetic surgery, management of immunosuppressed and cancer patients, intensive care stewardship, care of the elderly and neonates, will be seriously jeopardized. Due to the lack of data on resistance prevalence, the choice of an antibiotic for a bacterial infection is complicated. Therefore, there is an urgent need for data concerning the antimicrobial resistance situation in hospitals, the community, and in livestock. Action is needed on many fronts, new approaches for prevention and treatment of infections, development of rapid diagnostic tools, innovations into new drugs, and better molecular resistance surveillance.

REFERENCES

Aarestrup, F.M., Wiuff, C., Mølbak, K., Threlfall, E.J., 2003. Is it time to change the break points for fluoroquinolones for *Salmonella* spp.? Antimicrob. Agents Chemother. 47, 827–829.

Agersø, Y., Torpdahl, M., Zachariasen, C., Seyfarth, A., Hammerum, A.M., Nielsen, E.M., 2012. Tentative colistin epidemiological cut-off value for *Salmonella* spp. Foodborne. Pathog. Dis. 94, 367–369.

Aminov, R.I., 2010. A brief history of the antibiotic era: lessons learned and challenges for the future. Front. Microbiol. 1, 134.

Anonymous, 2009a. ECDC/EMEA joint technical report. The bacterial challenge: time to react. European Centre for Disease Prevention and Control (ECDC), Stockholm.

Anonymous, 2009b. Joint opinion on antimicrobial resistance focused on zoonotic infections. EFSA J. 7, 1372.

Anonymous, 2011a. Antimicrobial resistance surveillance in Europe 2010. Annual Report of the European Antimicrobial Resistance Surveillance Network (EARS-Net). European Centre for Disease Prevention and Control (ECDC), Stockholm.

Anonymous, 2011b. Fact sheet N°125:Enterohaemorrhagic *Escherichia coli* (EHEC). World health Organization (WHO).

Anonymous, 2013a. Global Risks 2013 An Initiative of the Risk Response Network, eighth ed. World Economic Forum, Switzerland.

Anonymous, 2013b. Scientific report of EFSA and ECDC. The European union summary report on trends and sources of zoonoses, zoonotic agents and food-borne outbreaks in 2011. EFSA J. 11 (4), 3129.

Bi, S., Yan, H., Chen, M., Zhang, Z., Shi, L., Wang, H., 2011. New variant *Salmonella* genomic island 1-U in *Proteus mirabilis* clinical and food isolates from South China. J. Antimicrob. Chemother. 66, 1178–1179.

Burch, D., 2011. Pig feed: potential effects of antimicrobial carry over. Pig Prog. 27 (9).

Butaye, P., Devriese, L.A., Haesebrouck, F., 1998. Effects of different test conditions on the minimal inhibitory concentration of growth promoting antibacterial agents with Enterococci. J. Clin. Microbiol. 36, 1907–1911.

Butaye, P., Devriese, L.A., Haesebrouck, F., 1999. Phenotypic distinction in *Enterococcus faecium* and *Enterococcus faecalis* strains between susceptibility and resistance to growth-enhancing antibiotics. Antimicrob. Agents Chemother. 43, 2569–2570.

Butaye, P., Devriese, L.A., Haesebrouck, F., 2000. Influence of different medium components on the *in vitro* activity of the growth-promoting antibiotic flavomycin against enterococci. J. Antimicrob. Chemother. 46, 713–716.

Butaye, P., Devriese, L.A., Haesebrouck, F., 2003. Antimicrobial growth promoters used in animal feed: a review of the less well known antibiotics and their effects on Gram-positive bacteria. Clin. Microbiol. Rev. 16, 175–188.

Butaye, P., Michael, G.B., Schwarz, S., Barrett, T.J., Brisabois, A., White, D.G., 2006. The clonal spread of multidrug-resistant non-typhi *Salmonella* serotypes. Microbes. Infect. 8, 1891–1897.

Crombé, F., Argudín, M.A., Vanderhaeghen, W., Hermans, K., Haesebrouck, F., Butaye, P., 2013. Transmission dynamics of methicillin-resistant *Staphylococcus aureus* in pigs. Front. Microbiol. 4, 57.

Davis, T.K., McKee, R., Schnadower, D., Tarr, P.I., 2013. Treatment of Shiga toxin-producing *Escherichia coli* infections. Infect. Dis. Clin. North. Am. 27, 577–597.

Douard, G., Praud, K., Cloeckaert, A., Doublet, B., 2010. The *Salmonella* genomic island 1 is specifically mobilized in trans by the IncA/C multidrug resistance plasmid family. PLoS. One. 5 (12), e15302.

Doublet, B., Golding, G.R., Mulvey, M.R., Cloeckaert, A., 2007. Potential integration sites of the *Salmonella* genomic island 1 in *Proteus mirabilis* and other bacteria. J. Antimicrob. Chemother. 59, 801–803.

Anonymous, 2013c. Use of colistin products in animals within the European Union: development of resistance and possible impact on human and animal health. EMA/755938/2012.

García Rodríguez, J.A., Barberán Lopez, J., González Nuñez, J., Orero González, A., Prieto Prieto, J., 2006. La otra historia de los antimicrobianos. Ed. Grupo Ars XXI de Comunicación, S.L., Barcelona.

Hermans, D., Pasmans, F., Messens, W., Martel, A., Van Immerseel, F., Rasschaert, G., et al., 2012. Poultry as a host for the zoonotic pathogen *Campylobacter jejuni*. Vector. Borne. Zoonotic. Dis. 12, 89–98.

Johnson, K.E., Thorpe, C.M., Sears, C.L., 2006. The emerging clinical importance of non-O157 Shiga toxin-producing *Escherichia coli*. Clin. Infect. Dis. 43, 1587–1595.

Kostrzewa, M., Sparbier, K., Maier, T., Schubert, S., 2013. MALDI-TOF MS: an upcoming tool for rapid detection of antibiotic resistance in microorganisms. Proteomics Clin. Appl. 7, 767–778.

Mellmann, A., Harmsen, D., Cummings, C.A., Zentz, E.B., Leopold, S.R., Rico, A., et al., 2011. Prospective genomic characterization of the German enterohemorrhagic *Escherichia coli* O104:H4 outbreak by rapid next generation sequencing technology. PLoS. One. 6, e22751.

Nakamura, A., Nakazawa, K., Miyakozawa, I., Mizukoshi, S., Tsurubuchi, K., Nakagawa, M., et al., 2000. Macrolide esterase-producing *Escherichia coli* clinically isolated in Japan. J. Antibiot. (Tokyo) 53, 516–524.

Nemeghaire, S., Roelandt, S., Argudín, M.A., Haesebrouck, F., Butaye, P., 2013. Characterization of methicillin-resistant *Staphylococcus aureus* from healthy carrier chickens. Avian Pathol. 42, 342–346.

Nemeghaire, S., Argudín, M.A., Feßler, A.T., Hauschild, T., Schwarz, S., Butaye, P., 2014a. The ecological importance of the *Staphylococcus sciuri* species group as a reservoir for resistance and virulence genes. Vet. Microbiol. 171, 342–356.

Nemeghaire, S., Argudín, M.A., Haesebrouck, F., Butaye, P., 2014b. Molecular epidemiology of methicillin-resistant *Staphylococcus sciuri* in healthy chickens. Vet. Microbiol. 171, 357–363.

Patel, J.B., Gorwitz, R.J., Jernigan, J.A., 2009. Mupirocin resistance. Clin. Infect. Dis. 49, 935–941.

Pletinckx, L.J., Verhegghe, M., Crombé, F., Dewulf, J., De Bleecker, Y., Rasschaert, G., et al., 2013. Evidence of possible methicillin-resistant *Staphylococcus aureus* ST398 spread between pigs and other animals and people resisding on the same farm. Prev. Vet. Med. 109, 293–303.

Seah, C., Alexander, D.C., Louie, L., Simor, A., Low, D.E., Longtin, J., et al., 2012. MupB, a new high-level mupirocin resistance mechanism in *Staphylococcus aureus*. Antimicrob. Agents Chemother. 56, 1916–1920.

Shryock, T.R., Mortensen, J.E., Baumholtz, M., 1998. The effects of macrolides on the expression of bacterial virulence mechanisms. J. Antimicrob. Chemother. 41, 505–512.

Singh, V., 2013. *Salmonella* serovars and their host specificity. J. Vet. Sci. Anim. Husb. 1, 301.

Siebor, E., Neuwirth, C., 2011. The new variant of *Salmonella* genomic island 1 (SGI1-V) from a *Proteus mirabilis* French clinical isolate harbours *bla*VEB-6 and *qnrA1* in the multiple antibiotic resistance region. J. Antimicrob. Chemother. 66, 2513–2520.

Siebor, E., Neuwirth, C., 2013. Emergence of *Salmonella* genomic island 1 (SGI1) among *Proteus mirabilis* clinical isolates in Dijon, France. J. Antimicrob. Chemother. 68, 1750–1756.

Smet, A., Rasschaert, G., Martel, A., Persoons, D., Dewulf, J., Butaye, P., et al., 2011. In situ ESBL conjugation from avian to human *Escherichia coli* during cefotaxime administration. J. Appl. Microbiol. 110, 541–549.

Threlfall, E.J., 2000. Epidemic *Salmonella typhimurium* DT 104—a truly international epidemic clone. J. Antimicrob. Chemother. 46, 7–10.

Threlfall, E.J., 2009. Resistant gut bacteria. In: Cook, G.C., Zumla, A.I. (Eds.), Manson's Tropical Diseases, Twenty Second Ed. Saunders Elsevier, pp. 943–952.

Threlfall, E.J., Ward, L.R., Frost, J.A., Willshaw, G.A., 2000. The emergence and spread of antibiotic resistance in food-borne bacteria. Int. J. Food. Microbiol. 62, 1–5.

Vanderhaeghen, W., Hermans, K., Haesebrouck, F., Butaye, P., 2010. Methicillin-resistant *Staphylococcus aureus* (MRSA) in food production animals. Epidemiol. Infect. 138, 606–625.

van Hoek, A.H., Mevius, D., Guerra, B., Mullany, P., Roberts, A.P., Aarts, H.J., 2011. Acquired antibiotic resistance genes: an overview. Front. Microbiol. 2, 203.

Chapter 2

Antimicrobial Resistance of Shiga Toxin-Producing *Escherichia coli*

Jinru Chen

Department of Food Science and Technology, The University of Georgia, Griffin, GA, USA

Chapter Outline

INTRODUCTION

Shiga Toxin-Producing *E. coli* and Enterohemorrhagic *E. coli*

Shiga toxin-producing *E. coli* (STEC) is a group of *E. coli* that is capable of producing at least one of the potent, proteinous cytotoxins known as Shiga toxins. The structure and biological activity of Shiga toxins of *E. coli* share striking resemblance to the cytotoxin produced by *Shigella dysenteriae* serotype 1 (O'Brien et al., 1982). A subset of STEC with great clinical significance and public health impact is enterohemorrhagic *E. coli* (EHEC), which causes severe clinic manifestations such as hemorrhagic colitis, hemolytic-uremic syndrome,

Antimicrobial Resistance and Food Safety. DOI: http://dx.doi.org/10.1016/B978-0-12-801214-7.00002-8

and thrombotic thrombocytopenic purpura. Since its debut as a notorious food-borne pathogen in 1982, EHEC has caused numerous outbreaks of infections worldwide. EHEC serotype O157:H7 is responsible for the majority of the outbreaks (Siegler, 1995) although non-O157 EHEC has also been involved, especially in Australia, Germany, Austria, and the United States (Elliott et al., 2001; Gerber et al., 2002).

Cattle have been identified as the symptomless carrier and primary reservoir of STEC, and calves seem to carry EHEC O157:H7 more frequently than adult cattle. As a result, foods originating from a bovine source have most frequently been implicated in outbreaks of EHEC infections. In addition to foods of bovine origin, other foods have also been linked to outbreaks of EHEC infections. EHEC outbreaks associated with fresh produce, such as spinach, lettuce, and alfalfa sprouts have been traced to contamination at farm level (CDC, 2007).

STEC has been isolated from different types of meat products including beef, lamb, pork, and poultry (Brooks et al., 2001; Doyle and Schoeni, 1987). It has also been isolated from unpasteurized cheese, raw and pasteurized milk, as well as mayonnaise (Watanabe et al., 1999; Werber et al., 2006). Vegetables and fruits have been contaminated with STEC during cultivation, harvesting, handling, processing, and distribution (CDC, 2007). Apple drops which are commonly used to make cider have been contaminated with STEC through contact with animal manure on the ground or during cider processing (CDC, 1997).

Antibiotic and Antimicrobial Use During Food Production and Processing

Antibiotics are natural or synthesized substances that have the ability to kill or inhibit the growth of bacteria. Antimicrobials are, nevertheless, products that act against a wide range of organisms including bacteria, viruses, fungi, protozoa, and helminths. Understandably, antibiotics are one type of antimicrobial, but not all antimicrobials are antibiotics.

The introduction of antibiotics into clinical practice in the 1930s and 1940s revolutionized human medicine (Cohen, 2000). Unfortunately, clinical success in the treatment of infectious diseases was soon followed by the emergence of antibiotic-resistant pathogens. The use of antibiotics in food animal production started shortly after World War II. It was suggested at the time that the addition of antibiotics into feed or water would promote faster growth in chickens (Moore et al., 1946). Since then, sub-therapeutic levels of antibiotics have been used in the production of many farm animals (Landers et al., 2012). The mechanism on how antibiotics improve animal growth has never been fully understood, although several theories have been proposed (Cromwell, 1991; John, 2006).

Antibiotics are used, during animal production, for different purposes, for example, therapeutic treatment of existing bacterial infections, metaphy-lactic prevention of infectious diseases, prophylactic prevention of infectious diseases during high-risk periods, or growth promotion (Schwarz and

Chaslus-Dancla, 2001). Antibiotics important for human medicine, including tetracycline, penicillin, erythromycin, and other important therapeutic drugs, have been used extensively in food animal production. Some studies suggest that antibiotic use in animal production selects antibiotic-resistant commensal bacteria and zoonotic enteropathogens, which will eventually diminish the therapeutic value of antibiotics in human medicine (Endtz et al., 1991; Levy et al., 1976; Linton et al., 1975; Low et al., 1997). Other studies, however, indicated that antibiotic resistance was easier to acquire in some bacterial species and did not have a convincing link to antibiotic use in food animals (Dargatz et al., 2000; Wells et al., 2001). Regardless of the connection between the two issues, epidemiological data have revealed an increased antibiotic resistance in many different pathogens that are threats to human health (CDC, 2013).

Various antimicrobial substances and interventions have been used to inhibit microbial growth and extend the shelf life of food products. As osmolytes, salt and sugar increase the osmotic pressure and reduce the water activity of food, therefore inhibiting the growth of various microorganisms. In addition to elevating the acidity of food, either naturally or artificially, organic acids have been used as cleaners to disinfect foods such as fresh produce and animal carcasses. Oxidizing agents, such as hydrogen peroxide and sodium hypochlorite, are among the most commonly used sanitizers to disinfect food, or food contact, surfaces. Adaptive exposure of bacterial cells to sub-lethal levels of an antimicrobial agent increases their resistance to a higher level of the same antimicrobial agent and offers cross-protection to bacteria against different types of antimicrobial intervention. Bacterial resistance to stress is also affected by other factors such as their physiological state as well as their ability to express extracellular substances and form biofilms.

RESISTANCE OF *E. COLI* O157:H7 TO ANTIBIOTICS

Antibiotic Resistance and Mobile DNA Elements

Bacteria susceptible to antibiotics usually gain their resistance through genetic mutation or genetic material exchange with an antibiotic-resistant donor. There are three different approaches for horizontal gene transfer, that is, transformation, conjugation, and transduction (Thomas and Nielsen, 2005), and conjugation is believed to be the major mechanism of antibiotic resistance gene exchange under *in vivo* conditions (Schwarz et al., 2006). Mobile DNA elements play an essential role in the dissemination of antibiotic resistance genes in natural environments.

As mobile DNA elements, temperate bacteriophage (Brabban et al., 2005) and transmissible plasmid (Makino et al., 1999; Miwa et al., 2002) have both served as carriers for antibiotic resistance genes in *E. coli* O157:H7. *E. coli* O157:H7 Sakai strain was found to carry 18 prophages that had the genes for virulence as well as multidrug resistance (Asadulghani et al., 2009). Many of the prophages were defective but inducible from *E. coli* O157:H7 chromosomes and some could even be transferred to other *E. coli* strains. *E. coli* O157:H7 strains isolated from a mass outbreak in Obihiro-City, Hokkaido, Japan in 1996 had two distinct *tet*^{r} R plasmids with sizes of 50 and 95 kb, respectively (Makino et al., 1999). The *tet*^{r} on the plasmids were 100% homologous to the *tet*^{r} of pSC101. In a separate study, plasmids of various sizes were identified from some of the *E. coli* O157:H7 strains isolated in Japan during the period June 1996 to June 1997. Three plasmids with a molecular weight of 8, 46, and 54 MDa were found to carry the gene for ampicillin resistance. The 54-MDa plasmid also had the gene for tetracycline resistance (Miwa et al., 2002).

In addition to transmissible plasmids and temperate bacteriophage, integrons have been recognized as carriers for antibiotic resistance genes in *E. coli* O157:H7 and other STEC. An integron itself is not mobile, and its horizontal transfer between bacterial cells must be facilitated by mobile DNA elements such as transposons and transmissible plasmids (Fluit and Schmitz, 2004). Integrons carry the genes for their basic components: integrase, promoter, gene cassette, and a 59-base recombination element. The cassette contains the genes for resistance to antibiotics or disinfectants. The integrase is an enzyme that catalyzes the integration and excision of antibiotic/disinfectant resistance gene cassette at a specific receptor site, *att*. The promoter is responsible for the expression of the resistance genes in a new host. The gene for promoter is upstream of the gene cassettes in the integron (Hall and Collis, 1998), and the 59-base recombination element is at the 3' end of the coding sequence, which interacts with the *att*-specific receptor site (Hall and Collis, 1995).

At least four classes of integrons have been identified so far, which are distinguished by their respective integrase (*int*) genes (Hall and Collis, 1998). Class 1 integron is most prevalent, and its gene cassettes encode resistance against a wide range of antibiotics including β-lactam, aminoglycosides, trimethoprim, chloramphenicol, rifampin, quaternary compounds, and erythromycin (Fluit and Schmitz, 1999; Hall and Collis, 1998). Class 1 integrons are associated with Tn21 and Tn21-related transposons, which are generally located on conjugative plasmids (Fluit and Schmitz, 1999, 2004). Class 2 integron is usually associated with the Tn7 family and contains three integrated gene cassettes (*dfr1-sat-aadA1*) adjacent to a defective integrase gene (*intI2*) at the 5' end of the coding sequence. The integrase gene of class 2 integron contains an internal stop codon, thus there is a limited variation of gene cassettes among class 2 integrons. The gene cassette *dfr1-sat-aadA1* encodes resistance to trimethoprim, streptothricin, and streptomycin, respectively. Class 2 integron has been found on both conjugative plasmids and bacterial chromosomes (Rodriguez et al., 2006). Class 3 integron has been far less reported than class 1 and class 2 integrons. Gene cassette associated with class 3 integron has *bla*$_{IMP}$ that confers resistance to broad-spectrum β-lactams (Correia et al., 2003). Class 4 integron has only been found in the *Vibrio cholerae* genome and is not known to be associated with antibiotic resistance (Mazel et al., 1998).

STEC has been found to carry both class I and class II integrons. In a study conducted by Nagachinta and Chen (2009), 14 out of 175 examined STEC (8%), including one O157:H7 strain, were found to carry class 1 integron and one STEC isolate (0.6%) tested positive for class 2 integron. The cassette region of the class 1 integrons had a uniform size of 1.1 kb and contained a nucleotide sequence identical to that of *aadA1*. The cassette region of the class 2 integron was at a size of 2.0 kb and carried nucleotide sequences homologous to those of *aadA1, sat1*, and *dfrA1*. Other researchers reported higher incidences of integron-positive STEC strains. Morabito et al. (2002) identified 17 integron-positive strains from the 59 STEC strains screened in their study (29%). The host strains more frequently had serotype O111 or O26 than O157. Most integrons had *aadA1* gene cassette either alone or together with *drfA1* gene cassette. One of the integrons contained *aadA2* and *drfA12*. Most of the integrons were carried by Tn21, making them transmissible to other *E. coli* strains. In the study of Murinda et al. (2005), the class 1 integron was found in 7 out of the 14 *E. coli* isolates from dairy cows with mastitis (50%) and 1 out of the 10 O157 isolates of fecal sources (10%). Similar to that reported by Morabito et al. (2002), the O157:H7 strains tested in this particular study also had a lower prevalence of integron carriage compared to *E. coli* with other serotypes.

Antibiotic Resistance Gene Dissemination

It is known that cattle are a natural reservoir for STEC, including strains with serotype O157:H7. A study showed that cow's rumen is a possible environment for antibiotic resistance gene exchange between pathogenic and commensal bacterial cells. In a study that used a commensal *E. coli* transformed with a con-jugative R plasmid as a donor and nalidixic acid-resistant *E. coli* O157:H7 as a recipient, the R plasmid was transferred at a higher frequency during a 6-h incubation period in rumen fluid (Mizan et al., 2002). The frequency of transfer was comparable to that which occurred in LB broth, indicating that the conditions in rumen fluid supported the transfer of the antibiotic resistance plasmid from commensal *E. coli* to *E. coli* O157. Blake et al. (2003) also observed the transfer of antibiotic resistance genes between commensal *E. coli* and *E. coli* O157, which took place in a simulated porcine ileum environment. A plasmid of environmental source and with genes for resistance to amoxicillin, streptomycin, sulfonamides, tetracycline, and inorganic mercury ions was successfully transferred to *E. coli* O157:H7 in a conjugation experiment (Van Meervenne et al., 2012). Transferred resistance genes were successfully expressed, evidenced by a marked increase in the resistance of the transconjugants to antibiotics.

Horizontal transfer of conjugative plasmids was deemed responsible for the dissemination of class 1 integron-mediated antibiotic resistance genes from STEC to *E. coli* K-12 MG1655 in the study of Nagachinta and Chen (2008, 2009). Antibiotic resistance traits not mediated by integrons were co-transferred with integron-mediated antibiotic resistance genes. The antibiotic resistance gene exchange took place in microbiological media (Nagachinta and Chen, 2009) as well as artificially inoculated bovine feces and storm water (Nagachinta and Chen, 2008). Yang et al. (2010) observed the transfer of class 1 integron from commensal *E. coli* isolated from a dog fecal sample to *E. coli* O157:H7 at a frequency of 10^{-7}. In a separate study, ampicillin resistance encoded by the TEM-1 β-lactamase gene on a plasmid-borne class 1 integron was transferred from an *E. coli* O157:H7 strain isolated from a 1–year-old girl in Japan to *E. coli* HB101 in a conjugation experiment (Ahmed, 2005).

Resistance to Antibiotics as Affected by Biocide Use and Bile Exposure

Sagripanti et al. (1997) reported that bacterial strains that more frequently survived the exposure to bactericidal agents were often those that were most commonly associated with hospital infections, suggesting that pathogens that have survived prior sanitizing treatments tend to be more robust and virulent. Braoudaki and Hilton (2004a,b) found that the development of resistance to high levels of triclosan, an antimicrobial agent commonly found in consumer products, decreased the susceptibility of *E. coli* O157:H7 to chloramphenicol, erythromycin, imipenem, tetracycline, and trimethoprim, as well as a number of biocides. In comparison, triclosan-adapted cells of *E. coli* K-12 only had cross-resistance to chloramphenicol and those of *E. coli* O55 only exhibited resistance to trimethoprim. Adaptive treatment with bile salts upregulated several genes in *E. coli* O157:H7, including those for an efflux system (*acrAB*), a two-component signal transduction system (*basRS pmrAB*), and lipid A modification (*arnBCADTEF* and *ugd*) (Kus et al., 2011). Increased expression of these genes let *E. coli* O157:H7 develop a *basS-* and *arnT*-dependent resistance to polymyxin.

RESISTANCE OF *E. COLI* O157:H7 AND OTHER STEC TO ANTIMICROBIAL INTERVENTIONS

Resistance to Oxidative Stress

In the presence of an oxidizing agent such as hydrogen peroxide or sodium hypochlorite, molecular oxygen gains one or more electrons and during this process active oxygen species are formed (Hill and Allen, 1978). These species are much more reactive than molecular oxygen and can cause serious damage to DNA, RNA, protein, and lipid molecules. Bacterial cells use both enzymatic and non-enzymatic processes to destroy active oxygen species and repair the damage caused by them. Oxidative stress occurs when a bacteria cell is overwhelmed by an excessive amount of active oxygen species. A study has shown that exposure of *E. coli* O157:H7 Sakai and TM14359 to oxidative stress caused a differential expression of over 380 genes (Wang et al., 2009). Several regulatory genes responsive to oxidative stress were upregulated, including the genes for putative oxidoreductases, cysteine biosynthesis, iron–sulfur cluster assembly, and antibiotic resistance. Oxidoreductases are a class of enzymes that catalyze the oxidation–reduction of chemical compounds.

Peroxidases and superoxide dismutases are two subclasses of oxidoreductases which have been shown to be associated with STEC resistance to hydrogen peroxide (Kim et al., 2006). Superoxide dismutases are enzymes that convert superoxide radicals into hydrogen peroxide and oxygen molecules with the input of protons, while peroxidases use nicotinamide adenine dinucleotide (NADH) or nicotinamide adenosine dinucleotide phosphate (NADPH) as an electron source to break down hydrogen peroxide. Catalase is a member of the hydrogen peroxidase family, which coverts hydrogen peroxide to oxygen and water (Wang, 1955). It is one of the enzymes that bacteria use to defend themselves against oxidative stress (Heimberger and Eisenstark, 1988). Cells of *E. coli* K-12 produce two catalases, KatG and KatE, as well as alkyl hydroperoxide reductase encoded by *ahpC*. In addition to these enzymes, *E. coli* O157:H7 strain ATCC 43895 (EDL933) also expresses KatP as a functional catalase (Uhlich, 2009). The plasmid-borne *katP* is regulated by OxyR and RpoS at stationary growth phase. During exponential growth, KatP scavenged more hydrogen peroxide than KatE or AhpC, but the largest amount of exogenous H_2O_2 is scavenged by KatG in *E. coli* O157:H7 and *katG* and *ahpC* together are sufficient for a full peroxide resistance in disk diffusion assays (Uhlich, 2009).

Other enzymes related to bacterial resistance to oxidative stress include glutathione synthetase, glutathione reductase, and glutathione peroxidase. Glutathione peroxidase catalyzes the reduction of hydrogen peroxide and hydroperoxide by glutathione, an important antioxidant tripeptide composed of glutamate, cysteine, and glycine (Saydam et al., 1997). Production of glutathione is catalyzed by glutathione synthetase. Glutathione reductase catalyzes the reduction of oxidized glutathione to reduced glutathione in the presence of NADPH. The glutathione synthesis pathway has been identified in *E. coli* O157:H7 strain EDL933 (Ohnishi et al., 2002).

Bearson et al. (2009) reported that glutamate- and arginine-dependent acid resistance systems were capable of providing protection to *E. coli* O157:H7 against oxidative stress caused by the exposure to diamide, a membrane-permeable thiol-specific oxidizing agent, or hydrogen peroxide. The protection against oxidative stress in the presence of glutamate required a low external pH such as 2.5. At pH 5.5, a similar protective effect did not occur.

Resistance to Osmotic Stress

As osmolytes, sugar and salt reduce water activity and increase the osmotic pressure of the environment, therefore, inhibiting the growth of microorganisms. STEC O157:H7 grows well in 2.5% NaCl and slowly in 6.5% NaCl. It does not grow at all in 8.5% NaCl (Glass et al., 1992). The minimum water activity required for STEC growth is about 0.95, similar to other *E. coli* cells.

In a high osmotic environment, *E. coli* cells use available osmoregulation systems to maintain their internal osmotic pressure by accumulating compatible solutes which are small organic molecules sharing a number of common properties (Galinski and Trüper, 1994). The uptake system for glycine betaine, an important compatible solute, is transcriptionally induced by an increased osmotic pressure (Gutierrez et al. 1995). Similar to the protective ability of betaine, synthesized or exogenously provided carnitine was found to be associated with aerobic growth of *E. coli* O157:H7 under osmotic stress (Verheul et al., 1998). Osomoprotectant transporters, ProP and ProU, for organic osmolytes such as betaine and carnitine are present and expressed in both commensal and pathogenic *E. coli* strains (Ly et al., 2004). The transport system is believed to play important roles in pathogen growth in low-moisture food (Verheul et al., 1998).

Osmotic stress in *E. coli* induces the expression of *trk* and *kdp* (Laimins et al., 1981), and both genes have been identified in multiple strains of *E. coli* O157:H7 (www.uniport.org). Trk is a low-affinity potassium uptake system, while Kdp is a high-affinity potassium uptake system. The expression of the two systems results in an increased uptake of potassium, which could lead to the restoration of normal osmotic pressure. Increased synthesis or uptake of glutamate that occurs concurrently with the uptake of potassium elevates the cytoplasmic concentration of potassium glutamate to a point that could impair enzyme function and inhibit bacterial growth. However, the increased cytoplasmic concentration of potassium glutamate causes the accumulation of thehalose, an α-linked disacchride of glucose which regulates the rate of release of potassium glutamate.

The role of several genes, including *recA*, *dps*, and *rpoS*, of *E. coli* O157:H7 was examined in desiccating and osmotically challenging environments (Stasic et al., 2012). It was found that RpoS was necessary for *E. coli* O157:H7 survival during desiccation and osmotic stress, and the contribution of Dps to survival was only significant in Luria-Bertani (LB) with 12% NaCl at 37°C. Using several *gfp(uv)* gene fusions, Gawande and Griffiths (2005) found that osmotic shock caused about a fourfold increase in green fluorescence of *E. coli* O157:H7 harboring *uspA::gfp(uv)* or *rpoS::gfp(uv)* at 37°C and room temperature, whereas osmotic shock at 5°C did not induce green fluorescence. Garmendia and Frankel (2005) studied the expression of *E. coli* O157 *espJ--tccP* operon, which encodes the type III secretion system effector proteins EspJ and TccP. It was found that *in vitro* expression of the genes is affected by the composition of growth medium and environmental signals such as temperature, pH, osmolarity, and O_2 pressure. Parker et al. (2012) have shown that many σ(S)-regulated genes, including *gadA*, *osmE*, *osmY*, and *katE*, were correlated with the resistance of *E. coli* O157:H7 to not only osmotic stress but also acid stress and oxidative stress.

Resistance to Acidic Stress

The optimum growth of STEC occurs at pH 6–7. The rate of growth declines rapidly at pH below this range (Buchanan and Klawitter, 1992; Vimont et al., 2007). Some strains of STEC are acid-resistant and non-pathogenic *E. coli* strains are, in general less acid-tolerant than pathogenic strains (Gorden and Small, 1993). Acid-resistant strains of *E. coli* O157:H7 can grow at pH 4.6 and some can even survive at pH as low as 2.5 (Vimont et al., 2007). Both organic acid and inorganic acid inhibit the growth of bacterial pathogens, but the mechanism of inhibition by the two acids is completely different. At pH below their pK_a values organic acids are present in the un-disassociated form which can diffuse across bacterial cell membrane, subsequently dissociate and acidify cytoplasm causing the ease of bacterial metabolism and death of bacterial cells. Thus, organic acid tends to be more effective than inorganic acid in inhibiting the growth of bacterial pathogens.

The involvement of an oxidative system, a glutamate decarboxylase system, and an arginine decarboxylase system in the acid resistance of *E. coli* has been studied (Lin et al., 1995; Chung et al., 2006). All three systems are regulated by the alternative sigma factor, σ^s (Lin et al., 1996; Bhagwat, 2003). The oxidative acid resistance system is induced during bacterial growth in rich media and under aerobic condition. The system protects *E. coli* cells from acidic pH above 3.0 (Bhagwat et al., 2006). The arginine decarboxylase- and glutamate decarboxylase-dependent acid resistance systems are induced in minimal growth media supplemented with glutamate or arginine and confer protection to cells at pH below 3.0 (Lin et al., 1996; Bhagwat et al., 2006). The amino acid decarboxylase systems maintain intracellular pH by consuming protons during amino acid decarboxylation, thus alleviating the acidification of cytoplasm and contributing to the alkalization of the environment (Cotter et al., 2001). Diez-Gonzalez and Karaibrahimoglu (2004) investigated the effect of growth conditions on acid resistance systems of *E. coli* O157:H7. It was found that although the lysine-dependent acid resistance system could be just as important as the arginine-dependent system, the contribution of both systems to overall acid resistance response of *E. coli* O157:H7 might be minor compared with the glutamate-dependent system. In addition, glutathione reductase, the enzyme important for bacterial resistance to oxidative stress has been shown to assist *E. coli* O157:H7 to combat acid stress according to a study of Bearson et al. (2003).

Effect of Repeated Exposure and Pro-Adaptation on Resistance

The adaptive responses of *E. coli* to environmental stress have been investigated intensively. Cells' response to one type of stress can sometimes increase their resistance to other adverse environmental conditions (Volker et al., 1992; Gunasekera et al., 2008) and repeated exposure to a specific type of stress can elevate the level of bacterial tolerance. A study has found that 50 consecutive exposures to disinfection significantly increased the resistance of *E. coli* O157:H7 to sanitizers including sodium dichloroisocyanurate, iodophor, and quaternary ammonium, but the adaptive treatments did not change the resistance of *E. coli* O157:H7 to chlorhexidiniacetas (Zhang et al., 2005). Further analysis revealed that the exposure to sodium dichloroisocyanurate and quaternary ammonium changed the restriction profiles of pO157 and PFGE patterns of chromosomal DNA, suggesting that both chromosome DNA and pO157 have genes related to resistance to the two sanitizers. Repeated disinfection with iodophor only changed the PFGE pattern of chromosomal DNA, and neither plasmid profile nor PFGE pattern was changed by the repeated disinfection with chlorhexidiniacetas. Rajkovic et al. (2009) found that exposure of multiple strains of *E. coli* O157:H7 for 20 consecutive cycles did not result in an increased resistance to chlorine dioxide. However, repeated use of a physical intervention with intense light pulse significantly increased the resistance of *E. coli* O157:H7 to the treatment with chlorine dioxide (Rajkovic et al., 2009).

Pre-exposure of *E. coli* O157:H7 to a sub-lethal concentration (0.1%) of peroxyacetic acid significantly increased its tolerance to oxidative stress (80 mM H_2O_2) (Zook et al., 2001). Peroxyacetic acid-adapted cells had a $D_{80\,mM}$ value of greater than 3 h while the non-adapted cells had a $D_{80\,mM}$ value of 0.19 h. Neo et al. (2013) found that acid-adapted cells were more resistant to peroxyacetic acid as well as chlorine treatment than non-adapted cells. Acid adaptation also increased the resistance of *E. coli* O157:H7 to Dynashock wave power ultrasound in cloudy apple juice (Gabriel, 2012). However, acid adaptation did not increase the resistance of *E. coli* O157:H7 to the antimicrobial hurdles involved in jerky processing (Calicioglu, 2002). Chapman et al. (2006) reported that an increased salt concentration prolonged the survival of *E. coli* O157 in a broth model simulating the aqueous phase of a food dressing or sauce containing acetic acid.

Heat Susceptibility as Affected by Adaptive Treatment

Splittstoesser et al. (1996) found that increasing the Brix from 11.8° to 16.5° had no effect on thermal resistance of *E. coli* O157:H7 in a model Empire apple juice, while increasing L-malic acid concentration from 0.2% to 0.8% or reducing the pH from 4.4 to 3.6 significantly increased the susceptibility of *E. coli* O157:H7 to heat treatment. The presence of 1,000 mg/L of benzoic and sorbic acid decreased the tolerance of *E. coli* O157:H7 to heat. The D_{50} was 5.2 min in the presence of sorbic acid and 0.64 min in the presence of benzoic acid compared with an average D_{52} of 18 min in juice without any chemical preservatives. Enache et al. (2006) found that cells of *E. coli* O157:H7 were more heat-resistant after being exposed to adaptation under high osmotic pressure or low pH in single-strength juice. In juice concentrate however, acid-adapted cells were less heat-resistant than non-adapted cells. In the study of Dock et al. (2000), modification of the pH and addition of chemical preservatives: potassium sorbate (0–0.2%), sodium benzoate (0–0.2%), and malic acid (0–1%) into apple juice significantly decreased the D values of *E. coli* O157:H7. In natural apple cider, the D_{50} was about 65 min, and this value decreased significantly in cider supplemented with malic acid, sorbate, and benzoate. Addition of both sorbate and malic acid did not further lower the D value. However, the addition of 0.2% benzoate and 1% malic acid lowered D_{50} to 0.3 min. Addition of all three preservatives (0.2% sorbate, 0.2% benzoate, and 1% malic acid) resulted in a D_{50} of 18 s. In a separate study, Huang and Juneja (2003) found that when sodium lactate was added to ground beef, heating temperature was the only factor affecting the D-values. The thermal resistance of *E. coli* O157:H7 was neither affected by the addition of sodium lactate nor the interactions between sodium lactate and temperature. Zook et al. (2001) found that preadaptation with peroxyacetic acid adaptation did not increase the thermal resistance of *E. coli* O157:H7 at 54°C.

Resistance of *E. coli* O157:H7 as Affected by Its Physiological State

The physiological state of bacterial cells affects their tolerance to environmental stress. Cells in the stationary phase are more resistant to stress than those in the exponential phase. Cherchi and Gu (2011) found that *E. coli* O157:H7 at stationary phase was more resistant to oxidative stress, particularly to chlorine treatment, and had the lowest rate of inactivation compared to cells in lag and exponential phase. The impact of cell age on osmotic stress was observed by Hajmeet et al. (2006) who reported that 24-hour-old cells were more resistant to salt treatment compared to 12-hour-old cells. Enache et al. (2006) studied the thermal resistance of different bacterial pathogens in white grape juice concentrate and found that the most heat-resistant cells of *E. coli* O157:H7 were those in the stationary phase. In a study of Lisle et al. (1998) starvation increased bacterial tolerance to stress, and chlorine resistance of *E. coli* O157:H7 progressively increased through the 29-day starvation treatment. A strain-specific resistance of wild-type strains of *E. coli* O157:H7 to high hydrostatic pressure, mild heat and other stresses was observed by Benito (1999). Stationary phase cells of pressure-resistant strains were also more resistant to acid, oxidative, and osmotic stress although similar protection was also observed with cells in the exponential growth phase. Arnold and Kaspar (1995) observed that acid tolerance in the stationary phase was not affected by pre-adaptation at milder pH levels.

Influence of Extracellular Polysaccharide Production and Biofilm Formation on Resistance

E. coli produces a number of cell surface components and some are polysaccharide in nature. Two good examples are cellulose and the EPS comprised of colanic acid. Using a colanic acid-proficient wild-type *E. coli* O157 strain and its colanic acid-deficient mutant, Mao et al. (2006) demonstrated that the colanic acid-deficient mutant was substantially less tolerant to acid (pH 4.5 and 5.5) and heat (55°C and 60°C) in comparison to its wild-type parent. In a subsequent study, Chen et al. (2004) found that cells of *E. coli* O157:H7 deficient in colanic acid production were more susceptible than its wild-type parent to treatments with salt ($P < 0.05$) and hydrogen peroxide ($P \leq 0.05$).

The contribution and importance of EPS in biofilm formation have been extensively investigated. Biofilm is a structure that has been shown to protect bacterial cells against treatment with antibiotics as well as other antimicrobial agents. Uhlich et al. (2006) observed that biofilm-associated cells of two *E. coli* O157:H7 strains had a greater resistance ($P < 0.05$) to hydrogen peroxide and quaternary ammonium sanitizer than their respective planktonic cells. The study also shows that the rdar phenotype of *E. coli* O157:H7 strain, producing both curli and cellulose as cell surface components, is important for resistance to sanitizers.

Cellulose is a long polymer of glucose with β-1-4 glycosidic bonds and is produced by many members of the *Enterobacteriaceae*. In a study by Park and Chen (2011) cellulose-producing STEC were treated with acetic and lactic acid (2% and 4%) under laboratory conditions. It was found that the amounts of cellulose produced by STEC affected their susceptibilities to the acid treatments. The residual amounts of cellulose on STEC positively correlated to the surviving populations of STEC after the treatments with the two acids ($r = 0.64–0.94$) with a significance of correlation ranging from 83% to 99%. A separate study showed that cells expressing a greater amount of cellulose formed more biofilms, were more difficult to inactivate, and had greater residual biofilm mass left after sanitizing treatments (Park and Chen, unpublished). Uhlich et al. (2010) found that *E. coli* O157 retained in biofilm formed by its companion strains had increased resistance to oxidative stress. Wang et al. (2013) also reported that bacterial EPS not only enhanced sanitizer resistance of the EPS-producing strains but also rendered protection to the strains co-existing with them, regardless of the species, in multi-species biofilms.

Spontaneous cellulose-deficient mutants have been intensively studied, and the conversion from cellulose-proficient wild-type to cellulose-deficient phenotype is believed to be the result of different levels of expression of the genes encoding the enzymes responsible for cellulose biosynthesis (Krystynowicz et al., 2005). In a study by Yoo et al. (2010), cellulose-deficient spontaneous mutants of two STEC strains were selected by repeatedly selecting dimmer colonies on LB no salt agar supplemented with calcofluor white stain because cellulose-proficient colonies would fluoresce on the medium under a long-wave UV light. Using the wild-type STEC strains and their spontaneous cellulose-deficient counterparts, Yoo and Chen (2010, 2012) found that cellulose production conferred cells tolerance to oxidative stress and osmotic stress, as well as acid stress.

CONCLUSION

Shiga toxin-producing *E. coli* and enterohemorrhagic *E. coli* are important foodborne bacterial pathogens. Their resistance to antimicrobial substances and interventions is an important food safety issue. Understanding of the mechanisms of antimicrobial resistance is critical and will lead to the development of effective control strategies, and subsequently the availability of safe food supplies.

REFERENCES

Ahmed, A.M., 2005. Genomic analysis of a multidrug-resistant strain of enterohaemorrhagic *Escherichia coli* O157:H7 causing a family outbreak in Japan. J. Med. Microbiol. 54, 867–872.

Arnold, K.W., Kaspar, C.W., 1995. Starvation- and stationary induced acid tolerance in *Escherichia coli* O157:H7. Appl. Environ. Microbiol. 61, 2037–2039.

Asadulghani, M., Ogura, Y., Ooka, T., Itoh, T., Sawaguchi, A., Iguchi, A., et al., 2009. The defective prophage pool of *Escherichia coli* O157: prophage–prophage interactions potentiate horizontal transfer of virulence determinants. PLoS. Pathog. Available from: http://dx.doi.org/doi:10.1371/journal.ppat.1000408.

Bearson, B., Bearson, S.M., Casey, T.A., 2003. Glutathione is Required for Acid Resistance in *E. coli* O157:H7. Am. Soc. Microbiol. 365 [Abstract].

Bearson, B.L., Lee, I.S., Caset, T.A., 2009. *Escherichia coli* O157:H7 glutamate- and arginine-dependent acid-resistance systems protect against oxidative stress during extreme acid challenge. Microbiology 155, 805–812.

Benito, A., 1999. Variation in resistance of natural isolates of *Escherichia coli* O157 to high hydrostatic pressure, mild heat, and other stresses. Appl. Environ. Microbiol. 65, 1564–1569.

Bhagwat, A.A., 2003. Regulation of the glutamate-dependent acid-resistance system of diarrheagenic *Escherichia coli* strains. FEMS Microbiol. Lett. 227, 39–45.

Bhagwat, A.A., Tan, J., Sharma, M., Kothary, M., Low, S., Tall, B.D., et al., 2006. Functional heterogeneity of RpoS in stress tolerance of enterohemorrhagic *Escherichia coli* strains. Appl. Environ. Microbiol. 72, 4978–4986.

Blake, D.P., Hillman, K., Fenlon, D.R., Low, J.C., 2003. Transfer of antibiotic resistance between commensal and pathogenic members of the Enterobacteriaceae under ileal conditions. J. Appl. Microbiol. 95, 428–436.

Brabban, A.D., Hite, E., Callaway, T.R., 2005. Evolution of foodborne pathogens via temperate bacteriophage-mediated gene transfer. Foodborne Pathog. Dis. 2, 287–303.

Braoudaki, M., Hilton, A.C., 2004a. Adaptive resistance to biocides in *Salmonella enterica* and *Escherichia coli* O157 and cross-resistance to antimicrobial agents. J. Clin. Microbiol. 42, 73–78.

Braoudaki, M., Hilton, A.C., 2004b. Low level of cross-resistance between triclosan and antibiotics in *Escherichia coli* K-12 and *E. coli* O55 compared to *E. coli* O157. FEMS Microbiol. Lett. 235, 305–309.

Brooks, H.J.L., Mollison, B.D., Bettelheim, K.A., Matejka, K., Paterson, K.A., Ward, V.K., 2001. Occurrence and virulence factors of non-O157 Shiga toxin-producing *Escherichia coli* in retail meat in Dunedin, New Zealand. Lett. Appl. Microbiol. 32, 118–122.

Buchanan, R.L., Klawitter, L.A., 1992. The effect of incubation-temperature, initial pH, and sodium-chloride on the growth-kinetics of *Escherichia coli* O157:H7. Food Microbiol. 9, 185–196.

Calicioglu, M., 2002. Inactivation of acid-adapted and nonadapted *Escherichia coli* O157:H7 during drying and storage of beef jerky treated with different marinades. J. Food Prot. 65, 1394–1405.

CDC, 1997. Outbreaks of *Escherichia coli* O157:H7 infection and cryptosporidiosis associated with drinking unpasteurized apple cider-Connecticut and New York, October 1996. Morb. Mortal. Wkly. Rep. 46, 4–8.

CDC, 2007. Preliminary FoodNet data on the incidence of infection with pathogens transmitted commonly through food—10 States, 2006. Morb. Mortal. Wkly. Rep. 56, 336–339.

CDC, 2013. Press release: Untreatable: reports by CDC details today's drug resistance health threats. <http://www.cdc.gov/media/releases/2013/p0916-untreatable.html> (accessed 06.06.14).

Chapman, B., Jensen, N., Ross, T., Cole, M., 2006. Salt, alone or in combination with sucrose, can improve the survival of *Escherichia coli* O157 (SERL 2) in model acidic sauces. Appl. Environ. Microbiol. 72, 165–172.

Chen, J., Lee, S.M., Mao, Y., 2004. Protective effect of exopolysaccharide colanic acid of *Escherichia coli* O157:H7 to osmotic and oxidative stress. Int. J. Food Microbiol. 93, 281–286.

Cherchi, C., Gu, A.Z., 2011. Effect of bacterial growth stage on resistance to chlorine disinfection. Water Sci. Technol. 64, 7–13.

Chung, H.J., Bang, W., Drake, M.A., 2006. Stress response of *Escherichia coli*. Compr. Rev. Food Sci. Food Saf. 5, 52–64.

Cohen, M.L., 2000. Changing patterns of infectious disease. Nature 406, 762–767.

Correia, M., Boavida, F., Grosso, F., Salgado, M.J., Lito, L.M., Cristino, J.M., et al., 2003. Molecular characterization of a new class 3 integron in *Klebsiella pneumoniae*. Antimicrob. Agents Chemother. 47, 2838–2843.

Cotter, P.D., Gahan, C.G., Hill, C., 2001. Glutamatedecarboxylase system protects *Listeria monocytogenes* in gastric fluid. Mol. Microbiol. 40, 465–475.

Cromwell, G.L., 1991. Antimicrobial agents. In: Miller, E.R., Ullrey, D.E., Lewis, A.J. (Eds.), Swine Nutrition Butterworth-Heinemann, Stoneham, MA, pp. 297–314.

Dargatz, D.A., Fedorka-Cray, P.J., Ladely, S.R., Ferris, K.E., 2000. Survey of *Salmonella* serotypes shed in feces of beef cows and their antimicrobial susceptibility patterns. J. Food Prot. 63, 1648–1653.

Diez-Gonzalez, F., Karaibrahimoglu, Y., 2004. Comparison of the glutamate-, arginine- and lysine-dependent acid resistance systems of *Escherichia coli* O157:H7. J. Appl. Microbiol. 96, 1237–1244.

Dock, L.L., Floros, J.D., Linton, R.H., 2000. Heat inactivation of *Escherichia coli* O157:H7 in apple cider containing malic acid, sodium benzoate, and potassium sorbate. J. Food Prot. 63, 1026–1031.

Doyle, M.P., Schoeni, J.L., 1987. Isolation of *Escherichia coli* O157:H7 from retail fresh meats and poultry. Appl. Environ. Microbiol. 53, 2394–2396.

Elliott, E.J., Robins-Browne, R.M., O'Loughlin, E.V., Bennett-Wood, V., Bourke, J., Henning, P., et al., 2001. Nationwide study of haemolytic uraemic syndrome: clinical, microbiological, and epidemiological features. Arch. Dis. Child. 85, 125–131.

Enache, E., Chen, Y., Awuah, G., Economides, A., Scott, V.N., 2006. Thermal resistance parameters for pathogens in white grape juice concentrate. J. Food Prot. 69, 564–569.

Endtz, H.P., Ruijs, G.J., van Klingeren, B., Jansen, W.H., van der Reyden, T., Mouton, R.P., 1991. Quinolone resistance in *Campylobacter* isolated from man and poultry following the introduction of fluoroquinolones in veterinary medicine. J. Antimicrob. Chemother. 27, 199–208.

Fluit, A.C., Schmitz, F.J., 1999. Class 1 integrons, gene cassettes, mobility, and epidemiology. Eur. J. Clin. Microbiol. Infect. Dis. 18, 761–770.

Fluit, A.C., Schmitz, F.J., 2004. Resistance integrons and super-integrons. Clin. Microbiol. Infec. 10, 272–288.

Gabriel, A.A., 2012. Microbial inactivation in cloudy apple juice by multi-frequency Dynashock power ultrasound. Ultrason. Sonochem. 19, 346–351.

Galinski, C., Trüper, A., 1994. Microbial behaviour in salt-stressed ecosystems. FEMS Microbiol. Rev. 15, 95–108.

Garmendia, J., Frankel, G., 2005. Operon structure and gene expression of the *espJ–tccP* locus of enterohaemorrhagic *Escherichia coli* O157:H7. FEMS Microbiol. Lett. 247, 137–145.

Gawande, P.V., Griffiths, M.W., 2005. Effects of environmental stresses on the activities of the *uspA*, *grpE* and *rpoS* promoters of *Escherichia coli* O157:H7. Int. J. Food Microbiol. 99, 91–98.

Gerber, A., Karch, H., Allerberger, F., Verweyen, H.M., Zimmerhackl, L.B., 2002. Clinical course and the role of Shiga toxin-producing *Escherichia coli* infection in the hemolytic-uremic syndrome in pediatric patients, 1997–2000, in Germany and Austria: a prospective study. J. Infect. Dis. 186, 493–500.

Glass, K.A., Loeffelholz, J.M., Ford, J.P., Doyle, M.P., 1992. Fate of *Escherichia coli* O157:H7 as affected by pH or sodium chloride and in fermented, dry sausage. Appl. Environ. Microbiol. 58, 2513–2516.

Gorden, J., Small, P.L., 1993. Acid resistance in enteric bacteria. Infect. Immun. 61, 364–367.

Gunasekera, T.S., Csonka, L.N., Paliy, O., 2008. Genome-wide transcriptional responses of *Escherichia coli* K-12 to continuous osmotic and heat stresses. J. Bacteriol. 190, 3712–3720.

Gutierrez, C., Abee, T., Booth, I.R., 1995. Physiology of the osmotic stress response in microorganisms. Int. J. Food Microbiol. 28, 233–244.

Hajmeet, M., Ceylan, E., Marsden, J.L., Fund, D.Y., 2006. Impact of sodium chloride on *Escherichia coli* O157:H7 and *Staphylococcus aureus* analysed using transmission electron microscopy. Food Microbiol. 23, 446–452.

Hall, R.M., Collis, C.M., 1995. Mobile gene cassettes and integrons: capture and spread of genes by site-specific recombination. Mol. Microbiol. 15, 593–600.

Hall, R.M., Collis, C.M., 1998. Antibiotic resistance in gram-negative bacteria: the role of gene cassettes and integrons. Drug Resist. Update 1, 109–119.

Heimberger, A., Eisenstark, A., 1988. Compartmentalization of catalases in *Escherichia coli*. Biochem. Biophys. Commun. 154, 392–397.

Hill, O., Allen, H., 1978. The chemistry of dioxygen and its reduction products. In: Fitzsimons, D.W. (Ed.), Oxygen Free Radicals and Tissue Damage Elsevier Science Publishing, Inc., New York, NY, pp. 5–12.

Huang, L., Juneja, V.K., 2003. Thermal inactivation of *Escherichia coli* O157:H7 in ground beef supplemented with sodium lactate. J. Food Prot. 66, 664–667.

John, F.P., 2006. History of antimicrobial usage in agriculture: an overview. In: Aarestrup, F.M. (Ed.), Antimicrobial Resistance in Bacteria of Animal Origin ASM Press, Washington, DC, pp. 19–29.

Kim, Y.H., Lee, Y., Kim, S., Yeom, J., Yeom, S., Seok Kim, B., et al., 2006. The role of periplasmic antioxidant enzymes (superoxide dismutase and thiol peroxidase) of the Shiga toxin-producing *Escherichia coli* O157:H7 in the formation of biofilms. Proteomics 6, 6181–6193.

Krystynowicz, A., Koziolkiewicz, M., Wiktorowska-Jezierska, A., Bielecki, S., Klemenska, E., Masny, A., et al., 2005. Molecular basis of cellulose biosynthesis disappearance in submerged culture of *Acetobacter xylinum*. Acta Biochim. Pol. 52, 691–698.

Kus, J.V., Gebremedhin, A., Dang, V., Tran, S.L., Serbanescu, A., Barnett Foster, D., 2011. Bile salts induce resistance to polymyxin in enterohemorrhagic *Escherichia coli* O157:H7. J. Bacteriol. 193, 4509–4515.

Laimins, L.A., Rhoads, D.B., Epstein, W., 1981. Osmotic control of *kdp* operon expression in *Escherichia coli*. Proc. Natl. Acad. Sci. USA 78, 464–468.

Landers, T.F., Coher, B., Larson, E.L., 2012. A review of antibiotic use in food animals: perspective, policy and potential. Public Health Rep. 127, 4–22.

Levy, S.B., FitzGerald, G.B., Macone, A.B., 1976. Spread of antibiotic-resistant plasmids from chicken to chicken and from chicken to man. Nature 260, 40–42.

Lin, J., Lee, I.S., Frey, J., Slonczewski, J.L., Foster, J.W., 1995. Comparative analysis of extreme acid survival in *Salmonella typhimurium*, *Shigella flexneri*, and *Escherichia coli*. J. Bacteriol. 177, 4097–4104.

Lin, J., Smith, M.P., Chapin, K.C., Baik, H.S., Bennett, G.N., Foster, J.W., 1996. Mechanisms of acid resistance in enterohemorrhagic *Escherichia coli*. Appl. Environ. Microbiol. 62, 3094–3100.

Linton, A.H., Howe, K., Osborne, A.D., 1975. The effects of feeding tetracycline, nitrovin and quindoxin on the drug-resistance of coli-aerogenes bacteria from calves and pigs. J. Appl. Bacteriol. 38, 255–275.

Lisle, J.T., Broadaway, S.C., Prescott, A.M., Pyle, B.H., Fricker, C., McFeters, G.A., 1998. Effects of starvation on physiological activity and chlorine disinfection resistance in *Escherichia coli* O157:H7. Appl. Environ. Microbiol. 64, 4658–4662.

Low, J.C., Angus, M., Hopkins, G., Munro, D., Rankin, S.C., 1997. Antimicrobial resistance of *Salmonella enterica* Typhimurium DT104 isolates and investigation of strains with transferable apramycin resistance. Epidemiol. Infect. 118, 97–103.

Ly, A., Henderson, J., Lu, A., Culham, D.E., Wood, J.M., 2004. Osmoregulatory systems of *Escherichia coli*: identification of betaine-carnitine-choline transporter family member BetU and distributions of *betU* and *trkG* among pathogenic and nonpathogenic isolates. J. Bacteriol. 186, 296–306.

Makino, S., Asakura, H., Obayashi, T., Shirahata, T., Ikeda, T., Takeshim, K., 1999. Molecular epidemiological study on tetracycline resistance R plasmids in enterohaemorrhagic *Escherichia coli* O157:H7. Epidemiol. Infect. 123, 25–30.

Mao, Y., Doyle, M.P., Chen, J., 2006. Role of colanic acid exopolysaccharide in the survival of enterohaemorrhagic *Escherichia coli* O157:H7 in simulated gastrointestinal fluids. Lett. Appl. Microbiol. 42, 642–647.

Mazel, D., Dychinco, B., Webb, V.A., Davies, J., 1998. A distinctive class of integron in the *Vibrio cholerae* genome. Science 280, 605–608.

Miwa, Y., Matsumoto, M., Hiramatsu, R., Yamazaki, M., Saito, H., Saito, M., et al., 2002. Drug resistance of enterohemorrhagic *Escherichia coli* O157 and a possible relation of plasmids to the drug-resistance. Kansenshogaku. Zasshi. 76, 285–290.

Mizan, S., Lee, M.D., Harmon, B.G., Tkalcic, S., Maurer, J.J., 2002. Acquisition of antibiotic resistance plasmids by enterohemorrhagic *Escherichia coli* O157:H7 within rumen fluid. J. Food Prot. 65, 1038–1040.

Moore, P.R., Evenson, A., Luckey, T.D., McCoy, E., Elvehjem, C.A., Hart, E.B., 1946. Use of sulfasuxidine, streptothricin, and streptomycin in nutritional studies with the chick. J. Biol. Chem. 165, 437–441.

Morabito, S., Tozzoli, R., Caprioli, A., Karch, H., Carattoli, A., 2002. Detection and characterization of class 1 integrons in enterohemorrhagic *Escherichia coli*. Microb. Drug Resist. 8, 85–91.

Murinda, S.E., Ebner, P.D., Nguyen, L.T., Mathew, A.G., Oliver, S.P., 2005. Antimicrobial resistance and class 1 integrons in pathogenic *Escherichia coli* from dairy farms. Foodborne Pathog. Dis. 2, 348–352.

Nagachinta, S., Chen, J., 2008. Transfer of class 1 integron-mediated antibiotic resistance genes from Shiga toxin-producing *Escherichia coli* to a susceptible *E. coli* K-12 strain in storm water and bovine feces. Appl. Environ. Microbiol. 74, 5063–5067.

Nagachinta, S., Chen, J., 2009. Integron-mediated antibiotic resistance in Shiga toxin-producing *Escherichia coli*. J. Food Prot. 72, 21–27.

Neo, S., Lim, P.Y., Phua, L.K., Khoo, G.H., Kim, S.J., Lee, S.C., et al., 2013. Efficacy of chlorine and peroxyacetic acid on reduction of natural microflora, *Escherichia coli* O157:H7, *Listeria monocyotgenes* and *Salmonella* spp. on mung bean sprouts. Food Microbiol. 36, 475–480.

O'Brien, A.D., LaVeck, G.D., Thompson, M.R., Formal, S.B., 1982. Production of *Shigella dysenteriae* type 1-like cytotoxin by *Escherichia coli*. J. Infect. Dis. 146, 763–769.

Ohnishi, M., Terajima, J., Kurokawa, K., Nakayama, K., Murata, T., Tamura, K., et al., 2002. Genomic diversity of enterohemorrhagic *Escherichia coli* O157 revealed by whole genome PCR scanning. Proc. Natl. Acad. Sci. USA 99, 17043–17048.

Park, Y.J., Chen, J., 2011. Inactivation of Shiga toxin-producing *Escherichia coli* (STEC) and degradation/removal of cellulose from STEC surfaces using selected enzymatic and chemical treatments. Appl. Environ. Microbiol. 77, 8532–8537.

Park, Y.J., Chen, J., Unpublished. Control of biofilms of Shiga toxin-producing *Escherichia coli* using treatments with organic acids and commercial sanitizers.

Parker, C.T., Kyle, J.L., Huynh, S., Carter, M.Q., Brandl, M.T., Mandrell, R.E., 2012. Distinct transcriptional profiles and phenotypes exhibited by *Escherichia coli* O157:H7 isolates related to the 2006 spinach-associated outbreak. Appl. Environ. Microbiol. 78, 455–463.

Rajkovic, A., Smigic, N., Uyttendaele, M., Medic, H., de Zutter, L., Devlieghere, F., 2009. Resistance of *Listeria monocytogenes*, *Escherichia coli* O157:H7 and *Campylobacter jejuni* after exposure to repetitive cycles of mild bactericidal treatments. Food Microbiol. 26, 889–895.

Rodriguez, I., Martin, M.C., Mendoza, M.C., Rodicio, M.R., 2006. Class 1 and class 2 integrons in non-prevalent serovars of *Salmonella enterica*: structure and association with transposons and plasmids. J. Antimicrob. Chemother. 58, 1124–1132.

Sagripanti, J.L., Eklund, C.A., Trost, P.A., Jinneman, K.C., Abeyta Jr, C., Kaysner, C.A., et al., 1997. Comparative sensitivity of 13 species of pathogenic bacteria to seven chemical germicides. Am. J. Infect. Control. 25, 335–339.

Saydam, N., Kirb, A., Demirb, Ö., Hazanc, E., Oto, Ö., Saydam, O., et al., 1997. Determination of glutathione, glutathione reductase, glutathione peroxidase and glutathione *S*-transferase levels in human lung cancer tissues. Cancer. Lett. 119, 13–19.

Schwarz, S., Chaslus-Dancla, E., 2001. Use of antimicrobials in veterinary medicine and mechanisms of resistance. Vet. Res. 32, 201–225.

Schwarz, S., Cloeckaert, A., Roberts, M.C., 2006. Mechanism and spread of bacterial resistance to antimicrobial agents. In: Aarestrup, F.M. (Ed.), Antimicrobial Resistance in Bacteria of Animal Origin ASM Press, Washington, DC, pp. 73–98.

Siegler, R.L., 1995. The hemolytic uremic syndrome. Pediatr. Clin. N. Am. 42, 1505–1529.

Splittstoesser, D.F., McLellan, M.R., Churey, J.J., 1996. Heat resistance of *Escherichia coli* O157:H7 in apple juice. J. Food Prot. 59, 226–229.

Stasic, A.J., Lee Wong, A.C., Kaspar, C.W., 2012. Osmotic and desiccation tolerance in *Escherichia coli* O157:H7 requires *rpoS* (σ(38)). Curr. Microbiol. 65, 660–665.

Thomas, C.M., Nielsen, K.M., 2005. Mechanisms of and barriers to, horizontal gene transfer between bacteria. Nat. Rev. Microbiol. 3, 711–721.

Uhlich, G.A., 2009. KatP contributes to OxyR-regulated hydrogen peroxide resistance in *Escherichia coli* serotype O157:H7. Microbiology 155, 3589–3598.

Uhlich, G.A., Cooke, P.H., Solomon, E.B., 2006. Analyses of the red-dry-rough phenotype of an *Escherichia coli* O157:H7 strain and its role in biofilm formation and resistance to antibacterial agents. Appl. Environ. Microbiol. 72, 2564–2572.

Uhlich, G.A., Rogers, D.P., Mosier, D.A., 2010. *Escherichia coli* serotype O157:H7 retention on solid surfaces and peroxide resistance is enhanced by dual-strain biofilm formation. Foodborne Pathog. Dis. 7, 935–943.

Van Meervenne, E., Van Coillie, E., Kerckhof, F.M., Devlieghere, F., Herman, L., De Gelder, L.S., et al., 2012. Strain-specific transfer of antibiotic resistance from an environmental plasmid to foodborne pathogens. J. Biomed. Biotechnol. 2012, 8. http://dx.doi.org/doi:10.1155/2012/834598 [Article ID 834598].

Verheul, A., Wouters, J.A., Rombouts, F.M., Abee, T., 1998. A possible role of ProP, ProU and CaiT in osmoprotection of *E. coli* by carnitine. J. Appl. Microbiol. 85, 1036–1046.

Vimont, A., Vernozy-Rozand, C., Montet, M.P., Bavai, C., Fremaux, B., Delignette-Muller, M.L., 2007. Growth of Shiga-toxin producing *Escherichia coli* (STEC) and bovine feces background microflora in various enrichment protocols. Vet. Microbiol. 123, 274–281.

Volker, U., Mach, H., Schmid, R., Hecker, M., 1992. Stress proteins and cross-protection by heat shock and salt stress in *Bacillus subtilis*. J. Gen. Microbiol. 138, 2125–2135.

Wang, J.H., 1955. On the detailed mechanism of a new type of catalase-like action. J. Am. Chem. Soc. 77, 4715–4719.

Wang, R., Kalchayanand, N., Schmidtm, J.W., Harhay, D.M., 2013. Mixed biofilm formation by Shiga toxin-producing *Escherichia coli* and *Salmonella enterica* serovar Typhimurium enhanced bacterial resistance to sanitization due to extracellular polymeric substances. J. Food Prot. 76, 1513–1522.

Wang, S., Deng, K., Zaremba, S., Deng, X., Lin, C., Wang, Q., et al., 2009. Transcriptomic response of *Escherichia coli* O157:H7 to oxidative stress. Appl. Environ. Microbiol. 75, 6110–6123.

Watanabe, Y., Ozasa, K., Mermin, J.H., Griffin, P.M., Masuda, K., Imashuku, S., et al., 1999. Factory outbreak of *Escherichia coli* O157:H7 infection in Japan. Emerg. Infect. Dis. 5, 424–428.

Wells, S.J., Fedorka-Cray, P.J., Dargatz, D.A., Ferris, K., Green, A., 2001. Fecal shedding of *Salmonella* spp. by dairy cows on farm and at cull cow markets. J. Food Prot. 64, 3–11.

Werber, D., Behnke, S.C., Fruth, A., Merle, R., Menzler, S., Glaser, S., et al., 2006. Shiga toxin-producing *Escherichia coli* infection in Germany—different risk factors for different age groups. Am. J. Epidemiol. 165, 425–434.

Yang, H., Byelashov, O.A., Geornaras, I., Goodridge, L.D., Nightingale, K.K., Belk, K.E., et al., 2010. Characterization and transferability of class 1 integrons in commensal bacteria isolated from farm and nonfarm environments. Foodborne Pathog. Dis. 7, 1441–1451.

Yoo, B.K., Chen, J., 2010. Role of extracellular polysaccharide polysaccharides in protecting the cells of Shiga toxin producing *Escherichia coli* against osmotic stress and chlorine treatment. J. Food Prot. 73, 2084–2088.

Yoo, B.K., Chen, J., 2012. Role of exopolysaccharides in protecting the cells of Shiga toxin producing *Escherichia coli* against oxidative and acidic stress. Food Control. 23, 289–292.

Yoo, B.K., Stewart, T., Guard-Bouldin, J., Musgrove, M., Gust, R., Chen, J., 2010. Selection and characterization of cellulose deficient mutants of Shiga toxin producing *Escherichia coli*. J. Food Prot. 73, 1038–1046.

Zhang, B., Liu, H.C., Zhang, Y., Li, F.P., Tu, G.P., 2005. Study on the resistance of *Escherichia coli* O157:H7 to disinfectants in relationship with its plasmid pO157 and chromosome DNA. Sichuan. Da. Xue. Xue. Bao. Yi. Xue. Ban. 36, 862–865.

Zook, C.D., Busta, F.F., Brady, L.J., 2001. Sublethal sanitizer stress and adaptive response of *Escherichia coli* O157:H7. J. Food Prot. 64, 767–769.

Chapter 3

Antibiotic Resistance in Pathogenic *Salmonella*

Steven C. Ricke and Juliany Rivera Calo
Center for Food Safety and Department of Food Science, University of Arkansas, Fayetteville, AR, USA

Chapter Outline

INTRODUCTION

Foodborne illnesses continue to be one of the primary public health concerns in the United States. Each year, it is estimated that over 1 million Americans contract *Salmonella* infections (Scallan et al., 2011). Annual costs for *Salmonella* control efforts are estimated to be several billion dollars and can include expenses such as long-term care (Scharff, 2012; McLinden et al., 2014). Usually foodborne *Salmonella* infections are self-limited in healthy individuals, only causing mild gastroenteritis with recovery occurring anywhere from 4 to 7 days after initial exposure. Symptoms of salmonellosis include diarrhea, fever, and abdominal cramps occurring several hours after consumption of food containing the pathogen (Ricke et al., 2013b). *Salmonella* infections are typically contracted through the consumption of contaminated food, water, or through direct contact with an infected host (Foley and Lynne, 2008; Foley et al., 2008, 2013; Ricke et al., 2013b).

Eggs, poultry, poultry products, meat, and meat products have all been classified as common food vehicles of salmonellosis in humans and can originate from several of the numerous *Salmonella* serovars (Ricke, 2003b; Patrick et al., 2004; Braden, 2006; Dunkley et al., 2009; Foley et al., 2011; 2013; Finstad et al., 2012;

Antimicrobial Resistance and Food Safety. DOI: http://dx.doi.org/10.1016/B978-0-12-801214-7.00003-X

Howard et al., 2012; Koo et al., 2012; Painter et al., 2013; Pires et al., 2014). *Salmonella* are also associated with contamination of fresh fruits and vegetables (Hanning et al., 2009; Erickson, 2010; Brandl et al., 2013). Prevention of *Salmonella* can be achieved by proper food handling, avoiding cross-contamination, sanitation, personal hygiene, and public education (Carrasco et al., 2012). Proper cooking following the recommended minimum internal temperatures and subsequent prompt cooling to 3–4°C or freezing within 2 h are recommended as means to eliminate *Salmonella* from foods (USDA/FSIS, 2013). In some cases, salmonellosis can become highly invasive and eventually systemic in the human host. In such cases more drastic measures such as antibiotic administration may be required to clear the pathogen. The prevalence and persistence of antibiotic resistance in certain *Salmonella* isolates can be a dilemma for clinical application of antibiotics and is considered a rising concern in the public health community. The objectives of this review are to discuss the characteristics and pathogenesis of foodborne *Salmonella* followed by an assessment of antibiotic resistance in this organism and how this continues to be an important factor for public health considerations.

SALMONELLA SPP. AS A PATHOGEN

Salmonella Characteristics

The genus name *Salmonella* was first suggested by Lignieres in 1900 in recognition of the work carried out by the American bacteriologist, D.E. Salmon, who, with T. Smith in 1886, described the hog cholera bacillus causing "swine plague" (Li et al., 2013; Ricke et al., 2013b). *Salmonella* species are Gram-negative, facultative anaerobic, flagellated, rod-shaped, motile bacteria of the enterobacteria group, of approximately 2–3 × 0.4–0.6 μm in size (Pui et al., 2011; Li et al., 2013). *Salmonella* can survive under various environmental conditions outside their respective living host (Altier, 2005; Park et al., 2008; Spector and Kenyon, 2012). Most *Salmonella* serovars grow at a temperature range of 5–47°C with an optimum temperature of 35–37°C, but some can grow at temperatures as low as 2–4°C and as high as 54°C (Pui et al., 2011). Given the ability of *Salmonella* spp. to grow under a variety of different growth conditions and their well-characterized genetics, they are a fairly ideal foodborne pathogen model for developing an understanding of the mechanism(s) involved in the efficacy of antimicrobial compounds.

Salmonella Classification

Classification of *Salmonella* is based on their susceptibility to different bacteriophages, as well as grouping of their respective somatic (O), flagellar (H), and capsular (Vi) antigenic patterns (Brenner et al., 2000; Tindall et al., 2005; Dunkley et al., 2009; Pui et al., 2011; Li et al., 2013; Ricke et al., 2013b). *Salmonella* H and O surface antigens have served as the primary typing vehicle for identifying the over 2,500 serovars of *Salmonella* currently known to exist

(Pui et al., 2011; Ricke et al., 2013b). Based on more recent molecular analysis the *Salmonella* genus has now been classified into two species: *S. enterica* and *S. bongori*, with the majority of the human disease-causing serovars being members of *S. enterica* (Foley et al., 2013; Ricke et al., 2013b). The species *S. enterica* is divided into six subspecies: I (*enterica*), II (*salamae*), IIIa (*arizonae*), IIIb (*diarizonae*), IV (*houtenae*), and VI (*indica*) based on biochemical and genetic properties, while *Salmonella bongori* contains the members of subspecies V (Pui et al., 2011; Li et al., 2013; Ricke et al., 2013b).

Epidemiologically *Salmonella* is classified into three groups based on the host preferences that may include (i) host-restricted, which are the serotypes capable of causing a typhoid-like disease in a single host species (e.g., *Salmonella* Typhi in humans), (ii) host-adapted, which are serotypes associated with one host species, but also able to cause disease in other hosts; some of these are human pathogens and may be contracted from foods, and (iii) unadapted serovars (no host preference), which are pathogenic for humans and other animals, and they include most foodborne serovars, for example *S.* Typhimurium and *S.* Enteritidis (Pui et al., 2011; Li et al., 2013; Ricke et al., 2013b).

PATHOGENESIS OF *SALMONELLA*

Salmonellosis

The primary sources of *Salmonella* are the gastrointestinal tracts of humans, domestic and wild animals, birds, and rodents (Dunkley et al., 2009; Foley et al., 2013). Because of this, they are widespread in the environment, including such diverse niches as water, soil, and even animal feeds where they do not usually multiply in a significant way but may survive for long periods of time (Maciorowski et al., 2004; Park et al., 2008; Pui et al., 2011). *Salmonella enterica* species are normally orally acquired pathogens that subsequently reach the intestinal tract where their presence can lead to one of four major syndromes: enteric fever (typhoid), enterocolitis (gastroenteritis) or diarrhea, bacteremia, and chronic asymptomatic carriage (Coburn et al., 2007). The organisms are ultimately excreted through the feces from which they can then be transmitted by insects and other vectors to a wide range of potential hosts and environmental niches (Park et al., 2008; Pui et al., 2011; CDC, 2012; Ricke et al., 2013b).

Depending on the host and serotype, *Salmonella* can cause diseases ranging from mild gastroenteritis to typhoid fever (Chiu et al., 2004; Foley et al., 2013; Ricke et al., 2013b). *Salmonella*-caused gastroenteritis is commonly referred to as salmonellosis, which can result in a much more systemic infection, depending on several factors including susceptibility of the host and the specific serovar (Coburn et al., 2007). *Salmonella enterica* serovars Typhimurium and Enteritidis cause gastroenteritis and these are two of the more common serovars responsible for human infections (CDC, 2011; Dunkley et al., 2009; Foley et al., 2011; Finstad et al., 2012; Howard et al., 2012). *S.* Enteritidis cases have often been associated with consumption of contaminated eggs (Guard-Petter, 2001; Gantois et al., 2009; Howard et al., 2012; Ricke et al., 2013a). *S.* Heidelberg and *S.* Kentucky have also become more common in poultry but only Heidelberg thus far has been linked consistently with human disease (Foley et al., 2011). *Salmonella enterica* serovar Typhi is the most invasive, causing typhoid fever in humans. Some *Salmonella* strains may exhibit low infectious doses (500 or less) while others require 100,000 or more organisms to cause infection depending on the virulence of the strain (Finstad et al., 2012). After ingestion with contaminated food, disease symptoms usually develop in 12–14 h, but this time has been reported to be shorter or it can be longer, lasting up to 72 h (Ricke et al., 2013b).

Pathogenic Mechanisms

Salmonella growth, survival, and virulence respond to a series of stress factors such as pH, oxygen availability, and osmolarity (Altier, 2005; Humphrey, 2004; Ricke, 2003a; Spector and Kenyon, 2012). When there is a lack of nutrients, such as carbon sources, this can trigger global stress responses in *Salmonella* resulting in a changed physiological state with a wide array of genes impacted including some virulence genes (Spector and Kenyon, 2012). For example, a connection between nutrient-poor conditions in the gastrointestinal ract and enhanced virulence has been reported in laying hens where extended feed removal led to increased expression of key virulence regulatory genes in *S.* Enteritidis followed by increased systemic invasion and appearance of *S.* Enteritidis in organs such as the spleen and the ovaries (Durant et al., 1999a; Ricke, 2003b; Dunkley et al., 2007; Ricke et al., 2013a).

Salmonella possess clusters of genes located within their chromosome that are associated with pathogenesis and are referred to as *Salmonella* pathogenicity islands, which can vary in size, number, and composition (Li et al., 2013). Some of the genes necessary for invasion of intestinal epithelial cells and induction of intestinal secretory and inflammatory responses are encoded in SPI-1 while SPI-2 is key for the establishment of systemic infection beyond the intestinal epithelium as well as encoding essential genes for intracellular replication (Cirillo et al., 1998; Hensel, 2000; Coburn et al., 2007; Foley et al., 2013). Adhesion of *Salmonella* to the intestinal epithelial surface is an important first step in pathogenesis and is central to its colonization of the intestine. Once *Salmonella* is attached to the intestinal epithelium, it can express multiprotein flagellar-like complexes, known as type III secretion system (T3SS), that facilitate adhesion, invasion, and macrophage survival (Coburn et al., 2007; Foley et al., 2013; Li et al., 2013). These T3SSs are associated with SPI-1 and SPI-2, generate effector proteins that among other functions form a channel in the host epithelial cell for translocating invasion proteins into the cell (Foley et al., 2013; Li et al., 2013).

Historically, researchers have used cultured mammalian cells as *in vitro* models to study the interaction and internalization of *Salmonella* under a range of different conditions (Giannella et al., 1973; Finlay et al., 1988; Durant et al., 1999b). Invasion in cultured epithelial cells is commonly used to measure the pathogenicity of *Salmonella* and compare different isolates (Giannella et al., 1973; van Asten et al., 2000, 2004; Shah et al., 2011). A series of classic studies using tissue culture revealed that *Salmonella* invasion into cultured mammalian cells can be influenced by several environmental stimuli such as osmolarity (Galán and Curtis, 1990; Tartera and Metcalf, 1993) carbohydrate availability (Schiemann, 1995), and oxygen availability (Ernst et al., 1990; Lee et al., 1990; Francis et al., 1992). In their research, Durant et al. (1999b) reported that *Salmonella* can encounter high concentrations of short-chain fatty acids (SCFA) in the lower regions of the gastrointestinal tract, in addition to changes in pH, oxygen tension, and osmolarity. Since a combination of environmental conditions are what regulate virulence gene expression, SCFA may contribute to the environmental stimuli that regulate cell-association and invasion of *S.* Typhimurium epithelial cells (Durant et al., 1999b). Species and even strain variation may also be a factor in observed differences in invasiveness (Shah et al., 2011; Heithoff et al., 2012).

SALMONELLA AND ANTIBIOTICS

General Concepts

Antimicrobials have a long history of effective chemotherapy in medical applications (Butler and Buss, 2006; Aminov, 2010). Antibiotics have been essentially defined by US regulatory agencies as drugs produced by a microorganism that possess the capability, in dilute solutions, to inhibit the growth of or to kill other microorganisms; while an antimicrobial compound is any substance that kills bacteria (bactericidal) or suppresses (bacteriostatic) their multiplication or growth, including antibiotics and synthetic agents (FDA, 2002; CDC, 2010a,b). Intuitively, antibiotic resistance is essentially the ability of bacteria to either inactivate or exclude antibiotics or be able to block the inhibitory or lethal effects of antibiotics (FDA, 2002).

Historically, a number of antibiotics have been used in a diverse range of agriculture production settings from livestock production to large-scale grain-based ethanol fermenters, but the emergence of multidrug-resistant (MDR) pathogens has increased concerns over the consequences and their impact on human health due to antibiotic resistance (Threlfall, 1992; Tollefson and Miller, 2000; Casewell et al., 2003; Jones and Ricke, 2003; Roe and Pillai, 2003; Talbot et al., 2006; Muthaiyan et al., 2011). Reducing antibiotic-resistant clinically important pathogens by removal of antibiotics in livestock systems is widely touted, yet questions remain not only due to the complex epidemiology associated with establishing the quantitative impact of transmission routes between humans and livestock, but the potentially inherent biological slowness of reversing antibiotic resistance in organisms (Sørum and L'Abée-Lundm, 2002; Barber et al., 2003; Su et al., 2004; Andersson and Hughes, 2010; Mather et al., 2013; Woolhouse and Ward, 2013).

However, interest has continued to increase in not only seeking a wide range of antimicrobial alternatives which exhibit mitigation properties against various foodborne pathogens, but also positive commercial growth in agriculture production systems that are antibiotic-free such as natural and organic livestock (Burt, 2004; Berghman et al., 2005; O'Bryan et al., 2008; Sirsat et al., 2009; Li et al., 2011; Muthaiyan et al., 2011; Bajpai et al., 2012; Solórzano-Santos and Miranda-Novales, 2012; Ricke et al., 2012; Van Loo et al., 2012).

Salmonellosis and Clinical Antibiotic Treatment

Antibiotic therapy is not recommended for simple cases of *Salmonella*-induced gastroenteritis and may only come into play for the severe systemic infection where antibiotic treatment is more likely warranted (Ricke et al., 2013b). However, for more systemic infections, treatment with antibiotics such as ciprofloxacin, ceftriaxone, or cefotaxime may be needed (Li et al., 2013). In invasive life-threatening infections, the use of such antimicrobial drugs is required, but the efficacy of these drugs is decreasing due to the ongoing presence of antimicrobial-resistant *Salmonella* strains in clinical settings (Angulo et al., 2000; Winokur et al., 2000; Varma et al., 2005a,b). Historically, administration of antibiotics such as ampicillin, chloramphenicol, and tetracycline has become compromised due to the emergence of MDR *Salmonella* strains to these agents (Su et al., 2004). Likewise, *Salmonella* species have now been isolated that are resistant to fluoroquinolones, such as ciprofloxacin, or extended-spectrum cephalosporins commonly used to treat adult patients infected with *Salmonella enterica* or to treat a severe case of gastroenteritis (Hohmann, 2001; Mølbak, 2005; Chen et al., 2007; Hur et al., 2012). The fact that emerging fluoroquinolone resistance prevalence has been identified in several *Salmonella* serovars and can occur as a multiple drug resistance phenotype, which has caused international concern because of the clinical importance of this particular antibiotic (Olsen et al., 2001; Mølbak, 2005; Alcaine et al., 2007; Chen et al., 2007; Hur et al., 2012).

Varma et al. (2005b) reported that antimicrobial resistance in nontyphoidal *Salmonella* is characterized by an increased frequency of bloodstream infections and hospitalizations of the corresponding patients. Bloodstream infections have been reported to occur more frequently among people infected with a nontyphoidal *Salmonella* isolate resistant to more than one antimicrobial agent (CDC, 2003a,b). Human-health consequences of increasing resistance may be substantial if antimicrobial-resistant infection increases the risk of bloodstream infection (Varma et al., 2005b). Previous research has revealed a strong association between resistance, bloodstream infection, and hospitalization (Holmberg et al., 1984, 1987; Lee et al., 1994; Helms et al., 2002; Varma et al., 2005b). More recently, it has been demonstrated that patients who become infected with resistant *Salmonella* isolates are more likely to experience severe outcomes when administered antimicrobial therapy (Krueger et al., 2014).

Mechanisms of Antimicrobial Action

To inhibit microbial growth, antimicrobials tend to act on several specific target sites such as inhibition of synthesis of specific proteins, DNA and RNA synthesis, interference with cell wall synthesis, disruption of membrane permeability structure, and inhibition of essential metabolite synthesis (Hughes and Mellows, 1978; Reynolds, 1989; Falla et al., 1996; Hooper, 2001). Consequently, due to their different biological structures and properties, target sites for each antimicrobial are quite likely to vary depending on the type of microorganism that is being exposed to the respective antimicrobial agent. For example, antibiotics such as penicillins and cephalosporins inhibit bacterial cell wall synthesis by interference with the enzymes associated with peptidoglycan layer synthesis; while tetracyclines and streptogramins work by disrupting protein synthesis (Tenover, 2006).

A microbial strain is considered resistant to a particular antimicrobial when the antimicrobial is no longer efficacious for treatment of a clinical disease caused by a particular bacterial pathogen (Alcaine et al., 2007). Over the past several years, details of antimicrobial resistance in *Salmonella* have been extensively documented elsewhere (Su et al., 2004; Alcaine et al., 2007; Hur et al., 2012) and will only be briefly discussed in the current review. In general, bacterial resistance to antimicrobial agents may be mediated by a variety of mechanisms including (i) energy-dependent removal of antimicrobials via membrane-bound efflux pumps, (ii) changes in bacterial cell permeability, which restricts the access of antibiotics to target sites, (iii) modifications of the site targeted by drug action, (iv) acquisition of a replacement for the target protein, and (v) inactivation or destruction of antimicrobials by mechanisms such as secretion of specific enzymes that target the respective antibiotic (Chen et al., 2004; Alcaine et al., 2007; Hur et al., 2012; Frye and Jackson, 2013). Alcaine et al. (2007) have defined the two mechanisms known to be critical for the spread of antimicrobial resistance in *Salmonella* as (i) horizontal transfer of antibiotic resistance genes and (ii) clonal spread of antimicrobial drug-resistant *Salmonella* isolates.

It has been suggested that high-level resistance of *Salmonella* to fluoroquinolones is due to the combination of two major resistance mechanisms, active efflux mediated by AcrAB-TolC and multiple target gene mutations (Velge et al., 2005; Hur et al., 2012). Several studies have demonstrated that resistance to fluoroquinolines in several *Salmonella* serovars results from one or more target gene mutations (Chu et al., 2005; Baucheron et al., 2002, 2004, 2013). In the case of tetracyclines, *Salmonella* acquires resistance to these antibiotics through an efflux mechanism by which the drug is pumped out from the cell (Alcaine et al., 2007; Hur et al., 2012). In their review, Hur et al. (2012) concluded that tetracycline and chloramphenicol resistance was highly associated with the acquisition of efflux pumps and their subsequent expression, which in turn reduced toxic levels of the drug in the bacterial cells.

Streptomycin and kanamycin belong to the aminoglycosides family, which bind to conserved sequences within the 16S rRNA of the 30S ribosomal subunit, leading to a codon misreading and translation inhibition (Alcaine et al., 2007). Aminoglycoside resistance in bacteria has been associated with the expression of aminoglycoside-modifying enzymes often encoded by genes carried on plasmids and can also be associated with transposons (Shaw et al., 1993). Ampicillin belongs to the beta-lactam family, where the ability to interfere with the penicillin-binding proteins (involved in the synthesis of peptidoglycan) confers antimicrobial effects on the target microorganism (Alcaine et al., 2007). In *Salmonella*, the mechanism of resistance to beta-lactams that has been most commonly described involves the secretion of beta-lactamases (Alcaine et al., 2007).

Chloramphenicol belongs to the phenicols family and its mode of action is via the prevention of peptide bond formation. There are two resistance mechanisms in *Salmonella* for this antibiotic: (i) enzymatic inactivation of the antibiotic and (ii) removal of the antibiotic via an efflux pump (Alcaine et al., 2007). The mechanism of action for tetracyclines involves inhibiting protein synthesis; and its resistance in *Salmonella* is conferred by production of an energy-dependent efflux pump, which excretes the antibiotic from the bacterial cell (Alcaine et al., 2007). Sulfamethoxazole belongs to the sulfonamides family of antibiotics, which act by inhibiting enzymes that are involved in the synthesis of tetrahydrofolic acid (Alcaine et al., 2007). The presence of an additional *sul* gene, which expresses an insensitive form of the target enzyme, dihyrdropteroate synthetase, is responsible for the resistance phenotype observed in *Salmonella* (Alcaine et al., 2007).

Salmonella MDR Strains

MDR *Salmonella* strains are defined as those strains which are resistant to two or more antimicrobial agents. For example, *Salmonella* Heidelberg isolates have been typically characterized by resistance to ampicillin, amoxicillin-clavulanic acid, ceftiofur, as well as cephalothin (CDC, 2004a; Alcaine et al., 2007). *Salmonella* that are MDR may carry their resistance determinants on chromosomal sites, resistance gene carrying plasmids or on both (Chen et al., 2004; Alcaine et al., 2007; Lindsey et al., 2009). *S.* Typhimurium MDR isolates possess two common resistance patterns, namely, (i) resistance to ampicillin, kanamycin, streptomycin, sulfamethoxazole, and tetracycline or (ii) resistance to ampicillin, chloramphenicol, streptomycin, sulfamethoxazole, and tetracycline, the resistance profile most commonly associated with *S.* Typhimurium DT104 (CDC, 2004a; Alcaine et al., 2007). The difference between these two resistance patterns for MDR serovars such as *S.* Typhimurium is essentially that one has kanamycin and the other has chloramphenicol. These antibiotics pertain to different families of antibiotics and each of them possesses a specific mechanism of action.

Overall, the human clinical nontyphoidal *Salmonella* isolates resistant to one or more drug classes declined from 1996 to 2003 in the United States

(CDC, 2004b). Moreover, in 2011 the CDC reported that resistance to one or more drug classes was lower than during the period of 2003–2007 (CDC, 2013). However, MDR nontyphoidal *Salmonella* remain a public health concern because these strains are not only occurring in clinical settings but are also being isolated from food animals, meat carcasses, and retail food samples (White et al., 2001; Chen et al., 2004; Davis et al., 2007; Zhao et al., 2007, 2008; M'ikanatha et al., 2010; Brichta-Harhay et al., 2011; Schmidt et al., 2012; Frye and Jackson, 2013; Krueger et al., 2014). Likewise, infections by MDR *Salmonella* remain a persistent concern due to potential treatment difficulties resulting in human infections with these strains being more severe and patients, in turn, being at greater risk of bacteremia, hospitalization, and death compared to patients infected with antibiotic-susceptible strains (Helms et al., 2002; CDC, 2004b; Mølbak, 2005; Talbot et al., 2006; Krueger et al., 2014).

Finally, concerns remain about the frequency of MDR in particular *Salmonella* serovars that may be of more clinical importance. Eight of the ten serotypes most commonly reported by the CDC include some isolates that displayed resistance to at least five or more antimicrobial drugs (CDC, 2004a,b; Alcaine et al., 2007). *Salmonella* serotypes reported with the highest MDR were Typhimurium, Heidelberg, and Newport (CDC, 2004a; Alcaine et al., 2007). In addition, the frequency of MDR in Typhimurium and Newport is reportedly increasing (Alcaine et al., 2007; Zhao et al., 2007; Hur et al., 2012). Serovars such as Agona, Anatum, Cholerasuis, Derby, Dublin, Kentucky, Pullorum, Schwarzengrund, Seftenberg, and Uganda also contain MDR *Salmonella* strains (Chen et al., 2004; Alcaine et al., 2007; Zhao et al., 2007; Hur et al., 2012).

FACTORS THAT INFLUENCE *SALMONELLA* ANTIBIOTIC RESISTANCE

While general aspects of antibiotic resistance in bacteria such as the influence of mobility of genetic elements encoding resistance among bacteria have been well documented (Mebrhatu et al., 2014), there are several less defined factors that could alter how microorganisms such as *Salmonella* respond to antibiotic exposure under a given situation. For example, the formation of biofilms by bacteria including *Salmonella* on different inert surfaces, host animal tissues, and food-processing environments may change the physiology of the organism such that it is less susceptible to antibiotic exposure (Høiby et al., 2010; Carrasco et al., 2012; Steenackers et al., 2012). Likewise, the ability of *Salmonella* to survive a wide range of environmental stresses and elicit cross resistances to a multitude of apparently unrelated stresses can have a confounding impact on predicting the susceptibility of organism when exposed to a normally lethal level of a particular antimicrobial agent (Spector and Kenyon, 2012).

In addition, there are host and pathogenesis factors that may also impact antibiotic effectiveness. For example, the ability of *Salmonella* to survive in host tissues may counter the efficacy of the antibiotic being administered to the host.

Even though chloramphenicol is used as treatment for systemic infections, a large number of cultures taken from the bone marrow over the first week of treatment remained positive for *Salmonella* (Gasem et al., 1995, 2003). This persistence of *Salmonella*, potentially surviving inside macrophages, could be responsible for a relapse of the infection (Sanders, 1965; Gasem et al., 2003). *Salmonella* in an invasive virulent physiological state may also be a fairly elusive target for the development of new antibiotics. To address this question, Becker et al. (2006) conducted proteomic assessment of *S.* Typhimurium-infected mice along with comparisons with previously published *in vivo* data on *Salmonella*-expressed phenotypes during infection in various hosts. They concluded that most potential metabolic essential gene targets for antibiotics were generally not expressed during infection either because of genetic redundancy in *Salmonella* or the nutrient-rich conditions of the host tissues.

CONCLUSIONS

Antibiotic resistance in foodborne *Salmonella* continues to be a recurring problem for public health efforts. This is not only reflective in the expansion of serovars displaying antibiotic resistance but the number of strains that possess multiple-drug resistance. Through the years, the pharmaceutical industry has invested heavily in the research and discovery of natural products with the end goal of producing therapeutic antibiotic formulations (Butler and Buss, 2006). The majority of the antibiotics in clinical use are either derived from natural product leads or are a result of their semisynthetic derivatives (Butler and Buss, 2006; Benowitz et al., 2010). The ongoing problem of increasing drug resistance has created a more urgent demand to develop new and improved antimicrobials against resistant microorganisms (Spellberg et al., 2004; Fischbach and Walsh, 2009). While traditional *in vitro* screening tools can still provide some answers to antimicrobial choices and development strategies, it is becoming clear that more realistic scenarios and *in vivo*/host screens may need to be employed to achieve accurate efficacy assessments. With the advent of multiple comprehensive system analyses approaches that include not just genomics but proteomics as well as metabolomics, such assessments are much more practical and offer opportunities to identify more precise pathogen targets for effective delivery of newly developed antimicrobials.

REFERENCES

Alcaine, S.D., Warnick, L.D., Wiedmann, M., 2007. Antimicrobial resistance in nontyphoidal *Salmonella*. J. Food Prot. 70, 780–790.

Altier, C., 2005. Genetic and environmental control of *Salmonella* invasion. J. Microbiol. 43, 85–92.

Aminov, R.I., 2010. A brief history of the antibiotic era: lessons learned and challenges for the future. Front. Microbiol. 1, 134.

Andersson, D.I., Hughes, D., 2010. Antibiotic resistance and its cost: is it possible to reverse resistance? Nat. Rev. Microbiol. 8, 260–271.

Angulo, F.J., Johnson, K.R., Tauxe, R.V., Cohen, M.L., 2000. Origins and consequences of antimi-crobial-resistant nontyphoidal *Salmonella*: implications for the use of fluoroquinolones in food animals. Microb. Drug Resist. 6, 77–83.

Bajpai, V.K., Baek, K.-H., Kang, S.C., 2012. Control of *Salmonella* in foods by using essential oils: a review. Food Res. Int. 45, 722–734.

Barber, D.A., Miller, G.Y., McNamara, P.E., 2003. Models of antimicrobial resistance and food-borne illness: examining assumptions and practical applications. J. Food Prot. 66, 700–709.

Baucheron, S., Imberechts, H., Chaslus- Dancla, E., Cloeckaert, A., 2002. The AcrB multidrug transporter plays a major role in high-level fluoroquinolone resistance in *Salmonella enterica* serovar Typhimurium phage type DT204. Microb. Drug Resist. 8, 281–289.

Baucheron, S., Chaslus-Dancla, E., Cloeckaert, A., 2004. Role of TolC and *parC* mutation in high-level fluoroquinolone resistance in *Salmonella enterica* serotype Typhimurium DT204. J. Antimicrob. Chemother. 53, 657–659.

Baucheron, S., Le Hello, S., Doublet, B., Giraud, E., Weill, F.-X., Cloeckaert, A., 2013. *ram R* mutations affecting fluoroquinolone susceptibility in epidemic multidrug-resistant *Salmonella enterica* serovar Kentucky ST198. Front. Microbiol. 4 (Article 213), 1–6.

Becker, D., Selbach, M., Rollenhagen, C., Ballmaier, M., Meyer, T.F., Mann, M., et al., 2006. Robust *Salmonella* metabolism limits possibilities for new antimicrobials. Nature 440, 303–307.

Benowitz, A.B., Hoover, J.L., Payne, D.J., 2010. Antibacterial drug discovery in the age of resis-tance. Microbe 5, 390–396.

Berghman, L.R., Abi-Ghanem, D., Waghela, S.D., Ricke, S.C., 2005. Antibodies: an alternative for antibiotics? Poult. Sci. 8, 660–666.

Braden, C.R., 2006. *Salmonella enterica* serotype Enteritidis and eggs: a national epidemic in the United States. Clin. Infect. Dis. 43, 512–517.

Brandl, M.T., Cox, C.E., Teplitski, M., 2013. *Salmonella* interactions with plants and their associ-ated microbiota. Phytopathology 103, 316–325.

Brenner, F.W., Villar, R.G., Angulo, F.J., Tauxe, R., Swaminathan, B., 2000. *Salmonella* nomencla-ture. J. Clin. Microbiol. 38, 2465–2467.

Brichta-Harhay, D.M., Arthur, T.M., Bosilevac, J.M., Kalchayanand, N., Shackelford, S.D., Wheeler, T.L., et al., 2011. Diversity of multidrug-resistant *Salmonella enterica* strains associ-ated with cattle at harvest in the United States. Appl. Environ. Microbiol. 77, 1783–1796.

Burt, S., 2004. Essential oils: their antibacterial properties and potential applications in foods—a review. Int. J. Food Microbiol. 94, 223–253.

Butler, M.S., Buss, A.D., 2006. The future scaffolds for novel antibiotics? Biochem. Pharmcol. 71, 919–929.

Carrasco, E., Morales-Rueda, A., García-Gimeno, R.M., 2012. Cross-contamination and recontami-nation by *Salmonella* in foods: a review. Food Res. Int. 45, 545–556.

Casewell, M., Friis, C., Marco, E., McMullin, P., Phillips, I., 2003. The European ban on growth-promoting antibiotics and emerging consequences for human and animal health. J. Antimicrob. Chemother. 52, 159–161.

Centers for Disease Control and Prevention (CDC), 2003a. FoodNet Surveillance Report for 2001: Final Report. Centers for Disease Control and Prevention, Atlanta, GA.

Centers for Disease Control and Prevention (CDC), 2003b. National Antimicrobial Resistance Monitoring System for Enteric Bacteria: Annual Report, 2001. Centers for Disease Control and Prevention, Atlanta, GA.

Centers for Disease Control and Prevention (CDC), 2004a. National Antimicrobial Resistance Monitoring System For Enteric Bacteria (NARMS): 2002 Human Isolates Final Report. US Department of Health and Human Services, Atlanta, GA.

Centers for Disease Control and Prevention (CDC), 2004b. Salmonella *Surveillance Annual Summary, 2003.* US Department of Health and Human Services, Atlanta, GA.

Centers for Disease Control and Prevention (CDC). 2010a. Antibiotic/antimicrobial resistance. <http://www.cdc.gov/drugresistance/glossary.html#antibiotic>.

Centers for Disease Control and Prevention (CDC). 2010b. Glossary: antibiotic/antimicrobial resistance. <http://www.cdc.gov/drugresistance/glossary.html>.

Centers for Disease Control and Prevention (CDC), 2011. *National* Salmonella *Surveillance Annual Data Summary, 2009.* US Department of Health and Human Services, CDC, Atlanta, GA.

Centers for Disease Control and Prevention (CDC), 2012. Salmonella *Infection (Salmonellosis) and Animals.* US Department of Health and Human Services, CDC, Atlanta, GA, <http://www.cdc.gov/healthypets/diseases/salmonellosis.htm> .

Centers for Disease Control and Prevention (CDC), 2013. National Antimicrobial Resistance Monitoring System for Enteric Bacteria (NARMS): Human Isolates Final Report, 2011. US Department of Health and Human Services, CDC, Atlanta, GA.

Chen, S., Zhao, S., White, D.G., Schroeder, C.M., Lu, R., Yang, H., et al., 2004. Characterization of multiple-antimicrobial-resistant *Salmonella* serovars isolated from retail meats. Appl. Environ. Microbiol. 70, 1–7.

Chen, S., Cui, S., McDermott, P.F., Zhao, S., White, D.G., Paulsen, I., et al., 2007. Contribution of target gene mutations and efflux to decreased susceptibility of *Salmonella enterica* serovar Typhimurium to fluoroquinolones and other antimicrobials. Antimicrob. Agents Chemother. 51, 535–542.

Chiu, C.-H., Su, L.-H., Chu, C., 2004. *Salmonella enterica* serotype Choleraesuis: epidemiology, pathogenesis, clinical disease and treatment. Clin. Microbiol. Rev. 2, 311–322.

Chu, C., Su, L.H., Chu, C.H., Baucheron, S., Cloeckaert, A., Chiu, C.H., 2005. Resistance to fluoroquinolones linked to *gyrA* and *parC* mutations and overexpression of acrAB efflux pump in *Salmonella enterica* serotype Choleraesuis. Microb. Drug Resist. 11, 248–253.

Cirillo, D.M., Valdivia, R.H., Monack, D.M., Falkow, S., 1998. Macrophage-dependent induction of the *Salmonella* pathogenicity island 2 type III secretions system and its role in intracellular survival. Mol. Microbiol. 30, 175–188.

Coburn, B., Grassl, G.A., Finlay, B.B., 2007. *Salmonella*, the host and disease: a brief review. Immunol. Cell. Biol. 85, 112–118.

Davis, M.A., Besser, T.E., Eckmann, K., MacDonald, J.K., Green, D., Hancock, D.D., et al., 2007. Multidrug-resistant *Salmonella* Typhimurium, Pacific Northwest, United States. Emerg. Infect. Dis. 13, 1583–1586.

Dunkley, K.D., McReynolds, J.L., Hume, M.E., Dunkley, C.S., Callaway, T.R., Kubena, L.F., et al., 2007. Molting in *Salmonella* Enteritidis challenged laying hens fed alfalfa crumbles I. *Salmonella* Enteritidis colonization and virulence gene *hilA* response. Poult. Sci. 86, 1633–1639.

Dunkley, K.D., Callaway, T.R., Chalova, V.I., McReynolds, J.L., Hume, M.E., Dunkley, C.S., et al., 2009. Foodborne *Salmonella* ecology in the avian gastrointestinal tract. Anaerobe 15, 26–35.

Durant, J.A., Corrier, D.E., Byrd, J.A., Stanker, L.H., Ricke, S.C., 1999a. Feed deprivation affects crop environment and modulates *Salmonella enteritidis* colonization and invasion of Leghorn hens. Appl. Environ. Microbiol. 65, 1919–1923.

Durant, J.A., Lowry, V.K., Nisbet, D.J., Stanker, L.H., Corrier, D.E., Ricke, S.C., 1999b. Short-chain fatty acids affect cell-association and invasion of HEp-2 cells by *Salmonella typhimurium.* J. Environ. Sci. Health B34, 1083–1099.

Erickson, M.C., 2010. Microbial risks associated with cabbage, carrots, celery, onions, and deli salads made with these produce items. Comp. Rev. Food Sci. Food Safety 9, 602–619.

Ernst, R.K., Dombroski, D.M., Merrick, J.M., 1990. Anaerobiosis, type 1 fimbriae, and growth phase are factors that affect invasion of HEP-2 cells by *Salmonella typhimurium*. Infect. Immun. 58, 2014–2016.

Falla, T.J., Karunaratne, D.N., Hancock, E.W.R., 1996. Mode of action of the antimicrobial peptide indolicidin. J. Biol. Chem. 271, 19298–19303.

FDA. 2002. The judicious use of antimicrobials for beef producers. Retrieved from: <http://www.fda.gov/downloads/AnimalVeterinary/SafetyHealth/AntimicrobialResistance/JudiciousUseofAntimicrobials/UCM095583.pdf> (accessed 05.12.12).

Finlay, B.B., Gumbiner, B., Falkow, S., 1988. Penetration of *Salmonella* through a polarized Madin-Darby canine kidney epithelial cell monolayer. J. Cell Biol. 107, 221–230.

Finstad, S., O'Bryan, C.A., Marcy, J.A., Crandall, P.G., Ricke, S.C., 2012. *Salmonella* and broiler processing in the United States: relationship to foodborne salmonellosis. Food Res. Int. 45, 789–794.

Fischbach, M.A., Walsh, C.T., 2009. Antibiotics for emerging pathogens. Science 325, 1089–1093.

Foley, S.L., Lynne, A.M., 2008. Food animal-associated *Salmonella* challenges: pathogenicity and antimicrobial resistance. J. Anim. Sci. 86, E173–E187.

Foley, S.L., Lynne, A.M., Nayak, R., 2008. *Salmonella* challenges: prevalence in swine and poultry and potential pathogenicity of such isolates. J. Anim. Sci. 86, E149–E162.

Foley, S., Nayak, R., Hanning, I.B., Johnson, T.L., Han, J., Ricke, S.C., 2011. Population dynamics of *Salmonella enterica* serotypes in commercial egg and poultry production. Appl. Environ. Microbiol. 77, 4273–4279.

Foley, S.L., Johnson, T.J., Ricke, S.C., Nayak, R., Danzelsen, J., 2013. *Salmonella* pathogenicity and host adaptation in chicken-associated serovars. Microbiol. Mol. Biol. Rev. 77, 582–607.

Francis, C.L., Starnbach, M.N., Falkow, S., 1992. Morphological and cytoskeletal changes in epithelial cells occur immediately upon interaction with *Salmonella typhimurium* grow under low-oxygen conditions. Mol. Microbiol. 6, 3077–3087.

Frye, J.G., Jackson, C.R., 2013. Genetic mechanisms of antimicrobial resistance identified in *Salmonella enterica, Escherichia coli*, and *Enteroccocus* spp. isolated from U.S. food animals. Front. Microbiol. 4, 135. <http://dx.doi.org/10.3389/fmicb.2013.00135>.

Galán, J.E., Curtis III, R., 1990. Expression of *Salmonella typhimurium* genes required for invasion is regulated by changes in DNA supercoiling. Infect. Immun. 58, 1879–1885.

Gantois, I., Ducatelle, R., Pasmans, F., Haesebrouck, F., Gast, R., Humphrey, T.J., et al., 2009. Mechanisms of egg contamination by *Salmonella enteritidis*. FEMS Microbiol. Rev. 33, 718–738.

Gasem, M.H., Dolmans, W.M.V., Isbandrio, B.B., Wahyono, H., Keuter, M., Djokomoeljanto, R., 1995. Culture of *Salmonella typhi* and *Salmonella paratyphi* from blood and bone marrow in suspected typhoid fever. Trop. Geogr. Med. 47, 164–167.

Gasem, M.H., Keuter, M., Dolmans, W.M.V., van der Ven-Jongekrijg, J., Djokomoeljanto, R., van der Meer, J.W.M., 2003. Persistence of salmonellae in blood and bone marrow: randomized controlled trial comparing ciprofloxacin and chloramphenicol treatments against enteric fever. Antimicrob. Agents Chemother. 47, 1727–1731.

Giannella, R.A., Washington, O., Gemski, P., Formal, S.B., 1973. Invasion of HeLa cells by *Salmonella typhimurium*: a model for study of invasiveness of *Salmonella*. J. Infect. Dis. 128, 69–75.

Guard-Petter, J., 2001. The chicken, the egg and *Salmonella enteritidis*. Environ. Microbiol. 3, 421–430.

Hanning, I.B., Nutt, J.D., Ricke, S.C., 2009. Salmonellosis: outbreaks in the United States due to fresh produce: sources and potential intervention measures. Foodborne Pathog. Dis. 6, 635–648.

Heithoff, D.M., Shimp, W.R., House, J.K., Xie, Y., Weimer, B.C., Sinsheimer, R.L., et al., 2012. Intraspecies variation in the emergence of hyperinfectious bacterial strains in nature. PLoS Pathog. 8, 1–17.

Helms, M., Vastrup, P., Gerner-Smidt, P., Mølbak, K., 2002. Excess mortality associated with antimicrobial drug-resistant *Salmonella typhimurium*. Emerg. Infect. Dis. 8, 490–495.

Hensel, M., 2000. *Salmonella* pathogenicity island 2. Mol. Microbiol. 36, 1015–1023.

Hohmann, E.L., 2001. Nontyphoidal salmonellosis. Clin. Infect. Dis. 32, 263–269.

Høiby, N., Bjarnsholt, T., Givskov, M., Molin, S., Ciofu, O., 2010. Antibiotic resistance of bacterial biofilms. Int. J. Antimicrob. Agents 35, 322–332.

Holmberg, S.D., Wells, J.G., Cohen, M.L., 1984. Animal-to-man transmission of antimicrobial-resistant *Salmonella*: investigations of US outbreaks, 1971–1983. Science 225, 833–834.

Holmberg, S.D., Solomon, S., Blake, P., 1987. Health and economic impacts of antimicrobial resistance. Rev. Infect. Dis. 9, 1065–1078.

Hooper, D.C., 2001. Mechanisms of action of antimicrobials: focus on fluoroquinolones. Clin. Infect. Dis. 32 (Suppl. 1), S9–S15.

Howard, Z.R., O'Bryan, C.A., Crandall, P.G., Ricke, S.C., 2012. *Salmonella enteritidis* in shell eggs: current issues and prospects for control. Food Res. Int. 45, 755–764.

Hughes, J., Mellows, G., 1978. On the mode of action of pseudomonic acid: inhibition of protein synthesis in *Staphylococcus aureus*. J. Antibiot. 31, 330–335.

Humphrey, T., 2004. *Salmonella*, stress responses and food safety. Nat. Rev. Microbiol. 2, 504–509.

Hur, J., Jawale, C., Lee, J.H., 2012. Antimicrobial resistance of *Salmonella* isolated from food animals: a review. Food Res. Int. 45, 819–830.

Jones, F.T., Ricke, S.C., 2003. Observations on the history of the development of antimicrobials and their use in poultry feeds. Poult. Sci. 82, 613–617.

Koo, O.K., Sirsat, S.A., Crandall, P.G., Ricke, S.C., 2012. Physical and chemical control of *Salmonella* in ready-to-eat products. Agric. Food Anal. Bacteriol. 2, 56–68.

Krueger, A.L., Greene, S.A., Barzilay, E.J., Henao, O., Vugia, D., Hanna, S., et al., 2014. Clinical outcomes of nalidixic acid, ceftriaxone, and multidrug-resistant nontyphoidal *Salmonella* infections compared with pansusceptible infections in FoodNet Sites, 2006–2008. Foodborne Pathog. Dis. 11, 335–341.

Lee, C.A., Jones, B.D., Falkow, S., 1990. Identification of a *Salmonella typhimurium* invasion locus by selection for hyperinvasive mutants. Proc. Natl. Acad. Sci. USA. 89, 1847–1851.

Lee, L.A., Puhr, N.D., Maloney, E.K., Bean, N.H., Tauxe, R.V., 1994. Increase in antimicrobial resistant *Salmonella* infections in the United States, 1989–1990. J. Infect. Dis. 170, 128–134.

Li, M., Muthaiyan, A., O'Bryan, C.A., Gustafson, J.E., Li, Y., Crandall, P.G., et al., 2011. Use of natural antimicrobials from a food safety perspective for control of *Staphylococcus aureus*. Curr. Pharm. Biotechnol. 12, 1240–1254.

Li, H., Wang, H., D'Aoust, J.Y., Maurer, J., 2013. *Salmonella* species. In: Doyle, M.P., Buchanan, R.L. (Eds.), Food Microbiology—Fundamentals and Frontiers, fourth ed. American Society of Micobiology Press, Washington, DC, pp. 225–261 (Chapter 10).

Lindsey, R.L., Fedorka-Cray, P.J., Frye, J.G., Meinersmann, R.J., 2009. Inc A/C plasmids are prevalent in multidrug-resistant *Salmonella enterica* isolates. Appl. Environ. Microbiol. 75, 1908–1915.

Maciorowski, K.G., Jones, F.T., Pillai, S.D., Ricke, S.C., 2004. Incidence and control of food-borne *Salmonella* spp. in poultry feeds—a review. World's Poult. Sci. J. 60, 446–457.

Mather, A.E., Reid, S.W.J., Maskell, D.J., Parkhill, J., Fookes, M.C., Harris, S.R., et al., 2013. Distinguishable epidemics of multidrug-resistant *Salmonella typhimurium* DT104 in different hosts. Science 341, 1514–1517.

McLinden, T., Sargeant, J.M., Thomas, M.K., Papadopolo, A., Fazil, A., 2014. Association between component costs, study methodologies, and foodborne illness related factors with the cost of nontyphoidal *Salmonella* illness. Foodborne Pathog. Dis. 11, 718–726.

Mebrhatu, M.T., Cenens, W., Aertsen, A., 2014. An overview of the domestication and impact of the *Salmonella* mobilome. Crit. Rev. Microbiol. 40, 63–75.

M'ikanatha, N.M., Sandt, C.H., Localio, A.R., Tewari, D., Rankin, S.C., Whichard, J.M., et al., 2010. Multidrug-resistant *Salmonella* isolates from retail chicken meat compared with human clinical isolates. Foodborne Pathog. Dis. 7, 929–934.

Mølbak, H., 2005. Human health consequences of antimicrobial drug-resistant *Salmonella* and other foodborne pathogens. Clin. Infect. Dis. 41, 1613–1620.

Muthaiyan, A., Limayem, A., Ricke, S.C., 2011. Antimicrobial strategies for limiting bacterial contaminants in fuel bioethanol fermentations. Prog. Energ. Combust. 37, 351–370.

O'Bryan, C.A., Crandall, P.G., Ricke, S.C., 2008. Organic poultry pathogen control from farm to fork. Foodborne Pathog. Dis. 5, 709–720.

Olsen, S.J., DeBess, E.E., McGivern, T.E., Marano, N., Eby, T., Mauvais, S., et al., 2001. A nosocomial outbreak of fluoroquinolone-resistant *Salmonella* infection. N. Engl. J. Med. 344, 1572–1579.

Painter, J.A., Hoekstra, R.M., Ayers, T., Tauxe, R.V., Braden, C.R., Angulo, F.J., et al., 2013. Attribution of foodborne illnesses, hospitalizations, and deaths to food commodities by using outbreak data, United States, 1998–2008. Emerg. Infect. Dis. 19, 407–415.

Park, S.Y., Woodward, C.L., Kubena, L.F., Nisbet, D.J., Birkhold, S.G., Ricke, S.C., 2008. Environmental dissemination of foodborne *Salmonella* in preharvest poultry production: reservoirs, critical factors and research strategies. Crit. Rev. Environ. Sci. Technol. 38, 73–111.

Patrick, M.E., Adcock, P.M., Gomez, T.M., Altekruse, S.F., Holland, B.H., Tauxe, R.V., et al., 2004. *Salmonella enteritidis* infections, United States, 1985–1999. Emerg. Infect. Dis. 10, 1–7.

Pires, S.M., Vieira, A.R., Hald, T., Cole, D., 2014. Source attribution of human salmonellosis: an overview of methods and estimates. Foodborne Pathog. Dis. 11, 667–676.

Pui, C.F., Wong, W.C., Chai, L.C., Tunung, R., Jeyaletchumi, P., Noor Hidayah, M.S., et al., 2011. *Salmonella*: a foodborne pathogen. Int. Food Res. J. 18, 465–473.

Reynolds, P.E., 1989. Structure, biochemistry and mechanism of action of glycopeptide antibiotics. Eur. J. Clin. Microbiol. Infect. Dis. 8, 943–950.

Ricke, S.C., 2003a. Perspectives on the use of organic acids and short chain fatty acids as antimicrobials. Poult. Sci. 82, 632–639.

Ricke, S.C., 2003b. The gastrointestinal tract ecology of *Salmonella enteritidis* colonization in molting hens. Poult. Sci. 82, 1003–1007.

Ricke, S.C., Van Loo, E.J., Johnson, M.G., O'Bryan, C.A. (Eds.), 2012. Organic Meat Production and Processing Wiley Scientific/IFT, New York, NY.

Ricke, S.C., Dunkley, C.S., Durant, J.A., 2013a. A review on development of novel strategies for controlling *Salmonella enteritidis* colonization in laying hens: fiber-based molt diets. Poult. Sci. 92, 502–525.

Ricke, S.C., Koo, O.-K., Foley, S., Nayak, R., 2013b. Salmonella. In: Labbé, R., García, S. (Eds.), Guide to Foodborne Pathogens, second ed. Wiley-Blackwell, Oxford, UK, pp. 112–137 (Chapter 7).

Roe, M.T., Pillai, S.D., 2003. Monitoring and identifying antibiotic resistance mechanisms in bacteria. Poult. Sci. 82, 622–626.

Sanders, W.L., 1965. Treatment of typhoid fever: a comparative trial of ampicillin and chloramphenicol. Br. Med. J. 5472, 1226–1227.

Scallan, E., Hoekstra, R.M., Angulo, F.J., Tauxe, R.V., Widdowson, M.-A., Roy, S.L., et al., 2011. Foodborne illness acquired in the United States—major pathogens. Emerg. Infect. Dis. 17, 7–15.

Scharff, R.L., 2012. Economic burden from health losses due to foodborne illness in the United States. J. Food. Prot. 75, 123–131.

Schiemann, D.A., 1995. Association with MDCK epithelial cells by *Salmonella typhimurium* is reduced during utilization of carbohydrates. Infect. Immun. 63, 1462–1467.

Schmidt, J.W., Brichta-Harhay, D.M., Kalchayanand, N., Bosilevac, J.M., Shackelford, S.D., Wheeler, T.L., et al., 2012. Prevalence, enumeration, serotypes, and antimicrobial resistance phenotypes of *Salmonella enterica* on carcasses at two large United States pork processing plants. Appl. Environ. Microbiol. 78, 2716–2726.

Shah, D.H., Zhou, X., Addwebi, T., Davis, M.A., Orfe, L., Call, D.R., et al., 2011. Cell invasion of poultry-associated *Salmonella enterica* serovar Enteritidis isolates is associated with pathogenicity, motility and proteins secreted by the type III secretion system. Microbiology 157, 1428–1445.

Shaw, K.J., Rather, P.N., Hare, R.S., Miller, G.H., 1993. Molecular genetics of aminoglycoside resistance genes and familial relationships of the aminoglycoside-modifying enzymes. Microbiol. Rev. 57, 138–163.

Sirsat, S.A., Muthaiyan, A., Ricke, S.C., 2009. Antimicrobials for pathogen reduction in organic and natural poultry production. J. Appl. Poult. Res. 18, 379–388.

Solórzano-Santos, F., Miranda-Novales, M.G., 2012. Essential oils from aromatic herbs as antimicrobial agents. Curr. Opin. Biotechnol. 23, 136–141.

Sørum, H.T., L'Abée-Lund, T.M., 2002. Antibiotic resistance in food-related bacteria—a result of interfering with the global web of bacterial genetics. Int. J. Food Microbiol. 78, 43–56.

Spector, M.P., Kenyon, W.J., 2012. Resistance and survival strategies of *Salmonella enterica* to environmental stresses. Food Res. Int. 45, 455–481.

Spellberg, B., Powers, J.H., Brass, E.P., Miller, L.G., Edwards Jr., J.E., 2004. Trends in antimicrobial drug development: implications for the future. Clin. Infect. Dis. 38, 1279–1286.

Steenackers, H., Hermans, K., Vanderleyden, J., De Keersmaecker, S.C.J., 2012. *Salmonella* biofilms: an overview on occurrence, structure, regulation and eradication. Food Res. Int. 45, 502–531.

Su, L.-H., Chiu, C.-H., Chu, C., Ou, J.T., 2004. Antimicrobial resistance in nontyphoid *Salmonella* serotypes: a global challenge. Clin. Infect. Dis. 39, 546–551.

Talbot, E.A., Gagnon, E.R., Greenblatt, J., 2006. Common ground for the control of multidrug-resistant *Salmonella* in ground beef. Clin. Infect. Dis. 42, 1455–1462.

Tartera, C., Metcalf, E.S., 1993. Osmolarity and growth phase overlap in regulation of *Salmonella typhi* adherence to and invasion of human intestinal cells. Infect. Immun. 61, 3084–3089.

Tenover, F.C., 2006. Mechanisms of antimicrobial resistance in bacteria. Am. J. Med. 119, S3–S10.

Threlfall, E.J., 1992. Antibiotics and the selection of food-borne pathogens. J. Appl. Microbiol. 73, 96S–102S.

Tindall, B.J., Grimont, P.A.D., Garrity, G.M., Euzeby, J.P., 2005. Nomenclature and taxonomy of the genus *Salmonella*. Int. J. Syst. Evol. Microbiol. 55, 521–524.

Tollefson, L., Miller, M.A., 2000. Antibiotic use in food animals: controlling the human health impact. J. AOAC Int. 83, 245–254.

USDA/FSIS. 2013. *Salmonella* questions and answers. <http://www.fsis.usda.gov/wps/portal/fsis/topics/food-safety-education/get-answers/food-safety-fact-sheets/foodborne-illness-and-disease/salmonella-questions-and-answers/> (accessed 23.11.13).

Van Asten, F.J., Hendriks, H.G., Koninkx, J.F., Van der Zeijst, B.A., Gaastra, W., 2000. Inactivation of the flagellin gene of *Salmonella enterica* serotype Enteritidis strongly reduces invasion into differentiated Caco-2 cells. FEMS Microbiol. Lett. 185, 175–179.

Van Asten, F.J., Hendriks, H.G., Koninkx, J.F., van Dijk, J.E., 2004. Flagella-mediated bacterial motility accelerates but is not required for *Salmonella* serotype Enteritidis invasion of differentiated Caco-2 cells. Int. J. Med. Microbiol. 294, 395–399.

Van Loo, E.J., Alali, W., Ricke, S.C., 2012. Food safety and organic meats. Annu. Rev. Food Sci. Technol. 3, 205–225.

Varma, J.K., Greene, K.D., Ovitt, J., Barrett, T.J., Medalla, F., Angulo, F.J., 2005a. Hospitalization and antimicrobial resistance in *Salmonella* outbreaks, 1984–2002. Emerg. Infect. Dis. 11, 943–946.

Varma, J.K., Mølbak, K., Barrett, T.J., Beene, J.L., Jones, T.F., Rabatsky-Her, T., et al., 2005b. Antimicrobial-resistant nontyphoidal *Salmonella* is associated with excess bloodstream infections and hospitalizations. J. Infect. Dis. 191, 554–561.

Velge, P., Cloeckaert, A., Barrow, P., 2005. Emergence of *Salmonella* epidemics: the problems related to *Salmonella enterica* serotype Enteritidis and multiple antibiotic resistance in other major serotypes. Vet. Res. 36, 267–288.

White, D.G., Zhao, S., Sudler, R., Ayers, S., Friedman, S., Chen, S., et al., 2001. The isolation of antibiotic-resistant *Salmonella* from retail ground meats. N. Engl. J. Med. 345, 1147–1154.

Winokur, P.L., Brueggemann, A., DeSalvo, D.L., Hoffmann, L., Apley, M.D., Uhlenhopp, E.K., et al., 2000. Animal and human multidrug-resistant, cephalosporin-resistant *Salmonella* isolates expressing a plasmid-mediated CMY-2 AmpC beta-lactamase. Antimicrob. Agents Chemother. 44, 2777–2783.

Woolhouse, M.E.J., Ward, M.J., 2013. Sources of antimicrobial resistance. Science 341, 1460–1461.

Zhao, S., McDermott, P.F., White, D.G., Qaiyumi, S., Friedman, S.L., Abbott, J.W., et al., 2007. Characterization of multidrug resistant *Salmonella* recovered from diseased animals. Vet. Microbiol. 123, 122–132.

Zhao, S., White, D.G., Friedman, S.L., Glenn, A., Blickenstaff, K., Ayers, S.L., et al., 2008. Antimicrobial resistance in *Salmonella enterica* serovar Heidelberg isolates from retail meats, including poultry, from 2002 to 2006. Appl. Environ. Microbiol. 74, 6656–6662.

Chapter 4

Antimicrobial Resistance and *Campylobacter jejuni* and *C. coli*

Mirko Rossi[1], Satu Olkkola[1], Mati Roasto[2], Rauni Kivistö[1] and Marja-Liisa Hänninen[1]

[1]*Department of Food Hygiene and Environmental Health, University of Helsinki, Finland,*
[2]*Institute of Veterinary Medicine and Animal Sciences, Estonian University of Life Sciences, Kreutzwaldi, Tartu, Estonia*

Chapter Outline

INTRODUCTION

Antimicrobial resistance (AMR) threatens the effective prevention and treatment of an increasing range of bacterial infections including zoonotic *Campylobacter* spp. (World Health Organization, 2014). AMR monitoring programs are active in many parts of the world but strengthening global AMR surveillance is critical to monitor the effectiveness of public health interventions and to detect new trends in resistance which threaten public health.

Campylobacter spp. are members of the epsilonproteobacterial group containing two major species of clinical importance: *C. jejuni* and *C. coli*. *C. jejuni* and *C. coli* are zoonotic pathogens which colonize a wide spectrum of different animal species. Most human infections are caused by *C. jejuni* (>90%) with a minority of cases associated with *C. coli* although geographic differences in the relative contribution of each species do exist (Allos and Blaser, 2010; EFSA and ECDC, 2013). *C. jejuni* and *C. coli* have been identified as major pathogens

Antimicrobial Resistance and Food Safety. DOI: http://dx.doi.org/10.1016/B978-0-12-801214-7.00004-1

of bacterial human gastrointestinal infections since the mid-1990s, when more advanced detection and recording methods for zoonotic infections were started in Europe and North America. In the EU, the number of reported, confirmed human *Campylobacter* infections was 214,000 in 2012 (an average 55 cases per 100,000) (EFSA and ECDC, 2014). In the United States, active surveillance through the Foodborne Diseases Active Surveillance Network (FoodNet) indicates 14 cases per 100,000 (CDC, 2013). Many more cases go undiagnosed or unreported, and campylobacteriosis is estimated to affect over 1.3 million persons every year in the USA (NCEZID, 2014). In temperate regions, campylobacteriosis occurs more frequently during the summer months than during winter. The organism is isolated more frequently from infants and young adults than other age groups and from males more often than females (Allos and Blaser, 2010). The occurrence of foodborne zoonoses and the antimicrobial susceptibility of the causative agents including *Campylobacter* in the EU is monitored at the member state level with the data collected and analyzed annually by the European Food Safety Authority (EFSA) and the European Centre for Disease Prevention and Control (EFSA and ECDC, 2014) to provide the Commission and the Member States with up-to-date information on the current situation. In North America, both the United States (National Antibiotic Resistance Monitoring; NARMS) (NARMS, 2014) and Canada (Canadian Integrated Program for Antimicrobial Resistance Surveillance; CIPARS) (CIPARS, 2007) have established their own monitoring programs for antimicrobial susceptibility patterns and trends of zoonotic pathogens including *Campylobacter* species. Therefore, these geographic regions have data which can be used for interpreting the results and to plan methods to combat AMR.

Most campylobacteriosis cases are seemingly sporadic and reported outbreaks have been associated with improperly heated or handled chicken meat, unpasteurized milk, or contaminated drinking water. Epidemiological studies performed in different geographical areas indicate that chicken meat is one of the major reservoirs and sources of human infections. In addition, contact with pet and domestic animals, drinking unpasteurized milk or drinking water are also risk factors for the infections (Allos and Blaser, 2010). A recent EU risk assessment report suggested that handling, preparation and eating contaminated chicken meat could explain 20–30% of the human cases and that overall 50–80% of the cases might be attributed to poultry as a reservoir (EFSA, 2011). However, it is important to recognize that the relative proportions of risk factors vary in different geographic areas.

During recent years mathematical source attribution models have been developed covering data from several geographic regions in Europe and New Zealand (Wilson et al., 2008; Sheppard et al., 2009; de Haan et al. 2013; Muellner et al., 2013). The models are based on population genetic analyses of genotyping data (most commonly multi locus sequence types) occurring in different sources to attribute clinical *C. jejuni* populations to different potential infection sources. These studies have shown a remarkable attribution of human infections to chicken and bovine/sheep isolates (Wilson et al., 2008; Sheppard et al., 2009; de Haan et al., 2013; Muellner et al., 2013). The general conclusion from the present research is that domestic animals and the food products from those animals represent the most common reservoir and source of human infections. As a consequence, antimicrobial use in animals reflects resistance patterns of both animal and human clinical *C. jejuni* and *C. coli* isolates. Due to geographic differences in the use of antimicrobials in animal production, the risk of acquiring resistant isolates from food sources varies considerably. An example is the different resistance patterns among patients who acquired infections during foreign travel versus infection from domestic sources. This is especially evident in countries where antimicrobial use in veterinary medicine is strictly controlled and levels of resistance are low (Engberg et al., 2004). This difference has been documented in several countries including the United Kingdom (Nichols et al., 2012), the United States (Ricotta et al., 2014), Switzerland (Niederer et al., 2012), and Finland (Hakanen et al., 2003; Feodoroff et al., 2009, 2010). This pattern needs to be considered when evaluating clinical data from countries where a high proportion of campylobacteriosis cases are associated with travel outside of the country (Engberg et al., 2004). Resistant isolates are spread to the environment though several routes including wastewater plants treating human and animal wastes. In consideration of the ecology and epidemiology of AMR, the most important sites where resistant *Campylobacter* strains are generated are the intestines of domestic animals, especially those species commonly colonized with *Campylobacter* and frequently treated with antimicrobials. In some regions, 80–90% of the commercial chicken flocks are colonized by *C. jejuni* at levels of 10^{-6} to 10^{-8} per gram of fecal material, indicating that chicken flocks are large reactors capable of producing resistant isolates if treated with antimicrobial agents (Luangtongkum et al., 2009; Hermans et al., 2012; Lawes et al., 2012). The low fitness cost of certain types of resistances such as fluoroquinolone (FQ) resistance may promote persistence and circulation of the resistant strains among the host animal and other animal species, that is, wild birds and wildlife. From these sources, resistant strains are transferred back into domestic animals and to humans. AMR has been an emerging issue during the last 35 years since *C. jejuni* and *C. coli* have been recognized as important causes of human gastroenteritis (Allos and Blaser, 2010). A recent WHO report on worldwide AMR focused on resistances spreading through food chain from animals to humans (World Health Organization, 2014).

RESISTANCE MECHANISMS

Campylobacter spp. use similar mechanisms for the acquisition of resistance as other Gram-negative bacteria. The most common antimicrobials known to be associated with acquired AMR are FQs, macrolides, tetracyclines (tet), aminoglycosides, and β-lactams (Ge et al., 2013). The cellular targets of resistance for these antimicrobials are presented in Figure 4.1.

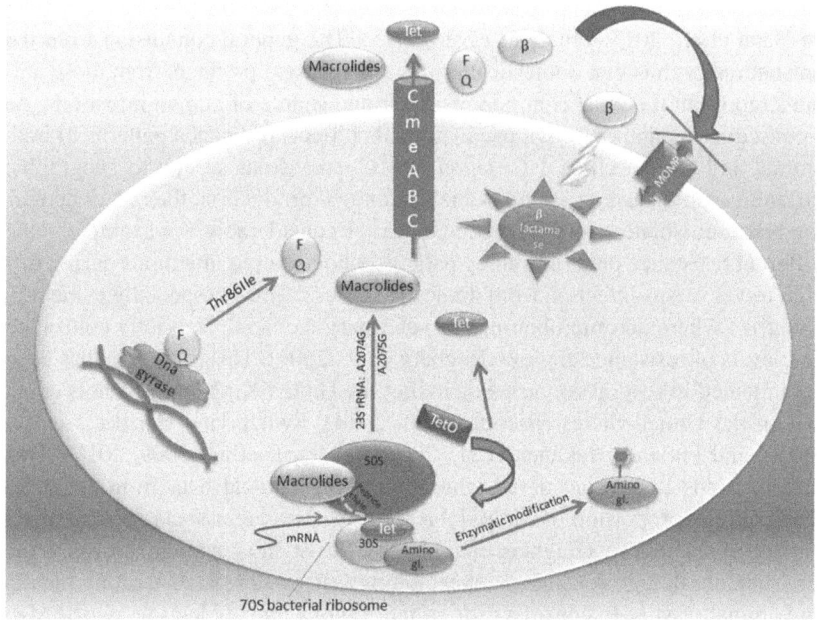

FIGURE 4.1 Schematic presentation of cellular targets of antimicrobial activities.

Fluoroquinolones

The use of FQs to treat human infections (especially diarrhea) started in the late 1980s and early 1990s and their use has expanded since that time (Engberg et al., 2001). Currently, FQs are one of the most frequently used groups of antimicrobials. FQs are broad-spectrum antibiotics, effective against infections caused by both Gram-negative and Gram-positive pathogens. Treatment of domestic food animals with FQs started in the early 1990s. Soon after their expanded use began in the 1990s, an increased FQ resistance among chicken and human isolates was identified (Endtz et al., 1991; Segreti et al., 1992; Smith et al., 1999; Piddock et al., 2003). The most commonly used FQ in human medicine is ciprofloxacin, while in animals enrofloxacin is more widely used. Some newer molecules, such as levofloxacin or moxifloxacin, have been used recently (Luangtongkum et al., 2009; Iovine, 2013). All have, however, similar resistance mechanisms. Since the 1990s, resistance to FQs has increased rapidly (CDC, 2013; EFSA and ECDC, 2013; NARMS, 2014) and numerous studies and/or reports of resistance monitoring programs have been published detailing the frequency of FQ resistance among both human and animal *C. jejuni* and *C. coli* isolates from different regions of the world.

FQs bind to a specific region known as the quinolone-resistance-determining region (QRDR) in the DNA gyrase gene (*gyrA*) and affect DNA supercoiling, decreasing transcription and causing bacterial death (bactericidal effect) (Wang et al., 1993; Han et al., 2012). Mutations in the QRDR region are associated with resistance. Mutation in codon Thr89Ile (Thr86Ala) is associated with high-level resistance (MIC>16 mg/L). Other mutations associated with increased MICs are in Asp90Asn, Asp90Tyr, and Ala70Thr and these mutations cause an intermediate level of resistance in *C. jejuni* and *C. coli* (Wang et al., 1993; Ge et al., 2005). A recent study showed that mutations in Asp90 do not cause changes in DNA supercoiling (Han et al., 2012). Genes for a total of 12 different efflux pumps have been identified but their role in resistance is less well known, with the exception of CmeABC which is responsible for tolerance and transport of physiologically important compounds, such as bile salts but also plays a role in the acquisition of FQ resistance (Luo et al., 2003; Ge et al., 2005; Akiba et al., 2006; Yan et al., 2006; Hannula and Hanninen, 2008a). CmeABC efflux is important in colonization (Lin et al., 2003). The CmeABC efflux pump is negatively regulated by the CmeR regulator (Guo et al., 2008). In addition, CmeR regulates a set of other activities in the bacterial cell making this regulator important for cell function as well as for regulating AMR. The transcriptional regulator CmeR binds specifically with an IR region (TGTAATAAAAATTACA) upstream of the CmeA gene (Guo et al., 2008). The length and sequence of this intergenic regulatory region differs between *C. jejuni* (94 bp) and *C. coli* (107 bp). It has been shown that mutations in the repressing site lead to an overexpression of the efflux pump and to enhanced resistance to several antimicrobials. Studies have identified mutations in the IR region which could change the affinity for CmeR binding (Hanninen and Hannula, 2007; Perez-Boto et al., 2010). Another pump consisting of CmeG also confers resistance to FQs and other antimicrobials. Mutations in the *cmeG* gene decreased MICs to ciprofloxacin by four-fold (Jeon et al., 2011).

FQ resistance develops rapidly after exposure (van Boven et al., 2003; Zhang et al., 2003) because only a point mutation in the QRDR is required for phenotypic resistance. Another reason which might explain rapid development of FQ resistance is the diversity of the population in relation to mutation(s) in the QRDR. Susceptible *C. jejuni* populations exist as a major population within the wild-type QRDR region and a minor population having resistance-associated mutation(s) in the QRDR region (Hanninen and Hannula, 2007; Luangtongkum et al., 2009; Andersson and Hughes, 2014). Exposure to inhibitory concentrations of FQs kills the susceptible part of the population but selects a subpopulation having the resistance-associated mutation. This mechanism of acquisition of AMR at subinhibitory levels is relevant not only to FQ resistance but is common for several other types of resistances as well. During antimicrobial treatments the concentration of antimicrobials needs to be above the mutant "selection window" to prevent development of resistance (Drlica and Zhao, 2007) (Figure 4.2). In the concentration area of the mutant selection window, preexisting mutants are selected and predominate the bacterial population in later colonization phases (Lin et al., 2003). Another mechanism is *de novo*

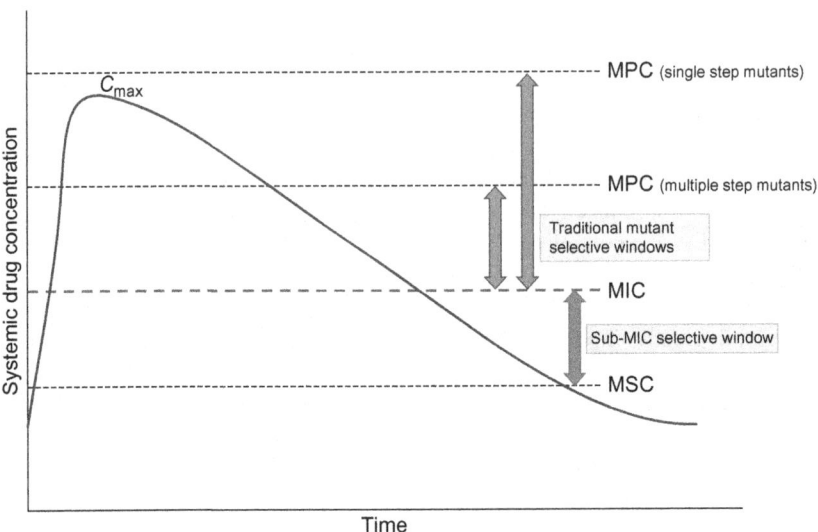

FIGURE 4.2 Mutant selection "window" for selection of resistance.

mutations which are less common among *C. jejuni*. Horizontal *in vivo* transfer of FQ resistance was not shown to be highly evident in an experimental study (Luangtongkum et al., 2009) even though the natural transformation capability of *C. jejuni* is high (Wang and Taylor, 1990).

Macrolides

Erythromycin (a macrolide) is the first-line choice for treatment of campylobacteriosis (Allos and Blaser, 2010). Resistance monitoring of *C. jejuni* and *C. coli* against this group of antimicrobials is common. This group contains several related compounds (e.g., azithromycin, clindramycin, and telithromycin) used in human medicine but the most commonly used macrolide in food production animals is tylosin, used widely in pig farming to treat dysentery in weaned pigs (Juntunen et al., 2010; Logue et al., 2010). Macrolides inhibit bacterial protein synthesis by binding the 50S ribosomal unit (Figure 4.1). Macrolides interfere with aminoacyl translocation, preventing the transfer of tRNA from the A site to the P site of the rRNA complex. Without this translocation, the A site remains occupied and thus prevents amino acids from attaching (Gibreel et al., 2005; Gibreel and Taylor, 2006; Payot et al., 2006; Corcoran et al., 2006). They have a molecular weight of >500 requiring membrane transportation. Acquisition of macrolide resistance in *Campylobacter* is strongly associated with a mutation in the 23S rRNA gene in positions 2074 and 2075. The mutation rate in the acquisition of macrolide resistance is low <10^{-10}/mL (Luangtongkum et al., 2009) indicating that resistance does not develop easily because high concentrations of *Campylobacter* do not often exist in the intestine. In addition to target specific mutation, the efflux pump (CmeABC) acts in macrolide resistance by increasing transport of macrolides and increases the MIC (Mamelli et al., 2005). Resistance development after tylosin treatment is slow and requires a longer treatment period. The resistant isolates persist for a few months but were not detected 7 months after treatment and monitoring at a large farrowing farm (Juntunen et al., 2010), indicating that resistant isolates disappear when the selection pressure is absent or removed. Similar results were found in experimental studies using chickens (Lin et al., 2007). Both human and animal *C. coli* isolates have higher resistance levels for erythromycin than *C. jejuni*. The most evident reason for the difference in resistance patterns between *C. coli* and *C. jejuni* may be the common use of tylosin in swine production.

Tetracyclines

Tetracyclines (tet) were brought into use in the1940s and have activity against both Gram-negative and Gram-positive organisms. Tets were used more commonly during the 1970s and 1980s. After 2000, tets were mainly used in veterinary medicine as therapeutic treatments. Recognition of increasing resistance and development of other antimicrobials has limited their later use (Akhtar, 1988; Dasti et al., 2007). Tetracyclines function by inhibiting protein synthesis.

Known mechanisms of tetracycline resistance in *Campylobacter* include the alteration of tetracycline ribosomal target (Figure 4.1) and interference with the efflux pump, CmeABC (Manavathu et al., 1988; Chopra and Roberts, 2001; Gibreel et al., 2004b). Tet binds to the 30S subunit of ribosomes to inhibit protein synthesis. Resistance is associated with the presence of the *tet*(O) gene which in most cases is located on a plasmid but also sometimes is located in different positions on the chromosome (Dasti et al., 2007; Perez-Boto et al., 2010). The *tet*(O) protein binds to the same ribosomal site as tetracycline and inhibits binding of tetracycline (Epe et al., 1987). Both the *tet*(O) gene and resistance seem to disappear if the use of tetracyclines decreases as seen among *C. coli* in Finnish swine farms (Juntunen et al., 2013). The original hosts of the *tet*(O) gene are intestinal enterococci which have the capability to transfer it to *C. jejuni* or *C. coli* during antimicrobial treatment (Avrain et al., 2004).

Aminoglycosides

Aminoglycosides are bacterial protein synthesis inhibitors which act by binding to the 30S ribosome (bactericidal effect) (Figure 4.1). Aminoglycosides contain a structural component, an amino-modified glycoside, which binds to the 30S ribosome of the molecule. This class, including gentamicin, amikacin, neomycin, tobramycin, kanamycin, and streptomycin, has traditionally been used in therapeutic treatments, especially streptomycin and neomycin which were commonly used in veterinary medicine (Iovine, 2013). The main mechanisms of resistance in *C. coli* and *C. jejuni* have not been thoroughly elucidated but in principle should be similar to other bacteria, that is, through action on aminoglycoside-modifying enzymes (i.e., 3'-aminoglycosidephos-photransferase encoded by *aph*-3) (Gibreel et al., 2004a). Streptomycin resistance is encoded by *aadE*, 6'-adenyl transferase (Pinto-Alphandary et al., 1990). These resistance genes are often located on plasmids or as recently shown, on a genomic island in *C. coli* (Quin et al., 2012) and are probable transferred from enterococci to *C. jejuni* and *C. coli*. Sometimes simultaneous resistance to tetracycline and streptomycin is detected and the encoding genes are often located on the same plasmid (Nirdnoy et al., 2005). Also, integrons seem to have a role in aminoglycoside resistance (Lee et al., 2002; O'Halloran et al., 2004). Furthermore, high-level streptomycin resistance has been shown to be associated with mutations in S12 ribosomal protein (encoded by the *rpsL* gene) in *C. coli* (Olkkola et al., 2010). Streptomycin susceptibility is not often monitored. Gentamicin is included in the USDA NARMS program as well as other programs (NARMS, 2014).

β-Lactam Group of Antimicrobials

β-Lactam antibiotics inhibit cell wall biosynthesis (protein synthesis inhibitors) (Figure 4.1). This group has been the most widely available antimicrobial on the market, covering approximately 65% of the total antimicrobial market worldwide (Chandel et al., 2008). β-Lactam antibiotics are used extensively in veterinary medicine. β-Lactam antibiotics include penicillins, oxacillin, amoxicillin, ampicillin, carbpenicillin, and others. Emerging resistance has compromised their use. Several resistance mechanisms exist but the most commonly studied is β-lactamase production. Studies have shown that ampicillin resistance is increasing among *Campylobacter* strains (NARMS, 2014) even though most national resistance monitoring programs do not include β-lactam resistance among their surveillance of *Campylobacter* strains. Studies have shown that β-lactamase production is common among animal and human *C. jejuni* and *C. coli* isolates (Tajada et al., 1996; Griggs et al., 2009; Juntunen et al., 2012; NARMS, 2014). Although β-lactamase production is most often associated with the resistance other mechanisms may exist which have not been well characterized. A β-lactamase gene *OXA-61* associated with ampicillin resistance has been described (Alfredson and Korolik, 2007) even though the *OXA-61* gene is present in strains which do not produce β-lactamase (Juntunen et al., 2012; Zeng et al., 2014). A recent study has revealed that a mutation (G to T transversion) in the promoter region is required to activate β-lactamase production (Zeng et al., 2014). In addition to the target-specific mechanism, an efflux pump CmeABC has been associated with the ampicillin resistance in *Campylobacter* (Iovine, 2013).

DEVELOPMENT OF RESISTANCE

The dynamics of the resistance development against the most commonly used antimicrobials in *C. jejuni* and *C. coli* is highly dependent of the antimicrobial class, exposure concentrations, and mutation frequency related to each anti-microbial agent. Concentrations of antimicrobials at the infection site of the body of therapeutically treated animals should be above the MIC of the susceptible populations (Farnell et al., 2005). This concentration is bactericidal to the susceptible populations but may select for a potential minority from the existing population with target-specific mutation(s), especially if the concentration is under the MIC value of the resistant subpopulation (Luangtongkum et al., 2009). The selection of resistant subpopulations occurs in the area of concentrations known as the "mutant selection window" (Figure 4.2) (Drlica and Zhao, 2007). *C. jejuni* mutants have been shown to be rapidly selected after FQ treatment of animals, such as chickens with enrofloxacin *in vivo* (McDermott et al., 2002; Payot et al., 2002; van Boven et al., 2003; Zhang et al., 2003; Griggs et al., 2005; Asai et al., 2007; Juntunen et al 2011). Less well known is the potential selection of resistance at sub-MIC concentrations in *C. jejuni*. Recently, resistance selection at sub-MIC concentrations has been shown to occur in *Escherichia coli* and *Salmonella enterica* serovar Typhimurium and is suspected to be common in other zoonotic pathogens as well (Andersson and Hughes, 2014).

All zoonotic pathogens experience sub-MIC exposures when antimicrobials are used as growth promoters. The acquisition of resistance at low antimicrobial concentrations might be important for the ecology of pathogen populations, since selection at sub-MIC concentrations (ng/mL) can explain the low fitness cost of FQ resistance in *C. jejuni* recognized *in vivo* both in chicken colonization experiments (Luo et al., 2003) as well as in field conditions in the USA where ciprofloxacin-resistant *C. jejuni* strains have persisted among poultry for several years after the withdrawal of enrofloxacin treatment of chickens in 2005 (Davidson, 2004; Price et al., 2007; Ge et al., 2013). Some experimental data have supported the development of clinically relevant ciprofloxacin-resistant *C. jejuni* strains after subinhibitory MIC level exposure of *C. jejuni* strains (Hannula and Hanninen, 2008b). Mutation frequency for selection of FQ ORDR mutants is in the range of 10^{-6} to 10^{-9} (Hänninen and Hannula, 2007) but is much lower for erythromycin ($<10^{-10}$/mL) (Luangtongkum et al., 2009). Mutation frequency is a central factor in the emergence of resistant populations.

Persistence of the tetracycline resistance gene *tet(O)* is probably not sustainable as a Finnish study showed no *tet(O)* genes nor tetracycline resistance were detected among *C. coli* isolates from nine swine farms included in the study. In a follow-up study at the same swine farm where a tetracycline treatment was used, no recovered isolates developed tetracycline resistance during the 2 weeks of therapy (Juntunen et al., 2013). These data suggest that tetracycline resistance requires the presence of selective pressure for the *tet(O)* gene to persist. Resistance can be transferred between animals living in close proximity on poultry and swine farms (Avrain et al., 2004; Luangtongkum et al., 2009).

GENETIC ADAPTIVE MECHANISMS IN RESISTANT BACTERIA

Antibiotic exposure causes some additional adaptive mutations in bacteria as described in *in vitro* studies of *C. jejuni* under either ciprofloxacin or erythromycin exposure. Exposure of *C. jejuni* strain 81-176 to ciprofloxacin caused in addition to the mutation in QRDR, point mutations in genes associated with, for example, motility (Hyytiainen et al., 2013). Exposure to antimicrobials can cause changes in the transcription of certain genes. Exposure of a *C. jejuni* wild-type strain NCTC 11168 to an inhibitory dose (16 times the MIC) of erythromycin produced changes in the expression of more than 250 genes including both up- and downregulating genes. The common gene groups affected were those associated with motility, regulation, and the tricarboxylic acid cycle. The corresponding responses were not identified in the resistant variant (Hao et al., 2013). Furthermore, exposure to subinhibitory concentrations resulted in transcription changes in fewer genes. In strain 81-176 exposure to ciprofloxacin also caused transcriptional changes in a large number of genes. The transcriptional analysis revealed that ciprofloxacin exposure caused changes in the expression of genes involved in DNA replication and repair in the wild-type as well as in the resistant variant (Hyytiainen et al., 2013).

CONSEQUENCES OF RESISTANCE ON BACTERIAL FITNESS

Most studies performed on the effect of AMP on bacterial fitness costs have been performed using FQ (ciprofloxacin) exposure. Fitness cost can be detected as a changed growth rate in laboratory media or survival in food matrices compared with the wild type or competition in animal gut colonization experiments. Luo et al. (2005) found that the strain NCTC 11168 having the Thr86Ile mutation in *gyrA* outcompeted the wild type in chicken intestinal colonization experiments even though both variants had similar colonization levels when inoculated separately. Zeitouni and Kempf (2011) conducted similar experiments using a different *C. jejuni* strain than Luo et al. and in addition tested a *C. coli* strain. Zeitouni and Kempf found a decreased competition capacity among ciprofloxacin-resistant isolates in chicken colonization experiments. Macrolide-resistant strains disappear rather rapidly after tylosin therapy at pig farms (Juntunen et al., 2011). Experimental studies have revealed that macrolide resistance (ery) decreases the survival in food matrices and results in a slower growth rate when compared to the wild-type strains (Gibreel et al., 2005; Han et al., 2009; Almofti et al., 2011). Also co-inoculation of a wild-type and an ery-resistant variant in mice resulted in lower colonization of the resistant strain compared to the susceptible strain (Zeitouni et al., 2012). In competition experiments the mutant strain grew at a significantly slower rate and survived less than the susceptible parent strain. The mutation imposed a fitness cost in the ery-resistant mutant compared to the susceptible parent strain. A limited number of studies have been undertaken in regards to the effects of tetracycline resistance on the fitness of *Campylobacter*.

RESISTANCE DETECTION

Standardized susceptibility testing methods have been developed for *C. jejuni* and *C. coli* (McDermott et al., 2004, 2005). These testing methods and the interpretation of the results are described in more detailed in a recent review (Ge et al., 2013). These protocols have improved the interpretation and comparison of the MIC results between laboratories. The protocols have been published as the standards of the Clinical and Laboratory Standards Institute (CLSI, 2014) in the United States and the European Committee on Antimicrobial Susceptibility Testing (EUCAST, 2014) in Europe. Both CLSI and EUCAST developed guidelines for standardized antimicrobial susceptibility testing and formulated interpretation criteria for categorizing organisms as susceptible, intermediate, or resistant. These interpretation criteria have been published by these organizations. Harmonization of the methods is in progress and is recommended also by a recent WHO report (World Health Organization, 2014).

The standardized methods include disk diffusion, broth microdilution, and agar dilution methods. Disk diffusion methods include the commercial E Test. The commercial broth microdilution includes the VetMic testing system (Ge et al., 2013). CLSI has standardized all three types of methods but EUCAST does not include the agar dilution method. The basic growth media is either Mueller-Hinton agar or broth (cation-adjusted) containing lysed horse blood (5%), with sheep blood (5%) used in the disk diffusion method of CLSI. The initial inoculum size is the same in all methods, a 0.5 MacFarland standard. An adjusted inoculum is of major importance to achieve comparable results. Agar dilution and broth microdilution are the primary choices of method. The agar plates and broth tubes always contain twofold dilutions of the antimicrobial agents and are inoculated with the standardized suspension of bacteria. MIC (minimum inhibitory concentration) is defined as the lowest antimicrobial concentration that inhibits bacterial growth. Reference strain *C. jejuni* ATCC 33560 is used as the standard. Several publications compare the performance of the methods (Ge et al., 2013). The purpose of the research study and the capacity of the laboratory conducting the research or surveillance determines the best method to be used. More detailed information is available (Ge et al., 2013).

GENETIC METHODS IN THE DETECTION OF RESISTANCE DETERMINING TARGETS

Genetic targets of the most important AMRs are described for *C. jejuni and C. coli*. FQ resistance can be determined by PCR amplification of the QRDR region of the *gyrA* gene and Sanger sequencing of the PCR product. Point mutations in amino acid Thr86Ile or Thr86Arg are commonly detected in bacteria having high MIC values for FQs. In addition, mutations can be found in amino acid 90 which leads to a lowerlevel of resistance (Wang et al., 1993; Piddock et al., 2003). High-level macrolide resistance (>512 µg/mL) can be detected as a point mutation of the 23S RNA gene in the region corresponding to nucleotides 2074 and 2075 (Lehtopolku et al., 2011). Tetracycline resistance can be determined by PCR amplification of the *tet*(O) gene (Gibreel et al., 2004b). This gene is located generally on a plasmid but sometimes on the chromosome. Susceptible isolates do not contain the *tet*(O) gene (Juntunen et al., 2013) indicating that presence or absence of PCR may be used to determine resistance or susceptibility. Aminoglycoside resistance has been studied less but is associated with the *aad2* gene located either on an integron or a plasmid (Lee et al., 2002; O'Halloran et al., 2004; Gibreel et al., 2004a) and also mutations in the ribosomal genes (*rpsL*) have been linked to a high level of resistance to streptomycin in *C. coli* (Olkkola et al., 2010). β-Lactamase resistance and its genetic mechanisms in *C. jejuni* and *C. coli* are not completely known. Detection of the β-lactamase gene *blaoxa-61* (cj0299) does not always indicate resistance because susceptible isolates have been found to carry the gene, but not produce β-lactamase (Juntunen et al., 2012; Zeng et al., 2014). A recent study found that *bla-oxa61* is actually a promoter controlling a mutation (G- >T transversion) associated with a high level of β-lactam resistance (Zeng et al., 2014).

GENOMIC ANALYSIS OF RESISTANCE

As NGS technologies and bioinformatics analysis tools evolve and prices drop, pathogens are targeted for whole-genome sequencing (WGS). This approach offers the opportunity to recognize all potential resistance mechanisms simultaneously and additionally has the potential to find new targets (Rossolini and Thaller, 2010). Platforms and pipelines have been developed for analysis of the resistome of bacterial strains (CGE, 2014) or Comprehensive Antibiotic Research Database (CARD) (McArthur et al., 2013). The usefulness of genome data in AMR studies is highlighted by the recent findings of novel genes associated with gentamicin resistance in *C. coli* (Chen et al., 2013). Although genomics will provide all resistance-associated target genes, the level of resistance will still need to be analyzed using traditional MICs because clinical breakpoints are needed when selecting suitable therapy.

CLINICAL BREAKPOINTS AND EPIDEMIOLOGICAL CUT-OFF VALUES

Clinical breakpoint values (CBRV) are determined based on the therapeutic concentrations of antimicrobials in the body. CBRVs are determined from the pharmacokinetic/pharmacodynamic data (PK/PD) of the antimicrobials in the target host (humans, animals), *in vitro* susceptibility testing of the isolates and data from clinical experiments (CLSI, 2014; EUCAST, 2014). Epidemiological cut-off values are developed for monitoring purposes of zoonotic foodborne pathogens to predict temporary and spatial fluctuations in the susceptibility of bacterial populations (EUCAST, 2014). Cut-off values (ECOFFs) are determined as MIC values from a large population of organisms isolated from different sources. Population-level analysis divides the population usually into two parts; one contains isolates with low MICs (wild-type) and another with higher MICs (non-wild-type) and usually there is a gap between the groups. ECOFF is a MIC clearly separating the groups. Therefore ECOFFs are lower than clinical breakpoints. In long-term monitoring, changes in the population structure closer to the wild-type population or to an increased number of isolates in the non-wild-type population can be followed and antimicrobial use in animals can be regulated. *C. jejuni* and *C. coli* are typical zoonotic pathogens which acquire resistance when animals are treated with antimicrobials at therapeutic levels for certain infectious diseases or subtherapeutic levels for preventive purposes and can be indicators for antimicrobial use in animal hosts.

Resistance Levels in Different Countries and Sources

National resistance monitoring programs, such as NARMS and CIPAS, started to evolve in 1990 (CIPARS, 2007; NARMS, 2014). The European Commission launched a strategy to combat the threat of AMR to human, animal, and plant health covering the areas of data collection and national surveillance (Directive 2003/99/EC). Even though data from different systems are rather comparable, NARMS uses clinical breakpoint values in the assessment of resistance, similarly the EU uses clinical breakpoints for the assessment of clinical data. Animal data in the EU are assessed using ECOFFS. In the EU data for *Campylobacter* spp., clinical resistance to common antimicrobials was detected frequently (EFSA and ECDC, 2014). A high proportion of human isolates (47.4% EU average; including isolates from both domestic and travel-associated patients) were resistant to ciprofloxacin with increasing trends observed in several Member States, followed by ampicillin (36.4%) and tetracycline (32.4%) resistances. High levels of multiresistance were observed in some countries. Resistance to erythromycin was low overall (3.1%), but moderately high in human *C. coli* isolates (15.1%). *Campylobacter* monitoring results from poultry in 2012 showed high levels of resistance to ciprofloxacin (44%), and tetracyclines (34%) while resistance to erythromycin was low (<1%). *C. coli* isolates from poultry had higher resistance levels than *C. jejuni*, especially erythromycin resistance which was on average 11% higher. Resistance levels in poultry meat samples followed the same trend as cecal samples. *C. coli* isolates from swine fecal samples had an average resistance to ciprofloxacin of 31%, tetracyclines of 76%, and erythromycin of 26% with an increase in resistance from 2006 to 2012. In cattle, the average resistance level reported for ciprofloxacin was 33% and for tetracyclines was 44%. However, a wide variation in all resistances between different EU member countries is evident. A common trend among both human and animal *C. jejuni* and *C. coli* isolates was lower levels of resistance in the Nordic countries (Finland, Norway, and Sweden) but higher in the southern European countries with >90% FQ resistance in the Spain and Portugal. Germany, France, and the United

Kingdom reported high levels of resistance among both animals and human samples (EFSA and ECDC, 2014). Other published data, for example from Spain, have indicated high levels of resistance especially to FQs (Melero et al., 2012) starting from the 1990s (Saenz et al., 2000). In the United States in 2010, the NARMS study reported a lower average resistance to ciprofloxacin (22%) but the same level of resistance for tetracyclines (43% of *C. jejuni* and 49% of *C. coli*) among human isolates (NARMS, 2014) as in the EU. Erythromycin resistance levels were higher, 43% and 40% among human and chicken *C. coli* isolates, respectively. Gentamicin resistance increased between 2007 and 2010 among both human and chicken *C. coli* isolates from <1% to 11% and 13%, respectively, but the levels remained low among the *C. jejuni* isolates (<1%) (NARMS, 2014). The *C. coli* isolates were shown to have identical pulsed-field gel electrophoresis patterns indicating clonal spread of this type (Chen et al., 2013). In the United States, FQ resistance among human isolates has increased from 13% in 1997 to 23% in 2009–2011 (CDC, 2013). The FQ resistance trend is especially interesting to follow-up in the United States due to the withdrawal of FQs from poultry production in 2005 (Davidson, 2004). Whether the lower level of FQ resistance observed among *C. jejuni* isolates in the United States when compared to the average FQ resistance levels in the EU is associated with the removal of FQ use in poultry production remains to be seen. Resistance levels among Asian countries are mostly reported from Thailand where, for example, ciprofloxacin resistance is very high (>90%) among animal isolates. Europeans as well, who acquire campylobacteriosis while travelling to Thailand, have highly resistant infecting strains (Engberg et al., 2001, 2004; Luangtongkum et al., 2009; Feodoroff et al., 2010). A recent report from three regions in Peru indicated high levels of resistance, especially to FQs (up to 90%) revealing that a worldwide strategy is needed to combat resistance (Pollett et al., 2012). Wide variations in the resistance levels of both human and animal *Campylobacter* isolates between countries can be explained by differing strategies and practices of antimicrobial use in animal production. Countries which strictly regulate and restrict use of antimicrobials have lower resistance levels than countries which have more liberal strategies. The Nordic countries (EFSA and ECDC, 2014), New Zealand (Heffernan et al., 2011), and Australia (Unicomb et al., 2006) are examples of countries where the use of antimicrobials in poultry production is strictly controlled and as a consequence a low level of resistance occurs. A recent comparative study on the association of antimicrobial use per kilogram of meat produced and the measured resistance levels of monitored bacteria including *Campylobacter* in the EU countries confirmed the association between the consumption of antimicrobials by meat-producing animals and resistance levels in pathogenic bacteria (Garcia-Migura et al., 2014). For example, in Finland, Denmark, Norway, and Sweden, an approximate level of 0.2 mg FQ per kg meat produced was consumed by meat production animals compared to the Czech Republic, the Netherlands, and France where 1.2–2.9 mg FQ per kg meat produced was consumed by meat production animals and a clear positive association between the use and levels of resistance was detected (Garcia-Migura et al., 2014). The weakness of the present reporting systems is that specific data on the use of different antimicrobials in the various animal species is lacking; even all the MIC monitoring data produced are animal species specific. Another uncertainty factor in the study of the resistance might be the fact that resistant *Campylobacter* isolates can persist and circulate for long periods even after the decreased use of antimicrobials.

REFERENCES

Akhtar, S., 1988. Antimicrobial sensitivity and plasmid-mediated tetracycline resistance in *Campylobacter jejuni* isolated in Bangladesh. Chemotherapy 34, 326–331.

Akiba, M., Lin, J., Barton, Y.W., Zhang, Q., 2006. Interaction of CmeABC and CmeDEF in conferring antimicrobial resistance and maintaining cell viability in *Campylobacter jejuni*. J. Antimicrob. Chemother. 57, 52–60.

Alfredson, D.A., Korolik, V., 2007. Identification of putative zinc hydrolase genes of the metallo-β-lactamase superfamily from *Campylobacter jejuni*. FEMS Immunol. Med. Microbiol. 49, 159–164.

Allos, B.M., Blaser, M.J., 2010. *Campylobacter jejuni* and related species. In: Mandell, G., Dolin, R., Bennett, J. (Eds.), Mandell, Douglas, and Bennett's Principles and Practice of Infectious Diseases, Churchill, Livingstone, Elsevier, pp. 2793–2802.

Almofti, Y.A., Dai, M., Sun, Y., Haihong, H., Yuan, Z., 2011. Impact of erythromycin resistance on the virulence properties and fitness of *Campylobacter jejuni*. Microb. Pathog. 50, 336–342.

Andersson, D.I., Hughes, D., 2014. Microbiological effects of sublethal levels of antibiotics. Nat. Rev. Microbiol. 12, 465–478.

Asai, T., Harada, K., Ishihara, K., Kojima, A., Sameshima, T., Tamura, Y., et al., 2007. Association of antimicrobial resistance in *Campylobacter* isolated from food-producing animals with antimicrobial use on farms. Jpn. J. Infect. Dis. 60, 290.

Avrain, L., Vernozy-Rozand, C., Kempf, I., 2004. Evidence for natural horizontal transfer of tetO gene between *Campylobacter jejuni* strains in chickens. J. Appl. Microbiol. 97, 134–140.

CDC, 2013. Incidence and trends of infection with pathogens transmitted commonly through food-foodborne diseases active surveillance network, 10 U.S. sites, 1996–2012. Morb. Mortal. Wkly. Rep. 62, 283–287.

CGE, 2014. Center for Genomic Epidemiology. <http://www.genomicepidemiology.org/>.

Chandel, A.K., Rao, L.V., Narasu, M.L., Singh, O.V., 2008. The realm of penicillin G acylase in β-lactam antibiotics. Enzyme Microb. Technol. 42, 199–207.

Chen, Y., Mukherjee, S., Hoffmann, M., Kotewicz, M.L., Young, S., Abbott, J., et al., 2013. Whole-genome sequencing of gentamicin-resistant *Campylobacter coli* isolated from U.S. retail meats reveals novel plasmid-mediated aminoglycoside resistance genes. Antimicrob. Agents Chemother. 57, 5398–5405.

Chopra, I., Roberts, M., 2001. Tetracycline antibiotics: mode of action, applications, molecular biology, and epidemiology of bacterial resistance. Microbiol. Mol. Biol. Rev. 65, 232–260. (Second page, table of contents.)

CIPARS, 2007. Canadian Integrated Program for Antimicrobial Resistance Surveillance. <http://www.phac-aspc.gc.ca/cipars-picra/index-eng.php>.

CLSI, 2014. Clinical and Laboratory Standard Institute. <http://clsi.org/>.

Corcoran, D., Quinn, T., Cotter, L., Fanning, S., 2006. An investigation of the molecular mechanisms contributing to high-level erythromycin resistance in *Campylobacter*. Int. J. Antimicrob. Agents 27, 40–45.

Dasti, J.I., Gross, U., Pohl, S., Lugert, R., Weig, M., Schmidt-Ott, R., 2007. Role of the plasmid-encoded tet(O) gene in tetracycline-resistant clinical isolates of *Campylobacter jejuni* and *Campylobacter coli*. J. Med. Microbiol. 56, 833–837.

Davidson, D., 2004. In the matter of enrofloxacin for poultry: withdrawal of approval of Bayer Corporation's new animal drug application 1 (NADA) 140-828 (Baytril). FDA Docket No 00N-1571 2004.

de Haan, C., Lampén, K., Corander, J., Hänninen, M., 2013. Multilocus sequence types of environmental *Campylobacter jejuni* isolates and their similarities to those of human, poultry and bovine *C. jejuni* isolates. Zoonoses Public Health 60, 125–133.

Drlica, K., Zhao, X., 2007. Mutant selection window hypothesis updated. Clin. Infect. Dis. 44, 681–688.

EFSA, 2011. Panel on biological hazards (BIOHAZ); scientific opinion on *Campylobacter* in broiler meat production: control options and performance objectives and/or targets at different stages of the food chain. EFSA J. 9, 141.

EFSA, ECDC, 2013. The European Union summary report on trends and sources of zoonoses, zoonotic agents and food-borne outbreaks in 2011. EFSA J. 11, 3129–3379.

EFSA, ECDC, 2014. The European Union summary report on trends and sources of zoonoses, zoonotic agents and food-borne outbreaks in 2012. EFSA J. 12, 3547–3859.

Endtz, H.P., Ruijs, G.J., van Klingeren, B., Jansen, W.H., van der Reyden, T., Mouton, R.P., 1991. Quinolone resistance in *Campylobacter* isolated from man and poultry following the introduction of fluoroquinolones in veterinary medicine. J. Antimicrob. Chemother. 27, 199–208.

Engberg, J., Aarestrup, F.M., Taylor, D.E., Gerner-Smidt, P., Nachamkin, I., 2001. Quinolone and macrolide resistance in *Campylobacter jejuni* and *C. coli*: resistance mechanisms and trends in human isolates. Emerg. Infect. Dis. 7, 24–34.

Engberg, J., Neimann, J., Nielsen, E.M., Aerestrup, F.M., Fussing, V., 2004. Quinolone-resistant *Campylobacter* infections: risk factors and clinical consequences. Emerg. Infect. Dis. 10, 1056–1063.

Epe, B., Woolley, P., Hornig, H., 1987. Competition between tetracycline and tRNA at both P and A sites of the ribosome of *Escherichia coli*. FEBS Lett. 213, 443–447.

EUCAST, E., 2014. European Committee on antimicrobial susceptibility testing. <www.eucast. org/>.

Farnell, M., Donoghue, A., Cole, K., Reyes-Herrera, I., Blore, P., Donoghue, D., 2005. *Campylobacter* susceptibility to ciprofloxacin and corresponding fluoroquinolone concentrations within the gastrointestinal tracts of chickens. J. Appl. Microbiol. 99, 1043–1050.

Feodoroff, B., Ellström, P., Hyytiäinen, H., Sarna, S., Hänninen, M., Rautelin, H., 2010. *Campylobacter jejuni* isolates in Finnish patients differ according to the origin of infection. Gut Pathog. 2, 22.

Feodoroff, F., Lauhio, A., Sarna, S., Hänninen, M., Rautelin, H., 2009. Severe diarrhoea caused by highly ciprofloxacin-susceptible *Campylobacter* isolates. Clin. Microbiol. Infect. 15, 188–192.

Garcia-Migura, L., Hendriksen, R.S., Fraile, L., Aarestrup, F.M., 2014. Antimicrobial resistance of zoonotic and commensal bacteria in Europe: the missing link between consumption and resistance in veterinary medicine. Vet. Microbiol. 170, 1–9.

Ge, B., McDermott, P.F., White, D.G., Meng, J., 2005. Role of efflux pumps and topoisomerase mutations in fluoroquinolone resistance in *Campylobacter jejuni* and *Campylobacter coli*. Antimicrob. Agents Chemother. 49, 3347–3354.

Ge, B., Wang, F., Sjölund-Karlsson, M., McDermott, P.F., 2013. Antimicrobial resistance in *Campylobacter*: susceptibility testing methods and resistance trends. J. Microbiol. Methods 95, 57–67.

Gibreel, A., Taylor, D.E., 2006. Macrolide resistance in *Campylobacter jejuni* and *Campylobacter coli*. J. Antimicrob. Chemother. 58, 243–255.

Gibreel, A., Sköld, O., Taylor, D.E., 2004a. Characterization of plasmid-mediated aphA-3 kanamycin resistance in *Campylobacter jejuni*. Microb. Drug Resist. 10, 98–105.

Gibreel, A., Tracz, D.M., Nonaka, L., Ngo, T.M., Connell, S.R., Taylor, D.E., 2004b. Incidence of antibiotic resistance in *Campylobacter jejuni* isolated in Alberta, Canada, from 1999 to 2002, with special reference to tet(O)-mediated tetracycline resistance. Antimicrob. Agents Chemother. 48, 3442–3450.

Gibreel, A., Kos, V.N., Keelan, M., Trieber, C.A., Levesque, S., Michaud, S., et al., 2005. Macrolide resistance in *Campylobacter jejuni* and *Campylobacter coli*: molecular mechanism and stability of the resistance phenotype. Antimicrob. Agents Chemother. 49, 2753–2759.

Griggs, D.J., Johnson, M.M., Frost, J.A., Humphrey, T., Jorgensen, F., Piddock, L.J., 2005. Incidence and mechanism of ciprofloxacin resistance in *Campylobacter* spp. isolated from commercial poultry flocks in the United Kingdom before, during, and after fluoroquinolone treatment. Antimicrob. Agents Chemother. 49, 699–707.

Griggs, D.J., Peake, L., Johnson, M.M., Ghori, S., Mott, A., Piddock, L.J., 2009. Beta-lactamase-mediated beta-lactam resistance in Campylobacter species: prevalence of Cj0299 (bla OXA-61) and evidence for a novel beta-lactamase in *C. jejuni*. Antimicrob. Agents Chemother. 53, 3357–3364.

Guo, B., Wang, Y., Shi, F., Barton, Y.W., Plummer, P., Reynolds, D.L., et al., 2008. CmeR functions as a pleiotropic regulator and is required for optimal colonization of *Campylobacter jejuni* in vivo. J. Bacteriol. 190, 1879–1890.

Hakanen, A.J., Lehtopolku, M., Siitonen, A., Huovinen, P., Kotilainen, P., 2003. Multidrug resistance in *Campylobacter jejuni* strains collected from Finnish patients during 1995–2000. J. Antimicrob. Chemother. 52, 1035–1039.

Han, F., Pu, S., Wang, F., Meng, J., Ge, B., 2009. Fitness cost of macrolide resistance in *Campylobacter jejuni*. Int. J. Antimicrob. Agents 34, 462–466.

Han, J., Wang, Y., Sahin, O., Shen, Z., Guo, B., Shen, J., et al., 2012. A fluoroquinolone resistance associated mutation in gyrA affects DNA supercoiling in *Campylobacter jejuni*. Front Cell Infect. Microbiol 1 (2), 21.

Hanninen, M.L., Hannula, M., 2007. Spontaneous mutation frequency and emergence of ciprofloxacin resistance in *Campylobacter jejuni* and *Campylobacter coli*. J. Antimicrob. Chemother. 60, 1251–1257.

Hannula, M., Hanninen, M.L., 2008a. Effect of putative efflux pump inhibitors and inducers on the antimicrobial susceptibility of *Campylobacter jejuni* and *Campylobacter coli*. J. Med. Microbiol. 57, 851–855.

Hannula, M., Hanninen, M.L., 2008b. Effects of low-level ciprofloxacin challenge in the in vitro development of ciprofloxacin resistance in *Campylobacter jejuni*. Microb. Drug Resist. 14, 197–201.

Hao, H., Yuan, Z., Shen, Z., Han, J., Sahin, O., Liu, P., et al., 2013. Mutational and transcriptomic changes involved in the development of macrolide resistance in *Campylobacter jejuni*. Antimicrob. Agents Chemother. 57, 1369–1378.

Heffernan, H., Lok Wong, T., Lindsay, J., Bowen, B., Woodhouse, R., 2011. A baseline survey of antimicrobial resistance in bacteria from selected New Zealand foods, 2009–2010. MAF Technical Paper 2011/53.

Hermans, D., Pasmans, F., Messens, W., Martel, A., Van Immerseel, F., Rasschaert, G., et al., 2012. Poultry as a host for the zoonotic pathogen *Campylobacter jejuni*. Vector-Borne and Zoonotic Diseases 12, 89–98.

Hyytiainen, H., Juntunen, P., Scott, T., Kytomaki, L., Venho, R., Laiho, A., et al., 2013. Effect of ciprofloxacin exposure on DNA repair mechanisms in *Campylobacter jejuni*. Microbiology 159, 2513–2523.

Iovine, N.M., 2013. Resistance mechanisms in *Campylobacter jejuni*. Virulence 4, 230–240.

Jeon, B., Wang, Y., Hao, H., Barton, Y.W., Zhang, Q., 2011. Contribution of CmeG to antibiotic and oxidative stress resistance in *Campylobacter jejuni*. J. Antimicrob. Chemother. 66, 79–85.

Juntunen, P., Heiska, H., Olkkola, S., Myllyniemi, A., Hänninen, M., 2010. Antimicrobial resistance in *Campylobacter coli* selected by tylosin treatment at a pig farm. Vet. Microbiol. 146, 90–97.

Juntunen, P., Olkkola, S., Hänninen, M., 2011. Longitudinal on-farm study of the development of antimicrobial resistance in *Campylobacter coli* from pigs before and after danofloxacin and tylosin treatments. Vet. Microbiol. 150, 322–330.

Juntunen, P., Heiska, H., Hänninen, M., 2012. *Campylobacter coli* Isolates from Finnish Farrowing Farms Using Aminopenicillins: High Prevalence of bla OXA-61 and β-Lactamase Production, But Low MIC Values. Foodborne Pathog. Dis. 9, 902–906.

Juntunen, P., Laurila, T., Heinonen, M., Hänninen, M., 2013. Absence of tetracycline resistance in Campylobacter coli isolates from Finnish finishing pigs treated with chlortetracycline. J. Appl. Microbiol. 114, 974–981.

Lawes, J., Vidal, A., Clifton-Hadley, F., Sayers, R., Rodgers, J., Snow, L., et al., 2012. Investigation of prevalence and risk factors for *Campylobacter* in broiler flocks at slaughter: results from a UK survey. Epidemiol. Infect. 140, 1725–1737.

Lee, M.D., Sanchez, S., Zimmer, M., Idris, U., Berrang, M.E., McDermott, P.F., 2002. Class 1 integron-associated tobramycin-gentamicin resistance in *Campylobacter jejuni* isolated from the broiler chicken house environment. Antimicrob. Agents Chemother. 46, 3660–3664.

Lehtopolku, M., Kotilainen, P., Haanpera-Heikkinen, M., Nakari, U.M., Hanninen, M.L., Huovinen, P., et al., 2011. Ribosomal mutations as the main cause of macrolide resistance in *Campylobacter jejuni* and *Campylobacter coli*. Antimicrob. Agents Chemother. 55, 5939–5941.

Lin, J., Sahin, O., Michel, L.O., Zhang, Q., 2003. Critical role of multidrug efflux pump CmeABC in bile resistance and in vivo colonization of *Campylobacter jejuni*. Infect. Immun. 71, 4250–4259.

Lin, J., Yan, M., Sahin, O., Pereira, S., Chang, Y.J., Zhang, Q., 2007. Effect of macrolide usage on emergence of erythromycin-resistant *Campylobacter* isolates in chickens. Antimicrob. Agents Chemother. 51, 1678–1686.

Logue, C., Danzeisen, G., Sherwood, J., Thorsness, J., Mercier, B., Axtman, J., 2010. Repeated therapeutic dosing selects macrolide-resistant *Campylobacter* spp. in a turkey facility. J. Appl. Microbiol. 109, 1379–1388.

Luangtongkum, T., Jeon, B., Han, J., Plummer, P., Logue, C.M., Zhang, Q., 2009. Antibiotic resistance in *Campylobacter*: emergence, transmission and persistence. Future Microbiol. 4, 189–200.

Luo, N., Pereira, S., Sahin, O., Lin, J., Huang, S., Michel, L., et al., 2005. Enhanced in vivo fitness of fluoroquinolone-resistant *Campylobacter jejuni* in the absence of antibiotic selection pressure. Proc. Natl. Acad. Sci. USA 102, 541–546.

Luo, N., Sahin, O., Lin, J., Michel, L.O., Zhang, Q., 2003. In vivo selection of *Campylobacter* isolates with high levels of fluoroquinolone resistance associated with gyrA mutations and the function of the CmeABC efflux pump. Antimicrob. Agents Chemother. 47, 390–394.

Mamelli, L., Prouzet-Mauleon, V., Pages, J.M., Megraud, F., Bolla, J.M., 2005. Molecular basis of macrolide resistance in *Campylobacter*: role of efflux pumps and target mutations. J. Antimicrob. Chemother. 56, 491–497.

Manavathu, E.K., Hiratsuka, K., Taylor, D.E., 1988. Nucleotide sequence analysis and expression of a tetracycline-resistance gene from *Campylobacter jejuni*. Gene 62, 17–26.

McArthur, A.G., Waglechner, N., Nizam, F., Yan, A., Azad, M.A., Baylay, A.J., et al., 2013. The comprehensive antibiotic resistance database. Antimicrob. Agents Chemother. 57, 3348–3357.

McDermott, P.F., Bodeis, S.M., English, L.L., White, D.G., Walker, R.D., Zhao, S., et al., 2002. Ciprofloxacin resistance in *Campylobacter jejuni* evolves rapidly in chickens treated with fluoroquinolones. J. Infect. Dis. 185, 837–840.

McDermott, P., Bodeis, S., Aarestrup, F.M., Brown, S., Traczewski, M., Fedorka-Cray, P., et al., 2004. Development of a standardized susceptibility test for *Campylobacter* with quality-control ranges for ciprofloxacin, doxycycline, erythromycin, gentamicin, and meropenem. Microb. Drug Resist. 10, 124–131.

McDermott, P.F., Bodeis-Jones, S.M., Fritsche, T.R., Jones, R.N., Walker, R.D., 2005. Broth microdilution susceptibility testing of *Campylobacter jejuni* and the determination of quality control ranges for fourteen antimicrobial agents. J. Clin. Microbiol. 43, 6136–6138.

Melero, B., Juntunen, P., Hanninen, M.L., Jaime, I., Rovira, J., 2012. Tracing *Campylobacter jejuni* strains along the poultry meat production chain from farm to retail by pulsed-field gel electrophoresis, and the antimicrobial resistance of isolates. Food Microbiol. 32, 124–128.

Muellner, P., Pleydell, E., Pirie, R., Baker, M., Campbell, D., Carter, P., et al., 2013. Molecular-based surveillance of campylobacteriosis in New Zealand—from source attribution to genomic epidemiology. Euro Surveill. 18, pii 20365.

NARMS, 2014. National antimicrobial resistance monitoring system. <http://www.ars.usda.gov/Main/docs.htm>.

NCEZID, 2014. National Center for Emerging and Zoonotic Infectious Diseases, *Campylobacter*—General Information. <http://www.cdc.gov/nczved/divisions/dfbmd/diseases/Campylobacter/>.

Nichols, G.L., Richardson, J.F., Sheppard, S.K., Lane, C., Sarran, C., 2012. *Campylobacter* epidemiology: a descriptive study reviewing 1 million cases in England and Wales between 1989 and 2011. BMJ Open 2 10.1136/bmjopen-2012-001179.

Niederer, L., Kuhnert, P., Egger, R., Buttner, S., Hachler, H., Korczak, B.M., 2012. Genotypes and antibiotic resistances of *Campylobacter jejuni* and *Campylobacter coli* isolates from domestic and travel-associated human cases. Appl. Environ. Microbiol. 78, 288–291.

Nirdnoy, W., Mason, C.J., Guerry, P., 2005. Mosaic structure of a multiple-drug-resistant, conjugative plasmid from *Campylobacter jejuni*. Antimicrob. Agents Chemother. 49, 2454–2459.

O'Halloran, F., Lucey, B., Cryan, B., Buckley, T., Fanning, S., 2004. Molecular characterization of class 1 integrons from Irish thermophilic *Campylobacter* spp. J. Antimicrob. Chemother. 53, 952–957.

Olkkola, S., Juntunen, P., Heiska, H., Hyytiäinen, H., Hänninen, M., 2010. Mutations in the rpsL gene are involved in streptomycin resistance in *Campylobacter coli*. Microb. Drug Resist. 16, 105–110.

Payot, S., Cloeckaert, A., Chaslus-Dancla, E., 2002. Selection and characterization of fluoroquinolone-resistant mutants of *Campylobacter jejuni* using enrofloxacin. Microb. Drug Resist. 8, 335–343.

Payot, S., Bolla, J., Corcoran, D., Fanning, S., Mégraud, F., Zhang, Q., 2006. Mechanisms of fluoroquinolone and macrolide resistance in *Campylobacter* spp. Microb. Infect. 8, 1967–1971.

Perez-Boto, D., Lopez-Portoles, J.A., Simon, C., Valdezate, S., Echeita, M.A., 2010. Study of the molecular mechanisms involved in high-level macrolide resistance of Spanish *Campylobacter jejuni* and *Campylobacter coli* strains. J. Antimicrob. Chemother. 65, 2083–2088.

Piddock, L.J., Ricci, V., Pumbwe, L., Everett, M.J., Griggs, D.J., 2003. Fluoroquinolone resistance in *Campylobacter* species from man and animals: detection of mutations in topoisomerase genes. J. Antimicrob. Chemother. 51, 19–26.

Pinto-Alphandary, H., Mabilat, C., Courvalin, P., 1990. Emergence of aminoglycoside resistance genes *aadA* and *aadE* in the genus *Campylobacter*. Antimicrob. Agents Chemother. 34, 1294–1296.

Pollett, S., Rocha, C., Zerpa, R., Patiño, L., Valencia, A., Camiña, M., et al., 2012. *Campylobacter* antimicrobial resistance in Peru: a ten-year observational study. BMC Infect. Dis. 12, 193.

Price, L.B., Lackey, L.G., Vailes, R., Silbergeld, E., 2007. The persistence of fluoroquinolone-resistant *Campylobacter* in poultry production. Environ. Health Perspect. 115, 1035–1039.

Quin, S., Wang, Y., Zhang, Q., Chen, X., Shen, Z., Deng, F., et al., 2012. Identification of a novel genomic island conferring resistance to multiple amninoglycoside antibiotics in *Campylobacter coli*. Antimicrobial. Agents Chemother 56, 5332–5339.

Ricotta, E.E., Palmer, A., Wymore, K., Clogher, P., Oosmanally, N., Robinson, T., et al., 2014. Epidemiology and antimicrobial resistance of international travel-associated *Campylobacter* infections in the United States, 2005–2011. Am. J. Public Health 104, e108–e114.

Rossolini, G.M., Thaller, M.C., 2010. Coping with antibiotic resistance: contributions from genomics. Genome Med. 2, 15.

Saenz, Y., Zarazaga, M., Lantero, M., Gastanares, M.J., Baquero, F., Torres, C., 2000. Antibiotic resistance in *Campylobacter* strains isolated from animals, foods, and humans in Spain in 1997–1998. Antimicrob. Agents Chemother. 44, 267–271.

Segreti, J., Gootz, T.D., Goodman, L.J., Parkhurst, G.W., Quinn, J.P., Martin, B.A., et al., 1992. High-level quinolone resistance in clinical isolates of *Campylobacter jejuni*. J. Infect. Dis. 165, 667–670.

Sheppard, S.K., Dallas, J.F., Strachan, N.J., MacRae, M., McCarthy, N.D., Wilson, D.J., et al., 2009. *Campylobacter* genotyping to determine the source of human infection. Clin. Infect. Dis. 48, 1072–1078.

Smith, K.E., Besser, J.M., Hedberg, C.W., Leano, F.T., Bender, J.B., Wicklund, J.H., et al., 1999. Quinolone-resistant *Campylobacter jejuni* infections in Minnesota, 1992–1998. N. Engl. J. Med. 340, 1525–1532.

Tajada, P., Gomez-Graces, J.L., Alos, J.I., Balas, D., Cogollos, R., 1996. Antimicrobial susceptibilities of *Campylobacter jejuni* and *Campylobacter coli* to 12 beta-lactam agents and combinations with beta-lactamase inhibitors. Antimicrob. Agents Chemother. 40, 1924–1925.

Unicomb, L.E., Ferguson, J., Stafford, R.J., Ashbolt, R., Kirk, M.D., Becker, N.G., Australian Campylobacter Subtyping Study Group, 2006. Low-level fluoroquinolone resistance among *Campylobacter jejuni* isolates in Australia. Clin. Infect. Dis. 42, 1368–1374.

van Boven, M., Veldman, K.T., de Jong, M.C., Mevius, D.J., 2003. Rapid selection of quinolone resistance in *Campylobacter jejuni* but not in *Escherichia coli* in individually housed broilers. J. Antimicrob. Chemother. 52, 719–723.

Wang, Y., Huang, W.M., Taylor, D.E., 1993. Cloning and nucleotide sequence of the *Campylobacter jejuni gyrA* gene and characterization of quinolone resistance mutations. Antimicrob. Agents Chemother. 37, 457–463.

Wang, Y., Taylor, D.E., 1990. Chloramphenicol resistance in *Campylobacter coli*: nucleotide sequence, expression, and cloning vector construction. Gene 94, 23–28.

Wilson, D.J., Gabriel, E., Leatherbarrow, A.J., Cheesbrough, J., Gee, S., Bolton, E., et al., 2008. Tracing the source of campylobacteriosis. PLoS Genetics 4, e1000203.

World Health Organization, 2014. Antimicrobial resistance: global report on surveillance.

Yan, M., Sahin, O., Lin, J., Zhang, Q., 2006. Role of the CmeABC efflux pump in the emergence of fluoroquinolone-resistant *Campylobacter* under selection pressure. J. Antimicrob. Chemother. 58, 1154–1159.

Zeitouni, S., Kempf, I., 2011. Fitness cost of fluoroquinolone resistance in *Campylobacter coli* and *Campylobacter jejuni*. Microb. Drug Resist. 17, 171–179.

Zeitouni, S., Collin, O., Andraud, M., Ermel, G., Kempf, I., 2012. Fitness of macrolide resistant *Campylobacter coli* and *Campylobacter jejuni*. Microb. Drug Resist. 18, 101–108.

Zeng, X., Brown, S., Gillespie, B., Lin, J., 2014. A single nucleotide in the promoter region modulates the expression of the beta-lactamase OXA-61 in *Campylobacter jejuni*. J. Antimicrob. Chemother. 69, 1215–1223.

Zhang, Q., Lin, J., Pereira, S., 2003. Fluoroquinolone-resistant *Campylobacter* in animal reservoirs: dynamics of development, resistance mechanisms and ecological fitness. Anim. Health Res. Rev. 4, 63–72.

Chapter 5

Antimicrobial Resistance in *Yersinia enterocolitica*

Anna Fàbrega[1], Clara Ballesté-Delpierre[1] and Jordi Vila[1,2]

[1]*ISGlobal, Barcelona Ctr. Int. Health Res. (CRESIB), Hospital Clínic – Universitat de Barcelona, Barcelona, Spain,* [2]*Department of Clinical Microbiology, Hospital Clínic, School of Medicine, University of Barcelona, Barcelona, Spain*

Chapter Outline

INTRODUCTION

Bacteria within the genus *Yersinia* are Gram-negative, pleomorphic bacilli belonging to the family Enterobacteriaceae. There are currently 15 species within the genus *Yersinia*, three of which are pathogenic to humans. *Y. pestis* is the etiological agent of bubonic plague, while *Y. enterocolitica* and *Y. pseudotuberculosis* are known primarily as foodborne pathogens mainly acquired through ingestion of contaminated pork and related products, and water (Bottone, 1997). However, transmission from animal to human and from person to person can also occur (Fredriksson-Ahomaa et al., 2006). *Y. pseudotuberculosis* is not frequently detected as a cause of human infection and, therefore, in this chapter we will mainly focus on *Y. enterocolitica*.

Classification of *Y. enterocolitica* strains into biotypes or serotypes (or biose-rotypes) is based on biochemical tests and the somatic O antigen (lipopolysaccha-ride or LPS), with six biotypes (1A, 1B, 2, 3, 4, and 5) and more than 57 serotypes (Wauters et al., 1987). Nonetheless, most of the strains belong to biotypes 2, 3, and 4 and to serotypes O:3, O:5,27, O:8, and O:9 (Fredriksson-Ahomaa et al., 2006). *Y. enterocolitica* is subdivided into the *Y. enterocolitica* subspecies *entero-colitica*, which includes mainly biotype 1B, and the *Y. enterocolitica* subspecies

Antimicrobial Resistance and Food Safety. DOI: http://dx.doi.org/10.1016/B978-0-12-801214-7.00005-3

palearctica, which includes the remaining biotypes (Neubauer et al., 2000). Both pathogenic and non-pathogenic *Y. enterocolitica* strains have been described, and their virulence potential relies on the presence or absence of several chromosome- and plasmid-encoded virulence genes (see subsection on "Virulence Factors") (Fabrega and Vila, 2012). Pathogenic strains can be classified, in turn, into two groups: high-pathogenicity and low-pathogenicity. The first group is mainly composed of bioserotype 1B/O:8 strains, which are associated with more severe clinical manifestations since they harbor additional virulence traits. In contrast, the second group leads to milder infections and comprises biotypes 2 to 5, highlighting serotypes O:3 and O:9 as the most representative among clinical isolates (Fabrega and Vila, 2012; Lamps et al., 2006). In the case of strains belonging to biotype 1A, they have typically been considered as non-pathogenic since they usually do not carry virulence genes (Lee et al., 1977; Stephan et al., 2013). However, recent evidence suggests that some of these strains are virulent and able to cause gastroenteritis indistinguishable from that caused by other biotypes traditionally considered pathogenic (Burnens et al., 1996; Morris et al., 1991). Thus, biotype 1A strains likely deserve to be included in the low-pathogenicity group.

Human Infections

Human infection with *Y. enterocolitica* was first reported by Schleifsten and Coleman in 1939. The infections caused by this pathogen are named yersiniosis and occur worldwide, despite the incidence being higher in cooler climates. In several countries human yersiniosis has become a reportable disease since *Y. enterocolitica* is the third most commonly reported etiological agent after *Campylobacter* and *Salmonella* (Drummond et al., 2012). The populations at increased risk for infection include children under 4 years of age and immunocompromised patients (Greene et al., 1993; Hoogkamp-Korstanje and Stolk-Engelaar, 1995; Jouquan et al., 1984). The majority of cases are sporadic, however, several outbreaks have been occasionally described in different parts of the world and associated with the consumption of pork- and milk-related products (Black et al., 1978; Drummond et al., 2012; Grahek-Ogden et al., 2007; Jones, 2003; Tacket et al., 1984).

The onset of illness is usually within 24–48h after ingestion of the contaminated product. The clinical manifestations usually last for 1–3 weeks or even longer. The symptoms reported for intestinal yersiniosis are fever, watery diarrhea (5–10 stools per day), and abdominal pain (Fabrega and Vila, 2012; Vantrappen et al., 1977). Mesenteric lymphadenitis and terminal ileitis are also frequent manifestations which mimic appendicitis, particularly in older children. Serious local complications include hemorrhage and necrosis of the small intestine, intestinal perforation, peritonitis, mesenteric vein thrombosis, or ileocecal intussusception (Abdel-Haq et al., 2000). Cases of sepsis, however, have also been reported and are associated with patients with iron overload or those receiving blood transfusion (Boelaert et al., 1987; Guinet et al., 2011). Among other less frequent clinical manifestations, *Y. enterocolitica* can cause sporadic cases of endocarditis, pneumonia in immunocompromised patients and also community-acquired pneumonia in immunocompetent people (Greene et al., 1993; Lupi et al., 2013; Wong et al., 2013). Additionally, postinfection sequelae as a consequence of *Y. enterocolitica* infection have been observed, including reactive arthritis and erythema nodosum as the most frequently detected (Bottone, 1997).

Diagnosis

The diagnosis of enteritis caused by *Y. enterocolitica* is mainly carried out by culture on laboratory plates. Cefsulodin–irgasan–novobiocin (CIN) agar is the media most frequently used and is based on *Y. enterocolitica*'s resistance to the abovementioned antibiotics (Schiemann, 1979). Plates are incubated at 22–28°C for 48 h, preferentially, or at 30°C for 24 h. *Y. enterocolitica* colonies appear as a dark red "bull's eye" with a translucent edge allowing easy identification. Corroboration is carried out by the specific biochemical properties of *Y. enterocolitica* such as motility at 25°C but not at 37°C, production of urease, lack of oxidase activity and lactose fermentation, as well as absence of either gas or hydrogen sulfide on Kligler's iron agar. Alternatively, the API 20E system can also be used to identify the *Y. enterocolitica* (Bottone, 1997). However, recent technology such as MALDI-TOF mass spectrometry has been used as a rapid, accurate method to also identify the species of the genus *Yersinia* based on protein profiles (Ayyadurai et al., 2010). Additional methods to identify pathogenic *Y. enterocolitica* include both detection of the presence of virulence genes and tests to assess phenotypic characteristics associated with the production of plasmid-encoded virulence factors (Fredriksson-Ahomaa and Korkeala, 2003; Lambertz et al., 2008).

Treatment

Antimicrobial therapy is not usually recommended for treating enterocolitis in immunocompetent hosts since most of the gastrointestinal infections are self-limiting. However, immunocompromised patients with invasive infection, who are at increased risk for developing bacteremia or even septicemia, need special attention and antibiotic treatment since the mortality rate in these cases can be as high as 50% (Fabrega and Vila, 2012).

According to the common profile of susceptibility among *Y. enterocolitica* strains (see section on "Antimicrobial Susceptibility"), the initial recommendations for antimicrobial chemotherapy from public institutions, such as the WHO, included tetracycline, chloramphenicol, gentamicin, and cotrimoxazole (Crowe et al., 1996). However, other compounds such as ciprofloxacin, ceftriaxone, and cefotaxime are also considered since they have shown excellent *in vitro* activity and have been successfully used to treat complicated infections (liver abscess, endocarditis, and septicemia) (Abdel-Haq et al., 2006; Chiu et al., 2003; Crowe et al., 1996; Hoogkamp-Korstanje et al., 2000; Jimenez-Valera et al., 1998;

Lupi et al., 2013). Current protocols include ciprofloxacin 500 mg/12 h for 3–5 days to treat enterocolitis, whereas ciprofloxacin or third-generation cephalosporins in combination with gentamicin should be used for 3 weeks or more for treating complicated and invasive infections.

Epidemiology

Y. enterocolitica is widely distributed in nature and can be isolated in water, soil, and from many domestic and wild animals. Bioserotype 4/O:3 is the most prevalent isolated from animals, particularly pigs, and since indistinguishable genotypes among human and porcine strains have been reported, pigs are considered the major reservoir (Bonardi et al., 2013; Bonke et al., 2011; Fredriksson-Ahomaa et al., 2006; Tadesse et al., 2013). This predominant bioserotype, however, has also been recovered from domestic animals, such as dogs and cats, suggesting that pets may also represent a source of human infection, especially for young children (Fredriksson-Ahomaa et al., 2006). On the contrary, the prevalence of strains belonging to biotypes 2 and 3 and serotypes O:5,27, O:8, and O:9 is much lower, despite several studies reporting sporadic cases from the slaughter of pigs, cows, sheep, goats, monkeys, and wild rodents with potential transmission to humans (Bonke et al., 2011; Fearnley et al., 2005; Fukushima et al., 1993; Hayashidani et al., 1995). However, among all the bioserotypes isolated from animals, including the most and least frequently detected, only a few have been associated with human infections, 4/O:3, 2/O:5,27, 2/O:9, and 3/O:3 being the most prevalent (Fredriksson-Ahomaa et al., 2006).

Even though bioserotype 4/O:3 is also the most prevalent type isolated from humans worldwide, a relatively significant prevalence of other types can also be detected in particular geographic areas as a cause of human yersiniosis. Bioserotype 1B/O:8 strains have been isolated in the United States and have also been sporadically recovered in Japan and Europe, where it has been increasingly reported in Poland (Bottone, 1997; Hayashidani et al., 1995; Rastawicki et al., 2009). In China, the dominant epidemic serotypes are O:5 and O:8, with biotype 1A being the dominant biotype accounting for 84.7% of isolates (Mu et al., 2013). Biotype 1A strains are also significantly recovered from Australia, India, and the United States (Bottone, 1997; Pham et al., 1991a; Sharma et al., 2004). Moreover, bioserotype 3/O:3 has also been responsible for human yersiniosis in Japan and China (Fukushima et al., 1997).

On analyzing the incidence of yersiniosis over time, statistically significant decreasing trends have been noted in six EU member states: Denmark, Germany, Lithuania, Slovenia, Spain, and Sweden; while on the contrary, increasing trends have been noted in Hungary, Romania, and Slovakia (EFSA (European Food Safety Authority) and ECDC (European Centre for Disease Prevention and Control), 2013). In the United States, the incidence of yersiniosis decreased more than 50% comparing data obtained in 2009 with the period 1996–1998 (CDC (Centers for Disease Control and Prevention), 2010).

Virulence Factors

The ability of *Y. enterocolitica* strains to cause disease is attributed to the presence of different virulence factors, either located in the chromosome or on a 70 kb virulence plasmid named pYV, which is only detected in virulent strains (Bottone, 1997; Cornelis et al., 1998; Fabrega and Vila, 2012).

Three proteins have been shown to take over the invasion process of the intestinal mucosa: Inv, the invasin detected in all isolates (Pepe and Miller, 1993), YadA, the plasmid-encoded adhesin (El Tahir and Skurnik, 2001) and Ail, involved in adhesion and invasion only found among pathogenic strains (Pierson and Falkow, 1993). Two type 3 secretion systems (T3SSs) have also been reported. The Ysc T3SS is encoded by the pYV plasmid and is important during systemic stages of infection (Gemski et al., 1980). The Ysa T3SS, only detected in highly pathogenic biotype 1B strains and involved in early stages of infection, is encoded in the chromosomal Ysa pathogenicity island (Haller et al., 2000; Venecia and Young, 2005). Additional virulence factors include the High-Pathogenicity Island, which is only present in the chromosome of biotype 1B strains and is involved in the production of the siderophore yersiniabactin (Carniel et al., 1996), the Yst enterotoxin, frequently detected in diarrheagenic biotype 1A strains (Singh and Virdi, 2004), and the *myf* operon, which encodes a fibrillar structure reportedly involved in adhesion (Iriarte et al., 1993). Lastly, flagella as well as LPS have also been reported to contribute to virulence (Bengoechea et al., 2004; Young et al., 2000).

ANTIMICROBIAL SUSCEPTIBILITY

The resistance profiles to different antimicrobial agents have been examined among strains collected from animal and environmental reservoirs, meat products, as well as those recovered from the clinical setting. Heterogeneity of the antimicrobial resistance pattern is shown to be depending on the bioserotype and geographical distribution (Tables 5.1 and 5.2).

The levels of resistance to β-lactams, which are the major family of antibiotics currently used, have been extensively studied in strains of both animal and human origin. In general terms, high percentages of resistance to ampicillin are detected: values >85% resistance have been reported for non-clinical strains and >95% in the case of clinical isolates (Abdel-Haq et al., 2000; Baumgartner et al., 2007; Bhaduri et al., 2009; Bonardi et al., 2014; Bonke et al., 2011; Capilla et al., 2003, 2004; Fredriksson-Ahomaa et al., 2010, 2012; Gousia et al., 2011; Mayrhofer et al., 2004; Novoslavskij et al., 2013; Pham et al., 1991a; Prats et al., 2000; Preston et al., 1994; Rastawicki et al., 2000; Tadesse et al., 2013; Terentjeva and Berzins, 2010). Nonetheless, lower rates of resistance have been reported (13–57.1%) for strains obtained from the animal and, to a lesser extent from the human, setting. Strains were collected in the United States, Switzerland, and Canada and variability was shown to depend on

TABLE 5.1 Frequencies of Resistance to the Most Important Antimicrobial Compounds for *Y. enterocolitica* Clinical Isolates According to Bioserotype and Geographical Area

| Area | Sample | Bioserotype | No of Strains | % of Resistance | | | | | | | | | References |
| | | | | β-Lactams[a] | | | | Aminoglycosides[b] | | | Quinolones[c] | | |
				AMP	AMC	CEF	CXM	GEN	KAN	STR	NAL	CIP	
Poland	Human stool	4/O:3	114	100	0	–	0	0	–	–	–	0	Rastawicki et al. (2000)
Switzerland	Human stool	4/O:3	52	100	1.9	98.1	0	0	–	9.6	–	0	Baumgartner et al. (2007)
		2/O:5	22	100	90.9	100	0	0	–	0	–	0	
		2/O:9	34	100	94.1	100	0	0	–	0	–	0	
Switzerland	Human stool	1A	51	100	92.2	100	0		5.9	0	2		Fredriksson-Ahomaa et al. (2012)
		2/O:5,27	4	100	75	100	0		0	0	0		
		2/O:9	22	100	81.8	100	0		0	0	0		
		3/O:3	2	100	100	100	0		0	0	0		
		4/O:3	47	100	0	100	0		0	14.9	4.3		
Spain	Human stool	4/O:3	46	100	–	–	–	0	0	97.8	–	0	Capilla et al. (2003)
Spain	Human stool	4/O:3	271	–	–	–	–	–	–	–	23	0	Capilla et al. (2004)
Spain	Human stool	4/O:3	20	100	–	100	–	0	–	90	5	0	Prats et al. (2000)

Country	Source	Serotype	n	AMP[a]	AMC[a]	CEF[a]	CXM[a]	GEN[b]	KAN[b]	STR[b]	NAL[c]	CIP[c]	References
Germany	Human stool	2/O:9	2	100	–	–	–	0	0	0	0	0	Bonke et al. (2011)
		4/O:3	23	100	–	–	–	0	0	21.7	0	0	
Sweden		4/O:3	45	95.5	–	–	–	0	0	2.2	0	0	
Croatia		4/O:3	7	100	–	–	–	0	0	28.6	0	0	
Lithuania	Human stool	4/O:3	19	100	–	–	–	–	–	5.2	–	0	Novoslavskij et al. (2013)
Australia	Human stool	4/O:3	64	100	0	–	–	0	–	–	–	0	Pham et al. (1991b)
		1A	24	100	100	–	–	0	–	–	–	0	
		3/O:5,27	12	100	100	–	–	0	–	–	–	0	
India	Human stool	1A	36	100	100	–	100	–	–	–	–	–	Sharma et al. (2004)
USA	Human stool	ND[d]	30	97.7	–	–	12	0	–	–	–	–	Abdel-Haq et al. (2006)
Canada	Human stool	O:3	945	99.9	26.3	99.9	–	–	–	–	–	–	Preston et al. (1994)
		O:5,27	58	96.6	98.3	98.3	–	–	–	–	–	–	
		O:8	54	13	1.8	96.3	–	–	–	–	–	–	

[a]AMP, ampicillin; AMC, amoxicillin/clavulanic acid; CEF, cefalothin; CXM, cefuroxime.
[b]GEN, gentamycin; KAN, kanamycin; STR, streptomycin.
[c]NAL, nalidixic acid; CIP, ciprofloxacin.
[d]ND, Not determined.

TABLE 5.2 Frequencies of Resistance to the Most Important Antimicrobial Compounds for *Y. enterocolitica* Non-Clinical Isolates According to Bioserotype and Geographical Area

Area	Sample Source	Bioserotype	No of Strains	% of Resistance									References
				β-Lactams[a]				Aminoglycosides[b]			Quinolones[c]		
				AMP	AMC	CEF	CXM	GEN	KAN	STR	NAL	CIP	
Austria	Meat (chicken, beef, pork, turkey)	4/O:3; ND[d]	118	–	–	–	–	0	0	0	0	0	Mayrhofer et al. (2004)
Greece	Meat (pork)	ND	25	96	–	–	60	36	36	–	0	24	Gousia et al. (2011)
Italy	Pig tonsils	4/O:3	22	–	0	–	–	5	5	64	5	0	Bonardi et al. (2014)
Latvia	Pig tonsils	4/O:3	71	100	0	100	–	–	–	5.6	0	0	Terentjeva and Berzins (2010)
Lithuania	Pig feces, carcasses	4/O:3	41	100	–	–	–	–	–	7.3	–	0	Novoslavskij et al. (2013)
Switzerland	Pig feces	4/O:3	10	100	10	100	0	0	0	0	–	0	Baumgartner et al. (2007)
	Pig feces	2/O:5	25	100	100	100	0	0	0	8	–	0	
	Pork	2/O:5	26	100	100	100	0	0	0	0	–	0	
	Pig feces	2/O:9	8	100	100	100	0	0	0	0	–	0	

Country	Source	Serotype	n										Reference
Switzerland	Pig feces, wild boar	2/O:5,27	9	33.3	–	–	–	0	0	0	0	0	Bonke et al. (2011)
	Pig feces	2/O:9	1	100	–	–	–	0	0	0	0	0	
	Pig feces	4/O:3	20	100	–	–	–	0	0	5	0	0	
Germany	Pig feces, pork	4/O:3	44	90.9	–	–	–	0	0	75	0	0	
Sweden	Pork	4/O:3	13	92.3	–	–	–	0	0	7.7	0	0	Sharma et al. (2004)
Finland	Pig feces	4/O:3	7	85.7	–	–	–	14.3	14.3	28.6	0	0	
India	Pig, aquatic	1A	10	100	100	–	100	–	–	–	–	–	Bhaduri et al. (2009)
USA	Pig feces	4/O:3	80	100	0	85	–	0	0	0	0	0	
	Pig feces	2/O:5	26	100	0	100	–	0	0	0	0	0	
USA	Pig feces	4/O:3	74	31	2.7	–	–	0	0	–	0	0	Tadesse et al. (2013)
	Pig feces	O:5	44	13.6	2.3	–	–	0	0	–	0	0	
	Pig feces	O:9	7	57.1	0	–	–	0	0	–	0	0	
	Pig feces	O:8	5	20	20	–	–	0	0	–	0	0	

[a]AMP, ampicillin; AMC, amoxicillin/clavulanic acid; CEF, cefalothin; CXM, cefuroxime.
[b]CEN, gentamicin; KAN, kanamycin; STR, streptomycin.
[c]NAL, nalidixic acid; CIP, ciprofloxacin.
[d]ND, Not determined.

the bioserotype analyzed (4/O:3, O:8, O:9, 2/O:5,27, O:5) (Bonke et al., 2011; Tadesse et al., 2013; Preston et al., 1994).

In the case of amoxicillin/clavulanic acid, heterogeneous susceptibility profiles have been seen among *Y. enterocolitica* strains. High levels of resistance (100% for non-clinical strains and >75% for clinical isolates) have been shown to occur in strains belonging to biotypes 1A, 2, and 3 collected from around the world (Baumgartner et al., 2007; Fredriksson-Ahomaa et al., 2012; Pham et al., 1991a; Sharma et al., 2004), including serotype O:5,27 strains from Canada (presumably related to biotypes 2 or 3 according to the most prevalent biotype–serotype associations) (Preston et al., 1994). Tadesse et al., however, reported very low levels of resistance (<3%) for strains serotyped as O:5 and O:9 (putatively belonging to biotypes 2 or 3) obtained from animals in the United States (Tadesse et al., 2013). On the other hand, bioserotype 4/O:3 consistently shows the lowest resistance values, <10%, among all the isolates regardless of the country of isolation (Baumgartner et al., 2007; Bhaduri et al., 2009; Bonardi et al., 2014; Fredriksson-Ahomaa et al., 2012; Pham et al., 1991a; Rastawicki et al., 2000; Tadesse et al., 2013; Terentjeva and Berzins, 2010).

Similar to the resistance levels observed for ampicillin, most of the *Y. enterocolitica* strains are also resistant to cefalothin, a first-generation cephalosporin. Several studies have reported percentages of resistance >85% in strains of animal origin and >98% in strains isolated from humans (Baumgartner et al., 2007; Fredriksson-Ahomaa et al., 2010; Prats et al., 2000; Preston et al., 1994; Terentjeva and Berzins, 2010). Fortunately, most of the strains studied remain susceptible to more recent cephalosporins, including the second-generation compound cefuroxime and third-generation cephalosporins such as ceftriaxone, ceftazidime, and cefotaxime (Baumgartner et al., 2007; Bhaduri et al., 2009; Bonke et al., 2011; Capilla et al., 2003; Fredriksson-Ahomaa et al., 2010; Gousia et al., 2011; Prats et al., 2000; Preston et al., 1994; Rastawicki et al., 2000; Tadesse et al., 2013; Terentjeva and Berzins, 2010). Nonetheless, a few exceptions have been reported. In 2000, Abdel-Haq et al. reported low levels of resistance to cefuroxime (12%), ceftazidime (11%), and cefotaxime (1%) in strains isolated from patients at the Children's Hospital of Michigan (USA) (Abdel-Haq et al., 2000). Recently, in a study conducted in 2011 in Greece by Gousia et al., a high proportion of *Y. enterocolitica* isolates from animals (60%) was resistant to cefuroxime, whereas lower values were observed for ceftriaxone (8%) (Gousia et al., 2011). Regrettably, in these two studies no information was provided regarding the bioserotype of the isolates. More importantly, Sharma et al. detected that all 36 clinical strains tested for antimicrobial susceptibility in India, only belonging to biotype 1A, were resistant to cefotaxime (Sharma et al., 2004).

To our knowledge little information is available on the susceptibility of *Y. enterocolitica* to cephamycins and carbapenems. Furthermore, data on cefoxitin susceptibility has only been reported in a few studies. First, in a study performed in Australia, the authors revealed that all the isolates belonging to bioserotype 4/O:3 were susceptible to cefoxitin whereas those belonging to bio-types 1A (serotypes O:5 or O:8) and 3 (serotype O:5,27) were resistant (Pham et al., 1991a). Similarly, a more recent study conducted in Switzerland showed that all bioserotype 4/O:3 clinical isolates were susceptible, while higher resist-ance rates were observed for biotype 1A (45%) and bioserotypes 2/O:5,27, 2/O:9, and 3/O:3 (>80%) (Fredriksson-Ahomaa et al., 2012). Information regarding imipenem resistance is also scarce. To our knowledge, only one study from Greece has been conducted, showing 8% of the strains of animal origin to be resistant to this drug (Gousia et al., 2011). With respect to clinical isolates, no resistance to imipenem has been reported (Pham et al., 1991a; Rastawicki et al., 2000).

Other antimicrobials which deserve to be highlighted are those belonging to the group of aminoglycosides. Full susceptibility to kanamycin and gentamicin has been reported in almost all studies regardless of the source of isolation (Abdel-Haq et al., 2000; Baumgartner et al., 2007; Bhaduri et al., 2009; Bonke et al., 2011; Capilla et al., 2003; Mayrhofer et al., 2004; Pham et al., 1991a; Preston et al., 1994; Rastawicki et al., 2000; Tadesse et al., 2013; Terentjeva and Berzins, 2010). A few studies represent an exception since they report levels of resistance ranging from 5% to 36% in animal strains belonging to bioserotype 4/O:3, when information concerning bioserotyping is available (Bonardi et al., 2014; Gousia et al., 2011; Tadesse et al., 2013); and in human strains belonging to biotype 1A, despite being reported in only one study (Fredriksson-Ahomaa et al., 2012). Higher heterogeneity has been observed for strepto-mycin among clinical and non-clinical strains. Several studies have revealed a lack of resistance, particularly concerning biotype 2 strains (Baumgartner et al., 2007; Bhaduri et al., 2009; Bonke et al., 2011; Mayrhofer et al., 2004), while others have reported levels of resistance ranging from 5.6% to 28.6% (Baumgartner et al., 2007; Bonke et al., 2011; Fredriksson-Ahomaa et al., 2010, 2012; Novoslavskij et al., 2013; Terentjeva and Berzins, 2010). The highest lev-els of resistance to streptomycin have been described for strains belonging to bioserotype 4/O:3. Among strains of animal origin, studies carried out in Italy and Germany showed percentages of resistance of 64% and 75%, respectively (Bonardi et al., 2014; Bonke et al., 2011). In the clinical setting, strains isolated in Spanish hospitals have reported the highest levels of resistance to strepto-mycin (≥90%) (Capilla et al., 2003; Prats et al., 2000); whereas the remain-ing studies, also performed in Europe, show a prevalence of resistance <30% (Baumgartner et al., 2007; Bonke et al., 2011; Novoslavskij et al., 2013). Cases in which high levels of resistance have been reported are attributed to horizon-tal transfer of plasmids carrying genes which confer resistance to streptomycin (see subsection on "Mechanisms of Resistance to Aminoglycosides (Capilla et al., 2003).

A much more optimistic situation has been reported for quinolones. Almost all the *Y. enterocolitica* strains of any origin studied have shown a lack of resist-ance to nalidixic acid or a very low level of resistance (<5%) (Bhaduri et al.,

2009; Bonardi et al., 2014; Bonke et al., 2011; Fredriksson-Ahomaa et al., 2012; Mayrhofer et al., 2004; Prats et al., 2000; Tadesse et al., 2013; Terentjeva and Berzins, 2010). Nonetheless, one study from Spain revealed that up to 23% of the clinical isolates were resistant to this drug (Capilla et al., 2004). In most of the situations in which resistance is reported the strains belong to bioserotype 4/O:3. Moreover, full susceptibility to the fluoroquinolone ciprofloxacin has also been reported in the studies analyzed in this review (Baumgartner et al., 2007; Bhaduri et al., 2009; Bonardi et al., 2014; Bonke et al., 2011; Novoslavskij et al., 2013; Preston et al., 1994; Tadesse et al., 2013; Terentjeva and Berzins, 2010), except one study from Greece in which Gousia et al. reported that as many as 24% of strains they isolated from animal samples were ciprofloxacin-resistant (Gousia et al., 2011). Even though nalidixic acid was not evaluated in this study, it is very likely that ciprofloxacin-resistant strains might also be resistant to this drug, as acquisition of mutations in the target genes encoding resistance to quinolones (the most commonly found mechanism of resistance) confers resistance to nalidixic acid prior to ciprofloxacin (see subsection on "Mechanisms of Resistance to Quinolones").

In addition to these three main families of antibacterial drugs, tetracyclines, chloramphenicol, and cotrimoxazole can also represent potential treatments for human yersiniosis. Low rates of tetracycline-resistant strains of both animal and human origin have been documented among studies conducted in Europe and Canada, ranging from full susceptibility, particularly observed in biotype 2 strains collected from animals and all biotypes of human isolates, to 12% of resistance for bioserotype 4/O:3 (Baumgartner et al., 2007; Bonke et al., 2011; Fredriksson-Ahomaa et al., 2010; Gousia et al., 2011; Mayrhofer et al., 2004; Novoslavskij et al., 2013; Pham et al., 1991a; Preston et al., 1994; Rastawicki et al., 2000; Terentjeva and Berzins, 2010). Nonetheless, higher rates of tetracycline resistance have been reported in two studies from the United States assessing animal settings: Bhaduri et al. showed that up to 69.2% of serotype O:5 strains were resistant versus 13.8% among serotype O:3 (Bhaduri et al., 2009), whereas Tadesse et al. revealed that intermediate percentages (28.6% and 20%) were observed for serotypes O:9 and O:8, respectively, despite the recovery of a reduced number of isolates (Tadesse et al., 2013).

Regarding chloramphenicol, most studies report very low levels of resistance, from 0% to 4% of the isolates (Baumgartner et al., 2007; Bhaduri et al., 2009; Bonke et al., 2011; Fredriksson-Ahomaa et al., 2010; Mayrhofer et al., 2004; Pham et al., 1991a; Preston et al., 1994; Tadesse et al., 2013; Terentjeva and Berzins, 2010). Nonetheless, a few exceptions have been reported, although all concerned bioserotype 4/O:3 strains. In the animal setting Bonardi et al. showed 55% of resistance to this drug in a report from Italy (Bonardi et al., 2014), whereas clinical strains analyzed from Spanish hospitals have shown higher levels of resistance: Prats et al. detected an increase in resistance from 20% to 60% comparing strains collected during the period 1985–1987 versus 1995–1998 (Prats et al., 2000), whereas Capilla et al. showed 100% of resistance and suggested the clonal expansion of a particular isolate in their study (Capilla et al., 2003).

Similarly, the levels of resistance reported for trimethoprim and cotrimoxazole indicate that most strains remain susceptible (Abdel-Haq et al., 2000; Baumgartner et al., 2007; Fredriksson-Ahomaa et al., 2010; Mayrhofer et al., 2004; Pham et al., 1991a; Preston et al., 1994; Rastawicki et al., 2000; Tadesse et al., 2013; Terentjeva and Berzins, 2010). Only a single study performed in Spain by Capilla et al. reported high levels of resistance to cotrimoxazol of approximately 91% among clinical isolates which belonged to bioserotype 4/O:3 (Capilla et al., 2003).

Lastly, all the studies reviewed agree that erythromycin is the only antibiotic to which 100% of the *Y. enterocolitica* strains recovered either from animal reservoirs and human feces are resistant, likely associated with the constitutive expression of the efflux pump AcrAB-TolC, the main efflux pump characterized to confer the multidrug resistance (MDR) phenotype among Enterobacteriaceae (Fredriksson-Ahomaa et al., 2010; Novoslavskij et al., 2013; Preston et al., 1994; Terentjeva and Berzins, 2010; Vila et al., 2011).

Few reports have analyzed the rates of antimicrobial resistance in *Y. enterocolitica* over time in order to assess whether these values have increased in more recent study periods. Preston et al. compared the antimicrobial profile for strains collected during the period 1972–1976 and in 1980, 1985, and 1990. Their results showed that, in general terms, no major change was observed. The most important difference was observed for tetracycline resistance which increased from 0.4% in 1985 to 2% in 1990 (Preston et al., 1994). More recently, Prats et al. evaluated antimicrobial resistance in *Y. enterocolitica* isolated over two periods of time, 1985–1987 and 1995–1998. Their results showed an increase in resistance regarding streptomycin (from 72% to 90%) and chloramphenicol (from 20% to 60%), and, to a lesser extent, nalidixic acid (from 0% to 5%) (Prats et al., 2000).

Concerning the prevalence of MDR isolates, the results published in the literature have been analyzed in this chapter according to the definition of MDR proposed by Magiorakos et al. in which resistance to three or more antimicrobial classes is required for an organism to be considered MDR (Magiorakos et al., 2012). Several studies, conducted in different countries such as Switzerland, Italy, and the United States, have reported low levels of MDR (<2%) among clinical and non-clinical strains, with most of them belonging to bioserotype 4/O:3 (Baumgartner et al., 2007; Bonardi et al., 2013; Fredriksson-Ahomaa et al., 2012; Tadesse et al., 2013). On the contrary, other reports have shown increased percentages of MDR reaching levels as high as 50–60% (Bonardi et al., 2014; Prats et al., 2000; Sihvonen et al., 2011). This situation may, at least in part, be due to the fact that strains of bioserotype 4/O:3 were the only or predominant group collected and tested in these studies, and/or that a great proportion of these MDR strains may be clonally related and disseminated as an outbreak. This latter consideration was assessed in the study performed by

Sihvonen et al. (2011) which revealed different percentages, 59.1% and 18.3%, for outbreak-related strains and for sporadic cases, respectively (Sihvonen et al., 2011). The antibiotics most frequently found in the MDR patterns include streptomycin, sulfonamides, and chloramphenicol, and to a lesser extent cotrimoxazol, tetracycline, and nalidixic acid. The first association, resistance to streptomycin, sulfonamides, and chloramphenicol, has been shown to be carried by a conjugative plasmid (Sihvonen et al., 2011). Moreover, the association of sulfonamide-streptomycin resistance has also been suggested to be part of a mobile genetic element such as an integron (Prats et al., 2000).

According to this information and on comparing the percentage of resistance regarding all bioserotypes, most of the resistant strains reported from both the animal and clinical settings belong to bioserotype 4/O:3. Moreover, resistant isolates reported in human settings are likely to originate in the animal environment in which several antimicrobial compounds or related drugs are used or have been used in the veterinary field (e.g., streptomycin, apramycin, tetracycline, chloramphenicol, thiamphenicol, florphenicol, quinolones). Nonetheless, the low percentages of resistance to second- and third-generation cephalosporins, as well as cephamycins, are detected among clinical isolates, suggesting that resistance might emerge directly in the clinical setting as a consequence of human antimicrobial therapies.

Although the use of ciprofloxacin, third-generation cephalosporins, and gentamicin is the current treatment of choice for yersiniosis, a modest trend toward increased antimicrobial resistance has been observed in *Y. enterocolitica* hence underscoring the need for surveillance of resistance as well as for efforts focused on encouraging a rational use of the antimicrobial compounds.

MECHANISMS OF ANTIMICROBIAL RESISTANCE

The success of antibiotics in inhibiting bacterial growth relies on the ability of the drug to recognize its target and achieve the appropriate concentration of the antibiotic for efficient inhibition. Mutations leading to antimicrobial resistance are usually located in the chromosome and their spread depends on the dissemination of a particular clone. Acquisition of antibiotic resistance genes can also be spread by clonal expansion, although since they are located in transferable elements (i.e., transposons and plasmids) they can be horizontally transferred between bacteria (Martinez and Baquero, 2014).

Most of the transferable genes are embedded in structures called gene cassettes, which usually contain a single coding sequence (most commonly an antibiotic resistance gene) and an integrase-specific recombination site (*attC*). Integrons are not mobile genetic elements, but are characterized by their ability to capture and accumulate gene cassettes by site-specific recombination. Most of the clinically relevant antimicrobial-resistance-encoding integrons belong to class 1, whereas integrons belonging to class 2 and class 3 similarly carry cassettes predominantly composed of antibiotic resistance genes and are also of clinical concern (Hall, 2012). However, integrons can only be mobilized, and are responsible for transmission of resistance genes, when incorporated into transposons. Transposons are segments of DNA flanked by terminal inverted repeats carrying a transposition region. In some cases this transposition region includes the transposase gene (*tnpA*), which allows self-replication and mobilization to a new position in the chromosome or plasmid (Griffiths et al., 1999). Furthermore, integrons, transposons or standalone resistance genes are often harbored in plasmids, the major source for dissemination of antimicrobial resistance genes. In particular, conjugative plasmids can be transferred among different pathogens at a high frequency, thereby favoring their rapid and broad dissemination.

Mechanisms of Resistance to β-lactams

As mentioned previously, *Y. enterocolitica* is intrinsically resistant to certain β-lactam antibiotics (e.g., ampicillin and the first-generation cephalosporin cefalothin). It is well established that this phenotype is due to the presence and expression of two chromosomally encoded β-lactamase genes, namely *blaA* and *blaB* (Cornelis and Abraham, 1975). The *blaA* gene is constitutively expressed and encodes the broad-spectrum class A penicillinase (Cornelis and Abraham, 1975). The presence of clavulanic acid (20 μM) completely inhibits the activity of this enzyme, whereas no change is detected in the presence of aztreonam (Pham et al., 1991a). Unlike BlaA, the BlaB enzyme is a class C cephalosporinase (AmpC-type β-lactamase), which can be inhibited by aztreonam and cloxacillin but not by clavulanic acid (Cornelis and Abraham, 1975; Pham et al., 1991a). Moreover, imipenem has been shown to induce BlaB activity (Pham and Bell, 1992).

In order to study the individual contribution of these two enzymes, Seoane et al. reported the antimicrobial susceptibility profile to different penicillins and cephalosporins by cloning the *blaA* and *blaB* genes from strains belonging to bioserotypes 4/O:3 and 2/O:5b, respectively, and expressing them individually in an *E. coli* recipient strain. The results showed an increase in resistance to all of the antibiotics tested for the two enzymes, with only one exception; there was no increase in the MIC of cefoxitin in the case of BlaA. The most representative results for BlaA were detected for ampicillin and carbenicillin, whereas for BlaB the highest increments were observed for ampicillin, cefalothin, and cefotaxime. BlaB conferred higher levels of resistance to almost all the cephalosporins tested (Table 5.3) (Seoane and Garcia-Lobo, 1991).

Among clinical and non-clinical *Y. enterocolitica* strains, the presence and differential expression of these two enzymes have been characterized and associated with different levels and spectra of resistance. These strain variations have been reported to largely depend on the bioserotype and the geographic origin of the isolate (Table 5.4). Analyses on the presence or absence of the *blaA* and *blaB* genes by PCR amplification among clinical and non-clinical strains

TABLE 5.3 Effect of BlaA and BlaB Expression on the MICs of Several β-lactams

Strain	Enzyme Produced	Antimicrobial Susceptibilities (MICs, µg/mL)							
		Penicillins[a]			Cephalosporins[b] (Generation)				
		AMP	BPC	CAR	CFL (1st)	CEF (1st)	MAN (2nd)	FOX (2nd)	CTX (3rd)
E. coli HB101		2	ND[c]	4	2	2	<1	2	<0.03
E. coli HB101	BlaA	1024	ND	>1024	16	128	32	2	0.12
E. coli HB101	BlaB	512	ND	16	128	1024	16	64	4

[a]AMP, ampicillin; BPC, benzylpenicillin; CAR, carbenicillin.
[b]CFL, cephaloridine; CEF, cefalothin; MAN, cefamandole; FOX, cefoxitin; CTX, cefotaxime.
[c]ND, Not determined.

collected from all over the world and belonging to various biotypes (1A, 1B and 2–5) and serotypes (O:3, O:9 and O:5,27) indicate that almost all the strains harbor both genes (Bonke et al., 2011; Sharma et al., 2006; Stock et al., 1999, 2000b). While Stock et al. initially observed variable results in strains belonging to biotypes 1A, 1B, and 3 (Stock et al., 2000b), this situation was later suggested to be a problem due to lack of amplification in the previous protocol (Sharma et al., 2006). Only one bioserotype 3/O:3 strain isolated from a chinchilla in Germany lacked both genes (Bonke et al., 2011).

The results obtained from such qualitative studies have been complemented by phenotypic analysis. Both the β-lactam hydrolyzing activity from crude enzyme extracts and the MIC of several β-lactams have been tested for inhibition with clavulanic acid, aztreonam or cloxacillin or induction with imipenem (0.5 mg/L) (Bonke et al., 2011; Pham et al., 1991a; Stock et al., 1999). Inhibition by clavulanic acid is an indicator of BlaA activity, whereas inhibition by aztreonam or cloxacillin indicates BlaB production, which may or may not be induced by imipenem. If a complete inhibition is observed for a given compound, the corresponding enzyme is assumed to be the only enzyme expressed (Stock et al., 1999).

Concerning expression of the class A β-lactamase, BlaA, a repeated pattern can be concluded from several studies including isolates from different geographic regions and different sources (Table 5.4). Most of the strains (93–100%) of clinical and non-clinical origin belonging to biotypes 1A, 1B, 2 (only serotype O:9), and 4 expressed the BlaA enzyme (Bonke et al., 2011; Pham et al., 1991a, 1995; Sharma et al., 2006; Stock et al., 1999, 2000b). Only

TABLE 5.4 Presence, Production, and Induction of *blaA*/BlaA and *blaB*/BlaB in *Y. enterocolitica* Clinical and Non-Clinical Isolates According to Bioserotype and Geographic Area

Sample	Geographic Area	Bioserotype	No of Strains	% of Presence, Production, and Induction					References
				blaA	BlaA	*blaB*	BlaB	BlaB Induction	
Clinical origin	Australia	1A	24	_[a]	100	–	100	100	Pham et al. (1991a), Pham and Bell (1992)
		3	12	–	0	–	100	100	
		4/O:3	64	–	100	–	0	0	
	Europe, Asia, Brazil, South Africa	4/O:3	31	–	100	–	100	100	Pham et al. (1995)
	Australia, New Zealand	4/O:3	10	–	100	–	0	0	
	Canada	4/O:3	3	–	100	–	0	0	
	Canada	4/O:3	4	–	100	–	100	100	
	Germany	2/O:9	2	100	100	100	100	–	Bonke et al. (2011)
	Germany	4/O:3	23	100	100	100	91.3	–	
	Sweden	4/O:3	45	100	100	100	91.1	–	
	Croatia	4/O:3	7	100	100	100	100	–	
	India	1A	35	100	100	100	94.3	94.3	Sharma et al. (2006)
	Europe	1A	15	100	100	100	100	100	
	USA	1A	1	100	100	100	100	100	

(Continued)

Clinical and non-clinical origin								
Germany	1A	14	0	92.9	100	100	100	Stock et al. (2000b)
	1B	12	16.7	100	100	100	91.7	
	3	22	40.9	45.5	100	100	45.5	
Germany	2	12	100	41.7	100	100	41.7	Stock et al. (1999)
	4	13	100	100	100	100	100	
	5	10	100	80	100	100	60	
Non-clinical origin								
India	1A	30	100	100	100	100	100	Sharma et al. (2006)
Switzerland	2/O:5,27	9	100	0	100	100	–	Bonke et al. (2011)
Switzerland	2/O:9	6	100	100	100	100	–	
Germany	3/O:3	2	50	50	50	50	–	
Germany, Sweden, Switzerland, Finland, Croatia	4/O:3	92	100	97.8	100	96.7	–	

[a]Not determined in this study.

three strains were categorized as non-producers: one belonging to biotype 1A collected in Germany and two belonging to biotype 4 collected from pigs in Germany and Finland (Bonke et al., 2011; Stock et al., 2000a). In contrast, none of the strains belonging to biotype 2 (serovar O:5,27 strains isolated from animals in Switzerland) and biotype 3 (collected from humans in Australia) expressed BlaA. The former group harbored the gene whereas in the latter group the presence of *blaA* was not determined (Bonke et al., 2011; Pham et al., 1991a). Approximately 50% of the strains belonging to biotypes 2 (undetermined serotype) and 3 isolated in Germany expressed *blaA*, however, the reasons were different: all the biotype 2 strains harbored the gene but only some strains expressed it, whereas all biotype 3 strains demonstrating BlaA activity were also positive for the presence of the gene (Bonke et al., 2011; Stock et al., 1999, 2000b). Lastly, 80% of the biotype 5 strains, predominantly of animal origin, expressed BlaA.

If expression of the BlaB enzyme is considered, a great percentage (91–100%) of the clinical and non-clinical strains belonging to biotypes 1A, 1B, 2 (both serotypes O:9 and O:5,27), 3, and 5 produced this β-lactamase (Bonke et al., 2011; Pham et al., 1991a, 1995; Sharma et al., 2006; Stock et al., 1999, 2000b). Two clinical 1A strains isolated in India, which were positive for the presence of the gene, and a single biotype 3 strain collected from a chinchilla in Germany, which did not harbor *blaB*, did not express the BlaB enzyme (Bonke et al., 2011). The remaining strains belonging to the most commonly isolated bioserotype 4/O:3 showed different results according to the geographical region from which they were isolated. Strains collected in Australia, New Zealand, and Canada did not express the BlaB cephalosporinase (the presence or absence of *blaB* was not determined), whereas most of the strains isolated in Europe, Asia, Brazil, South Africa, and Canada did express the enzyme (Bonke et al., 2011; Pham et al., 1991a, 1995; Stock et al., 1999). Six clinical isolates did not express the protein; two isolated in Germany and four in Sweden (Bonke et al., 2011) (Table 5.4).

The results concerning induction of BlaB activity in the presence of imipenem are also shown in Table 5.4. The results have revealed that, when BlaB is produced it is almost always induced in the presence of imipenem regardless of the biotype. All strains belonging to biotypes 1A and 4, which are positive for BlaB production, show imipenem induction (Pham and Bell, 1993; Sharma et al., 2006; Stock et al., 2000b). More than 91% of the biotype 1B strains also produce BlaB which can be further induced (Stock et al., 2000b). Different results have been observed regarding strains belonging to biotypes 2 and 3. Stock et al. detected an association between BlaB inducibility and BlaA production: less than 50% of the strains only producing BlaB showed an inducible phenotype, suggesting that this enzyme could not be further induced because maximal levels were constitutively produced to compensate for the lack of BlaA (Stock et al., 1999, 2000b). Nonetheless, no clear results have been observed for biotype 5 strains and in other studies. Stock et al. showed that 60% of biotype

5 isolates showed BlaB induction regardless of the predominant expression of the enzymes (Stock et al., 1999), whereas Pham et al. reported induction of BlaB for all biotype 3 strains even in those producing only BlaB (Pham and Bell, 1993).

In view of these results, the high percentage of resistance to ampicillin (>85%) can be attributed to the fact that *Y. enterocolitica* strains always produce at least one of the two chromosomal β-lactamases, and both contribute to ampicillin resistance (Seoane and Garcia-Lobo, 1991). However, a few strains have been characterized to express both proteins and were susceptible or intermediate susceptible based on the MIC of ampicillin (Bonke et al., 2011). Cefalothin resistance (>85%) can be easily attributed to BlaB production, a trait generally detected in most of the strains (>91%). Nonetheless, there is no information regarding the MIC of cefalothin for the subset of 4/O:3 strains which do not express the cephalosporinase BlaB (Pham et al., 1991a, 1995). Although these strains may be considered cefalothin-susceptible, there are no data available to confirm this.

Conversely, there is no easy explanation for *Y. enterocolitica* susceptibility to amoxicillin-clavulanic acid. On one hand, Pham et al. have reported that all bioserotype 4/O:3 strains that only express BlaA, collected from Australia, New Zealand, and Canada, are susceptible to this combination since BlaA can be inhibited by clavulanic acid; whereas those expressing both β-lactamases are resistant and show increased MICs of cefoxitin (Pham et al., 1991a, 1995; Pham and Bell, 1993). On the other hand, other reports have shown that bioserotype 4/O:3 strains expressing both BlaA and BlaB are susceptible to amoxicillin-clavulanic acid, with no explanation being suggested (Bonardi et al., 2014; Fredriksson-Ahomaa et al., 2012; Rastawicki et al., 2000; Terentjeva and Berzins, 2010).

Thus, even though the phenotype of β-lactam resistance in most strains can be explained by the ability to express one or both of the chromosomal β-lactamases, there are particular situations in which this association does not provide a reasonable explanation. The levels of expression of these two enzymes, attributed to different regulatory effects which may depend on the intrinsic properties of the strain or a particular bioserotype, might account for such differences (Bonke et al., 2011; Pham et al., 1991b; Stock et al., 1999). Moreover, it is worth mentioning that, to the best of our knowledge, to date no isolate has been reported to carry an extended-spectrum β-lactamase (ESBL) which may contribute to resistance to such compounds.

Mechanisms of Resistance to Aminoglycosides

Aminoglycosides require an energy-dependent uptake system to reach their target site, a highly conserved motif of the 16S rRNA, which is part of the 30S ribosomal small subunit and essential for correct protein synthesis. There are two major mechanisms of resistance to aminoglycosides characterized so far: (i) reduction in the intracellular concentration of the antimicrobial via either decreased aminoglycoside uptake or increased active efflux; and (ii) inactivation of the antimicrobial by enzymatic modification preventing binding between the drug and the ribosomal RNA (Galimand et al., 2003; Magnet et al., 2001; Taber et al., 1987).

These structural modifications represent the most common mechanism of aminoglycoside resistance and are the result of particular plasmid-borne genes encoding enzymes belonging to three different classes: aminoglycoside nucleotidyltransferases (ANTs), aminoglycoside phosphotransferases (APHs), and aminoglycoside acetyltransferases (AACs). Each particular group includes enzymes with different regional specificities for aminoglycoside modifications: there are four nucleotidyltransferases (ANT(6), ANT(4″), ANT(3″), and ANT(2″)), seven phosphotransferases (APH(3′), APH(2″), APH(3″), APH(6), APH(9), APH(4), and APH(7″)), and four acetyltransferases (AAC(2′), AAC(6′), AAC(1), and AAC(3)). There also exists a bifunctional enzyme, AAC(6′)–APH(2″), that can sequentially acetylate and phosphorylate its aminoglycoside substrates (Kotra et al., 2000). The occurrence of the genes *ant(3)-Ia* and *ant(3)-Ib*, which confer resistance to streptomycin-spectinomycin, have been described as the most commonly found in clinical strains (Capilla et al., 2003; Levesque et al., 1995). These genes are integrated in mobile genetic elements as part of gene cassettes, integrons, and transposons and, thereby, can be disseminated among bacteria (Kotra et al., 2000). Moreover, in addition to their potential spread, cross-resistance has also been reported to be a problem that could limit the effective use of aminoglycoside compounds. For example, the use of apramycin in pig production has been correlated to the increased occurrence of cross-resistance with gentamicin and other aminoglycosides in *E. coli* strains collected from pigs carrying the *aac(3)-IV* gene in conjugative plasmids (Jensen et al., 2006). This phenomenon should be taken into account when determining the impact of antimicrobial use in animals on the dissemination of resistance into the clinical setting.

Mechanisms of Resistance to Quinolones

Overall, resistance to quinolones can be mediated by both chromosome- and plasmid-related mechanisms. Chromosomal mutations are the most prevalent mechanism contributing to quinolone resistance. These mutations can affect the genes encoding the type-II topoisomerase protein targets (*gyrA* and *gyrB* for the DNA gyrase, and *parC* and *parE* for the topoisomerase IV) and hence decrease the binding affinity towards the drug. In addition, chromosomal mutations resulting in decreased expression of porins, increased expression of efflux pumps, or the interplay of both, can be acquired leading to decreased accumulation of quinolones inside the bacteria. In terms of plasmid-mediated quinolone resistance, the following mechanisms have been found: (i) the expression of an aminoglycoside-modifying enzyme (AAC(6′)Ib-cr) which has the ability to acetylate an amino group in the piperazine ring of the quinolone structure; this enzyme affects all quinolones with the exception of those which have the amino group blocked, such as levofloxacin; (ii) the *qnr* gene family, which encode a peptide able to protect the complex between the DNA and the DNA gyrase or topoisomerase IV, thereby preventing quinolone binding; and (iii) two efflux pumps, OqxAB and QepA. All of these proteins lead to relatively small increases in the MICs of quinolones, but these changes are sufficient to facilitate the selection of mutants, mainly in the *gyrA* gene, with higher levels of resistance (Fabrega et al., 2009).

Since wild-type strains of *Y. enterocolitica* are usually susceptible to quinolones, few reports have focused on the study and characterization of the mechanisms involved. In a study conducted in Spain, 23% of the *Y. enterocolitica* strains were resistant to nalidixic acid, all of which had a mutation in the *gyrA* gene and, in some cases, overexpression of an efflux pump was also detected (Capilla et al., 2004). Similar results have recently been published by Drummond et al. in a study in which three nalidixic acid-resistant strains were isolated from human samples. All three strains carried a *gyrA* mutation in addition to increased efflux (Drummond et al., 2013). Although the authors in these two studies did not identify the efflux pump involved, AcrAB/TolC would be the most likely candidate according to its relevance in Enterobacteriaceae and taking into account that quinolones can be found among its substrates (Okusu et al., 1996; Vila et al., 2011). Moreover, four mutations, one in each of the *gyrA* and *gyrB* genes and two in *parC*, were found in addition to the overexpression of AcrAB-TolC in a quinolone-resistant *Y. enterocolitica* mutant obtained *in vitro*, showing a ciprofloxacin MIC of 64 µg/mL as well as resistance to other unrelated antibiotics (tetracycline, erythromycin, chloramphenicol, and cefoxitin). This combination of mechanisms explained the high level of resistance to ciprofloxacin and the MDR pattern in this mutant. The overexpression of AcrAB-TolC was found to be related to the increased expression of *marAYe*, an ortholog of the global regulator *marA* shown to activate expression of the *acrAB* and *tolC* genes (Fabrega et al., 2009, 2010). As far as we know, no plasmid-mediated quinolone resistance has been reported so far in *Y. enterocolitica*.

CONCLUDING REMARKS

Yersiniosis is a foodborne disease acquired via the ingestion of food or water mainly contaminated with *Y. enterocolitica*. Despite not being among the most frequent food-related pathogens, it is of great concern in terms of food safety as it can replicate at refrigeration temperature. The clinical manifestations and severity of the disease are diverse but, fortunately, the pathogen generally colonizes the gastrointestinal tract and causes local infection, which resolves naturally without antibiotic treatment. However, in some cases, mainly among inmunocompromised patients and children under 4 years of age, complications and/or systemic infection may occur, leading to the need for antimicrobial therapy.

Bioserotype 4/O:3 is widely distributed worldwide in both the clinical and non-clinical settings. Although resistance to most antimicrobial compounds still remains low in *Y. enterocolitica* clinical isolates, strains belonging to this bioserotype are more likely to be resistant to antimicrobials than are strains of other biotypes. Thus, in life-threatening situations the recommended antibacterial therapy includes either fluoroquinolones or third-generation cephalosporins (the antibiotics showing the highest levels of susceptibility) in combination with gentamicin for which the pathogen is still broadly susceptible. In this way, possible treatment failure as well as infrequent resistance to certain compounds should be overcome. Nevertheless, since pigs are the main reservoir of *Y. enterocolitica* and cross-resistance with antibiotics used in the veterinary field occurs, continuous surveillance of the antimicrobial susceptibility profile of *Y. enterocolitica* is needed in the swine industry. In addition, clonal dissemination of strains carrying multiple-resistance genes in the hospital environment also deserves special attention.

REFERENCES

Abdel-Haq, N.M., Asmar, B.I., Abuhammour, W.M., Brown, W.J., 2000. *Yersinia enterocolitica* infection in children. Pediatr. Infect. Dis. J. 19, 954–958.

Abdel-Haq, N.M., Papadopol, R., Asmar, B.I., Brown, W.J., 2006. Antibiotic susceptibilities of *Yersinia enterocolitica* recovered from children over a 12-year period. Int. J. Antimicrob. Agents 27, 449–452.

Ayyadurai, S., Flaudrops, C., Raoult, D., Drancourt, M., 2010. Rapid identification and typing of *Yersinia pestis* and other *Yersinia* species by matrix-assisted laser desorption/ionization time-of-flight (MALDI-TOF) mass spectrometry. BMC Microbiol. 10, 285.

Baumgartner, A., Kuffer, M., Suter, D., Jemmi, T., Rohner, P., 2007. Antimicrobial resistance of *Yersinia enterocolitica* strains from human patients, pigs and retail pork in Switzerland. Int. J. Food Microbiol. 115, 110–114.

Bengoechea, J.A., Najdenski, H., Skurnik, M., 2004. Lipopolysaccharide O antigen status of *Yersinia enterocolitica* O:8 is essential for virulence and absence of O antigen affects the expression of other *Yersinia* virulence factors. Mol. Microbiol. 52, 451–469.

Bhaduri, S., Wesley, I., Richards, H., Draughon, A., Wallace, M., 2009. Clonality and antibiotic susceptibility of *Yersinia enterocolitica* isolated from U.S. market weight hogs. Foodborne Pathog. Dis. 6, 351–356.

Black, R.E., Jackson, R.J., Tsai, T., Medvesky, M., Shayegani, M., Feeley, J.C., et al., 1978. Epidemic *Yersinia enterocolitica* infection due to contaminated chocolate milk. N. Engl. J. Med. 298, 76–79.

Boelaert, J.R., van Landuyt, H.W., Valcke, Y.J., Cantinieaux, B., Lornoy, W.F., Vanherweghem, J.L., et al., 1987. The role of iron overload in *Yersinia enterocolitica* and *Yersinia pseudotuberculosis* bacteremia in hemodialysis patients. J. Infect. Dis. 156, 384–387.

Bonardi, S., Bassi, L., Brindani, F., D'Incau, M., Barco, L., Carra, E., et al., 2013. Prevalence, characterization and antimicrobial susceptibility of *Salmonella enterica* and *Yersinia enterocolitica* in pigs at slaughter in Italy. Int. J. Food Microbiol. 163, 248–257.

Bonardi, S., Alpigiani, I., Pongolini, S., Morganti, M., Tagliabue, S., Bacci, C., et al., 2014. Detection, enumeration and characterization of *Yersinia enterocolitica* 4/O:3 in pig tonsils at slaughter in Northern Italy. Int. J. Food Microbiol. 177, 9–15.

Bonke, R., Wacheck, S., Stuber, E., Meyer, C., Martlbauer, E., Fredriksson-Ahomaa, M., 2011. Antimicrobial susceptibility and distribution of beta-lactamase A (*blaA*) and beta-lactamase B (*blaB*) genes in enteropathogenic *Yersinia* species. Microb. Drug Resist. 17, 575–581.

Bottone, E.J., 1997. *Yersinia enterocolitica*: the charisma continues. Clin. Microbiol. Rev. 10, 257–276.

Burnens, A.P., Frey, A., Nicolet, J., 1996. Association between clinical presentation, biogroups and virulence attributes of *Yersinia enterocolitica* strains in human diarrhoeal disease. Epidemiol. Infect. 116, 27–34.

Capilla, S., Goni, P., Rubio, M.C., Castillo, J., Millan, L., Cerda, P., et al., 2003. Epidemiological study of resistance to nalidixic acid and other antibiotics in clinical *Yersinia enterocolitica* O:3 isolates. J. Clin. Microbiol. 41, 4876–4878.

Capilla, S., Ruiz, J., Goni, P., Castillo, J., Rubio, M.C., Jimenez de Anta, M.T., et al., 2004. Characterization of the molecular mechanisms of quinolone resistance in *Yersinia enterocolitica* O:3 clinical isolates. J. Antimicrob. Chemother. 53, 1068–1071.

Carniel, E., Guilvout, I., Prentice, M., 1996. Characterization of a large chromosomal "high-pathogenicity island" in biotype 1B *Yersinia enterocolitica*. J. Bacteriol. 178, 6743–6751.

CDC (Centers for Disease Control and Prevention), 2010. Preliminary FoodNet data on the incidence of infection with pathogens transmitted commonly through food—10 states, 2009. Morb. Mortal. Wkly Rep. 59, 418–422.

Chiu, S., Huang, Y.C., Su, L.H., Lin, T.Y., 2003. *Yersinia enterocolitica* sepsis in an adolescent with Cooley's anemia. J. Formos. Med. Assoc. 102, 202–204.

Cornelis, G., Abraham, E.P., 1975. Beta-lactamases from *Yersinia enterocolitica*. J. Gen. Microbiol. 87, 273–284.

Cornelis, G.R., Boland, A., Boyd, A.P., Geuijen, C., Iriarte, M., Neyt, C., et al., 1998. The virulence plasmid of *Yersinia*, an antihost genome. Microbiol. Mol. Biol. Rev. 62, 1315–1352.

Crowe, M., Ashford, K., Ispahani, P., 1996. Clinical features and antibiotic treatment of septic arthritis and osteomyelitis due to *Yersinia enterocolitica*. J. Med. Microbiol. 45, 302–309.

Drummond, N., Murphy, B.P., Ringwood, T., Prentice, M.B., Buckley, J.F., Fanning, S., 2012. *Yersinia enterocolitica*: a brief review of the issues relating to the zoonotic pathogen, public health challenges, and the pork production chain. Foodborne Pathog. Dis. 9, 179–189.

Drummond, N., Stephan, R., Haughton, P., Murphy, B.P., Fanning, S., 2013. Further characterization of three *Yersinia enterocolitica* strains with a nalidixic acid-resistant phenotype isolated from humans with diarrhea. Foodborne Pathog. Dis. 10, 744–746.

EFSA (European Food Safety Authority), ECDC (European Centre for Disease Prevention and Control), 2013. The European Union summary report on trends and sources of zoonoses, zoonotic agents and food-borne outbreaks in 2011. EFSA J. 11, 3129.

El Tahir, Y., Skurnik, M., 2001. YadA, the multifaceted *Yersinia* adhesin. Int. J. Med. Microbiol. 291, 209–218.

Fabrega, A., Vila, J., 2012. *Yersinia enterocolitica*: pathogenesis, virulence and antimicrobial resistance. Enferm. Infecc. Microbiol. Clin. 30, 24–32.

Fabrega, A., Madurga, S., Giralt, E., Vila, J., 2009. Mechanism of action of and resistance to quinolones. Microb. Biotechnol. 2, 40–61.

Fabrega, A., Roca, I., Vila, J., 2010. Fluoroquinolone and multidrug resistance phenotypes associated with the overexpression of AcrAB and an orthologue of MarA in *Yersinia enterocolitica*. Int. J. Med. Microbiol. 300, 457–463.

Fearnley, C., On, S.L., Kokotovic, B., Manning, G., Cheasty, T., Newell, D.G., 2005. Application of fluorescent amplified fragment length polymorphism for comparison of human and animal isolates of *Yersinia enterocolitica*. Appl. Environ. Microbiol. 71, 4960–4965.

Fredriksson-Ahomaa, M., Korkeala, H., 2003. Low occurrence of pathogenic *Yersinia enterocolitica* in clinical, food, and environmental samples: a methodological problem. Clin. Microbiol. Rev. 16, 220–229.

Fredriksson-Ahomaa, M., Stolle, A., Korkeala, H., 2006. Molecular epidemiology of *Yersinia enterocolitica* infections. FEMS Immunol. Med. Microbiol. 47, 315–329.

Fredriksson-Ahomaa, M., Meyer, C., Bonke, R., Stuber, E., Wacheck, S., 2010. Characterization of *Yersinia enterocolitica* 4/O:3 isolates from tonsils of Bavarian slaughter pigs. Lett. Appl. Microbiol. 50, 412–418.

Fredriksson-Ahomaa, M., Cernela, N., Hachler, H., Stephan, R., 2012. *Yersinia enterocolitica* strains associated with human infections in Switzerland 2001–2010. Eur. J. Clin. Microbiol. Infect. Dis. 31, 1543–1550.

Fukushima, H., Gomyoda, M., Aleksic, S., Tsubokura, M., 1993. Differentiation of *Yersinia enterocolitica* serotype O:5,27 strains by phenotypic and molecular techniques. J. Clin. Microbiol. 31, 1672–1674.

Fukushima, H., Hoshina, K., Itogawa, H., Gomyoda, M., 1997. Introduction into Japan of pathogenic *Yersinia* through imported pork, beef and fowl. Int. J. Food Microbiol. 35, 205–212.

Galimand, M., Courvalin, P., Lambert, T., 2003. Plasmid-mediated high-level resistance to aminoglycosides in Enterobacteriaceae due to 16S rRNA methylation. Antimicrob. Agents Chemother. 47, 2565–2571.

Gemski, P., Lazere, J.R., Casey, T., 1980. Plasmid associated with pathogenicity and calcium dependency of *Yersinia enterocolitica*. Infect. Immun. 27, 682–685.

Gousia, P., Economou, V., Sakkas, H., Leveidiotou, S., Papadopoulou, C., 2011. Antimicrobial resistance of major foodborne pathogens from major meat products. Foodborne Pathog. Dis. 8, 27–38.

Grahek-Ogden, D., Schimmer, B., Cudjoe, K.S., Nygard, K., Kapperud, G., 2007. Outbreak of *Yersinia enterocolitica* serogroup O:9 infection and processed pork, Norway. Emerg. Infect. Dis. 13, 754–756.

Greene, J.N., Herndon, P., Nadler, J.P., Sandin, R.L., 1993. Case report: *Yersinia enterocolitica* necrotizing pneumonia in an immunocompromised patient. Am. J. Med. Sci. 305, 171–173.

Griffiths, A.J.F., Gelbart, W.M., Miller, J.H., Lewontin, R.C., 1999. Modern Genetic Analysis. WH Freeman, New York, NY.

Guinet, F., Carniel, E., Leclercq, A., 2011. Transfusion-transmitted *Yersinia enterocolitica* sepsis. Clin. Infect. Dis. 53, 583–591.

Hall, R.M., 2012. Integrons and gene cassettes: hotspots of diversity in bacterial genomes. Ann. N. Y. Acad. Sci. 1267, 71–78.

Haller, J.C., Carlson, S., Pederson, K.J., Pierson, D.E., 2000. A chromosomally encoded type III secretion pathway in *Yersinia enterocolitica* is important in virulence. Mol. Microbiol. 36, 1436–1446.

Hayashidani, H., Ohtomo, Y., Toyokawa, Y., Saito, M., Kaneko, K., Kosuge, J., et al., 1995. Potential sources of sporadic human infection with *Yersinia enterocolitica* serovar O:8 in Aomori Prefecture, Japan. J. Clin. Microbiol. 33, 1253–1257.

Hoogkamp-Korstanje, J.A., Stolk-Engelaar, V.M., 1995. *Yersinia enterocolitica* infection in children. Pediatr. Infect. Dis. J. 14, 771–775.

Hoogkamp-Korstanje, J.A., Moesker, H., Bruyn, G.A., 2000. Ciprofloxacin v placebo for treatment of *Yersinia enterocolitica* triggered reactive arthritis. Ann. Rheum. Dis. 59, 914–917.

Iriarte, M., Vanooteghem, J.C., Delor, I., Diaz, R., Knutton, S., Cornelis, G.R., 1993. The Myf fibrillae of *Yersinia enterocolitica*. Mol. Microbiol. 9, 507–520.

Jensen, V.F., Jakobsen, L., Emborg, H.D., Seyfarth, A.M., Hammerum, A.M., 2006. Correlation between apramycin and gentamicin use in pigs and an increasing reservoir of gentamicin-resistant *Escherichia coli*. J. Antimicrob. Chemother. 58, 101–107.

Jimenez-Valera, M., Gonzalez-Torres, C., Moreno, E., Ruiz-Bravo, A., 1998. Comparison of ceftriaxone, amikacin, and ciprofloxacin in treatment of experimental *Yersinia enterocolitica* O9 infection in mice. Antimicrob. Agents Chemother. 42, 3009–3011.

Jones, T.F., 2003. From pig to pacifier: chitterling-associated yersiniosis outbreak among black infants. Emerg. Infect. Dis. 9, 1007–1009.

Jouquan, J., Mottier, D., Pennec, Y., Boles, J.M., Voisin, P., Colloc, M.L., et al., 1984. *Yersinia enterocolitica* septicemia occurring in a hospital milieu. A case with septic shock in an immunocompromised patient. Sem. Hop. 60, 47–48.

Kotra, L.P., Haddad, J., Mobashery, S., 2000. Aminoglycosides: perspectives on mechanisms of action and resistance and strategies to counter resistance. Antimicrob. Agents Chemother. 44, 3249–3256.

Lambertz, S.T., Nilsson, C., Hallanvuo, S., Lindblad, M., 2008. Real-time PCR method for detection of pathogenic *Yersinia enterocolitica* in food. Appl. Environ. Microbiol. 74, 6060–6067.

Lamps, L.W., Havens, J.M., Gilbrech, L.J., Dube, P.H., Scott, M.A., 2006. Molecular biogrouping of pathogenic *Yersinia enterocolitica*: development of a diagnostic PCR assay with histologic correlation. Am. J. Clin. Pathol. 125, 658–664.

Lee, W.H., McGrath, P.P., Carter, P.H., Eide, E.L., 1977. The ability of some *Yersinia enterocolitica* strains to invade HeLa cells. Can. J. Microbiol. 23, 1714–1722.

Levesque, C., Piche, L., Larose, C., Roy, P.H., 1995. PCR mapping of integrons reveals several novel combinations of resistance genes. Antimicrob. Agents Chemother. 39, 185–191.

Lupi, A., Poletti, F., Mondino, V., Canale, C., Leonardo, L., Rognoni, A., et al., 2013. Subacute endocarditis caused by *Yersinia enterocolitica*: a case report. Scand. J. Infect. Dis. 45, 329–333.

Magiorakos, A.P., Srinivasan, A., Carey, R.B., Carmeli, Y., Falagas, M.E., Giske, C.G., et al., 2012. Multidrug-resistant, extensively drug-resistant and pandrug-resistant bacteria: an international expert proposal for interim standard definitions for acquired resistance. Clin. Microbiol. Infect. 18, 268–281.

Magnet, S., Courvalin, P., Lambert, T., 2001. Resistance-nodulation-cell division-type efflux pump involved in aminoglycoside resistance in *Acinetobacter baumannii* strain BM4454. Antimicrob. Agents Chemother. 45, 3375–3380.

Martinez, J.L., Baquero, F., 2014. Emergence and spread of antibiotic resistance: setting a parameter space. Ups. J. Med. Sci. 119, 68–77.

Mayrhofer, S., Paulsen, P., Smulders, F.J., Hilbert, F., 2004. Antimicrobial resistance profile of five major food-borne pathogens isolated from beef, pork and poultry. Int. J. Food Microbiol. 97, 23–29.

Morris Jr., J.G., Prado, V., Ferreccio, C., Robins-Browne, R.M., Bordun, A.M., Cayazzo, M., et al., 1991. *Yersinia enterocolitica* isolated from two cohorts of young children in Santiago, Chile: incidence of and lack of correlation between illness and proposed virulence factors. J. Clin. Microbiol. 29, 2784–2788.

Mu, Y.J., Zhao, J.Y., Guo, Q.S., Guo, X.C., Jing, H.Q., Xia, S.L., 2013. Investigation on distribution of *Yersinia enterocolitica* in Henan province between 2005 and 2011. Zhonghua Yu Fang Yi Xue Za Zhi 47, 612–615.

Neubauer, H., Aleksic, S., Hensel, A., Finke, E.J., Meyer, H., 2000. *Yersinia enterocolitica* 16S rRNA gene types belong to the same genospecies but form three homology groups. Int. J. Med. Microbiol. 290, 61–64.

Novoslavskij, A., Kudirkiene, E., Marcinkute, A., Bajoriuniene, A., Korkeala, H., Malakauskas, M., 2013. Genetic diversity and antimicrobial resistance of *Yersinia enterocolitica* isolated from pigs and humans in Lithuania. J. Sci. Food Agric. 93, 1858–1862.

Okusu, H., Ma, D., Nikaido, H., 1996. AcrAB efflux pump plays a major role in the antibiotic resistance phenotype of *Escherichia coli* multiple-antibiotic-resistance (Mar) mutants. J. Bacteriol. 178, 306–308.

Pepe, J.C., Miller, V.L., 1993. *Yersinia enterocolitica* invasin: a primary role in the initiation of infection. Proc. Natl. Acad. Sci. USA 90, 6473–6477.

Pham, J.N., Bell, S.M., 1992. Beta-lactamase induction by imipenem in *Yersinia enterocolitica*. Pathology 24, 201–204.

Pham, J.N., Bell, S.M., 1993. The prevalence of inducible beta-lactamase in clinical isolates of *Yersinia enterocolitica*. Pathology 25, 385–387.

Pham, J.N., Bell, S.M., Lanzarone, J.Y., 1991a. A study of the beta-lactamases of 100 clinical isolates of *Yersinia enterocolitica*. J. Antimicrob. Chemother. 28, 19–24.

Pham, J.N., Bell, S.M., Lanzarone, J.Y., 1991b. Biotype and antibiotic sensitivity of 100 clinical isolates of *Yersinia enterocolitica*. J. Antimicrob. Chemother. 28, 13–18.

Pham, J.N., Bell, S.M., Hardy, M.J., Martin, L., Guiyoule, A., Carniel, E., 1995. Comparison of beta-lactamase production by *Yersinia enterocolitica* biotype 4, serotype O:3 isolated in eleven countries. Contrib. Microbiol. Immunol. 13, 180–183.

Pierson, D.E., Falkow, S., 1993. The *ail* gene of *Yersinia enterocolitica* has a role in the ability of the organism to survive serum killing. Infect. Immun. 61, 1846–1852.

Prats, G., Mirelis, B., Llovet, T., Munoz, C., Miro, E., Navarro, F., 2000. Antibiotic resistance trends in enteropathogenic bacteria isolated in 1985–1987 and 1995–1998 in Barcelona. Antimicrob. Agents Chemother. 44, 1140–1145.

Preston, M.A., Brown, S., Borczyk, A.A., Riley, G., Krishnan, C., 1994. Antimicrobial susceptibility of pathogenic *Yersinia enterocolitica* isolated in Canada from 1972 to 1990. Antimicrob. Agents Chemother. 38, 2121–2124.

Rastawicki, W., Gierczynski, R., Jagielski, M., Kaluzewski, S., Jeljaszewicz, J., 2000. Susceptibility of Polish clinical strains of *Yersinia enterocolitica* serotype O3 to antibiotics. Int. J. Antimicrob. Agents 13, 297–300.

Rastawicki, W., Szych, J., Gierczynski, R., Rokosz, N., 2009. A dramatic increase of *Yersinia enterocolitica* serogroup O:8 infections in Poland. Eur. J. Clin. Microbiol. Infect. Dis. 28, 535–537.

Schiemann, D.A., 1979. Synthesis of a selective agar medium for *Yersinia enterocolitica*. Can. J. Microbiol. 25, 1298–1304.

Schleifstein, J., Coleman, M.B., 1939. An unidentified microorganism resembling *B. lignieri* and *Past. pseudotuberculosis* and pathogenic for man. NY State J. Med. 39, 1749–1753.

Seoane, A., Garcia-Lobo, J.M., 1991. Cloning of chromosomal beta-lactamase genes from *Yersinia enterocolitica*. J. Gen. Microbiol. 137, 141–146.

Sharma, S., Ramnani, P., Virdi, J.S., 2004. Detection and assay of beta-lactamases in clinical and non-clinical strains of *Yersinia enterocolitica* biovar 1A. J. Antimicrob. Chemother. 54, 401–405.

Sharma, S., Mittal, S., Mallik, S., Virdi, J.S., 2006. Molecular characterization of beta-lactamase genes *blaA* and *blaB* of *Yersinia enterocolitica* biovar 1A. FEMS Microbiol. Lett. 257, 319–327.

Sihvonen, L.M., Toivonen, S., Haukka, K., Kuusi, M., Skurnik, M., Siitonen, A., 2011. Multilocus variable-number tandem-repeat analysis, pulsed-field gel electrophoresis, and antimicrobial susceptibility patterns in discrimination of sporadic and outbreak-related strains of *Yersinia enterocolitica*. BMC Microbiol. 11, 42.

Singh, I., Virdi, J.S., 2004. Production of *Yersinia* stable toxin (YST) and distribution of *yst* genes in biotype 1A strains of *Yersinia enterocolitica*. J. Med. Microbiol. 53, 1065–1068.

Stephan, R., Joutsen, S., Hofer, E., Sade, E., Bjorkroth, J., Ziegler, D., et al., 2013. Characteristics of *Yersinia enterocolitica* biotype 1A strains isolated from patients and asymptomatic carriers. Eur. J. Clin. Microbiol. Infect. Dis. 32, 869–875.

Stock, A.M., Robinson, V.L., Goudreau, P.N., 2000a. Two-component signal transduction. Annu. Rev. Biochem. 69, 183–215.

Stock, I., Heisig, P., Wiedemann, B., 1999. Expression of beta-lactamases in *Yersinia enterocolitica* strains of biovars 2, 4 and 5. J. Med. Microbiol. 48, 1023–1027.

Stock, I., Heisig, P., Wiedemann, B., 2000b. Beta-lactamase expression in *Yersinia enterocolitica* biovars 1A, 1B, and 3. J. Med. Microbiol. 49, 403–408.

Taber, H.W., Mueller, J.P., Miller, P.F., Arrow, A.S., 1987. Bacterial uptake of aminoglycoside antibiotics. Microbiol. Rev. 51, 439–457.

Tacket, C.O., Narain, J.P., Sattin, R., Lofgren, J.P., Konigsberg Jr., C., Rendtorff, R.C., et al., 1984. A multistate outbreak of infections caused by *Yersinia enterocolitica* transmitted by pasteurized milk. JAMA 251, 483–486.

Tadesse, D.A., Bahnson, P.B., Funk, J.A., Morrow, W.E., Abley, M.J., Ponte, V.A., et al., 2013. *Yersinia enterocolitica* of porcine origin: carriage of virulence genes and genotypic diversity. Foodborne Pathog. Dis. 10, 80–86.

Terentjeva, M., Berzins, A., 2010. Prevalence and antimicrobial resistance of *Yersinia enterocolitica* and *Yersinia pseudotuberculosis* in slaughter pigs in Latvia. J. Food Prot. 73, 1335–1338.

Vantrappen, G., Ponette, E., Geboes, K., Bertrand, P., 1977. *Yersinia* enteritis and enterocolitis: gastroenterological aspects. Gastroenterology 72, 220–227.

Venecia, K., Young, G.M., 2005. Environmental regulation and virulence attributes of the Ysa type III secretion system of *Yersinia enterocolitica* biovar 1B. Infect. Immun. 73, 5961–5977.

Vila, J., Fabrega, A., Roca, I., Hernandez, A., Martinez, J.L., 2011. Efflux pumps as an important mechanism for quinolone resistance. Adv. Enzymol. Relat. Areas Mol. Biol. 77, 167–235.

Wauters, G., Kandolo, K., Janssens, M., 1987. Revised biogrouping scheme of *Yersinia enterocolitica*. Contrib. Microbiol. Immunol. 9, 14–21.

Wong, K.K., Fistek, M., Watkins, R.R., 2013. Community-acquired pneumonia caused by *Yersinia enterocolitica* in an immunocompetent patient. J. Med. Microbiol. 62, 650–651.

Young, G.M., Badger, J.L., Miller, V.L., 2000. Motility is required to initiate host cell invasion by *Yersinia enterocolitica*. Infect. Immun. 68, 4323–4326.

Chapter 6

Antimicrobial Resistance in *Vibrio* Species

Craig Baker-Austin

Centre for Environment Fisheries and Aquaculture Science, Weymouth, Dorset, UK

Chapter Outline

INTRODUCTION

Vibrios are rod-shaped bacteria, and are a functionally and phylogenetically diverse grouping of Gram-negative microbes found widely in aquatic, estuarine, and marine habitats. Approximately a dozen *Vibrio* species are known to cause disease in humans (Austin, 2005), and infection is usually initiated from exposure to seawater or consumption of raw or undercooked seafood (Altekruse et al., 2000; Dechet et al., 2008). Although a wide range of different bacterial species contain multiple chromosomes, *Vibrio* species are noted in that they possess two circular chromosomes (Yamaichi et al., 1999). Bacteria of the genus *Vibrio* are commonly found in tropical and temperate coastal and estuarine waters. Vibrios are among the most common bacteria that inhabit surface waters throughout the world and are responsible for a number of severe infections both in humans and animals (Vezzulli et al., 2013). Vibriosis is characterized by diarrhea, primary septicemia, wound infections, or other extraintestinal infections (Altekruse et al., 2000; Daniels et al., 2000). Select strains of *V. cholerae*, *V. parahaemolyticus*, *V. vulnificus*, and *V. alginolyticus* are perhaps considered the most serious human pathogens from this genus. Two *Vibrio* species in particular, *V. vulnificus* and *V. parahaemolyticus* are significant foodborne human pathogens, and most frequently infections occur via the consumption of naturally contaminated shellfish produce. It is worth noting that these pathogens represent a significant cause of morbidity and mortality. For example, an estimated 80,000 people contract *Vibrio* infections each year in the United States, with a sizeable fraction originating from foodborne sources, such as consumption of raw or undercooked seafood produce. Recent data from the Centers

Antimicrobial Resistance and Food Safety. DOI: http://dx.doi.org/10.1016/B978-0-12-801214-7.00006-5

for Disease Control and Prevention (CDC) in the United States have indicated that there has been a significant increase in reported infections associated with vibrios, particularly in the last two decades. The annual incidence of reported vibriosis per 100,000 population has increased significantly in the United States from 1996 to 2010 (Newton et al., 2012), highlighting the importance of these pathogens from a clinical context. Calculations based upon probable incidence of vibriosis have estimated that *V. vulnificus* and *V. parahaemolyticus* are the first and third most costly marine-borne pathogens, costing $233 and $20 million, respectively (Ralston et al., 2011). From a foodborne perspective *V. vulnificus* and *V. parahaemolyticus* represent the major pathogens from the *Vibrio* genus in terms of clinical impact and relevance, and as such this chapter is mostly concerned with these species. These taxa do not sustain prolonged presence in clinical or agricultural settings, where they would likely undergo human-induced selection for antibiotic resistance. As such, these bacteria represent a particularly interesting group of pathogens to study antibiotic resistance, as they provide a "snapshot" of resistance presumably acquired from environmental rather than clinical settings (Baker-Austin et al., 2008). Despite their public-health significance, strains of *V. parahaemolyticus* and *V. vulnificus* have not been extensively monitored for antimicrobial resistance, in contrast to enteric pathogens such as *Salmonella* or *Campylobacter* (Han et al., 2007). Given their increasing incidence, global distribution, and severity of disease progression (especially *V. vulnificus*) it is critical to gain a better understanding of the antimicrobial susceptibility patterns of *V. vulnificus* and *V. parahaemolyticus* originating from the environment (Shaw et al., 2014). Data from such sources is invaluable, particularly from routine antimicrobial screening of large numbers of environmental and clinical *Vibrio* strains as it can provide effective baseline data for treatment purposes.

EPIDEMIOLOGY AND UNDERREPORTING

A key issue regarding the study of antimicrobial resistance in pathogenic vibrios is related to datagaps associated with the reporting of infections. Effective prevention of waterborne and foodborne pathogens requires detailed epidemiological data to identify likely routes of exposure, risk characteristics (e.g., individuals at particular risk of infections) as well as possible mechanisms to disrupt disease transmission. A major challenge regarding the epidemiology of many foodborne pathogens is the role of underreporting in masking the true clinical burden of disease. Indeed, even dedicated surveillance systems set up to monitor specific pathogens record only 1–10% of foodborne cases (Huss et al., 2004). The problem of underreporting depends on *Vibrio* species studied as well as country of reporting, and has significant ramifications for the understanding of the basic epidemiology of these pathogens. An example of this problem is the enormous underreporting for foodborne *V. parahaemolyticus* in the United States. Much of this masking of reported infections is likely due to the low hospitalization and fatality rates associated with this pathogen (Scallan et al., 2011). Thus, because the vast majority of *V. parahaemolyticus* infections are self-limiting, the bulk of cases are unlikely to be reported to healthcare professionals. Even when help is sought by the patient, healthcare professionals may either misdiagnose or fail to recognize the illness (Ralston et al., 2011), and commonly in the case of foodborne *V. parahaemolyticus* infections less reliable biochemical rather than molecular testing is performed. In other cases, the healthcare provider may feel that the illness is not serious enough to warrant in-depth laboratory tests that would identify the causative pathogen. Whether or not the result of positive tests is cascaded further to regional and national databases is also questionable, as is whether the data are then made publically available to the *Vibrio* research community (Figure 6.1). A recent report

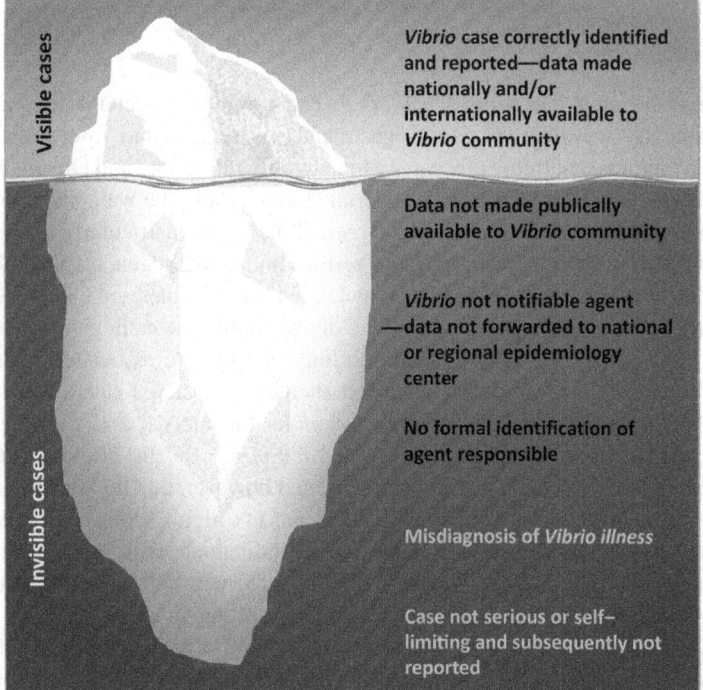

FIGURE 6.1   The enigma of under-reporting. The vast majority of foodborne *Vibrio* infections are not correctly identified, or correctly diagnosed using appropriate testing methods. Even where vibriosis is suspected, there are also gaps in identification and reporting of cases to appropriate agencies, especially in some countries and regions where vibriosis is considered a rare or non-endemic infection. These factors account for the fact that reliable epidemiological data and crucially antibiotic resistance datasets are available for only a small number of overall cases. As such, these cases may represent the "tip of the iceberg" of actual *Vibrio*-associated disease burden from foodborne cases. These limitations provide significant challenges in developing a baseline of antimicrobial resistance data for these foodborne pathogens.

from the United States indicated significant (almost 300 × under-reporting) of *V. parahaemolyticus* infections to the CDC, and almost 40,000 estimated cases each year (Scallan et al., 2011). The severity of *Vibrio*-associated infections is also likely to be related to whether or not cases are reported, and is likely to have some bearing on what proportion of the "iceberg" is observed by the *Vibrio* research, regulatory and public health community. Certainly, this masking of disease burden is undoubtedly a problem for foodborne *V. parahaemolyticus* disease, but is markedly less apparent for pathogens with a more severe clinical manifestation such as *V. vulnificus* (FAO, 2005; Jones and Oliver, 2009), although misdiagnosis with other pathogenic agents that cause sepsis is also likely. Irrespective, there are probably two times the number of actual *V. vulnificus* cases in the United States compared to those reported to the CDC (Mead et al., 1999), suggestive that the true clinical burden associated with these pathogens is significantly higher than previously thought.

Vibrio vulnificus

The Gram-negative bacterium *V. vulnificus* is a common inhabitant of estuarine environments. *V. vulnificus* causes gastroenteritis and wound infections with severe symptoms that frequently carry a high fatality rate (Jones and Oliver, 2009). *V. vulnificus* is part of the normal bacterial flora of estuarine waters and occurs in high numbers in molluscan shellfish around the world, particularly in warmer months (Oliver, 2006). *V. vulnificus* is a serious human pathogen, responsible for over 95% of seafood-related deaths in the United States (Jones and Oliver, 2009), and carries one of the highest fatality rate of any foodborne pathogen (Jones and Oliver, 2009; Rippey, 1994). Indeed, the high fatality rate associated with this pathogen (51.6% from recently reviewed shellfish-associated cases; Jones and Oliver, 2009) is comparable to a range of major biosafety 3 and 4 pathogenic agents, such as Ebola and Marburg virus and *Yersinia pestis*, the etiological agent responsible for bubonic plague. It must be stated however, that this striking case-fatality rate is normally associated with infections in a highly select demographic: *V. vulnificus* disease disproportionally affects individuals with underlying risk conditions, and the majority of cases are associated with immunocompromised males or patients with underlying conditions resulting in elevated serum iron levels, primarily alcohol-associated liver cirrhosis (Oliver and Kaper, 2001). Because of the pronounced influence of host susceptibility in the epidemiology of *V. vulnificus* disease, infections associated with this pathogen tend to be isolated and sporadic and as such *V. vulnificus* is not associated with typical foodborne outbreaks. Isolated incidents of *V. vulnificus* infections occur globally, with cases frequently reported in the United States, Europe, and the Far East (Dalsgaard et al., 1996; Chuang et al., 1992; Hlady and Klontz 1996; Baker-Austin et al., 2010). Unfortunately, molecular methods for the detection and enumeration of pathogenic *V. vulnificus* are hampered by the genetically diverse nature of this pathogen, the range of different biotypes capable of infecting humans, and the fact that *V. vulnificus* contains pathogenic as well as non-pathogenic variants. Infections are characterized by a short incubation period between exposure and the onset of symptoms (Baker-Austin et al., 2008), typically within 24 h of exposure (Jones and Oliver, 2009). For these two reasons, immediate and appropriate intervention with antimicrobials is absolutely essential. If treatment is delayed greater than 72 h the fatality rate of primary septicemia-associated *V. vulnificus* infections is 100% (Klontz et al., 1988).

Historically, *V. vulnificus* was not well recognized for its antibiotic resistance, and some previous studies have shown sensitivity to tetracyclines, aminoglycosides, third-generation cephalosporins, chloramphenicol, and newer fluoroquinolones (Morris and Tenney, 1985; Tang et al., 2002; Baker-Austin et al., 2008). More recently, several studies have analyzed the resistance of environmentally derived *V. vulnificus* or attempted to verify the effectiveness of commonly prescribed antibiotics in treating *V. vulnificus* infections. Pan et al. (2012) investigated antimicrobial resistance capabilities of *V. vulnificus* strains isolated in shrimps from retail markets in Hangzhou, China. They carried out antibiotic susceptibility testing of *V. vulnificus* strains against a battery of 21 antimicrobial agents alongside virulence testing of individual isolates. Some strains showed varying resistance to cefepime (3.03%), tetracycline (6.06%), aztreonam (24.24%), streptomycin (45.45%), gentamicin (93.94%), tobramycin (100%), and cefazolin (100%). Ji et al. (2011) analyzed the resistance characteristics of *V. vulnificus* strains isolated from Chinese retail shrimp samples taken from a range of retail establishments. The antimicrobial susceptibility of isolates to 12 antibiotics was subsequently ascertained following molecular typing to determine potential virulence. Most of the 169 isolates remained susceptible to the majority of antimicrobials tested, whereas some strains demonstrated resistance to amikacin, ampicillin, tetracycline, and gentamicin. In a similar study of isolated *Vibrio* strains from Thailand cultured shrimp produce, resistance to ampicillin and oxytetracycline (72% and 3%) was found in *V. parahaemolyticus* isolates (Yano et al., 2013). The authors further suggest that Shrimp may represent a vehicle for transfer of *Vibrio* species with antimicrobial resistance. Baker-Austin et al. (2008) screened 151 coastal isolates from the south-east seaboard of the United States and ten primary septicemia *V. vulnificus* isolates against 26 antimicrobial agents representing diverse modes of action. Unexpectedly, the frequency of multiple resistances to antibiotics from all sources was high, particularly during summer months, and a substantial proportion of isolates (17.3%) were resistant to eight or more antimicrobial agents. Of concern, they noted resistance to antibiotics routinely prescribed for *V. vulnificus* infections, such as doxycycline, tetracycline, aminoglycosides, and cephalosporins. This study was significant in that highly resistant strains were equally identified in industrially contaminated as well as pristine environments, and provided evidence that environmentally derived resistance traits may have direct clinical ramifications. Likewise, Kim et al. (2011) assessed the occurrence and antibiotic resistance of *V. vulnificus* strains isolated from seafood and environmental samples obtained from several fish markets and estuarine sites in Korea from May to December of 2009. Although the authors analyzed a relatively small number of isolates ($n = 31$), they observed resistance of isolates against three antibiotic classes: cephems, aminoglycosides, and folate pathway inhibitors. These data, as with studies in the United States and China, suggest that environmental sources of strains may represent an unexpected reservoir of resistant bacteria that can enter clinical settings. Not all published studies have found significant resistance to antimicrobials in *V. vulnificus* however. Han et al. (2007) carried out a large-scale examination of the antimicrobial susceptibilities of *V. vulnificus* in Louisiana Gulf and retail oysters. Their findings indicated that of a large library of strains analyzed ($n = 151$), the majority remained susceptible to most of the antimicrobials used, including ampicillin, cefotaxime, ceftazidime, chloramphenicol, ciprofloxacin, gentamicin, imipenem, and tetracycline. More recently, Shaw et al. (2014) analyzed antimicrobial resistance capabilities of *V. vulnificus* strains isolated from water samples in Chesapeake Bay, USA. Most isolates were susceptible to antibiotics recommended for treating *Vibrio* infections, although around 77% of isolates expressed intermediate resistance to chloramphenicol. Intermediate resistance to some aminoglycosides should be noted because these antibiotics are used to treat some pediatric *Vibrio* illnesses (Shaw et al., 2014).

The CDC recommend a range of antibiotic regimens to treat *V. vulnificus* infections. Doxycycline (100 mg PO/IV twice a day for 7–14 days) and a third-generation cephalosporin (e.g., ceftazidime 1–2 g IV/IM every 8 h) combination is generally recommended as an initial treatment regimen. A single agent regimen with a fluoroquinolone such as levofloxacin, ciprofloxacin, or gatifloxacin, has been reported to be at least as effective in an animal model as combination drug regimens with doxycycline and a cephalosporin (CDC, 2014). Unfortunately, there are few longstanding data on the efficacy of these treatment regimens *in vivo*. Neupane et al. (2010) analyzed the efficacy of several CDC-recommended drug regimens using mouse models infected with *V. vulnificus*. They found that ciprofloxacin was the most effective monotherapeutic drug, but a higher survival rate (50%) was achieved using the combination therapy of intraperitoneal doxycycline plus ceftriaxone. Recent data from case reviews in Taiwan (Chen et al., 2012) suggest that fluoroquinolones or third-generation cephalosporins plus minocycline are the best option for antibiotic treatment of some infections caused by *V. vulnificus*. Jang et al. (2014) studied a range of antimicrobial combinations and determined that a combination of ciprofloxacin and cefotaxime is an effective option for the treatment of *V. vulnificus* sepsis in humans. A current study by the author is an analysis of antibiotic treatment data in both the United States and EU to assess potential efficacy as well as trends in antimicrobial regimen usage regarding this pathogen. Because of the rapidity and severity of *V. vulnificus* infections, studies such as these outlined here are essential to determine the most effective treatment options for foodborne *V. vulnificus* infections.

Vibrio parahaemolyticus

V. parahaemolyticus is a Gram-negative, halophilic bacterium found commonly in temperate and warm estuarine waters worldwide (Powell et al., 2013). The presence of *V. parahaemolyticus* in the marine environment is closely related to water temperature, with strains readily isolated when environmental temperatures exceed 15°C (Baker-Austin et al., 2010). Globally, *V. parahaemolyticus* is the most prevalent food-poisoning bacterium associated with seafood consumption and typically causes acute gastroenteritis. As with *V. vulnificus* more serious infections tend to be associated with individuals that are suffering from serious underlying conditions, such as immune dysfunction. However, unlike *V. vulnificus*, *V. parahaemolyticus* has a well-established and -characterized basis for virulence and pathogenicity. The vast majority of strains associated with disease carry one (or rarely two) hemolysin genes. These include the thermostable direct hemolysin (TDH) (Bej et al., 1999; Nishibuchi et al., 1992) responsible for Kanagawa hemolysis and the TDH-related hemolysin (TRH) (Honda et al., 1988). Both genes are subsequently considered good predictive indicators of potential virulence (Baker-Austin et al., 2010), and utilized extensively for virulence and diagnostic testing. Until the mid-1990s, *V. parahaemolyticus* infections were typically sporadic, with few largescale foodborne outbreaks. In 1996, a sudden increase in *V. parahaemolyticus* infections emerged in Calcutta, India, with a distinctive serotype (O3:K6, Powell et al., 2013). This highly pathogenic strain has been responsible for large foodborne outbreaks across Asia, Africa, and America (Nair et al., 2007). More recently, a new and highly pathogenic clone, termed the Pacific Northwest clone (PNW) has radiated out of the Pacific region and has caused large outbreaks on both the Atlantic coasts of the United States and Europe (Martinez-Urtaza et al., 2013). Recently, there has been a significant increase in *V. parahaemolyticus* incidence in the United States. This pathogen increased in incidence from 1996 to 2010 in both CDC and FDA datasets used to assess the epidemiology associated with these bacteria (Newton et al., 2012). There is limited, yet increasingly reliable, evidence that the incidence of *V. parahaemolyticus* has also increased recently in Europe (Baker-Austin et al., 2010). Because *V. parahaemolyticus* infections tend to be self-limiting, antibiotic treatment is not usually required to treat these foodborne pathogens, with the exception of severe cases where underlying risk conditions are involved (Baker-Austin et al., 2008).

There have been a variety of studies assessing the environmental, clinical, and genomic basis of antibiotic resistance in *V. parahaemolyticus*. Sahilah et al. (2014) assessed resistance traits in strains isolated from cockles in Malaysia. Although no potentially pathogenic strains (e.g., *tdh* or *trh* positive isolates) were analyzed, they found surprising resistance traits in their library of strains. Antibiotic resistance analysis showed the *V. parahaemolyticus* isolates were highly resistant to bacitracin (92%, 34/37) and penicillin (89%, 33/37) followed by resistance toward ampicillin (68%, 25/37), cefuroxime (38%, 14/37), amikacin (6%, 2/37) and ceftazidime (14%, 5/37). None of the *V. parahaemolyticus* isolates demonstrated resistance to the antibiotics chloramphenicol, ciprofloxacin, ceftriaxone, enrofloxacin, norfloxacin, streptomycin, and vancomycin. Likewise, Elexson et al. (2014) analyzed resistance capabilities of *V. parahaemolyticus* strains isolated from seafood produce. Thirty-six *V. parahaemolyticus* isolates from seafood were tested for their susceptibility using 18 different antimicrobial agents. Two of these analyzed *V. parahaemolyticus* strains demonstrated resistance to bacitracin, chloramphenicol, rifampin, ampicillin, vancomycin, nalidixic acid, penicillin, and spectinomycin. Fourteen *V. parahaemolyticus* isolates were found to be resistant to bacitracin, tetracycline, rifampin, ampicillin, vancomycin, penicillin, and spectinomycin. This study, therefore, indicates wide-ranging resistance traits against a variety of structurally diverse (and clinically relevant) antimicrobials. In addition, the authors noted that some analyzed strains grown in biofilms demonstrated increased resistance capabilities compared to planktonically grown isolates. Jun et al. (2012) studied antimicrobial resistance capabilities of *V. parahaemolyticus* strains isolated from Korean seafood produce as well as environmental and clinical strains. Although this study analyzed a relatively small number of strains ($n = 24$) they carried out both culture-based as well as molecular characterization. Twenty-two commercial antibiotics were used to assess the antibiotic susceptibility of isolates, and, somewhat surprisingly, all the strains demonstrated resistance to more than four antibiotics. The strains harboring antibiotic-resistance genes such as *tetA* (25%) and *strB* (4.16%) were detected via PCR. De Melo et al. (2011) analyzed the antibiotic resistance patterns of *V. parahaemolyticus* strains isolated from Brazilian shellfish produce. Antibiotic susceptibility testing was performed using seven different antibiotics by the agar diffusion technique. Five strains (50%) presented multiple antibiotic resistance to ampicillin (90%) and amikacin (60%), while two strains (20%) displayed intermediate-level resistance to amikacin. Han et al. (2007) carried out one of the largest analyses of antimicrobial resistance in *V. parahaemolyticus* to date using Louisiana Gulf and retail oysters. Their data indicated low levels of resistance to the range of antimicrobials tested. Reduced susceptibility was detected only in *V. parahaemolyticus* for ampicillin (81%; minimum inhibitory concentration (MIC) ≥16 µg/mL). Additionally, *V. parahaemolyticus* displayed significantly higher MICs for cefotaxime, ciprofloxacin, and tetracycline than *V. vulnificus* (Han et al., 2007). In a study focusing on *V. parahaemolyticus* isolated in China, a total of 327 strains were isolated from food and 110 strains were isolated from active surveillance hospitals or food outbreaks during 2005 to 2008 for antimicrobial susceptibility testing. The authors noted significant differences in antibiotic resistance capabilities of strains based on serotype studied. Recently, Shaw et al. (2014) analyzed antimicrobial resistance capabilities of *V. parahaemolyticus* strains isolated from water samples in Chesapeake Bay, Maryland, as well as a small subset of clinically derived strains from the state of Maryland. A high percentage of resistance was observed against some of the penicillins (penicillin (68%); ampicillin (53%)), while a low percentage of resistance was seen against piperacillin (4%) and streptomycin (4%). All tested *V. parahaemolyticus* isolates were susceptible to 11 of the 26 tested antibiotics and four (carbapenems, tetracyclines, quinolones, and folate pathway inhibitors) of the eight tested antimicrobial classes (Shaw et al., 2014).

DATA LIMITATIONS AND EMERGING METHODS FOR ASSESSING ANTIBIOTIC RESISTANCE

There are now a range of standardized methods for assessing antibiotic resistance in pathogenic *Vibrio* species, such as *V. parahaemolyticus* and *V. vulnificus*. These include broth microdilution tests, disk diffusion and gradient-based testing approaches, among others. Each method has strengths and weaknesses. Alongside these testing platforms, the Clinical Laboratory Testing Institute (CLSI) provides tables that list the antimicrobial agents appropriate for testing *Vibrio* spp., with recommendations for agents that are important to test routinely, and those that may be tested or reported selectively based on the institution's formulary (Reller et al., 2009). This, therefore, provides a clear methodological framework for testing these pathogens, with clear guidelines for test conditions and interpretative criteria based upon a careful review of published microbiological data (distributions and MICs), and the extant clinical literature regarding therapy options (CLSI, 2010). Unfortunately, many of the research studies published in the peer-reviewed literature, and many reviewed here, do not use CLSI-based methods for antimicrobial susceptibility testing. There is a frequent problem therefore, in being able to adequately compare results from different research studies with regard to patterns of antibiotic resistance across these taxa, determine potential temporal and spatial changes in resistance characteristics as well as downstream characterization of resistance traits and mechanisms. In many instances, this deviation may be a different set of antibiotics used in commercially available antimicrobial testing platforms, the use of different MIC breakpoints or growth and culturing conditions. A key limitation in this regard is a lack of commercially available testing methods designed specifically for use on nonfastidious Gram-negative pathogenic bacteria such as vibrios. The availability of such testing platforms, in particular high-throughput microdilution tests would greatly improve the environmental and clinical surveillance data regarding antibiotic resistance in these pathogens, and is urgently required for research laboratory use.

However, there are also discrepancies between recommended antimicrobial testing guidelines used for these pathogens and clinical treatment options used in hospital settings. For example, doxycycline, a tetracycline derivative, is a CDC-recommended treatment drug for *V. vulnificus* infections (http://www.cdc.gov/vibrio/vibriov.html), yet no CLSI guidelines exist for this particular drug with regards to pathogenic vibrios. Significantly, resistance to doxycycline has been demonstrated previously in clinically isolated *V. vulnificus* strains

(Baker-Austin et al., 2008), highlighting the need for greater discourse and collaborative research efforts between environmental and clinical microbiologists as well as industry, government, and healthcare professionals involved in treating as well as assessing resistance in these pathogens.

There are several exciting methods currently undergoing development that hold the potential to quickly and reliably identify and characterize antibiotic-resistant foodborne bacteria. These approaches circumvent traditional culture-based testing methods and instead rely on rapid, high-throughput molecular platforms such as next-generation sequencing (NGS). Without question, such methods are particularly suited to analyzing highly invasive and progressive foodborne pathogens such as *V. vulnificus*, where identification of the etiological agent as well as possible resistance profile(s) may provide tangible benefits in clinical settings. Currently, conventional diagnostic testing for pathogens is narrow in scope and fails to detect the etiologic agent in a significant percentage of cases (Barnes et al., 1998). Coupled to this, conventional culture-based approaches to isolate and characterize a given pathogen, such as *V. vulnificus*, are time-consuming and will not provide antibiotic susceptibility data in a timeframe that can be used in a clinical context. Unbiased NGS approaches enable comprehensive pathogen detection in the clinical microbiology laboratory and have numerous applications for public health surveillance, outbreak investigation, and the diagnosis of infectious diseases (Naccache et al., 2014). Several methods that can provide data in a matter of hours, rather than days, are now a tangible reality. These methods are currently in their infancy, but a basic potential overview is presented here (Figure 6.2). An obvious advantage to this type

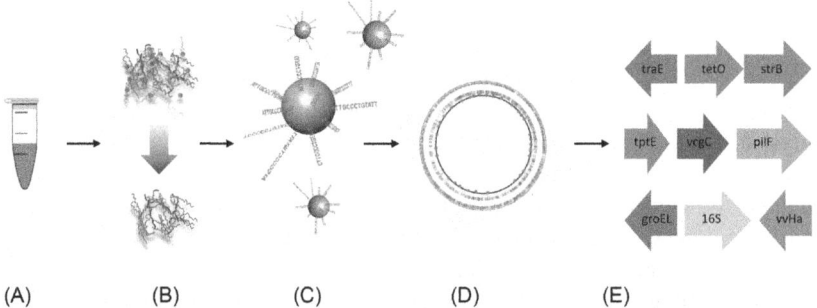

(A)   (B)   (C)   (D)   (E)

FIGURE 6.2 A clinical sample such as blood from an individual with a serious infection (e.g., primary septicemia) and where a systemic *Vibrio* infection is suspected is taken (A), and DNA extracted using a rapid automated extraction approach (B). Nucleic acid library construction is performed and DNA is analyzed using an appropriate rapid NGS platform (C), and genome data generated (D). Raw data reads are preprocessed by removal of adapter, low-quality, and low-complexity sequences, followed by computational subtraction of human reads. Data are scrutinized directly using an appropriate bioinformatic pipeline (e.g., SURPI, sequence-based ultrarapid pathogen identification) (Naccache et al., 2014), and genes of interest, such as virulence, antibiotic resistance, and species-specific markers are identified (E).

of methodology is the wealth of additional information that can be simultaneously generated, including potential resistance characteristics based on the presence of antimicrobial resistance genes, persistence factors, mobile genetic elements, virulence-associated genes and crucially, bacterial species-specific markers for unambiguous pathogen detection purposes. Such approaches are powerful, in that they are capable of identifying all potential pathogens in a single assay without a priori knowledge of the target. Given sufficiently long read lengths, multiple hits to the microbial genome, and a well-annotated reference database, nearly all microorganisms can be uniquely identified on the basis of their specific nucleic acid sequence (Naccache et al., 2014). Given that most of the major antibiotic resistance mechanisms as well as mobile genetic elements associated with resistance have been genomically and bioinformatically characterized, that allows rapid identification of gene hits against reference databases is now a tangible reality in point of care testing for important foodborne pathogens such as *V. vulnificus*. Another potential advantage of the technique is its capacity to identify co-infections, which is of great help to adapt therapeutics (Lecuit and Eliot, 2014).

CONCLUSIONS

To date, research in the field of antibiotic resistance with regards to the pathogens *V. vulnificus* and *V. parahaemolyticus* has been limited, fragmentary and most studies are unfortunately incomparable due to the use of non-standardized methodologies. There have been few large-scale studies, and most published reports are based on a small number of strains analyzed against a limited range of antimicrobial agents. As such, there is a need for additional systematic studies to develop a baseline of resistance data in these bacteria, where point of care can be better improved with the judicious use of appropriate and effective antimicrobials. Somewhat perversely, there has been more research work dedicated to studying antibiotic resistance in the pathogen *V. parahaemolyticus* than in *V. vulnificus*. Both of these bacteria represent important foodborne pathogens, but the striking severity, case fatality rate coupled to the rapid progression of infections associated with *V. vulnificus* highlight the need for increased research focus on this particular species in both environmental and clinical settings. Crucially, the identification of antibiotic resistance across a range of structurally diverse antimicrobials in these pathogens—and in particular to strains isolated from environments that are likely to have not undergone human-induced selection for antimicrobials—is somewhat troubling. Similar studies that have shown antibiotic resistance in environmental bacteria isolated from pristine habitats (D'Costa et al., 2006) suggest that antibiotic resistance is a complex biological phenomenon not restricted to clinical and community settings. Alongside traditional testing methods used to determine resistance traits in these pathogens, advances in molecular methods are helping to provide faster, more efficient and reliable treatment options. These new advances in testing methodologies, particularly utilizing high-throughput genomic analysis will not supersede but will augment the traditional approaches used for ascertaining antibiotic resistance in these important foodborne pathogens.

REFERENCES

Altekruse, S.F., Bishop, R.D., Baldy, L.M., et al., 2000. *Vibrio* gastroenteritis in the US Gulf of Mexico region: the role of raw oysters. Epidemiol. Infect. 124 (3), 489–495.

Austin, B., 2005. Bacteria pathogens of marine fish. In: Belkin, S., Colwell, R.R. (Eds.), Oceans and Health: Pathogens in the Marine Environment Kluwer, New York, NY, pp. 391–413.

Baker-Austin, C., McArthur, J.V., Tuckfield, R.C., Najarro, M., Lindell, A.H., et al., 2008. Antibiotic resistance in the shellfish pathogen *Vibrio parahaemolyticus* isolated from the coastal water and sediment of Georgia and South Carolina, USA. J. Food Prot. 71, 2552–2558.

Baker-Austin, C., Stockley, L., Rangdale, R., Martinez-Urtaza, J., 2010. Environmental occurrence and clinical impact of *Vibrio vulnificus* and *Vibrio parahaemolyticus*: a European perspective. Environ. Microbiol. Rep. 2, 7–18.

Baker-Austin, C., Trinanes, J., Hartnell, R., Taylor, N., Siitonen, A., Martinez-Urtaza, J., 2013. Emerging *Vibrio* risk at high-latitudes in response to ocean warming. Nat. Clim. Change 3, 73–77. Available from: http://dx.doi.org/10.1038/nclimate1628.

Barnes, G.L., Uren, E., Stevens, K.B., Bishop, R.F., 1998. Etiology of acute gastroenteritis in hospitalized children in Melbourne, Australia, from April 1980 to March 1993. J. Clin. Microbiol. 36, 133–138.

Bej, A.K., Patterson, D.P., Brasher, C.W., Vickery, M.C.L., Jones, D.D., Kaysner, C., 1999. Detection of total and hemolysin producing *Vibrio parahaemolyticus* in shellfish using multiplex PCR amplification of *tl*, *tdh*, and *trh*. J. Microbiol. Methods 36, 215–225.

Chen, S.C., Lee, Y.T., Tsai, S.J., Chan, K.S., Chao, W.N., Wang, P.H., et al., 2012. Antibiotic therapy for necrotizing fasciitis caused by *Vibrio vulnificus*: retrospective analysis of an 8 year period. J. Antimicrob. Chemother. 67, 488–493.

Chuang, Y.C., Yuan, C.Y., Liu, C.Y., Lan, C.K., Huang, A.H., 1992. *Vibrio vulnificus* infection in Taiwan: report of 28 cases and review of clinical manifestations and treatment. Clin. Infect. Dis. 15, 271–276.

CLSI, 2010. Methods for Antimicrobial Dilution and Disk Susceptibility Testing of Infrequently Isolated or Fastidious Bacteria; Approved Guideline-Second Edition (M45-A2). CLSI, Wayne, PA.

D'Costa, V., McGrann, K.M., Hughes, D.W., Wright, G.D., 2006. Sampling the antibiotic resistome. Science 311, 5759–5763.

Dalsgaard, A., Frimodt-Moller, N., Bruun, B., Hoi, L., Larsen, J.L., 1996. Clinical manifestations and molecular epidemiology of *Vibrio vulnificus* infections in Denmark. Eur. J. Clin. Microbiol. Infect. Dis. 15, 227–232.

Daniels, N.A., MacKinnon, L., Bishop, R., et al., 2000. *Vibrio parahaemolyticus* infections in the United States, 1973–1998. J. Infect. Dis. 181 (5), 1661–1666.

Dechet, A.M., Yu, P.A., Koram, N., Painter, J., 2008. Nonfoodborne *Vibrio* infections: an important cause of morbidity and mortality in the United States, 1997–2006. Clin. Infect. Dis. 46 (7), 970–976.

Elexson, N., Afsah-Hejri, L., Rukayadi, Y., Soopna, P., Lee, H.Y., Zainazor, T.C.T., et al., 2014. Effect of detergents as antibacterial agents on biofilm of antibiotics-resistant *Vibrio parahaemolyticus* isolates. Food Control 35 (1), 378–385.

Han, F., Walker, R.D., Janes, M.E., Prinyawiwatkul, W., Ge, B., 2007. Antimicrobial susceptibility of *Vibrio parahaemolyticus* and *Vibrio vulnificus* from Louisiana Gulf and Retail Raw Oysters. Appl. Environ. Microbiol. 73, 7096–7098.

Hlady, W.G., Klontz, K.C., 1996. The epidemiology of *Vibrio* infections in Florida, 1981–1993. J. Infect. Dis. 173, 1176–1183.

Huss, H.H., Ababouch, L., Gram, L., 2004. Assessment and Management of Seafood Quality and Safety. FAO Fisheries Technical Paper, No. 444. Food and Agriculture Organization of the United Nations, Rome, Italy, 240pp.

Ji, H., Chen, Y., Guo, Y., Liu, X., Wen, J., Liu, H., 2011. Occurrence and characteristics of *Vibrio vulnificus* in retail marine shrimp in China. Food Control 22, 1935–1940.

Jones, M.K., Oliver, J.D., 2009. *Vibrio vulnificus*: disease and pathogenesis. Infect. Immun. 77, 1723–1733.

Jun, J.W., Kim, J.H., Choresca Jr., C.H., Shin, S.P., Han, J.E., Han, S.Y., et al., 2012. Isolation, molecular characterization, and antibiotic susceptibility of *Vibrio parahaemolyticus* in Korean seafood. Foodborne Pathog. Dis. 9, 224–231.

Kim, J.H., Choresca Jr., C.H., Shin, S.P., Han, J.E., Jun, J.W., Park, S.C., 2011. Occurrence and antibiotic resistance of *Vibrio vulnificus* in seafood and environmental waters in Korea. J. Food Saf. 31, 518–524.

Klontz, K.C., Lieb, S., Schriber, M., Janoswski, H.T., Baldy, L.M., Gunn, R.A., 1988. Syndromes of *Vibrio vulnificus* infections. Clinical and epidemiologic features in Florida cases, 1981–1987. Ann. Intern. Med. 109, 318–323.

Lecuit, M., Eliot, M., 2014. The diagnosis of infectious diseases by whole genome next generation sequencing: a new era is opening. Front. Cell. Infect. Microbiol. 4, 25.

Martinez-Urtaza, J., Baker-Austin, C., Jones, J., Newton, A., Gonzalez-Aviles, G., DePaola, A., 2013. Spread of *Vibrio parahaemolyticus* outbreak clone. N. Engl. J. Med. 369 (16), 1573–1574.

Mead, P.S., Slutsker, L., Dietz, V., McGaig, L.F., Bresee, J.S., Shapiro, C., et al., 1999. Food-related illness and death in the United States. Emerg. Infect. Dis. 5, 607–625.

Morris Jr., J.G., Tenney, J., 1985. Antibiotic therapy for *Vibrio vulnificus* infection. J. Am. Med. Assoc. 253, 1121–1122.

Naccache, S.N., Federman, S., Veeraraghavan, N., Zaharia, M., Lee, D., Samayoa, E., et al., 2014. A cloud-compatible bioinformatics pipeline for ultrarapid pathogen identification from next-generation sequencing of clinical samples. Genome Res. 24 (7), 1180–1192.

Nair, G.B., Ramamurthy, T., Bhattacharya, S.K., Dutta, B., Takeda, Y., Sack, D.A., 2007. Global dissemination of *Vibrio parahaemolyticus* serotype O3:K6 and its serovariants. Clin. Microbiol. Rev. 20, 39–48.

Neupane, G.P., Kim, D.M., Kim, S.H., Lee, B.K., 2010. *In vitro* synergism of ciprofloxacin and cefotaxime against nalidixic acid-resistant *Salmonella enterica* serotypes Paratyphi A and Paratyphi B. Antimicrob. Agents Chemother. 54, 3696–3701.

Newton, A., Kendall, M., Vugia, D.J., Henao, O.L., Mahon, B.E., 2012. Increasing rates of vibriosis in the United States, 1996–2010: review of surveillance data from 2 systems. Clin. Infect. Dis. 54, S391–S395.

Nishibuchi, M., Fasano, A., Russell, R.G., Kaper, J.B., 1992. Enterotoxigenicity of *Vibrio parahaemolyticus* with and without genes encoding thermostable direct hemolysin. Infect. Immun. 60, 3539–3545.

Oliver, J.D., 2006. Vibrio vulnificus. In: Thompson, F.L., Austin, B., Swing, J. (Eds.), Biology of Vibrios Amer. Soc. Microbiol. Press, Washington, DC, pp. 349–366.

Oliver, J.D., Kaper, J., 2001. Vibrio species. In: Doyle M.P., et al. (Eds.), Food Microbiology: Fundamentals and Frontiers. pp. 263–300.

Pan, J.H., Zhang, Y.J., Jin, D.Z., Ding, G.Q., Luo, Y., Zhang, J.Y., et al., 2012. Molecular characterization and antibiotic susceptibility of *Vibrio vulnificus* in retail shrimps in Hangzhou, People's Republic of China. J. Food Prot. 76 (12), 2063–2068.

Powell, A., Baker-Austin, C., Wagley, S., Bayley, A., Hartnell, R., 2013. Pandemic *Vibrio parahaemolyticus* isolated from UK shellfish produce and water. Microb. Ecol. 65, 924–927.

Ralston, E.P., Kite-Powell, H., Beet, A., 2011. An estimate of the cost of acute health effects from food- and water-borne marine pathogens and toxins in the USA. J. Water Health 9, 680–694.

Reller, L.B., Weinstein, M., Jorgensen, J.H., Ferraro, M.J., 2009. Antimicrobial susceptibility testing: a review of general principles and contemporary practices. Clin. Infect. Dis. 49 (11), 1749–1755.

Rippey, S.R., 1994. Infectious diseases associated with molluscan shellfish consumption. Clin. Microbiol. Rev. 7, 419–425.

Sahilah, A.M., Laila, R.A.S., Sallehuddin, H.M., Osman, H., Aminah, A., Ahmad Azuhairi, A., 2014. Antibiotic resistance and molecular typing among cockle (*Anadara granosa*) strains of *Vibrio parahaemolyticus* by polymerase chain reaction (PCR)-based analysis. World J. Microbiol. Biotechnol. 30 (2), 649–659.

Scallan, E., Hoekstra, R.M., Angulo, F.J., Tauxe, R.V., Widdowson, M.A., et al., 2011. Foodborne illness acquired in the United States—major pathogens. Emerg. Infect. Dis. 17, 7–15.

Shaw, K.S., Rosenberg Goldstein, R.E., He, X., Jacobs, J.M., Crump, B.C., et al., 2014. Antimicrobial susceptibility of *Vibrio vulnificus* and *Vibrio parahaemolyticus* recovered from recreational and commercial areas of Chesapeake Bay and Maryland coastal bays. PLoS One 9 (2), e89616.

Tang, H.J., Chang, M.C., Ko, W.C., Huang, K.Y., Lee, C.L., Chuang, Y.C., 2002. *In vitro* and *in vivo* activities of newer fluoroquinolones against *Vibrio vulnificus*. Antimicrob. Agents Chemother. 46, 3580–3584.

Vezzulli, L., Colwell, R.R., Pruzzo, C., 2013. Ocean warming and spread of pathogenic vibrios in the aquatic environment. Microb. Ecol. 65 (4), 817–825.

Yamaichi, Y., Iida, T., Park, K.S., Yamamoto, K., Honda, T., 1999. Physical and genetic map of the genome of *Vibrio parahaemolyticus*: presence of two chromosomes in *Vibrio* species. Mol. Microbiol. 31 (5), 1513–1521.

Yano, Y., Hamano, K., Satomi, M., Tsutsui, I., Ban, M., Aue-umneoy, D., 2013. Prevalence and antimicrobial susceptibility of *Vibrio* species related to food safety isolated from shrimp cultured at inland ponds in Thailand. Food Control 38, 30–36.

Chapter 7

Antimicrobial Resistance in *Shigella* Species

Keith A. Lampel

Food and Drug Administration, Laurel, MD, USA

Chapter Outline

INTRODUCTION

Shigella spp. are members of the family Enterobacteriaceae and are considered a pathovar of *Escherichia coli*. These bacteria are Gram-negative rods, non-motile, primarily lactose negative (some *S. sonnei* are slow metabolizers), and other important biochemical characteristics include no production of H_2S, are unable to decarboxylate lysine, oxidase negative, and do not produce gas from glucose with the exception of *S. flexneri* 6 and *S. boydii* 14. There are four species of shigellae: *S. dysenteriae* (group A, 15 serotypes); *S. flexneri* (group B, 14 classical serotypes and subserotypes); *S. boydii* (group C, 19 serotypes); and *S. sonnei* (group D, 1 serotype). All *Shigella* species are considered to be pathogenic and cause shigellosis (bacillary dysentery), which is characterized by a myriad disease symptoms ranging from mild diarrhea to severe dysenteric syndrome, the latter may include abdominal pain, tenesmus, and bloody, mucoid stools. Symptoms appear 1–2 days after the ingestion of *Shigella* and usually resolve themselves within 5–7 days. Sequelae of patients with shigellosis can be manifested as hemolytic-uremic syndrome (HUS), reactive arthritis, seizures, and toxic megacolon. HUS occurs when Shiga toxins enter the bloodstream and damage vascular endothelial cells, particularly those in the kidney. HUS is characterized by hemolytic anemia, thrombocytopenia, and acute renal failure. Reactive arthritis is strongly associated with individuals of the HLA-B27 histocompatibility group (Simon et al., 1981). Mortality due to shigellosis is relatively low except in malnourished children, immunocompromised individuals, and the elderly.

Antimicrobial Resistance and Food Safety. DOI: http://dx.doi.org/10.1016/B978-0-12-801214-7.00007-7

In regard to the ability of *Shigella* spp. to cause HUS, historically only one serotype was associated with this sequelae, *S. dysenteriae* type 1. However, recent publications have noted that all strains of *Shigella* have been isolated that harbor the Shiga toxin gene, *stx* (Gupta et al., 2007; Gray et al., 2014), a genetic factor that enables *Shigella*, as well as some Shiga toxin-producing *E. coli* (STEC) strains, to cause HUS. As will be discussed below, the recent acquisition of the *stx1* gene by *Shigella* may present a parallel concern with other *stx*-carrying pathogens, notably the Shiga toxin-producing *E. coli* (STEC). Treatment of patients with infection by an STEC, such as *E. coli* O157:H7, becomes a complicated issue since treatment with an antibiotic may induce the *stx* gene, and therefore, produce the Shiga toxin which primarily attacks the kidney. As indicated above, children under the age of 5 are most vulnerable to the destructive action of this toxin.

The spread of *Shigella* is by the fecal–oral route with water, feces, food, fingers, flies, and fomites as the common transmission means. Humans are considered the only host for *Shigella* spp. although there have been some reports that higher primates have exhibited shigellosis. However, these occurrences have been associated with zoo environments and most likely due to close proximity of handlers that may have had the disease. Of importance with the spread of pathogens is that the infectious dose is low, ranging from 100–200 cells. Shigellosis, primarily by *S. flexneri*, is endemic in many developing countries and accounts for high rates of morbidity and mortality.

The World Health Organization (WHO) estimates that nearly 90–120 million people are infected annually with *Shigella* and most of the estimated 600,000 deaths occur within developing countries (2009 WHO http://www.who.int/vaccine_research/diseases/diarrhoeal/en/index6.html); 60% of the deaths reported occurred in children under the age of 5 with malnutrition as a potential contributing factor. Other impacted groups of people include the immunocompromised and the elderly, but most people are vulnerable to *Shigella* spp. Patients with severe shigellosis are subjected to an oral hydration regimen and antibiotics with ciprofloxacin recommended as the first-line drug of choice (Table 7.1); resistance to ampicillin, trimethoprim-sulfamethoxazole (TMP-SMX), and nalidixic acid by *Shigella* spp. has made these antibiotics noneffective for treatment (Table 7.2). Currently, there are no vaccines available for immunization although efforts continue to develop a *Shigella* vaccine.

ANTIBIOTIC RESISTANCE

Shigella spp., as with other bacterial pathogens, have acquired the genetic determinants to become resistant to multiple types of antibiotics. This phenomenon poses a serious concern for public health as the number of effective antibiotics against disease-causing agents decreases, particularly with the emergence of multiple drug resistance (MDR) strains. As with other bacterial pathogens, the occurrence of MDR strains is not localized to one specific geographical area but appears to be distributed worldwide.

TABLE 7.1 WHO Recommended Antibiotic Treatment for *Shigella* Infections

Antimicrobial	Treatment	
	Children	Adults
Ciprofloxacin[a]	15 mg/kg 2 times per day for 3 days, by mouth	500 mg 2 times per day for 3 days, by mouth
Pivmecillinam[b]	20 mg/kg 4 times per day for 5 days, by mouth	100 mg 4 times per day for 5 days, by mouth
Ceftriaxone[b]	50–100 mg/kg 50–100 mg/kg— once a day IM for 2–5 days	–
Azithromycin[b]	6–20 mg/kg once a day for 1–5 days, by mouth	1–1.5 g once a day for 1–5 days, by mouth

Source: From World Health Organization (2005).
[a]*First-line treatment.*
[b]*Second-line treatment.*

TABLE 7.2 Antibiotics Used to Treat *Shigella* Infections

Class	
Beta lactams	Ampicillin[a], amoxicillin[b], cephalosporins[b] (first and second generation), cefixime, ceftriaxone, pivmecillinam
Anti-metabolites	Tetracycline[a], TMP-SMX (cotrimoxazole)[a]
Quinolones	Nalidixic acid[a], ciprofloxacin, norfloxacin, ofloxacin
Aminoglycosides	Gentamicin[b], kanamycin[b]
Marcolides	Azithromycin
Others	Sulfonamides, furazolidone[b]

[a]*These antibiotics have been identified by the WHO as not effective against Shigella infections and are recommended not to be used.*
[b]*For effective treatment against Shigella spp. in patients, the antibiotic should be able to penetrate the intestinal mucosa; these antibiotics penetrate the intestinal mucosae poorly: nitrofurantoin, furazolidone, oral aminoglycosides (e.g., gentamicin, kanamycin), first- and second-generation cephalosporins (cefazolin, cephalotin, cefaclor, cefoxitin), and amoxicillin.*

Since shigellosis can be a self-limited disease, patients with mild cases can usually forgo treatment with antibiotics and expect full recovery. In the past, the empiric drug of choice to shorten the duration of the disease and reduce the shedding of the pathogen in stool was either ampicillin or TMP-SMX. However, if strains are known to be resistant to these antibiotics, alternative therapy may include fluoroquinolones, ceftriaxone, and azithromycin. At one brief point in time, nalidixic acid was given to patients with shigellosis but resistance to this drug developed rapidly in parts of Asia and other areas worldwide.

As with many bacterial species, *Shigella* acquire genetic loci encoding for antibiotic resistance through different mobile genetic elements, such as plasmids, transposons, and integrons (see Chapter 1; for general reviews, Balcazar, 2014; Gillings, 2014; Marti et al., 2014). As an example, genetic factors that confer quinolone resistance in *Shigella* spp. include mutations in the *gyrA* and *parC* genes, as well as *gnr*, *aac(6')-Ib-cr* and *qepA* genes, are carried on plasmids. These determinants are designated as plasmid-mediated quinolone resistance (PMQR). Another class of mobile genetic elements includes integrons. Typically, class 2 integrons are more commonly found in *Shigella* spp. than class 1 integrons. In addition to specific integrase genes, these genetic elements carry gene cassettes; typical for class 1 integrons *dfrA17-aadA5* and *dfrA12-orfF-aadA2* and for class 2 integrons *dfrA1*, *sat1*, and *aadA1*. The genes carried by integrons confer antibiotic resistance to gentamicin, kanamycin, streptomycin, tobramycin, sulfafurazole, trimethoprim, ampicillin, chloramphenicol, and tetracycline.

A possible complicating factor to the worldwide distribution of MDR *Shigella* spp. is the flux in the epidemiological landscape of this pathogen, i.e. the change of serogroup, and in some cases serotype, dominance in different countries and regions around the world. In the 1930s and 1940s, the dominant serogroup, *S. dysenteriae*, was being displaced by *S. flexneri*, which is commonly found today in developing countries. This continuum of serogroup change is highlighted by noting that just over a decade ago, the proportions of *S. flexneri*, *S. sonnei*, *S. boydii*, and *S. dysenteriae* were estimated to be 60% (predominantly serotype 2a, followed by 1b, 3a, 4a, and 6), 15%, 6%, and 6% (30% of *S. dysenteriae* cases were type 1), respectively. However, in industrialized countries, the ratio is 16% (predominantly *S. flexneri* serotype 3a, followed by 1b, 1c, 2a, and Y variant), 77%, 2%, and 1%, respectively. Therefore, the picture for *S. sonnei* is different with respect to *S. flexneri*, where it is responsible for nearly 75% of all shigellosis cases in these countries but only 15% in the developing countries. *Shigella boydii* is rarely found outside the Indian subcontinent. In addition to the change of serotype dominance within geographical regions around the world, the increase in antibiotic resistance in *Shigella* spp. has been observed for decades in certain regions, such as southeast Asia (Bennish et al., 1992; Dutta et al., 2001), Africa (Bogaerts et al., 1997), and South America (Suárez et al., 2000; Lima et al., 1995).

Part of the equation to shift serogroup dominance by *S. sonnei* in some countries, for example, Israel, Argentina, Thailand (Bangtrakulnonth et al., 2008), Iran (Ranjbar et al., 2008), and Vietnam (Vinh et al., 2009) may be due to higher socio-economic gains by the general population within each country. The consequence may have more than likely led to improved sanitation systems and better hygiene practices (von Seidlein et al., 2006).

Travelers may be an underappreciated means of dissemination of *Shigella* spp. as well as asymptomatic people. Within the former group, strains may carry antibiotic resistance markers not frequently seen in one country and currently are isolated from patients either as foreign visitors or returning residents. This phenomenon will be highlighted in regional coverage below.

As an example of the impact of travelers and the presence of MDR *Shigella* spp. in a "naïve population", one study noted that quinolone resistance in the United States is relatively rare as compared to Asian countries. However, an increase in fluoroquinolone-resistant *Shigella* strains has been reported in New York City (NYC), with one-half of these cases arising from travelers from Asia or people who may have come into contact with travelers from the Indian subcontinent (Wong et al., 2010). Furthermore, in a study conducted in NYC from 2006 to 2009, ciprofloxacin resistance was noted in six *Shigella* isolates (1% of the total) of which four were *S. sonnei* and two were *S. flexneri*. Other findings include that nalidixic acid resistance was found in 23 isolates (5%) of which 13 were *S. sonnei*, nine *S. flexneri* strains, and one *S. dysenteriae*. For MDR strains, the six ciprofloxacin-resistant isolates were also resistant to the quinolones nalidixic acid (6; 100%), ofloxacin (5; 83%), or levofloxacin (2; 33%). One person who had a ciprofloxacin-resistant strain had traveled outside of the United States. Of patients that had nalidixic-acid-resistant strains, 17 patients (77%) either spoke a language from the Indian subcontinent or traveled to China, India, or Vietnam. Of note, the first recorded ciprofloxacin-resistant *Shigella* strain (*S. flexneri*) was isolated from one 11-month-old male after travel to China (Wong et al., 2010).

With different patterns of antibiotic resistance among *Shigella* spp. being reported around the world, the choice of antimicrobial drugs to be dispensed has been affected by this phenomenon over the years. As noted by the WHO, resistance to ampicillin, co-trimoxazole, and nalidixic acid by *Shigella* spp. has increased significantly such that these antibiotics are no longer recommended. The WHO now recommends ciprofloxacin for all patients, including children, who present bloody diarrhea. Additional drugs of choice for all age patients were quinolones, specifically fluoroquinolones, as well as pivmecillinam (amdinocillin pivoxil) and ceftriaxone; azithromycin was recommended as an alternate antibiotic for adults. The drawbacks to these drugs include price, means of administration (injectables), and an unknown level of efficacy for some (see Tables 7.1 and 7.2; WHO, 2005). Reviews of the treatment of Shigella infections with antibiotics are available (http://www. thecochranelibrary.com/userfiles/ccoch/file/Water%20safety/CD006784.pdf and Klontz and Singh, 2015).

NORTH AMERICA

In the United States, the Centers for Disease Control and Prevention (CDC) monitors the rise of antibiotic resistance with *Shigella* spp. Similar to situations around the world, *Shigella* spp. have become commonly resistant to ampicillin, nalidixic acid, and TMP-SMX such that alternative drugs, for example, ciprofloxacin and azithromycin, are now recommended. For example, in the United States in 2008, the CDC received 552 *Shigella* isolates through its National Antimicrobial Resistance Monitoring System (NARMS), 89.9% were *S. sonnei* and 8.7% were *S. flexneri* strains. Among this collection, 62.5% were resistant to ampicillin, 41.1% to sulfamethoxazole, and 22.8% to both antibiotics (CDC, 2010). In addition, the CDC has reported that approximately 6% of all *Shigella* isolated in 2011 are resistant to both ampicillin and TMP-SMX (http://www.cdc.gov/drugresistance/threat-report-2013/pdf/ar-threats-2013-508.pdf#page=75).

In 2012, an outbreak that affected 43 people was caused by *S. sonnei* (MMWR 2013). This was an unusual outbreak in that it was the first documented case in the United States in which a strain was transmitted with a decreased susceptibility to azithromycin. In addition, four isolates were sent to the CDC, of which two were isolated from asymptomatic people, one of whom was a food handler. The importance of this outbreak is therefore two-fold; first the identification of an asymptomatic carrier and second the identification of a *Shigella* strain with decreased susceptibility to azithromycin, potentially limiting the options to treat MDR *Shigella* strains (Karlsson et al., 2013).

The Foodborne Diseases Active Surveillance Network (FoodNet) is a collaborative effort between the CDC, the USDA (Food Safety and Inspection Service), the US FDA, and 10 public US state laboratories that are responsible for tracking diseases caused by several foodborne bacterial microbes. NARMS laboratories found that 1,118 (out of a total of 1,376; 81%) *Shigella* isolates from the 10 state laboratories during 2000–2010 were resistant to the following antibiotics: 826 (74%) were resistant to ampicillin, 649 (58%) to streptomycin, 402 (36%) to TMP-SMX, 355 (32%) to sulfamethoxazole-sulfisoxazole, 312 (28%) to tetracycline, 19 (2%) to nalidixic acid, and 6 (0.5%) to ciprofloxacin. Of the isolates resistant to TMP-SMX, 75% were related to people who indicated that they had traveled internationally (Shiferaw et al., 2012).

In a report by Folster et al. (2011), 20 *Shigella* isolates (ten *S. flexneri*, six *S. sonnei*, three *S. boydii*, and one *S. dysenteriae*) were characterized as to the molecular mechanism of decreased susceptibility to ciprofloxacin. Of the 20 strains isolated by the NARMS laboratories, all had shown an MIC of $\geq 0.25 \mu g/mL$ to ciprofloxacin and 15 isolates (75%) had an MIC of $\geq 32 \mu g/mL$ to nalidixic acid. Particular attention was given to the travel history of these patients who could provide such information; eight of those ten individuals indicated that they had traveled to India or Thailand. Several studies had demonstrated that *Shigella* spp. worldwide, but particularly in Asia, have developed resistance to fluoroquinolones, including ciprofloxacin (Dutta et al., 2003; Mensa et al., 2008; Pu et al., 2009). As Folster et al. (2011) suggested, *Shigella* isolates that are resistant to ciprofloxacin in the United States are associated with travelers returning from Asian countries. In addition to these general findings, Folster et al. (2011) also reported on the genetic basis for the quinolone resistance in the 20 isolates. Those strains that acquired mutations in the chromosomal quinolone resistance-determining regions (QRDR), that is, DNA gyrase (*gyrA*) and topoisomerase IV (*parC*) genes and/or in plasmid (plasmid-mediated quinolone resistance (PMQR) determinants) associated genes, target protection (*qnr*), aminoglycoside acetyltransferase variants [*aac(6')-Ib-cr*], or efflux (*qepA*) genes showed increased resistance to the quinolones nalidixic acid and ciprofloxacin. The transfer of plasmid-mediated resistance to these quinolones may account for some of the observations of increased resistance of *Shigella* spp. as well as other members of the Enterobacteriaceae.

SOUTH AMERICA

Early studies in South America of antibiotic-resistant *Shigella* spp. indicated a concern of the number of isolates that were refractory to antibiotic treatment. In one report in 2005 (Fullá et al., 2005), 178 *Shigella* strains isolated from Chilean children under the age of 5 with diarrhea from a town outside Santiago were analyzed for MIC of several antibiotics. Of these strains, resistance was noted for ampicillin (82%), TMP-SMX (65%), tetracycline (53%), and chloramphenicol (49%). In addition, they reported that 51% were MDR strains. At this time, no *Shigella* isolates were resistant to ciprofloxacin, nalidixic acid, or cefotaxime.

A recent publication from Brazil reported on the analysis of 1,339 children with diarrheal disease for pathogen identification and antibiotic resistance patterns (da Cruz et al., 2014). Of the 30 isolates collected from this population, *S. flexneri* was predominant and accounted for 60%, followed by 22.2% for *S. sonnei*, and *S. boydii* and *S. dysenteriae* were 6.6% each, which is the general trend in Brazil. The antimicrobial profile of this collection of strains showed that 80% (24/30) were resistant to tetracycline, 40% (12/30) to ampicillin, 30% (9/30) to chloramphenicol and the same number to gentamicin; 13% (4/30) were multiple drug resistant to amkiacin and clavulanic acid. The number of strains susceptible to ciprofloxacin and ceftriaxone was 3% (1/30) whereas all isolates were sensitive to naladixic acid and kanamycin (da Cruz et al., 2014). Some of these findings are in contrast to earlier reports (Bastos and Loureiro, 2011). The 2011 report had characterized 122 *Shigella* spp. (81 *S. flexneri*, 40 *S. sonnei*, one *S. boydii*) from patients that had diarrhea in a district in northern Brazil. They reported that most were resistant to tetracycline (93.8%), chloramphenicol (63.9%), and TMP-SMX (63.1%) but none of the strains were resistant to cefotaxime, ceftazidime, ciprofloxacin, nalidixic acid, and nitrofurantoin.

In Uruguay, between June 2001 and January 2008, 48 out of 249 children with bloody diarrhea were infected with *Shigella*. Of these *Shigella* isolates, 34 (70%) were resistant to ampicillin and 13 (27%) to TMP-SMX. All *Shigella* isolates tested were susceptible to ciprofloxacin, nalidixic acid, and ceftriaxone. As reflected in most regions around the world, treatment of patients with shigellosis with TMP-SMX or ampicillin would not be appropriate (Mota et al., 2010).

Shigella spp. (414 isolates) were recovered from stool samples collected from people with diarrheal diseases from a rural locale in Peru in one study.

Most of the *Shigella* isolates were identified as *S. flexneri* (278/414; 67.1%), followed by *S. sonnei* (49/414; 11.8%), *S. boydii* (47/414; 11.4%), and *S. dysenteriae* (10/414; 2.4%); there were some untypable strains that accounted for 7.2% (30/414). Interestingly, 103 of the 414 (nearly 25%) isolates were from asymptomatic people. Antibiotic resistance was determined for 403 isolates (403/414; 97.3%) and most were found to be resistant to tetracycline (83%), ampicillin (73%), TMP-SMX (79%), erythromycin (69%), and chloroamphenicol (62%). However, *Shigella* spp. were found to be sensitive to ceftriaxone (97%), azithromycin (84% sensitive, 11% intermediate, and 5% resistant), nalidixic acid (95%, 3% intermediate, 2% resistant), and ciprofloxacin (97% sensitive, 3% intermediate, 1% resistant). One of the public health concerns for regions of the world, such as this site in Peru, is that normal treatment of children with dysentery would have included some of the antibiotics listed above, and most likely would have been ineffective. Alternative drugs, such as ciprofloxacin, azithromycin, ceftriaxone, and nalidixic acid may be reasonable therapy alternatives, but these antimicrobials are not readily available in this region (Kosek et al., 2008).

Lastly, since it is well recognized that the emergence of MDR *Shigella* spp. can have a profound effect on the treatment of disease, some countries have created and used electronic databases to track antibiotic resistance by this pathogen (Stelling et al., 2010). The application of WHONET and SaTScan, an electronic laboratory-based disease surveillance system that integrated statistical cluster detection methods can be used to identify disease outbreaks and perhaps provide a faster and appropriate antimicrobial therapy response.

EUROPE

In a review by Gu et al. (2012), emergence of *Shigella* strains resistant to nalidixic acid and ciprofloxacin in Asia, Africa, Europe, and the United States were reported. Comparisons were made between MDR *S. flexneri* and *S. sonnei* as well as between continents, Asia–Africa and Europe–America, in the years 1998–2009. The resistant rates reported for each region reflect a general trend; in Africa and Asia, higher rates to nalidixic acid (33.6%) and ciprofloxacin (5.0%) were recorded which were 10.5 and 16.7 times higher, respectively, for Europe–America. Also noted was that the resistance to these antibiotics increased annually in the Africa–Asia regions, whereas in the same time period, the trend in Europe–America was much lower. Interestingly, there were stark differences in antibiotic resistance by *S. flexneri* and *S. sonnei*; in Europe–America, *S. flexneri* had higher resistance to nalidixic acid and ciprofloxacin than *S. sonnei*. However, in Africa–Asia, the same trend was noted for ciprofloxacin but *S. sonnei* had higher resistance to nalidixic acid than *S. flexneri*. Their conclusion, which reflects a general pattern, is that resistance to quinolones by *Shigella* spp. is on the rise at an alarming rate.

During a similar period of time, 1990–2007, antimicrobial resistance by *S. sonnei* was monitored in Belgium (Vrints et al., 2009). A total of 3,186 strains were analyzed and data showed that resistance to tetracycline (59.3%), strepto-mycin (77.2%), trimethoprim (75.2%), sulfonamides (71.6%), and TMP-SMX (85.9%; only tested from 2000 to 2007) increased significantly during this time frame. Of particular concern was that 12.8% of these strains were resistant to nalidixic acid in 2004 and the increase in resistance to TMP-SMX, as the latter was used clinically to treat patients since 1970. The increase in the number of strains that were MDR was from 50% in 1990 to the latest reported number, 82%, in 2007. However, all strains were sensitive to ciprofloxacin and gentamicin.

As many Europeans and other nationalities travel to and from their respec-tive homelands, more reports are published that connect the emergence of resistance to antimicrobials by *Shigella* spp. to global travel. One study con-ducted in Finland (Haukka and Siitonen, 2008) found 1,814 *Shigella* strains (all four serogroups were included) of which 88% (1,592) were associated with foreign travel. Of these, the highest number of MDR strains originated from China and India but the authors also noted that travelers visiting other regions in Asia also contributed to the overall number. It was also noted that the first highly ciprofloxacin-resistant strain, *S. flexneri*, was isolated in 2004. Nalidixic acid resistance was found in both *S. flexneri* and *S. sonnei* strains where all of the former species had reduced susceptibility to ciprofloxacin but only 77% of the latter showed the same phenotype. Other Finnish travelers visited Egypt, Turkey, India, Tunisia, and Estonia and the pattern of antimicrobial resistance of isolates from these patients reflected the same resistance profiles found in the country of origin. These would include ampicillin, chloramphenicol, strepto-mycin, sulfonamide, tetracycline, and trimethoprim, whereas strains from Asia also were resistant to mecillinam, nalidixic acid, ciprofloxacin, gentamicin, and cefotaxime. As for the latter region, the number of strains resistant to nalidixic acid increased from 0% to 6% between 1990 and 1997, and subsequently, addi-tional increases were observed from 2003 to 2005.

Other European countries have experienced an increasingly significant num-bers of *Shigella* strains that were primarily due to travel or foods. As an exam-ple, outbreaks that occurred concurrently in Denmark and Australia were due to contaminated baby corn or sugar snaps with *S. sonnei* as the etiological agent. Both countries reported identical pulsed-field gel electrophoresis patterns as well as that all strains were resistant to tetracycline, ampicillin, sulfonamides, cephalothin, and streptomycin, but were found to be susceptible to nalidixic acid, ciprofloxacin, chloramphenicol, mecillinam, and gentamicin (Lewis et al., 2009).

In the Czech Republic, one patient who had traveled to China, Nepal, and India was ill with diarrhea. *Shigella sonnei* isolate displayed extended-spectrum β-lactamases (ESBL) which included resistance to cefotaxime and ceftazidime (Hrabák et al., 2008). In England and Wales, concern was raised in 2002 of the increase in all *Shigella* spp. that were resistant to nalidixic acid and, concur-rently, the decrease in susceptibility to ciprofloxacin. In earlier studies from 1978–1983, the resistance to nalidixic acid was reported as 0.5–0.6% for all

Shigella spp. tested, whereas in 2002, the increase in *Shigella* spp. resistant to nalidixic acid ranged from 10% to 13% (Cheasty et al., 2004).

Several other European studies focused on the integron analysis to monitor antibiotic resistance. A study published in 2003 (DeLappe et al., 2003) showed that several *S. sonnei* isolates carried a class 2 integron, associated with streptomycin, sulfonamides, tetracycline, and trimethoprim and one of these isolates also had a class 1 integron. A similar study was conducted in Hungary from 1998 to 2008 (Nógrády et al., 2013) with *S. flexneri* and *S. sonnei* isolates. Their conclusion showed that 67% (168/252) of the isolates were resistant to TMP-SMX, 47% displayed resistance to streptomycin, 32% to ampicillin, and 28% to tetracycline. It was also reported that 36% of the isolates were MDR, and that 33% of the *S. sonnei* (75) and 14% of the *S. flexneri* (3) strains carried either class 1 or class 2 drug resistance integrons or both.

AFRICA

In a 5-year study from Ethiopia, 214 *Shigella* isolates were tested for antibiotic resistance to gentamicin (7.9%), ciprofloxacin (8.9%), chloramphenicol (52.8%), TMP-SMX (73.4%), ampicillin (79.9%), and tetracycline (86%) (Yismaw et al., 2008). These data parallel results of another report from Ethiopia (Debas et al., 2011) in which 32 *Shigella* isolates analyzed over a period of 5 months in 2009 were all sensitive to ciprofloxacin. Furthermore, 90.6% and 75% of these 32 isolates were also sensitive to norfloxacin and gentamicin, respectively. In Mozambique, of 109 *Shigella* isolates (86% *S. flexneri* and 14% *S. sonnei*) that were isolated from 2001 to 2003, 56% were resistant to ampicillin, 84% to TMP-SMX, and 25% to both antibiotics (Mandomando et al., 2009). Of note, as in other regions around the world, narrow-spectrum quinolones have been used in treatment of shigellosis in the last few decades, but resistance to nalidixic acid has become more frequent in Africa. As an example of potential devastating public health effects due to MDR, most of the *S. dysenteriae* type I responsible for epidemics between 1993 and 1995 in refugee camps in Rwanda, Tanzania, and the Democratic Republic of the Congo were resistant to all the commonly used antibiotics including nalidixic acid (Kerneis et al., 2009).

Over a 2½-year period (2004–2006) in Senegal, 165 *Shigella* isolates were tested for resistance to sulfonamides, TMP-SMX, streptomycin, and tetracycline; resistance rates were 90%, 90%, 96%, and 94%, respectively. Most of the *Shigella* spp. were identified as *S. flexneri* (49%) and *S. sonnei* (45%), and of these, greater than half of the former strains were resistant to amoxicillin (59%), amoxicillin-clavulanic acid (58%), and chloramphenicol (52%), whereas *S. sonnei* isolates showed very different resistance rates to these antibiotics, 4%, 1%, and 4%, respectively. Only one *S. sonnei* was resistant to nalidixic acid and showed a reduced susceptibility to ciprofloxacin (Sire et al., 2008).

Most antibiotic resistance studies reported on *Shigella* spp. were isolated from stool samples. In some cases, *Shigella* spp. were isolated from food samples and assessed for antibiotic resistance. Guchi and Ashenafi (2010) found *Shigella* spp. in lettuce and green peppers from Nigeria. Their focus was on MDR strains and they found that of the 24 isolates tested, six isolates were resistant to three antibiotics, four to four antibiotics, seven to five antibiotics, six to six antibiotics, and three to seven antibiotics, the latter were the sole isolates that demonstrated resistance to ciprofloxacin.

Although *S. flexneri* is the dominant *Shigella* spp. isolated in Egypt, El-Gendy et al. (2012) investigated the antimicrobial profile of 40 *S. dysenteriae* and 30 *S. boydii* isolates with *S. dysenteriae*, the second most common isolate in three of the four studies from this group of children less than 5 years of age from 1998 to 2007. Resistances to tetracycline, ampicillin, chloramphenicol, sulfatrimethoxazole, cephalothin, and nalidixic acid by *S. dysenteriae* were 58%, 52%, 42%, 25%, 12%, and 5%, respectively, whereas for *S. boydii* they were 77%, 60%, 17%, 63%, 40%; all strains were susceptible to nalidixic acid. Half of the *S. dysenteriae* and *S. boydii* strains were MDR, ranging from three to six antibiotics; none were resistant to ciprofloxacin.

Other studies emanating from Africa parallel the data of the aforementioned examples. For example, in Nairobi, Kenya, 224 *Shigella* spp. isolated from stool samples showed similar results to those reported from Egypt; *Shigella flexneri* (64%) was the most common etiological agent followed by *S. dysenteriae* (11%), *S. sonnei* (9%), and *S. boydii* (5%) (Njuguna et al., 2013). From their data, more than 90% of all *Shigella* isolates were resistant to TMP-SMX and sulfisoxazole with a marked decrease in resistance to nalidixic acid (3%), ciprofloxacin (1%), and ceftriaxone (1%).

ASIA

Resistance of *Shigella* spp. to multiple antibiotics in Asian countries is similar to other regions around the world. Of concern to developing countries, the use of inexpensive antibiotics, such as ampicillin, chloramphenicol, TMP-SMX, and tetracycline has become ineffective to treat *Shigella* infections due to resistance to these antimicrobial agents. For a time, nalidixic acid was an effective treatment but resistance to this drug is now so prominent that other fluoroquinolone antibiotics, such as ciprofloxacin, have replaced nalidixic acid as a first-line therapy. But, as noted in countries in previous sections, resistance to ciprofloxacin is increasing, notably in Asian countries, such as India, where a 48% increase has been reported between 2002 and 2007 (Srinivasa et al., 2009). And as noted above in some countries, travelers to India have unintentionally imported *Shigella* strains resistant to ciprofloxacin (Haukka and Siitonen, 2008).

The emergence of fluoroquinolone-resistant *Shigella* strains in India has been recorded for over a decade. Two common serotypes in developing countries, *S. dysenteriae* type 1 and *S. flexneri* type 2a, were isolated that were not only resistant to fluoroquinolones but other antibiotics, such as ampicillin, tetracycline, streptomycin, and chloramphenicol but sensitive to azithromycin and ceftriaxone (Pazhani et al., 2008). Recently, ceftriaxone-resistant isolates of *S. flexneri* have now emerged in India (Gupta et al., 2010). In addition to this phenotype, this isolate was also resistant to ciprofloxacin; these two antibiotics are the drugs of choice to treat MDR *Shigella* strains. However, this strain was susceptible to ceftazidime. In other studies from India, similar patterns were observed. Eighty-four *Shigella* isolates, which included all four serotypes, were resistant to ampicillin (79%), nalidixic acid (51%), and ciprofloxacin (50%) and two of these strains had resistance to ciprofloxacin (Mandal et al., 2012). Another report (Bhattacharya et al., 2012), analyzed 88 *Shigella* strains from pediatric patients. They reported that all of the *Shigella* strains tested were resistant to ampicillin, and additional resistance to nalidixic acid (96.6%), tetracycline (92%), norfloxacin (81.8%), TMP-SMX (79.5%), ciprofloxacin (76.1%), and ofloxacin (71.6%) was found. In addition, resistance to the four third-generation cephalosporins (cefixime, ceftriaxone, cefotaxime, and ceftazidime) tested was detected in 17% of the *Shigella* strains. The scope of antibiotic resistance pattern changes in India was studied in two islands off the coast, Andaman and Nicobar Islands, from 2000 to 2011 (Bhattacharya et al., 2014). Two study periods were conducted, from 2000 to 2005 and from 2006 to 2011. Essentially, the authors showed that the frequency of antibiotic resistance in *Shigella* spp. increased significantly from the earlier study period to the latter with 20 of the 22 antibiotics tested.

Recent reports from other Asian countries, for example, China, Taiwan, Bangladesh, Iran, Japan, and Turkey have observed similar patterns of MDR *Shigella* strains. In Henan province in China, more than 99% of the *Shigella* spp. isolated from 2001 to 2008 were resistant to tetracycline, nalidixic acid, and pipemidic acid, and greater than 80% were found to be resistant to chloramphenicol, amoxicillin, and TMP-SMX and less than 5% were resistant to furazolidone, cefotaxime, and gentamicin. In addition, *S. flexneri* and *S. sonnei* were resistant to amoxicillin, ampicillin, chloramphenicol, ciprofloxacin, and TMP-SMX (Yang et al., 2013). Mehata and Duan (2011) found in 100 *Shigella* spp. that all (100%) were resistant to nalidixic acid and piperacilline, 85% to amoxicilline-clavum, 25% to ciprofloxacin, and 1.1% to polymixin B.

One hundred and ninety-five *S. sonnei* isolates associated with travelers to Japan were analyzed and a "core drug-resistance pattern" composed of streptomycin, sulfisoxazole, tetracycline, and TMP-SMX was found in 108 of the isolates. Interestingly, two isolates of travelers from China and India had resistance to cefotaxime and nalidixic acid (Izumiya et al., 2009). In neighboring Bangladesh, 3,789 *Shigella* spp. were collected and tested for fluorquinolone resistance. From 2004 to 2010, the increase in ciprofloxacin resistance was significant, from 0% to 44% in this 7-year period of time. Of the 2,181 *S. flexneri* isolates, which were the predominant species, 14.3% were found to be resistant to ciprofloxaxin (Azmi et al., 2014). In their analyses of these strains, a PFGE pattern matched isolates from Bangladesh with strains from China, suggesting a link between the two countries in regard to transmission of *Shigella* spp.

During a 7-year period from 2003 to 2009 in Turkey, it was found that of the 238 clinical *Shigella* spp. analyzed, 69.9% were resistant to TMP-SMX, 35.8% to ampicillin, and 4.7% to nalidixic acid; no isolate was found to be resistant to ciprofloxacin. In this same study, the number of MDR strains identified did not increase significantly (24.0% vs. 28.1%) between the current timeframe (2003–2009) and a previous study period between 1987 and 1994 (Özmert et al., 2011). Although other countries have experienced more of an increase in MDR *Shigella* strains than reported in this study, the concern still remains that nearly 30% of *Shigella* spp. in Turkey are MDR.

FINAL THOUGHTS

The rapid emergence and spread of MDR *Shigella* spp. worldwide has altered the treatment regimen of patients with shigellosis. In the past, first-line antibiotics, such as quinolones, nalidixic acid, TMP-SMX, and ampicillin were routinely administered as an effective means to target *Shigella* spp., thereby reducing the risk of further complications and duration of the disease. However, many of these first-line drugs have become ineffective and are no longer recommended for therapy (Tables 7.1 and 7.2). Due to the increase in the prevalence of MDR *Shigella* spp., the current choice of first-line antibiotic treatment is ciprofloxacin (for all age groups, including children); second-line antibiotics are pivmecillinam (where available), ceftriaxone, or azithromycin. For many developing countries, treatment of patients with infections by *Shigella* has become more complicated, not only because of the number of antibiotics presently ineffective, but also the cost and availability of the recommended drugs.

As noted in many countries, there has been a recorded rise in the number of ciprofloxacin-resistant *Shigella* spp. Since this antibiotic is the recommended first-line drug to treat shigellosis, a major public health concern is that it may become noneffective to counter new *Shigella* infections. Undoubtedly, this would leave healthcare providers with relatively limited choices for drug therapy, perhaps limited to only broad-spectrum antibiotics such as cephalosporins (e.g., ceftriaxone). However, the cost of such antibiotics may be too prohibitive in many developing countries and reduce the likelihood that they would be administered during treatment.

The United States recently announced (2014) an initiative to address the increase in antibiotic-resistant human pathogens and provide a road-map against the emergence and spread of these microbes (www.whitehouse.gov/sites/default/files/docs/carb_national_strategy.pdf). Although there are a number of measures to mitigate the rise in MDR pathogens, a common strategy emphasizes the judicious use of antibiotics, such as gentamicin and ciprofloxacin. The over-use and misuse of antibiotics has been an underlying basis for the emergence of MDR pathogens, including *Shigella*. An update of scientific opinions on the current concerns with antimicrobial resistance in select bacterial pathogens, such as *Shigella*, was made available by the WHO in 2014

(WHO, Antimicrobial resistance: Global Report on Surveillance. 2014 Summary. www.who.int/drugresistance/documents/surveillancereport/en/).

One of the striking causes of the dissemination of MDR strains of *Shigella* is attributed to the international traveler. As noted in several studies, there is a very strong connection between the emergence of unusual antibiotic resistance patterns that are not endemic to that country, and patients that have traveled to certain regions of the world, notably from countries in east and south Asia. Since a number of countries around the world have experienced an influx of MDR *Shigella* from international travelers, the spread of MDR strains within the "naïve" country will likely impact the therapeutic regimen to treat patients. This phenomenon places an additional burden on healthcare providers to identify the appropriate antibiotic treatment. In addition, the patient is at higher risk for a longer recovery period and perhaps further sequelae from the *Shigella* disease.

The extent of the spread of MDR pathogens has been noted by the WHO and countries around the world. The importance of controlling and preventing the spread of these resistant bacteria, and at the same time providing effective care, has gained international recognition. To mitigate the presence of MDR pathogens, scientists are focusing on the development of novel therapies and antibiotics, and at the same time efforts are being made to reduce the generally indiscriminate use of antibiotics. These objectives certainly apply to people infected with MDR *Shigella* spp., many of whom live in developing countries that encounter unacceptable levels of morbidity and mortality.

REFERENCES

Azmi, I.J., Khajanchi, B.K., Akter, F., Hasan, T.N., Shahnaij, M., Akter, M., et al., 2014. Fluoroquinolone resistance mechanisms of *Shigella flexneri* isolated in Bangladesh. PLoS One 9 (7), e102533.

Balcazar, J.L., 2014. Bacteriophages as vehicles for antibiotic resistance genes in the environment. PLoS Pathog. 10, 1–4.

Bangtrakulnonth, A., Vieira, A.R., Wong, D.M.A.L., Pornreongwong, S., Pulsrikarn, C., Sawanpanyalert, P., et al., 2008. *Shigella* from humans in Thailand during 1993 to 2006: spatial-time trends in species and serotype distribution. Foodborne Pathog. Dis. 5, 773–784.

Bastos, F.C., Loureiro, E.C., 2011. Antimicrobial resistance of *Shigella* spp. isolated in the State of Pará, Brazil. Rev. Soc. Bras. Med. Trop. 44, 607–610.

Bennish, M.L., Salam, M.A., Hossein, M.A., Myaux, J., Khan, E.H., Chakraborty, J., et al., 1992. Antimicrobial resistance among *Shigella* isolates in Bangladesh, 1983–1990: increasing frequency of strains multiply resistant to ampicillin, trimethoprim-sulfamethoxazole and nalidixic acid. Clin. Infect. Dis. 14, 1055–1060.

Bhattacharya, D., Bhattacharya, H., Thamizhmani, R., Sayi, D.S., Reesu, R., Anwesh, M., et al., 2012. Shigellosis in Bay of Bengal Islands, India: clinical and seasonal patterns, surveillance of antibiotic susceptibility patterns, and molecular characterization of multidrug-resistant *Shigella* strains isolated during a 6-year period from 2006 to 2011. Eur. J. Clin. Microbiol. Infect. Dis. 33, 157–170.

Bhattacharya, D., Bhattacharya, H., Sayi, D.S., Bharadwaj, A.P., Singhania, M., Sugunan, A.P., et al., 2014. Changing patterns and widening of antibiotic resistance in *Shigella* spp. over a decade (2000–2011), Andaman Islands, India. Epidemiol. Infect. 24, 1–8.

Bogaerts, J., Verhaegen, J., Munyabikali, J.P., Mukantabana, B., Lemmens, P., Vandeven, J., et al., 1997. Antimicrobial resistance and serotypes of *Shigella* isolates in Kigali, Rwanda (1983 to 1993): increasing frequency of multiple resistance. Diagn. Microbiol. Infect. Dis. 28, 165–171.

Centers for Disease Control and Prevention, 2010. National Antimicrobial Resistance Monitoring System for enteric bacteria (NARMS): Human Isolates Final Report, 2008. U.S. Department of Health and Human Services, Atlanta, GA.

Cheasty, T., Day, M., Threlfall, E.J., 2004. Increasing incidence of resistance to nalidixic acid in Shigellas from humans in England and Wales: implications for therapy. Clin. Microbiol. Infect. 10, 1033–1035.

da Cruz, C.B., de Souza, M.C., Serra, P.T., Santos, I., Balieiro, A., Pieri, F.A., et al., 2014. Virulence factors associated with pediatric shigellosis in Brazilian Amazon. Biomed. Res. Int. (Article ID 539697).

Debas, G., Kibret, M., Biadglegne, F., Abera, B., 2011. Prevalence and antimicrobial susceptibility patterns of shigella species at Felege Hiwot Referral Hospital, Northwest Ethiopia. Ethiop. Med. J. 49, 249–256.

DeLappe, N., O'Halloran, F., Fanning, S., Corbett-Feeney, G., Cheasty, T., Cormican, M., 2003. Antimicrobial resistance and genetic diversity of *Shigella sonnei* isolates from western Ireland, an area of low incidence of infection. J. Clin. Microbiol. 41, 1919–1924.

Dutta, S., Chatterjee, A., Dutta, P., Rajendran, K., Roy, S., Pramanik, K.C., et al., 2001. Sensitivity and performance characteristics of a direct PCR with stool samples in comparison to conventional techniques for diagnosis of *Shigella* and enteroinvasive *Escherichia coli* infection in children with acute diarrhoea in Calcutta, India. J. Med. Microbiol. 50, 667–674.

Dutta, S., Dutta, P., Matsushita, S., Bhattacharya, S.K., Yoshida, S., 2003. *Shigella dysenteriae* serotype 1, Kolkata, India. Emerg. Infect. Dis. 9, 1471–1474.

El-Gendy, A.M., Mansour, A., Weiner, M.A., Pimentel, G., Armstrong, Q.W., Young, S., et al., 2012. Genetic diversity and antibiotic resistance in *Shigella dysenteriae* and *Shigella boydii* strains isolated from children aged <5 years in Egypt. Epidemiol. Infect. 140, 299–310.

Folster, J.P., Pecic, G., Bowen, A., Rickert, R., Carattoli, A., Whichard, J.M., 2011. Decreased susceptibility to ciprofloxacin among *Shigella* isolates in the United States, 2006 to 2009. Antimicrob. Agents Chemother. 55, 1758–1760.

Fullá, N., Prado, V., Durán, C., Lagos, R., Levine, M.M., 2005. Surveillance for antimicrobial resistance profiles among *Shigella* species isolated from a semirural community in the northern administrative area of Santiago, Chile. Am. J. Trop. Med. Hyg. 72, 851–854.

Gillings, M.R., 2014. Integrons: past, present, and future. Microbiol. Mol. Biol. Rev. 78, 257–277.

Gray, M.D., Lampel, K.A., Strockbine, N.A., Fernandez, R.E., Melton-Celsa, A.R., Maurelli, A.T., 2014. Clinical isolates of Shiga toxin 1a–producing *Shigella flexneri* with an epidemiological link to recent travel to Hispañiola. Emerg. Infect. Dis. 20, 1669–1677.

Gu, B., Cao, Y., Pan, S., Zhuang, L., Yu, R., Peng, Z., et al., 2012. Comparison of the prevalence and changing resistance to nalidixic acid and ciprofloxacin of *Shigella* between Europe-America and Asia-Africa from 1998 to 2009. Int. J. Antimicrob. Agents 40, 9–17.

Guchi, B., Ashenafi, M., 2010. Microbial load, prevalence and antibiograms of *Salmonella* and *Shigella* in lettuce and green peppers. Ethiop. J. Health Sci. 20, 41–48.

Gupta, S., Mishra, B., Muralidharan, S., Srinivasa, H., 2010. Ceftriaxone resistant *Shigella flexneri*, an emerging problem. Indian J. Med. Sci. 64, 553–556.

Gupta, S.K., Strockbine, N., Omondi, M., Hise, K., Fair, M.A., Mintz, E., 2007. Emergence of Shiga toxin 1 genes within *Shigella dysenteriae* type 4 isolates from travelers returning from the island of Hispanola. Am. J. Trop. Med. Hyg. 76, 1163–1165.

Haukka, K., Siitonen, A., 2008. Emerging resistance to newer antimicrobial agents among *Shigella* isolated from Finnish foreign travelers. Epidemiol. Infect. 136, 476–482.

Hrabák, J., Empel, J., Gniadkowski, M., Halbhuber, Z., Rébl, K., Urbásková, P., 2008. CTX-M-15-producing *Shigella sonnei* strain from a Czech patient who traveled in Asia. J. Clin. Microbiol. 46, 2147–2148.

Izumiya, H., Tada, Y., Ito, K., Morita-Ishihara, T., Ohnishi, M., Terajima, J., et al., 2009. Characterization of *Shigella sonnei* isolates from travel-associated cases in Japan. J. Med. Microbiol. 58, 1486–1491.

Karlsson, M.S., Bowen, A., Reporter, R., Folster, J.P., Grass, J.E., Howie, R.L., et al., 2013. Outbreak of infections caused by *Shigella sonnei* with reduced susceptibility to azithromycin in the United States. J. Antimicrob. Agents Chemother. 57, 1559–1560.

Kerneis, S., Guerin, P.J., von Seidlein, L., Legros, D., Grais, R.F., 2009. A look back at an ongoing problem: *Shigella dysenteriae* type 1 epidemics in refugee settings in Central Africa (1993–1995). PLoS One 4, e4494.

Klontz, K.C, Singh, N., 2015. Treatment of drug-resistant *Shigella* infections. Expert Rev. Anti Infect. Ther. 13, 69–80.

Kosek, M., Yori, P.P., Pan, W.K., Olortegui, M.P., Gilman, R.H., Perez, J., et al., 2008. Epidemiology of highly endemic multiply antibiotic-resistant Shigellosis in children in the Peruvian Amazon. Pediatrics 122, e541–e549. (erratum: Pediatrics, 122, 1163–1164).

Lewis, H.C., Ethelberg, S., Olsen, K.E., Nielsen, E.M., Lisby, M., Madsen, S.B., et al., 2009. Outbreaks of *Shigella sonnei* infections in Denmark and Australia linked to consumption of imported raw baby corn. Epidemiol. Infect. 137, 326–334.

Lima, A.A., Lima, N.L., Pinho, M.C., Barros Juñior, E.A., Teixeira, M.J., Martins, M.C., et al., 1995. High frequency of strains multiply resistant to ampicillin, trimethoprim-sulfamethoxazole, streptomycin, chloramphenicol, and tetracycline isolated from patients with shigellosis in northeastern Brazil during the period 1988–1993. J. Antimicrob. Agents Chemother. 39, 256–259.

Mandal, J., Ganesh, V., Emelda, J., Mahadevan, S., Parija, S.C., 2012. The recent trends of shigellosis: a JIPMER perspective. J. Clin. Diagn. Res. 6, 1474–1477.

Mandomando, I., Jaintilal, D., Pons, M.J., Valles, X., Espasa, M., Mensa, L., et al., 2009. Antimicrobial susceptibility and mechanisms of resistance in *Shigella* and *Salmonella* isolates from children under five years of age with diarrhea in rural Mozambique. J. Antimicrob. Agents Chemother. 53, 2450–2454.

Marti, E., Variatza, E., Balcazar, J.L., 2014. The role of aquatic ecosystems as reservoirs of antibiotic resistance. Trends Microbiol. 22, 36–41.

Mehata, S., Duan, G.C., 2011. Molecular mechanism of multi-drug resistance in *Shigella* isolates from rural China. Nepal Med. Coll. J. 13, 27–29.

Mensa, L., Marco, F., Vila, J., Gascon, J., Ruiz, J., 2008. Quinolone resistance among *Shigella* spp. isolated from travellers returning from India. Clin. Microbiol. Infect. 14, 279–281.

MMWR, 2013. Notes from the field: outbreak of infections caused by *Shigella sonnei* with decreased susceptibility to azithromycin—Los Angeles, California, 2012. Morb. Mortal. Wkly. Rep. 62, 171.

Mota, M.I., Gadea, M.P., González, S., González, G., Pardo, L., Sirok, A., et al., 2010. Bacterial pathogens associated with bloody diarrhea in Uruguayan children. Rev. Argent. Microbiol. 42, 114–117.

Njuguna, H.N., Cosmas, L., Williamson, J., Nyachieo, D., Olack, B., Ochieng, J.B., et al., 2013. Use of population-based surveillance to define the high incidence of shigellosis in an urban slum in Nairobi, Kenya. PLoS One 8, e58437.

Nógrády, N., Király, M., Borbás, K., Tóth, Á., Pászti, J., Tóth, I., 2013. Antimicrobial resistance and genetic characteristics of integron-carrier shigellae isolated in Hungary (1998–2008). J. Med. Microbiol. 62, 1545–1551.

Özmert, E.N., İnce, O.T., Örün, E., Yalçın, S., Yurdakök, K., Gür, D., 2011. Clinical characteristics and antibiotic resistance of *Shigella gastroenteritis* in Ankara, Turkey between 2003 and 2009, and comparison with previous reports. Int. J. Infect. Dis. 15, e849–e853.

Pazhani, G.P., Niyogi, S.K., Singh, A.K., Sen, B., Taneja, N., Kundu, M., et al., 2008. Molecular characterization of multidrug-resistant *Shigella* species isolated from epidemic and endemic cases of shigellosis in India. J. Med. Microbiol. 57, 856–863.

Pu, X.Y., Pan, J.C., Wang, H.Q., Zhang, W., Huang, Z.C., Gu, Y.M., 2009. Characterization of fluoroquinolone-resistant *Shigella flexneri* in Hangzhou area of China. J. Antimicrob. Agents Chemother. 63, 917–920.

Ranjbar, R., Soltan Dallal, M.M., Talebi, M., Pourshafie, M.R., 2008. Increased isolation and characterization of *Shigella sonnei* obtained from hospitalized children in Tehran, Iran. J. Health Popul. Nutr. 26, 426–430.

Shiferaw, B., Solghan, S., Palmer, A., Joyce, K., Barzilay, E.J., Krueger, A., et al., 2012. Antimicrobial susceptibility patterns of *Shigella* isolates in Foodborne Diseases Active Surveillance Network (FoodNet) sites, 2000–2010. Clin. Infect. Dis. 54, S458–S463.

Simon, D.G., Kaslow, R.A., Rosenbaum, J., Kaye, R.L., Calin, A., 1981. Reiter's syndrome following epidemic shigellosis. J. Rheumatol. 8, 969–973.

Sire, J.M., Macondo, E.A., Perrier-Gros-Claude, J.D., Siby, T., Bahsoun, I., Seck, A., et al., 2008. Antimicrobial resistance in *Shigella* species isolated in Dakar, Senegal (2004–2006). Jpn. J. Infect. Dis. 61, 307–309.

Srinivasa, H., Baijayanti, M., Raksha, Y., 2009. Magnitude of drug resistant shigellosis: a report from Bangalore. Indian. J. Med. Microbiol. 27, 358–360.

Stelling, J., Yih, W.K., Galas, M., Kulldorff, M., Pichel, M., Terragno, R., Collaborative Group WHONET-Argentina, 2010. Automated use of WHONET and SaTScan to detect outbreaks of *Shigella* spp. using antimicrobial resistance phenotypes. Epidemiol. Infect. 138, 873–883.

Suárez, M.E., Carvajal, L., Culasso, C., 2000. Resistencia de *Shigella* spp. a los antimicrobianos en Cordoba, Argentina, durante el period 1990–1997. Pan Am. J. Public Health 7, 113–117.

Vinh, H., Nhu, N.T., Nga, T.V., Duy, P.T., Campbell, J.I., Hoang, N.V., et al., 2009. A changing picture of shigellosis in southern Vietnam: shifting species dominance, antimicrobial susceptibility and clinical presentation. BMC Infect. Dis. 9, 204.

von Seidlein, L., Kim, D.R., Ali, M., Lee, H., Wang, X., Thiem, V.D., et al., 2006. A multicentre study of *Shigella* diarrhea in six Asian countries: disease burden, clinical manifestations, and microbiology. PLoS. Med. 3, e353.

Vrints, M., Mairiaux, E., Van Meervenne, E., Collard, J.-M., Bertrand, S., 2009. Surveillance of antibiotic susceptibility patterns among *Shigella sonnei* strains isolated in Belgium during the 18-year period 1990 to 2007. J. Clin. Microbiol. 47, 1379–1385.

Wong, M.R., Reddy, V., Hanson, H., Johnson, K.M., Tsoi, B., Cokes, C., et al., 2010. Antimicrobial resistance trends of *Shigella* serotypes in New York City, 2006–2009. Microb. Drug Resist. 16, 155–161.

World Health Organization, WHO, 2005. Guidelines for the control of shigellosis, including epidemics due to *Shigella dysenteriae* type 1, Geneva, Switzerland. <http://whqlibdoc.who.int/publications/2005/9241592330.pdf>.

Yang, H., Sun, W., Duan, G., Zhu, J., Zhang, W., Xi, Y., et al., 2013. Serotype distribution and characteristics of antimicrobial resistance in *Shigella* isolated from Henan province, China, 2001–2008. Epidemiol. Infect. 141, 1946–1952.

Yismaw, G., Negeri, C., Kassu, A., 2008. A five-year antimicrobial resistance pattern of *Shigella* isolated from stools in the Gondar University Hospital, northwest Ethiopia. Trop. Doct. 38, 43–45.

Chapter 8

Antimicrobial Resistance in *Listeria* spp.

Nathan A. Jarvis, Philip G. Crandall, Corliss A. O'Bryan and Steven C. Ricke

Center for Food Safety, Food Science Department, University of Arkansas, Fayetteville, AR, USA

Chapter Outline

INTRODUCTION

The bacterium *Listeria monocytogenes* causes infections in humans with high death rates in susceptible populations and high costs associated with each case. The main route of infection is believed to be by consumption of contaminated foods (Allerberger and Wagner, 2010). In the normal population one estimate is that there will be 0.7 cases of listeriosis per 100,000 persons (Hof, 2004). In contrast, estimates increase greatly for susceptible populations with suppressed immune function (Hof, 2004). To illustrate, for every 100,000 persons in the following risk groups the estimate is that there would be two cases of listeriosis in people over the age of 70, 12 cases among pregnant women, 100 for kidney transplant recipients, 600 in AIDS patients, and 1,000 in acute leukemia patients (Hof, 2004). Their decreased immune function results in fatality rates as high as 30% (Painter and Slutsker, 2007), and hospitalization is required in up to 94% of cases (Scallan et al., 2011) because infections can results in sepsis, meningitis, endocarditis, or bacteremia (Painter and Slutsker, 2007). Unfortunately, some symptoms of listeriosis are difficult to distinguish from less serious clinical causes such as headaches, mental confusion, fatigue, and neck stiffness (Hale et al., 1994; Painter and Slutsker, 2007). This means both

Antimicrobial Resistance and Food Safety. DOI: http://dx.doi.org/10.1016/B978-0-12-801214-7.00008-9

that case number estimates may be low and that sick people do not receive the proper antimicrobials as quickly as they require.

Listeriosis cases have failed to decline in recent years despite efforts to eliminate *L. monocytogenes* from food products; this is similar to other pathogens (*Campylobacter* and *Escherichia coli* O157) and in contrast to decreases in salmonellosis and increases in vibriosis (Crim et al., 2014). This represents an opportunity for researchers and industry to identify why current measures are not continuing to reduce listeriosis cases.

Fortunately, in contrast to many other foodborne pathogens, *L. monocytogenes* strains remain without widespread resistance to antibiotics. One of the first reports of multi-antimicrobial-resistant *L. monocytogenes* was made in 1988 when a human *L. monocytogenes* meningitis isolate was shown to be resistant to chloramphenicol, erythromycin, streptomycin, and tetracycline (Poyart-Salmeron et al., 1990). Over the years there has been an increase in the reports of antimicrobial resistance in *L. monocytogenes* and other *Listeria* spp., though the subject has been less extensively investigated as compared to other foodborne pathogens (reviewed in Charpentier and Courvalin, 1999; Lungu et al., 2011). A potential trend toward resistance to fluoroquinolones and tetracycline was observed in a large collection of *L. monocytogenes* clinical isolates (Morvan et al., 2010), although other studies have not demonstrated increasing resistances (Hansen et al., 2005; Safdar and Armstrong, 2003). Co-selection for gentamicin resistance has also been shown to occur with repeated exposure to sublethal concentrations of benzalkonium chloride or ciprofloxacin and results from mutations to *L. monocytogenes'* efflux systems (Rakic-Martinez et al., 2011). Additionally, cephalosporins are well known for being ineffective against *L. monocytogenes* (Espaze and Reynaud, 1988). The collection of natural resistances and potential for acquisition make *L. monocytogenes* a unique system to study the interactions between antimicrobials and genes required for virulence. We briefly show that not only do innate resistances to certain classes of antimicrobials and acquisition of select transferable mechanisms play a significant role in *L. monocytogenes'* survival but also the relationship of antimicrobial resistance to virulence and the selection for resistant phenotypes impact food safety.

MECHANISMS OF ANTIBIOTIC RESISTANCE

Innate Resistance

L. monocytogenes possesses an innate or natural resistance to a variety of antimicrobials, including many β-lactams, most of the cephalosporins, cationic antimicrobial peptides, and some lantibiotics, particularly nisin in about 50% of isolates (Begley et al., 2005; Espaze and Reynaud, 1988; Hof, 2004; Kallipolitis et al., 2003; Roy, 2009). Troxler et al. (2000) found that most *Listeria* spp. have natural resistance to modern cephalosporins. However, they found differences in resistance for other antimicrobials including quinolones, co-trimoxazole, tri-methoprim, and fosfomycin. If patients are suffering from undiagnosed listeri-osis, this means that they may receive routinely prescribed antimicrobials which are ineffective in stemming the infection (Hof, 2004). A notable case is that of a *Listeria grayi* isolate responsible for bacteremia in a stem cell transplant recipient which was resistant to vancomycin (Salimnia et al., 2010). Described mechanisms for these innate resistances most often include cell wall acting gene products, two-component systems (TCSc), and efflux pumps.

The *penicillin-binding proteins (PBPs)* are enzymes responsible for extend-ing the glycan chains in peptidoglycan and cross-linking the peptides between chains (Zawadzka-Skomial et al., 2006). Vicente et al. (1990) identified five PBPs in *L. monocytogenes*, concluding PBP3 to be most significant. Guinane et al. (2006) later identified seven possible PBPs in *L. monocytogenes* strain EGD-e using *in silico* analysis. Vicente et al. (1990) found that there was a strong correlation between the binding affinity of PBP3 to an antimicrobial and whether *L. monocytogenes* was susceptible or not. While this seemed to indicate that PBP3 was key to β-lactam resistance, this has been challenged by more recent work using the nisin-controlled gene expression system (Krawczyk-Balska et al., 2012b). These researchers identified the putative gene encoding PBP3 (*lmo1438*) and demonstrated that its overexpression or deletion did not significantly influence minimum inhibitory concentrations (MIC). However, the PBP3 overexpressing strain was more sensitive, as demonstrated by partial autolysis, to subinhibitory concentrations of some β-lactams (Krawczyk-Balska et al., 2012b).

PBPs lmo0441, lmo2229, and lmo2754 contribute not only to penicillin resistance but also to cephalosporin resistance (Guinane et al., 2006), although Zawadzka-Skomial et al. (2006) found no effects on β-lactam resistance when disrupting *lmo2229*. Because many PBPs are responsible for cell wall synthesis it is intuitive that the loss of these genes could result in growth attenuation. Zawadzka-Skomial et al. (2006) found that a mutation in *lmo2229* resulted in no change in resistance to penicillin, imipenem, cephalothin, CTX, vancomycin, or nisin, but did result in increased resistance to moenomycin, which they inter-preted as indicating that the mutation affected the glycosyltransferase domain of PBP4 (Zawadzka-Skomial et al., 2006).

Other *cell wall acting gene products* influence innate resistance to β-lactams and cephalosporins. The gene *oatA* encodes an *O*-acetyltransferase which *O*-acetylates muramic acid in peptidoglycan thus conferring resistance to cefo-taxime and gallidermin and is also important for pathogenesis in mice, prob-ably via protection from macrophage killing (Aubry et al., 2011). The gene *mprF* plays a role in modifying phosphatidylglycerol and diphosphatidylglyc-erol with lysine in the cell membranes which bestows resistance to gallidermin and α-defensins (Thedieck et al., 2006). The protein MprF also plays a role in growth in brain–heart-infusion rich medium (Premaratne et al., 1991; Thedieck et al., 2006). The protein PrfA is a transcriptional regulator which regulates a large number of virulence genes in *L. monocytogenes* (Cossart, 2011). While PrfA was predicted to regulate *mprF* (Glaser et al., 2001) experiments have failed to confirm this so far (Thedieck et al., 2006). However, the VirRS regulon was shown to govern *mprF* (Mandin et al., 2005). Collins et al. (2010a) demonstrated that both *virR* and *mprF* mutants had reduced nisin and bacitracin MICs, as did mutants of *anrB*, *telA*, and *lisK*.

The ABC transporter gene which contributes to nisin resistance, *anrAB*, has a palindromic-binding sequence upstream which suggests that it is regulated by VirR (Collins et al., 2010a). The protein AnrAB not only contributes to nisin resistance but also to bacitracin and some β-lactam resistances (Collins et al., 2010a). Many of these sensitivities acquired through mutations in the above genes make sense in that they are cell-wall-modifying genes which are providing natural resistance to cell-wall-acting antimicrobials. For example, lantibiotics, which have thioether cross-links and the unusual amino acids 2,3-didehydroalanine and (Z)-2,3-didehydrobutyrine, act through interactions with lipid II preventing cell wall biosynthesis. Some lantibiotics, such as nisin, also create pores in the cell membrane (Begley et al., 2005; Draper et al., 2008; Islam et al., 2012). An interesting twist to the cell-wall-mediated antimicrobial resistance is the *gad* system. Begley et al. (2005) reported that approximately 50% of *L. monocytogenes* possess *gadD1* which functions to convert glutamate to γ-amino butyrate and produces ATP in the process. Mutants of *gadD1* exhibit increased sensitivity to both sublethal and lethal concentrations of nisin (Begley et al., 2005). They proposed that the natural resistance to nisin via *gadD1* is due to the increased intracellular pools or production ability of ATP. In essence, the gene allows the bacterium to keep the energy resources sufficiently high to survive even when losing ATP through the nisin-created pores.

Amino acid acetylation of phophatidylglycerol can confer resistance to a variety of antimicrobial substances, most notably cationic antimicrobial peptides (Roy, 2009). Phosphatidylglycerol synthases take amino acids from aa-tRNA and acetylate the respective phospholipids (Roy, 2009). This increases the overall positive charge of the lipid bilayer (Roy, 2009) and thus the repulsiveness to certain compounds. Lysinolated phosphatidylglycerol has been detected in *L. monocytogenes* (Fischer and Leopold, 1999; Roy, 2009; Thedieck et al., 2006). Lysinolated-cardiolipin has also been reported in *L. monocytogenes*, *Listeria innocua*, *Listeria seeligeri*, and *Listeria welshimeri* (Fischer and Leopold, 1999; Roy, 2009; Thedieck et al., 2006). Fischer and Leopold (1999) demonstrated that cardiolipin and lysinolated-cardiolipin increased as the cells entered the stationary phase for both *L. innocua* and *L. welshimeri*.

Other examples of genes contributing to innate resistance include *telA*, a gene homologous to tellurite resistance loci (Collins et al., 2010b), *anrAB*, a putative multidrug resistance transporter gene and contributes to bacitracin and some β-lactam resistances (Collins et al., 2010a), *dlt*, a gene regulated by VirR and which leads to the addition of D-alanine to lipoteichoic acids (Abachin et al., 2002; Mandin et al., 2005), and *lmo1416*, a homolog of *Enterococcus faecium vanZ* (Gottschalk et al., 2008). The TCSs CesRK and LisRK have also been identified as being involved in the innate resistance of *L. monocytogenes* to cephalosporins (Cotter et al., 2002; Kallipolitis et al., 2003). Transcriptional studies have confirmed the crucial role of LisRK and CesRK in the resistance of *L. monocytogenes* to β-lactams and have demonstrated that two other TCSs, LiaSR and VirRS, are also linked to this resistance (Nielsen et al., 2012).

Efflux pumps are proteins which can remove toxic substrates from the cytosol of both Gram-positive and -negative bacteria, including antimicrobials, outside of the cell (Van Bambeke et al., 2000). However, there is increasing evidence that they can contribute to cell signaling functions as well (Crimmins et al., 2008; Witte et al., 2013; Woodward et al., 2010). The genes for two separate efflux pumps, MdrL and Lde, seem to be present in almost all *L. monocytogenes* strains (Mereghetti et al., 2000; Romanova et al., 2006) indicating the possibility for innate resistance. Macrolide antimicrobials and cefotaxime, as well as heavy metals and ethidium bromide, are pumped outside the cell by MdrL (Mata et al., 2000). Fluoroquinolone resistance in addition to DNA intercalating dyes acridine orange and ethidium bromide resistances have been associated with the Lde pump (Godreuil et al., 2003).

Transferable and Acquired Antimicrobial Resistance

Plasmids, which replicate separately from the bacterial chromosome and can carry genes which can confer antibiotic resistance, were first reported in *L. monocytogenes* in 1982 (Margolles and de los Reyes-Gavilán, 1998; Pérez-Díaz et al., 1982; Stanisich, 1988). Plasmid pDB2011 from *L. innocua*, isolated from prepackaged sprouts in Switzerland, contains three antibiotic genes: *spc* (spectinomycin—adenyltransferase), *dfrD* (trimethoprim—dihydrofolate reductase), and *erm*(A) (erythromycin) (Bertsch et al., 2013a) and is a rolling-circle plasmid similar to pIP823 which has only the *dfrD* gene (Bertsch et al., 2013a; Charpentier and Courvalin, 1997; Charpentier et al., 1999). However, *erm*(A) encodes a protein which is 28 amino acids shorter than that encoded on transposon Tn*554*. This isolate with the shorter protein failed to be resistant to erythromycin, phenotypically confirming the truncated protein (Bertsch et al., 2013a). Whether *L. innocua* can serve as a reservoir for antibiotic resistance that is subsequently transferable to *L. monocytogenes* is still debated (Bertsch et al., 2013a). Replication of the plasmid pDB2011 in *Lactococcus lactis* subsp. *cremoris* MG1364, *E. coli* MC1061 and DH5α has been confirmed, but it failed to be transferred to *L. monocytogenes* 10403S (Bertsch et al., 2013a). Plasmid pIP813 confers tetracycline resistance via *tet*(L) to *L. monocytogenes*, although self-transferability has not been demonstrated in the lab (Poyart-Salmeron et al., 1992). Plasmid pWDB100 confers resistance to tetracycline, chloramphenicol, and erythromycin, may be related to pIP811, and was discovered in *L. monocytogenes* isolates from a patient with endocarditis (Hadorn et al., 1993; Poyart-Salmeron et al., 1990). Benzalkonium chloride resistance has also been shown to be conferred to *L. monocytogenes* strains by way of a plasmid, pLM80, isolated from the 1998–1999 hot dog outbreak strain, which is a part of an IS*1216* composite transposon (Dutta et al., 2013; Elhanafi et al., 2010). This plasmid was present in almost all, 70 of 71, benzalkonium chloride-resistant isolates (Dutta et al., 2013). Work by Kuenne et al. (2010) compared the sequences of 12 *L. monocytogenes* plasmids, one *L. innocua* plasmid, and one *L. grayi* plasmid and concluded that all *L. monocytogenes* plasmids contained genes for heavy metal resistance and multidrug transporters.

While plasmids are found in *L. monocytogenes*, they are found in much higher frequency in other foodborne bacteria such as enterococci and staphylococci. It is not clear why this is the case, however, a unique CRISPR system which is not associated with the *cas* genes and requires polynucleotide phosphorylase to function has recently been described, and may be responsible for some of the decreased long-term maintenance of plasmids (Sesto et al., 2014). Unstable resistance mechanisms have also been reported in both *Listeria ivanovii* and *L. monocytogenes* where the resistance mechanism disappeared after a year of storage (Bertsch et al., 2014; Hadorn et al., 1993).

A transposon, Tn*6188*, from an environmental isolate, has been demonstrated to provide benzalkonium resistance (Müller et al., 2013). It encodes all three transposase genes, a drug transporter similar to a multidrug family of *Sporosarcina newyorkensis*, and a transcriptional regulator. Transposon *6198* confers tetracycline and trimethoprim resistance in *L. monocytogenes* TTH-2007, a clinical isolate (Bertsch et al., 2013b). They showed that the transposon could be transferred to both *Enterococcus faecalis* and another strain of *L. monocytogenes*. *Streptococcus* Tn*1545*, which carries *tet*(M), was transferred from *E. faecalis* to *L. monocytogenes* in the GI tract of gnotobiotic mice with a transfer efficiency of 1.1×10^{-8} transconjugants per donor cell (Doucet-Populaire et al., 1991); they achieved ten-fold higher transconjugants when the mice were given water with low levels of tetracycline. In a continuous colonic fermentation, Haug et al. (2011) show transfer of marked plasmid, pRE25*, with erythromycin resistance, from *E. faecalis* to *L. monocytogenes* and other fecal bacteria. Katharios-Lanwermeyer et al. (2012) demonstrated conjugative transfer of resistance to benzalkonium chloride and cadmium from *L. innocua* and *L. welshimeri* to *L. monocytogenes* were equally efficient at 37°C as they were at 25°C. However, transfer between the nonpathogenic *Listeria* spp. was higher at 25°C than at 37°C which the authors speculate may be due to an environmental adaptation (Katharios-Lanwermeyer et al., 2012). This leaves open the possibility that nonpathogenic *Listeria* spp. could transfer resistance *in vivo* to *L. monocytogenes* (Katharios-Lanwermeyer et al., 2012). While this work did not address resistance to commonly used antimicrobials for clinical cases, further work is warranted to see if similar effects are seen with transposons and plasmids conferring such resistances. This work demonstrates the ability of *L. monocytogenes* to receive transposons originating from other genera and species.

RELATIONSHIP OF ANTIMICROBIAL RESISTANCE TO VIRULENCE

Subinhibitory concentrations of antimicrobials can not only induce virulence gene expression but also can inhibit it, depending on the conditions and organism (Laureti et al., 2013). *L. monocytogenes* virulence genes can be differentially regulated by some antimicrobials (Hof et al., 1997). Knudsen et al. (2012) found that exposure of *L. monocytogenes* to subinhibitory concentrations of ampicillin and tetracycline for 3 h influenced expression of virulence and stress genes. However, this exposure did not translate into increased Caco-2 cell invasion as compared to cells not exposed to the antimicrobials (Knudsen et al., 2012). In addition, there were some suggestions that subinhibitory concentrations of β-lactams or gentamicin may actually decrease the production or activity of listeriolysin, one of the proteins *L. monocytogenes* uses to escape into the host cell cytosol from the phagosome (Nichterlein et al., 1997, 1996). Listeriolysin production also is decreased by subinhibitory concentrations of ampicillin (Hof et al., 1997).

L. monocytogenes is well known for being resistant to cephalosporins yet this class of antimicrobials has some ability to downregulate *L. monocytogenes'* virulence genes (Hof et al., 1997). This would suggest that researchers need to expand their toolbox when assessing antimicrobials' effects on *L. monocytogenes*. Similarly, trans-cinnamaldehyde, carvacrol, and thymol have all been shown to downregulate some virulence genes of *L. monocytogenes in vitro*, in addition to influencing motility (Upadhyay et al., 2012); this carried over to reduced adhesion and invasion of Caco-2 and human brain microvascular endothelial cells.

An iron-binding, DPS-domain-containing, gene *fri* (*lmo0943*) contributes to resistance to β-lactams and cephalosporins (Krawczyk-Balska et al., 2012a) and this gene is induced by subinhibitory concentrations of penicillin G. It is possible that this gene contributes to antimicrobial resistance by reducing the possibility of oxidative damage by the binding of iron (Kohanski et al., 2007). Krawczyk-Balska and Lipiak (2013) later showed that the Fri protein is responsible for controlling cell membrane structure and autolysin presence, which corroborates research on β-lactam activity in other bacteria.

Kastbjerg et al. (2010) evaluated the influence of 11 common disinfectants on the gene expression of *prfA*, *plcA*, *inlA*, and *hly* using a *lacZ* reporter system. In general, chlorine compounds and one peroxy compound led to reduced expression of these genes. In contrast, one peroxy compound first downregulated and second upregulated these genes (Kastbjerg et al., 2010). Triclosan/ethanol and quaternary ammonium compounds resulted in increased expression of all genes, except for *plcA* which was not influenced by triclosan/ethanol (Kastbjerg et al., 2010). The authors concluded that the results reveal the necessity of insuring that lethal concentrations of disinfectant are used in order to avoid gene-induced resistance in surviving *L. monocytogenes*. Additional unintended consequences of using sublethal disinfectants on *L. monocytogenes* will be discussed in a later section.

Virulence Genes' Effects on Antimicrobial Susceptibility

Some antimicrobial susceptibilities are influenced by experimental conditions such that *in vivo* results are opposite to those of *in vitro* results (Scortti et al., 2006). It is therefore critical to be mindful of the limitations and applications of the methods used. Fosfomycin is a phosphonic-acid-based antimicrobial which inhibits peptidoglycan production by preventing the synthesis of *N*-acetylmuramic acid (Michalopoulos et al., 2011; Raz, 2012). Scortti et al. (2006) report that *L. monocytogenes* is resistant to fosfomycin *in vitro* but is susceptible to this antibiotic in infected mice. Due to a similar action in *E. coli* which is caused by UhpT sugar phosphate transporter, the authors suspected that the listerial hexose phosphate transporter, Hpt, which is regulated by PrfA might be responsible for the "*in vivo–in vitro* paradox" (Scortti et al., 2006). Being regulated by PrfA, *hpt* might not be expressed *in vitro* (Cossart, 2011). Fosfomycin is transported into the cell by two systems, one of which is the Hpt system (Michalopoulos et al., 2011). Knocking out the *hpt* gene and using *prfA* mutants, Scortti et al. (2006) showed that *L. monocytogenes* was susceptible to fosfomycin in mice and J774 macrophages. They (Scortti et al., 2006) also demonstrated that spontaneous mutants in either *prfA* or *hpt* had increased *in vivo* fosfomycin resistance and were also associated with decreased pathogenicity due to the limitations imposed by the mutations in the virulence genes. Further work with fosfomycin therefore ought to use *in vitro* conditions which induce *prfA* expression. Lastly, the presence of fosfomycin resulted in increased gene expression of genes under the regulation of the three-component system LiaFSR *in vitro* (Fritsch et al., 2011).

Another example of the *prfA*-dependent susceptibility is that of a potato defensin demonstrated by López-Solanilla et al. (2003). While they did not identify the molecular mechanism, they showed that causing *prfA* to be constitutively expressed resulted in susceptibility in the same way that growth at 37°C caused susceptibility (*prfA* is thermoregulated to be expressed at this temperature) (Cossart, 2011). Again, similar to *hpt* resulting in *in vivo* susceptibility to fosfomycin, perhaps a gene regulated by PrfA is sensitizing *L. monocytogenes* to the peptides.

The *htrA* gene encodes for a serine protease which is presumed to be responsible for degradation of misfolded proteins produced during stress responses (Wilson et al., 2006). It is partially responsible for response to osmotic stresses and is under the regulation of the LisRK TCS (Sleator and Hill, 2005). Without *htrA*, *L. monocytogenes* is less virulent in mice (Wilson et al., 2006). In other organisms, *htrA* mutants are more susceptible to puromycin because the antimicrobial halts protein chain lengthening and results in incomplete proteins (Wilson et al., 2006). Wilson et al. (2006) demonstrated that a *htrA* mutant exhibited impaired growth at 40°C both with and without puromycin but not at 30°C; they are also more susceptible to paraquat (Wilson et al., 2006).

These examples demonstrate a current gap in the knowledge of *L. monocytogenes* antimicrobial susceptibility and warrant further investigation. New protocols need to be developed which can mimic *in vivo* conditions to the extent that resistance to antimicrobials, such as fosfomycin, can be measured *in vitro*. A starting place would be overexpression of *prfA* so as to promote the expression of genes regulated by PrfA, as has been done by some researchers (López-Solanilla et al., 2003). It is possible that other antimicrobials behave similarly. It is also possible that antimicrobials which are effective *in vitro*, and which are prescribed to treat listeriosis, are not as effective *in vivo* due to similar effects. This would result in antimicrobial treatments being ineffective and possibly negative outcomes for the patients. It is therefore imperative that research used to define antimicrobial therapies employ adequate models for *in vivo* conditions.

SELECTION FOR RESISTANT PHENOTYPES

Subinhibitory concentrations of sanitizers on food-processing equipment or in the food-processing environment may select for resistant phenotypes of *L. monocytogenes*. While it is true that lethal chemical concentrations should be used on food equipment, product lines, and the other areas associated with processing, it should be remembered that even lethal concentrations can be diluted if there is remaining moisture on the equipment following cleaning. Sanitizers remaining on food contact surfaces may slowly degrade. Subinhibitory concentrations of cleaning and sanitizing compounds are therefore inevitable in food-processing environments. Selection for resistant phenotypes at sublethal and subinhibitory values may represent a larger long-term problem than the selections that are typically considered under lethal conditions (Sandegren, 2014). These low levels, even well below the MIC, have the potential not only to upregulate resistant genes and increase effects on virulence but also to select for resistant mutants (Gullberg et al., 2011). Furthermore, any biofilms that exist may represent heterogeneous systems where even the best delivery method of an antimicrobial results in concentration gradients across the biofilm (Laureti et al., 2013). Some researchers have therefore proposed the rotation of disinfectants or sanitizers to reduce selection of resistant phenotypes (Langsrud et al., 2003) although this has failed to be shown to be effective for *L. monocytogenes* for the tested disinfectants thus far evaluated (Lundén et al., 2003).

It has been shown with *E. coli* that within 10 h of sublethal exposure to ciprofloxacin in a microfluidic device, resistant cells emerge via single nucleotide polymorphisms and overtake the antimicrobial-sensitive cells (Zhang et al., 2011). This research is particularly important in that it demonstrates that the *gradient* of antimicrobial concentration is important in allowing for the emergence of a resistant phenotype. Antimicrobial-resistant cells actually moved toward higher concentrations of antimicrobial during the course of the experiment. If similar effects occurred in *L. monocytogenes* in food-processing plants then, returning to the work of Kastbjerg et al. (2010), it would take the sanitizers from only one cleaning shift at a plant running down the drains, to result in the emergence of resistant phenotypes. Similar experiments are therefore warranted with *L. monocytogenes*.

Gullberg et al. (2011) demonstrated, for both *E. coli* and *Salmonella* Typhimurium, that even very low levels of antimicrobials can result in resistant strains outcompeting sensitive strains in coculture. This is due to how the relative growth rates of the two respective phenotypes change as the concentration of antimicrobial is increased: the sensitive phenotypes' growth rate declines linearly as compared to that of the resistant phenotype which declines in a convex fashion, with very little initial decline followed by exponential reductions (Gullberg et al., 2011). This results in a point where the antimicrobial-resistant cells divide faster than those of the sensitive phenotype. They also demonstrate that the resistant strain can overgrow the sensitive strain by a factor of ten-fold in as few as 20 generations, depending on the antimicrobial identity and concentration (Gullberg et al., 2011). In *L. monocytogenes*, fluoroquinolone resistance has been reported to be mediated through either mutations in the DNA type II topoisomerase gyrase or the type IV topoisomerase, or through active efflux by Lde, a 12-transmembrane segmented protein (Godreuil et al., 2003).

L. monocytogenes' ubiquity in the natural environment (Sauders et al., 2012) requires that the potential for resistance development prior to food products entering processing facilities be considered. Thiele-Bruhn (2003) notes that antimicrobial presence in manure applied to crop fields can contribute to selective pressure on pathogenic microorganisms (Laureti et al., 2013; Thiele-Bruhn, 2003). Some studies have also demonstrated that the presence of antimicrobials can influence metabolic processes such as iron reduction and ammonia oxidation (Ding and He, 2010). Because *Listeria* spp. are often associated with soil (Sauders et al., 2012), it is possible that these residual antimicrobials in manure may be a source of resistance development. Even under organic or other conditions where antimicrobial use for animal treatment is strictly regulated, minute concentrations of antimicrobials may provide selection pressures. Microcosms with soil collected from various environmental sources can be constructed to evaluate the influence of antimicrobials on the microorganisms present (Popowska et al., 2010). Using such a model system would provide valuable information about the potential for *L. monocytogenes* to gain resistance through either mutations or horizontal gene transfer in the natural environment. If the effect on *L. monocytogenes* by very low levels of antimicrobials is similar to what has been observed with other organisms then it is possible that even the complete withdrawal of all antimicrobials from animal use would not appreciably decrease the selective pressure for antimicrobial resistance. Lastly, considering the work of Ding and He (2010), it is relevant to also consider how antimicrobials might influence the metabolic abilities of *L. monocytogenes*. One example might be the action of nisin which forms pores and results in ATP loss. Since some strains possess *gadD1*, they might be able to survive with the additional ATP production. Once the metabolic effect of some antimicrobials is known it may be possible to pair compounds to increase their efficacies or provide opportunities for desirable organisms to overtake the undesirables.

RELEVANCE TO FOOD SAFETY AND PUBLIC HEALTH

Foods associated with listeriosis are predominantly ready-to-eat foods; those with extended (usually refrigerated) shelf-life and capable of supporting growth of an infective dose of *L. monocytogenes* (Lianou and Sofos, 2007) being of particular importance. Antimicrobial resistance is currently not a widespread therapeutic problem, but that may change if resistance develops across most strains and serotypes via adaptation or horizontal transfer of key resistance genes. This is mostly likely to occur via background levels of antimicrobials or co-selection. It should be emphasized that there are no reports of human infection with foodborne *L. monocytogenes* that is resistant to the antimicrobials normally used to treat listeriosis (Barbosa et al., 2013; Bertsch et al., 2014; Morvan et al., 2010). Challenges still exist in determining why listeriosis cases have not continued to decline in recent years.

Future research will hopefully attempt to target some of the many natural resistances that *L. monocytogenes* possesses. It seems likely that the use of combinations of antimicrobials in both processing environments and clinical cases will rise in order to try to sidestep some of the resistance mechanisms. It is now apparent that horizontal gene transfer is possible, although not frequent among *Listeria* spp. Further research is warranted to determine how this might be prevented. In addition, whether there is a serotype difference would be helpful in assessing the risks, since only a few of the serotypes account for the majority of listeriosis cases (Gellin and Broome, 1989). The ability of antimicrobials to influence virulence gene expression is particularly interesting; further research ought to assess how long the altered gene expression lingers and whether it alters virulence in tissue culture and in *in vivo* models.

As mentioned previously, co-selection for gentamicin and benzalkonium chloride resistance is due to efflux pump mutations (Rakic-Martinez et al., 2011), but co-selection for resistance genes has been seen in other organisms as well (Ciric et al., 2011). This observation is important since quaternary ammonium compounds are routinely used for the sanitation of food production environment surfaces (McDonnell and Russell, 1999; Merianos, 1991; Rakic-Martinez et al., 2011). As a result, their frequent use may select for strains possessing reduced susceptibility to key therapeutic agents.

However, the presence of biofilms, if only on nonfood-contact surfaces, suggests that there will always be gradients of antimicrobials which can provide selective pressure. Therefore it would be wise for surveillance programs to continue to incorporate assessment of antimicrobial resistances. Lastly, we need to develop further models which will induce *in vivo* conditions so that antimicrobial susceptibility evaluation will most accurately reflect clinical conditions.

CONCLUSIONS

For over 30 years listeriosis has been an issue of public health. Despite listeriosis case reporting and serotyping, *L. monocytogenes* antimicrobial resistance data remain limited as compared to other foodborne pathogens. Limited resistance gene acquisition is encouraging and is cause for hope for future reductions in fatalities. Lastly, because invasive listeriosis almost always requires courses of antimicrobials for patient survival, it is important to continue to evaluate antimicrobials which will circumvent the natural resistances that *L. monocytogenes* possesses.

ACKNOWLEDGMENTS

Mr. Nathan Jarvis is supported by the University of Arkansas through a Distinguished Doctoral Fellowship.

REFERENCES

Abachin, E., Poyart, C., Pellegrini, E., Milohanic, E., Fiedler, F., Berche, P., et al., 2002. Formation of D-alanyl-lipoteichoic acid is required for adhesion and virulence of *Listeria monocytogenes*. Mol. Microbiol. 43, 1–14.

Allerberger, F., Wagner, M., 2010. Listeriosis: a resurgent foodborne infection. Clin. Microbiol. Infect. 16, 16–23.

Aubry, C., Goulard, C., Nahori, M.-A., Cayet, N., Decalf, J., Sachse, M., et al., 2011. OatA, a peptidoglycan *O*-acetyltransferase involved in *Listeria monocytogenes immune* escape, is critical for virulence. J. Infect. Dis. 204, 731–740.

Barbosa, J., Magalhães, R., Santos, I., Ferreira, V., Brandão, T.R.S., Silva, J., et al., 2013. Evaluation of antibiotic resistance patterns of food and clinical *Listeria monocytogenes* isolates in Portugal. Foodborne Pathog. Dis. 10, 861–866.

Begley, M., Sleator, R.D., Gahan, C.G., Hill, C., 2005. Contribution of the three bile-associated loci, *bsh*, *pva*, and *btlB*, to gastrointestinal persistence and bile tolerance of *Listeria monocytogenes*. Infect. Immun. 73, 894–904.

Bertsch, D., Anderegg, J., Lacroix, C., Meile, L., Stevens, M.J.A., 2013a. pDB2011, a 7.6 kb multidrug resistance plasmid from *Listeria innocua* replicating in Gram-positive and Gram-negative hosts. Plasmid 70, 284–287.

Bertsch, D., Uruty, A., Anderegg, J., Lacroix, C., Perreten, V., Meile, L., 2013b. Tn6198, a novel transposon containing the trimethoprim resistance gene *dfrG* embedded into a Tn916 element in *Listeria monocytogenes*. J. Antimicrob. Chemother. 68, 986–991.

Bertsch, D., Muelli, M., Weller, M., Uruty, A., Lacroix, C., Meile, L., 2014. Antimicrobial susceptibility and antibiotic resistance gene transfer analysis of foodborne, clinical, and environmental *Listeria* spp. isolates including *Listeria monocytogenes*. Microbiologyopen 3, 118–127.

Charpentier, E., Courvalin, P., 1997. Emergence of the trimethoprim resistance gene *dfrD* in *Listeria monocytogenes* BM4293. Antimicrob. Agents Chemother. 41, 1134–1136.

Charpentier, E., Courvalin, P., 1999. Antibiotic resistance in *Listeria* spp. Antimicrob. Agents Chemother. 43, 2103–2108.

Charpentier, E., Gerbaud, G., Courvalin, P., 1999. Conjugative mobilization of the rolling-circle plasmid pIP823 from *Listeria monocytogenes* BM4293 among gram-positive and gram-negative bacteria. J. Bacteriol. 181, 3368–3374.

Ciric, L., Mullany, P., Roberts, A.P., 2011. Antibiotic and antiseptic resistance genes are linked on a novel mobile genetic element: Tn*6087*. J. Antimicrob. Chemother. 66, 2235–2239.

Collins, B., Curtis, N., Cotter, P.D., Hill, C., Ross, R.P., 2010a. The ABC transporter AnrAB contributes to the innate resistance of *Listeria monocytogenes* to nisin, bacitracin, and various β-lactam antibiotics. Antimicrob. Agents Chemother. 54, 4416–4423.

Collins, B., Joyce, S., Hill, C., Cotter, P.D., Ross, R.P., 2010b. TelA contributes to the innate resistance of *Listeria monocytogenes* to nisin and other cell wall-acting antibiotics. Antimicrob. Agents Chemother. 54, 4658–4663.

Cossart, P., 2011. Illuminating the landscape of host–pathogen interactions with the bacterium *Listeria monocytogenes*. Proc. Natl. Acad. Sci. USA 108, 19484–19491.

Cotter, P.D., Guinane, C.M., Hill, C., 2002. The LisRK signal transduction system determines the sensitivity of *Listeria monocytogenes* to nisin and cephalosporins. Antimicrob. Agents Chemother. 46, 2784–2790.

Crim, S.M., Iwamoto, M., Huang, J.Y., Griffin, P.M., Gilliss, D., Cronquist, A.B., et al., 2014. Incidence and trends of infection with pathogens transmitted commonly through food—foodborne diseases active surveillance network, 10 U.S. sites, 2006–2013. MMWR. Morb. Mortal. Wkly. Rep. 63, 328–332.

Crimmins, G.T., Herskovits, A.A., Rehder, K., Sivick, K.E., Lauer, P., Dubensky, T.W., et al., 2008. *Listeria monocytogenes* multidrug resistance transporters activate a cytosolic surveillance pathway of innate immunity. Proc. Natl. Acad. Sci. USA 105, 10191–10196.

Ding, C., He, J., 2010. Effect of antibiotics in the environment on microbial populations. Appl. Microbiol. Biotechnol. 87, 925–941.

Doucet-Populaire, F., Trieu-Cuot, P., Dosbaa, I., Andremont, A., Courvalin, P., 1991. Inducible transfer of conjugative transposon Tn1545 from *Enterococcus faecalis* to *Listeria monocytogenes* in the digestive tracts of gnotobiotic mice. Antimicrob. Agents Chemother. 35, 185–187.

Draper, L.A., Ross, R.P., Hill, C., Cotter, P.D., 2008. Lantibiotic immunity. Curr. Protein. Pept. Sci. 9, 39–49.

Dutta, V., Elhanafi, D., Kathariou, S., 2013. Conservation and distribution of the benzalkonium chloride resistance cassette *bcrABC* in *Listeria monocytogenes*. Appl. Environ. Microbiol. 79, 6067–6074.

Elhanafi, D., Dutta, V., Kathariou, S., 2010. Genetic characterization of plasmid-associated benzalkonium chloride resistance determinants in a *Listeria monocytogenes* strain from the 1998–1999 outbreak. Appl. Environ. Microbiol. 76, 8231–8238.

Espaze, E.P., Reynaud, A.E., 1988. Antibiotic susceptibilities of *Listeria*: in vitro studies. Infection 16 (Suppl. 2), S160–S164.

Fischer, W., Leopold, K., 1999. Polar lipids of four *Listeria* species containing L-lysylcardiolipin, a novel lipid structure, and other unique phospholipids. Int. J. Syst. Bacteriol. 49 (Pt 2), 653–662.

Fritsch, F., Mauder, N., Williams, T., Weiser, J., Oberle, M., Beier, D., 2011. The cell envelope stress response mediated by the LiaFSR$_{Lm}$ three-component system of *Listeria monocytogenes* is controlled via the phosphatase activity of the bifunctional histidine kinase LiaS$_{Lm}$. Microbiology 157, 373–386.

Gellin, B.G., Broome, C.V., 1989. Listeriosis. JAMA 261, 1313–1320.

Glaser, P., Frangeul, L., Buchrieser, C., Rusniok, C., Amend, A., Baquero, F., et al., 2001. Comparative genomics of *Listeria* species. Science 294, 849–852.

Godreuil, S., Galimand, M., Gerbaud, G., Jacquet, C., Courvalin, P., 2003. Efflux pump Lde is associated with fluoroquinolone resistance in *Listeria monocytogenes*. Antimicrob. Agents Chemother. 47, 704–708.

Gottschalk, S., Bygebjerg-Hove, I., Bonde, M., Nielsen, P.K., Nguyen, T.H., Gravesen, A., et al., 2008. The two-component system CesRK controls the transcriptional induction of cell envelope-related genes in *Listeria monocytogenes* in response to cell wall-acting antibiotics. J. Bacteriol. 190, 4772–4776.

Guinane, C.M., Cotter, P.D., Ross, R.P., Hill, C., 2006. Contribution of penicillin-binding protein homologs to antibiotic resistance, cell morphology, and virulence of *Listeria monocytogenes* EGDe. Antimicrob. Agents Chemother. 50, 2824–2828.

Gullberg, E., Cao, S., Berg, O.G., Ilbäck, C., Sandegren, L., Hughes, D., et al., 2011. Selection of resistant bacteria at very low antibiotic concentrations. PLoS Pathog. 7, e1002158.

Hadorn, K., Hächler, H., Schaffner, A., Kayser, F.H., 1993. Genetic characterization of plasmid-encoded multiple antibiotic resistance in a strain of *Listeria monocytogenes* causing endocarditis. Eur. J. Clin. Microbiol. Infect. Dis. 12, 928–937.

Hale, E., Habte-Gabr, E., McQueen, R., Gordon, R., 1994. Co-trimoxazole for the treatment of listeriosis and its successful use in a patients with AIDS. J. Infect. 28, 110–113.

Hansen, J.M., Gerner-Smidt, P., Bruun, B., 2005. Antibiotic susceptibility of *Listeria monocytogenes* in Denmark 1958–2001. APMIS 113, 31–36.

Haug, M.C., Tanner, S.A., Lacroix, C., Stevens, M.J.A., Meile, L., 2011. Monitoring horizontal antibiotic resistance gene transfer in a colonic fermentation model. FEMS Microbiol. Ecol. 78, 210–219.

Hof, H., 2004. An update on the medical management of listeriosis. Expert. Opin. Pharmacother. 5, 1727–1735.

Hof, H., Nichterlein, T., Kretschmar, M., 1997. Management of listeriosis. Clin. Microbiol. Rev. 10, 345–357.

Islam, M.R., Nagao, J.-I., Zendo, T., Sonomoto, K., 2012. Antimicrobial mechanism of lantibiotics. Biochem. Soc. Trans. 40, 1528–1533.

Kallipolitis, B.H., Ingmer, H., Gahan, C.G., Hill, C., Søgaard-Andersen, L., 2003. CesRK, a two-component signal transduction system in *Listeria monocytogenes*, responds to the presence of cell wall-acting antibiotics and affects β-lactam resistance. Antimicrob. Agents Chemother. 47, 3421–3429.

Kastbjerg, V.G., Larsen, M.H., Gram, L., Ingmer, H., 2010. Influence of sublethal concentrations of common disinfectants on expression of virulence genes in *Listeria monocytogenes*. Appl. Environ. Microbiol. 76, 303–309.

Katharios-Lanwermeyer, S., Rakic-Martinez, M., Elhanafi, D., Ratani, S., Tiedje, J.M., Kathariou, S., 2012. Coselection of cadmium and benzalkonium chloride resistance in conjugative transfers from nonpathogenic *Listeria* spp. to other Listeriae. Appl. Environ. Microbiol. 78, 7549–7556.

Knudsen, G.M., Holch, A., Gram, L., 2012. Subinhibitory concentrations of antibiotics affect stress and virulence gene expression in *Listeria monocytogenes* and cause enhanced stress sensitivity but do not affect Caco-2 cell invasion. J. Appl. Microbiol. 113, 1273–1286.

Kohanski, M.A., Dwyer, D.J., Hayete, B., Lawrence, C.A., Collins, J.J., 2007. A common mechanism of cellular death induced by bactericidal antibiotics. Cell 130, 797–810.

Krawczyk-Balska, A., Lipiak, M., 2013. Critical role of a ferritin-like protein in the control of *Listeria monocytogenes* cell envelope structure and stability under β-lactam pressure. PLoS ONE 8, e77808.

Krawczyk-Balska, A., Marchlewicz, J., Dudek, D., Wasiak, K., Samluk, A., 2012a. Identification of a ferritin-like protein of *Listeria monocytogenes* as a mediator of β-lactam tolerance and innate resistance to cephalosporins. BMC Microbiol. 12, 278.

Krawczyk-Balska, A., Popowska, M., Markiewicz, Z., 2012b. Re-evaluation of the significance of penicillin binding protein 3 in the susceptibility of *Listeria monocytogenes* to β-lactam antibiotics. BMC Microbiol. 12, 57.

Kuenne, C., Voget, S., Pischimarov, J., Oehm, S., Goesmann, A., Daniel, R., et al., 2010. Comparative analysis of plasmids in the genus *Listeria*. PLoS ONE 5, 7.

Langsrud, S., Sundheim, G., Borgmann-Strahsen, R., 2003. Intrinsic and acquired resistance to quaternary ammonium compounds in food-related *Pseudomonas* spp. J. Appl. Microbiol. 95, 874–882.

Laureti, L., Matic, I., Gutierrez, A., 2013. Bacterial responses and genome instability induced by subinhibitory concentrations of antibiotics. Antibiotics 2, 100–114.

Lianou, A., Sofos, J.N., 2007. A review of the incidence and transmission of *Listeria monocytogenes* in ready-to-eat products in retail and food service environments. J. Food. Prot. 70, 2172–2198.

López-Solanilla, E., González-Zorn, B., Novella, S., Vázquez-Boland, J.A., Rodríguez-Palenzuela, P., 2003. Susceptibility of *Listeria monocytogenes* to antimicrobial peptides. FEMS Microbiol. Lett. 226, 101–105.

Lundén, J., Autio, T., Markkula, A., Hellström, S., Korkeala, H., 2003. Adaptive and cross-adaptive responses of persistent and non-persistent *Listeria monocytogenes* strains to disinfectants. Int. J. Food. Microbiol. 82, 265–272.

Lungu, B., O'Bryan, C.A., Muthaiyan, A., Milillo, S.R., Johnson, M.G., Crandall, P.G., et al., 2011. *Listeria monocytogenes*: antibiotic resistance in food production. Foodborne Pathog. Dis. 8, 569–578.

Mandin, P., Fsihi, H., Dussurget, O., Vergassola, M., Milohanic, E., Toledo-Arana, A., et al., 2005. VirR, a response regulator critical for *Listeria monocytogenes* virulence. Mol. Microbiol. 57, 1367–1380.

Margolles, A., de los Reyes-Gavilán, C.G., 1998. Characterization of plasmids from *Listeria monocytogenes* and *Listeria innocua* strains isolated from short-ripened cheeses. Int. J. Food. Microbiol. 39, 231–236.

Mata, M.T., Baquero, F., Pérez-Díaz, J.C., 2000. A multidrug efflux transporter in *Listeria monocytogenes*. FEMS Microbiol. Lett. 187, 185–188.

McDonnell, G., Russell, A.D., 1999. Antiseptics and disinfectants: activity, action, and resistance. Clin. Microbiol. Rev. 12, 147–179.

Mereghetti, L., Quentin, R., Marquet-Van Der Mee, N., Audurier, A., 2000. Low sensitivity of *Listeria monocytogenes* to quaternary ammonium compounds. Appl. Environ. Microbiol. 66, 5083–5086.

Merianos, J.J., 1991. Quaternary ammonium antimicrobial compounds. In: Block, S.S. (Ed.), Disinfection, Sterilization, and Preservation Lea & Feigner, Malvern, PA, pp. 225–255.

Michalopoulos, A.S., Livaditis, I.G., Gougoutas, V., 2011. The revival of fosfomycin. Int. J. Infect. Dis. 15, e732–e739.

Morvan, A., Moubareck, C., Leclercq, A., Hervé-Bazin, M., Bremont, S., Lecuit, M., et al., 2010. Antimicrobial resistance of *Listeria monocytogenes* strains isolated from humans in France. Antimicrob. Agents Chemother. 54, 2728–2731.

Müller, A., Rychli, K., Muhterem-Uyar, M., Zaiser, A., Stessl, B., Guinane, C.M., et al., 2013. Tn6188—a novel transposon in *Listeria monocytogenes* responsible for tolerance to benzalkonium chloride. PLoS ONE 8, e76835.

Nichterlein, T., Domann, E., Kretschmar, M., Bauer, M., Hlawatsch, A., Hof, H., et al., 1996. Subinhibitory concentrations of β-lactams and other cell-wall antibiotics inhibit listeriolysin production by *Listeria monocytogenes*. Int. J. Antimicrob. Agents 7, 75–81.

Nichterlein, T., Domann, E., Bauer, M., Hlawatsch, A., Kretschmar, M., Chakraborty, T., et al., 1997. Subinhibitory concentrations of gentamicin reduce production of listeriolysin, the main virulence factor of *Listeria monocytogenes*. Clin. Microbiol. Infect. 3, 270–272.

Nielsen, P.K., Andersen, A.Z., Mols, M., van der Veen, S., Abee, T., Kallipolitis, B.H., 2012. Genome-wide transcriptional profiling of the cell envelope stress response and the role of LisRK and CesRK in *Listeria monocytogenes*. Microbiology 158, 963–974.

Painter, J., Slutsker, L., 2007. Listeriosis in humans. In: Ryser, E.T., Marth, E.H. (Eds.), *Listeria*, Listeriosis, and Food Safety CRC Press, Boca Raton, FL, pp. 85–109.

Pérez-Díaz, J.C., Vicente, M.F., Baquero, F., 1982. Plasmids in *Listeria*. Plasmid 8, 112–118.

Popowska, M., Miernik, A., Rzeczycka, M., Łopaciuk, A., 2010. The impact of environmental contamination with antibiotics on levels of resistance in soil bacteria. J. Environ. Qual. 39, 1679–1687.

Poyart-Salmeron, C., Carlier, C., Trieu-Cuot, P., Courtieu, A.L., Courvalin, P., 1990. Transferable plasmid-mediated antibiotic resistance in *Listeria monocytogenes*. Lancet 335, 1422–1426.

Poyart-Salmeron, C., Trieu-Cuot, P., Carlier, C., MacGowan, A., McLauchlin, J., Courvalin, P., 1992. Genetic basis of tetracycline resistance in clinical isolates of *Listeria monocytogenes*. Antimicrob. Agents Chemother. 36, 463–466.

Premaratne, R.J., Lin, W.J., Johnson, E.A., 1991. Development of an improved chemically defined minimal medium for *Listeria monocytogenes*. Appl. Environ. Microbiol. 57, 3046–3048.

Rakic-Martinez, M., Drevets, D.A., Dutta, V., Katic, V., Kathariou, S., 2011. *Listeria monocytogenes* strains selected on ciprofloxacin or the disinfectant benzalkonium chloride exhibit reduced susceptibility to ciprofloxacin, gentamicin, benzalkonium chloride, and other toxic compounds. Appl. Environ. Microbiol. 77, 8714–8721.

Raz, R., 2012. Fosfomycin: an old-new antibiotic. Clin. Microbiol. Infect. 18, 4–7.

Romanova, N.A., Wolffs, P.F.G., Brovko, L.Y., Griffiths, M.W., 2006. Role of efflux pumps in adaptation and resistance of *Listeria monocytogenes* to benzalkonium chloride. Appl. Environ. Microbiol. 72, 3498–3503.

Roy, H., 2009. Tuning the properties of the bacterial membrane with aminoacylated phosphatidyl-glycerol. IUBMB Life. 61, 940–953.

Safdar, A., Armstrong, D., 2003. Antimicrobial activities against 84 *Listeria monocytogenes* isolates from patients with systemic listeriosis at a comprehensive cancer center (1955–1997). J. Clin. Microbiol. 41, 483–485.

Salimnia, H., Patel, D., Lephart, P.R., Fairfax, M.R., Chandrasekar, P.H., 2010. *Listeria grayi*: van-comycin-resistant, gram-positive rod causing bacteremia in a stem cell transplant recipient. Transpl. Infect. Dis. 12, 526–528.

Sandegren, L., 2014. Selection of antibiotic resistance at very low antibiotic concentrations. Ups. J. Med. Sci. 119, 103–107.

Sauders, B.D., Overdevest, J., Fortes, E., Windham, K., Schukken, Y., Lembo, A., et al., 2012. Diversity of *Listeria* species in urban and natural environments. Appl. Environ. Microbiol. 78, 4420–4433.

Scallan, E., Hoekstra, R.M., Angulo, F.J., Tauxe, R.V., Widdowson, M.-A., Roy, S.L., et al., 2011. Foodborne illness acquired in the United States—major pathogens. Emerg. Infect. Dis. 17, 7–15.

Scortti, M., Lacharme-Lora, L., Wagner, M., Chico-Calero, I., Losito, P., Vázquez-Boland, J.A., 2006. Coexpression of virulence and fosfomycin susceptibility in *Listeria*: molecular basis of an antimicrobial *in vitro–in vivo* paradox. Nat. Med. 12, 515–517.

Sesto, N., Touchon, M., Andrade, J.M., Kondo, J., Rocha, E.P.C., Arraiano, C.M., et al., 2014. A PNPase dependent CRISPR system in *Listeria*. PLoS Genet. 10, e1004065.

Sleator, R.D., Hill, C., 2005. A novel role for the LisRK two-component regulatory system in liste-rial osmotolerance. Clin. Microbiol. Infect. 11, 599–601.

Stanisich, V.A., 1988. 2 Identification and analysis of plasmids at the genetic level. Methods Microbiol. 21, 11–47.

Thedieck, K., Hain, T., Mohamed, W., Tindall, B.J., Nimtz, M., Chakraborty, T., et al., 2006. The MprF protein is required for lysinylation of phospholipids in listerial membranes and confers resistance to cationic antimicrobial peptides (CAMPs) on *Listeria monocytogenes*. Mol. Microbiol. 62, 1325–1339.

Thiele-Bruhn, S., 2003. Pharmaceutical antibiotic compounds in soils—a review. J. Plant Nutr. Soil Sci. 166, 145–167. doi:10.1002/jpln.200390023.

Troxler, R., von Graevenitz, A., Funke, G., Wiedemann, B., Stock, I., 2000. Natural antibiotic susceptibility of *Listeria* species: *L. grayi, L. innocua, L. ivanovii, L. monocytogenes, L. seeligeri* and *L. welshimeri* strains. Clin. Microbiol. Infect. 6, 525–535.

Upadhyay, A., Johny, A.K., Amalaradjou, M.A.R., Ananda Baskaran, S., Kim, K.S., Venkitanarayanan, K., 2012. Plant-derived antimicrobials reduce *Listeria monocytogenes* virulence factors *in vitro*, and down-regulate expression of virulence genes. Int. J. Food. Microbiol. 157, 88–94.

Van Bambeke, F., Balzi, E., Tulkens, P.M., 2000. Antibiotic efflux pumps. Biochem. Pharmacol. 60, 457–470.

Vicente, M.F., Pérez-Dáz, J.C., Baquero, F., Angel de Pedro, M., Berenguer, J., 1990. Penicillin-binding protein 3 of *Listeria monocytogenes* as the primary lethal target for β-lactams. Antimicrob. Agents Chemother. 34, 539–542.

Wilson, R.L., Brown, L.L., Kirkwood-Watts, D., Warren, T.K., Lund, S.A., King, D.S., et al., 2006. *Listeria monocytogenes* 10403S HtrA is necessary for resistance to cellular stress and virulence. Infect. Immun. 74, 765–768.

Witte, C.E., Whiteley, A.T., Burke, T.P., Sauer, J.-D., Portnoy, D.A., Woodward, J.J., 2013. Cyclic di-AMP is critical for *Listeria monocytogenes* growth, cell wall homeostasis, and establishment of infection. MBio 4 e00282–13.

Woodward, J.J., Iavarone, A.T., Portnoy, D.A., 2010. c-di-AMP secreted by intracellular *Listeria monocytogenes* activates a host type I interferon response. Science 328, 1703–1705.

Zawadzka-Skomial, J., Markiewicz, Z., Nguyen-Distèche, M., Devreese, B., Frère, J.-M., Terrak, M., 2006. Characterization of the bifunctional glycosyltransferase/acyltransferase penicillin-binding protein 4 of *Listeria monocytogenes*. J. Bacteriol. 188, 1875–1881.

Zhang, Q., Lambert, G., Liao, D., Kim, H., Robin, K., Tung, C., et al., 2011. Acceleration of emergence of bacterial antibiotic resistance in connected microenvironments. Science 333, 1764–1767.

Chapter 9

Antibiotic Resistance in Enterococci: A Food Safety Perspective

Anuradha Ghosh[1] and Ludek Zurek[1,2]

[1]Department of Diagnostic Medicine and Pathobiology, Kansas State University, Manhattan, KS, USA,
[2]Department of Entomology, Kansas State University, Manhattan, KS, USA

Chapter Outline

GENERAL INTRODUCTION

Enterococci are common members of the microbiota in the gastrointestinal tract of mammals and other animals including birds, insects, and reptiles and can also be found in soil, water, and food (Byappanahalli et al., 2012; Fisher and Phillips, 2009). In adult people, enterococci account for about 1% of the intestinal microbiota (Sghir et al., 2000). *Enterococcus faecalis* and *Enterococcus faecium* are most prevalent in the human gastrointestinal tract and are often used as a fecal indicator in water quality testing; *Enterococcus mundtii* and *Enterococcus casseliflavus* are commonly associated with plants (Byappanahalli et al., 2012). Several species appear to be host-specific such as *Enterococcus columbae* associated with pigeons, *Enterococcus asini* with donkeys, and *Enterococcus canintestini/canis* with dogs (Simjee et al., 2006). The cell numbers of *E. faecalis* in human feces range from 10^5 to 10^7 per gram, and those of *E. faecium* from 10^4 to 10^5 per gram. Overall, the detection of *E. faecium* and *E. faecalis* is less frequent from livestock than from human feces (Franz et al., 1999). For many

Antimicrobial Resistance and Food Safety. DOI: http://dx.doi.org/10.1016/B978-0-12-801214-7.00009-0

years, enterococci were considered harmless to humans and medically unimportant. Because they produce bacteriocins, enterococci have been used widely over recent decades in the food and feed industries as probiotics and due to their role in flavor development and fermentation also as a starter culture in production of fermented salami and several types of ripened cheese (Foulquie Moreno et al., 2006; Franz et al., 2011). Although enterococci, including *E. faecium* have a long history of safe use in food and feed, they have not been included in the list of microorganisms for Qualified Presumption of Safety (QPS) status published by the European Food Safety Authority (EFSA, 2007). Over the past three decades, enterococci have become the third most common nosocomial human pathogens, with a mortality rate up to 23% (Hidron et al., 2008; Karchmer, 2000). The reason for their spread in the hospital environment is partly due to their survival in high temperatures and in chemical disinfectants such as chlorine, glutaraldehyde, and alcohol (Bradley and Fraise, 1996). While the probiotic benefits of some strains are well established, their emergence as nosocomial and opportunistic pathogens along with their potential to horizontally transfer antimicrobial resistance and virulence traits to other bacteria requires a well-contemplated risk/benefit analysis (Foulquie Moreno et al., 2006). Studies from the food safety perspective have demonstrated that *E. faecalis* and *E. faecium* are present as contaminants in cheese, fish, sausages, minced beef, pork, and ready-to-eat food (Foulquie Moreno et al., 2006; Klein, 2003; Pesavento et al., 2014). This chapter introduces enterococci from both perspectives, as probiotics and as nosocomial pathogens, and describes their emergence as a foodborne pathogen with implications to human health.

TAXONOMY AND CHARACTERISTICS

In 1989, species in the genus *Streptococcus* were divided into three separate genera: *Streptococcus*, *Lactococcus*, and *Enterococcus* (Facklam and Collins, 1989). "Fecal streptococci" or "Lancefield's group D streptococci" associated with the gastrointestinal tract of humans and animals as well as with some fermented food products and in a range of other habitats constituted the new genus *Enterococcus*. This genus was placed within the family *Enterococcaceae* with genera *Bavariicoccus*, *Catellicoccus*, *Melissococcus*, *Pilibacter*, *Tetragenococcus*, *Vagococcus* (http://www.bacterio.net/enterococcaceae.html accessed in July 2014). Members of the genus *Enterococcus* are Gram-positive, catalase-negative, facultative anaerobic cocci with low (<50 mol%) G + C content and occur singly, in pairs, or as short chains. They are able to grow at 10°C and 45°C, in 6.5% NaCl, at pH 9.6, and hydrolyze esculin in the presence of 40% bile. They are chemoorganotrophic and produce L-lactic acid from hexoses by homofermentative lactic acid fermentation (Facklam and Collins, 1989). Enterococci contain C14:0, C16:1, C16:0, C18-unsaturated and c19-delta as the most prevalent fatty acids (Lang et al., 2001). At present, the genus consists of 53 species (http://www.bacterio.net/enterococcus.html accessed on

July 2014). Many of the recently described enterococcal species vary in their physiological properties from those of typical enterococci such as *Enterococcus durans, E. faecalis, E. faecium, Enterococcus gallinarum, Enterococcus hirae, E. casseliflavus*, and *E. mundtii*. There are also several species including *Enterococcus cecorum, E. columbae, Enterococcus dispar, Enterococcus pseudoavium, Enterococcus saccharolyticus*, and *Enterococcus sulfureus* that do not react with the group D antiserum.

Sequencing of the 16S rRNA gene with the universal primers for the species-level differentiation failed to distinguish phylogenetically closely related enterococcal species (Poyart et al., 2000). Recently, several molecular methods have been developed for species identification including, randomly amplified polymorphic DNA (RAPD) analysis and PCR amplification of the ribosomal intergenic spacer (ITS-PCR). However, these techniques offer various sensitivity and are also difficult to adopt and apply in a routine laboratory practice due to their high cost and requirement for highly skilled personnel (Kirschner et al., 2001). PCR amplification and sequencing of the internal DNA fragment of the manganese-dependent superoxide dismutase gene (*sodA*) resolved the majority of *Enterococcus* species and seem to be a viable low-cost approach for species identification (Frolkova et al., 2012; Poyart et al., 2000).

STATUS AS PROBIOTICS

Although the use of enterococci as human probiotics remains a controversial issue, some strains of *E. faecium* and *E. faecalis* are available on the market; however, to a much less extent compared to lactobacilli or bifidobacteria. Either whole cells, cell fragments, or cell lysates have been used in commercial probiotics based on *E. faecalis* and *E. faecium*. Only very few studies assessed the effectiveness of probiotic strains such as *E. faecium* CRL 183, SF68 (NCIMB 10415), L-X, L3, M-74, and *E. faecalis* DSM 16440, DSM 16431 in humans for the treatment of diarrhea, irritable bowel syndrome, in lowering serum cholesterol and for immune regulation (Christoffersen et al., 2012; Franz et al., 2011; Yamaguchi et al., 2013). Consequently, there is not enough data to make any firm conclusions on the efficacy of enterococci as probiotics for people.

In contrast, *Enterococcus* along with *Lactobacillus, Bacillus*, and *Saccharomyces* are the most common bacterial genera used as probiotics in livestock and poultry. Several studies have been performed to assess the effect of probiotic treatments on the immune system, gastrointestinal microbiota, and production efficiency of food animals. After feeding sows and piglets with the probiotic strain *E. faecium* SF68, it was shown that the total anaerobe and coliform bacterial populations were not significantly affected, in either sows or piglets; however, a remarkable decline in the frequency of β-hemolytic O141 serovars of *Escherichia coli* was observed in the intestinal contents of piglets. This also offered a possible explanation for the reduction in cytotoxic T-cells in these animals (Scharek et al., 2005). Other interdisciplinary studies on the modes of action of probiotics in swine showed that *E. faecium* NCIMB 10415 reduced the virulence gene expression of putative pathogenic *E. coli* strains (Taras et al., 2006) as well as the rate of natural *Chlamydia* infections (Pollmann et al., 2005). Although the exact mode of action of probiotics remains unclear, a correlation among their administration, reduction in pathogen load, and host inflammatory responses was observed. Broiler chickens challenged with *E. coli* K88 and treated with a probiotic strain of *E. faecium* had greater body weight, lower concentrations of *E. coli* and *Clostridium perfringens*, and increased populations of *Lactobacillus* and *Bifidobacterium* compared to those of the control group (Cao et al., 2013). In dairy cows, positive effects of direct-fed microbial supplementation (containing yeast and two *E. faecium* strains incorporated into a corn meal carrier) were observed in terms of enhanced ruminal digestibility of forage and increased milk yield (Nocek and Kautz, 2006) and higher percentage of milk fat and protein (Oetzel et al., 2007). Emmanuel et al. (2007) showed that feeding *E. faecium* strain EF212 in combination with yeast induced an inflammatory response in feedlot steers fed high-grain diets. Szabo et al. (2009) demonstrated that weaning piglets with *Salmonella enterica* serovar Typhimurium DT104 infections when orally treated with the probiotic strain NCIMB 10415 shed increased concentrations of *Salmonella* in their feces. Greater production of specific antibodies against *Salmonella* compared to the animals in the control group was also observed.

CLINICAL STATUS

Human Infections

Enterococci are opportunistic human pathogens that have become commonly resistant to many antimicrobial agents. Although *E. faecalis* and *E. faecium* are the most common clinical isolates, other species to cause human infections include *Enterococcus avium*, *E. gallinarum*, *E. casseliflavus*, *E. hirae*, *E. durans*, *Enterococcus raffinosus*, and *E. mundtii* (de Perio et al., 2006). Enterococci can cause endocarditis, urinary tract infections, intra-abdominal, pelvic, and wound infections, superinfections in immunocompromised patients, and bacteremia (often together with other organisms). Typically, *E. faecalis* is the only *Enterococcus* species isolated from the obturated root canal and causes the majority of human enterococcal endodontic infections (Love, 2001; Murray, 1990). Enterococci are second to staphylococci as a leading cause of nosocomial infections, accounting for about 12% of hospital-associated infections annually in the United States (Hidron et al., 2008). Some nosocomial infections are treated with single-drug therapy, most often with penicillin, ampicillin, or vancomycin. However, enterococci carry a large number of inherent and acquired resistance traits, including resistance to cephalosporins, chloramphenicol, erythromycin, clindamycin, tetracycline, aminoglycosides, oxacillin, quinupristin/dalfopristin, vancomycin, and others (Frye and Jackson, 2013; Murray, 1990). Newer drugs such as linezolid, daptomycin, and tigecycline (alone or in combination with older drugs) are used to treat enterococcal infections; however, some cases of resistance to these antibiotics have already been reported (Arias and Murray, 2008, 2012). Antibiotic-resistant clinical isolates often carry pathogenicity islands (PAI) encoding several genes that contribute to pathogenesis and are responsible for rapid evolution of nonpathogenic strains into pathogenic forms by horizontal transfer. For example, a 150 kb PAI encoding multiple virulence factors, the cytolysin toxin, enterococcal surface protein (Esp), aggregation substance, and other traits for metabolic functions has been characterized in *E. faecalis* (McBride et al., 2009). Intra- and interspecies dissemination of resistance and virulence traits takes place via horizontal gene transfer which is mainly mediated by conjugative pheromone-responsive and broad host range incompatibility group-18-type plasmids and other mobile genetic elements (Palmer et al., 2010).

The vancomycin-resistant enterococci (VRE) pose a major therapeutic challenge due to their intrinsic resistance to commonly used antibiotics and their ability to acquire resistance to many available antibiotics, either by mutation or by acceptance of foreign genetic material. Since the first outbreak of nosocomial VRE in the mid- and late 1980s, the prevalence of VRE increased from 0.4% to 23.2% in intensive care facilities and from 0.3% to 15.4% in nonintensive care settings (Martone, 1998). Meta-analysis of published case reports indicated that across the United States, the VRE acquisition rate in intensive care units was 10.2% (Ziakas et al., 2013). Interestingly, the ratio of *E. faecalis* to *E. faecium* infections changed in favor of *E. faecium* in the hospital environment during the mid- to late 1990s (reviewed in Top et al., 2008) with the worldwide emergence of the specific *E. faecium* clonal complex CC17. These nosocomial strains are characterized by ampicillin and quinolone resistance and the majority contains putative virulence genes *esp* and hyl_{Efm} (Willems et al., 2005). Considering nosocomial co-endemicity of VRE and meticillin-resistant *Staphylococcus aureus* (MRSA), horizontal transfer of *vanA*- and/or *vanB*-containing transposons could facilitate transformation of MRSA into vancomycin-resistant *S. aureus* (VRSA). In 2002, the first case of VRSA was reported from a diabetic patient in Michigan. A potential source of *vanA* in this VRSA was vancomycin-resistant *E. faecalis* that was also isolated from this patient (Flannagan et al., 2003). Plasmid analysis showed that the VRE isolate carried Tn*1546*-like transposon with *vanA* that could be transposed into the plasmid that was present in the vancomycin-susceptible MRSA strains. The Centers for Disease Control and Prevention (CDC) has recently confirmed the 13th case of the *vanA*-induced VRSA infection in the United States (Limbago et al., 2014). In comparison to all previous VRSA cases that belonged to CC5, a lineage associated primarily with health care, the most recent (13th) United States VRSA isolate was a community-associated clone.

Animal Infections

Some enterococcal species have been associated with infections in food animals and poultry. In chickens, *E. durans* has been implicated in causing bacteremia and encephalomalacia (Abe et al., 2006; Cardona et al., 1993). Cases of diarrhea associated with *E. durans* in calves (Rogers et al., 1992) and with *Enterococcus villorum* in piglets (Vancanneyt et al., 2001) have also been reported. Furthermore, in broilers, septicemia and endocarditis caused by *E. hirae* (Chadfield et al., 2005; Kolbjornsen et al., 2011), as well as pulmonary hypertension syndrome caused by *E. faecalis* (Tankson et al., 2001), have been documented. Steentjes et al. (2002) reported unilateral amyloid arthropathy in broiler breeders associated with *E. faecalis*. In addition, *E. faecium*, *E. faecalis*, and *E. saccharolyticus* are known to cause bovine mastitis (Wyder et al., 2011).

FOOD ANIMALS AS A RESERVOIR OF ANTIBIOTIC-RESISTANT STRAINS

Research on food animals and poultry in Europe and other parts of the world has provided a large amount of information on the ecology of antibiotic-resistant enterococci in the agricultural environment. In this section, we discuss some of the studies that highlight how use of antimicrobials on farms creates a reservoir of antibiotic-resistant strains. A comprehensive study carried out in Europe revealed that the distribution of *Enterococcus* species differs among countries. For example, in Spain and the United Kingdom, *E. faecalis* and *E. faecium* were most commonly isolated from clinical and environmental sources, while in Sweden and Denmark, *E. hirae* was highly prevalent among food animals and slaughtered animals (Kuhn et al., 2003). In the same study, enterococci were found in 77% of the Spanish samples from farmlands fertilized with pig manure and on crops grown on such farmlands, but in the United Kingdom, enterococci were found only in 21% of such samples. In farmland and crops receiving no animal fertilizer, enterococci were also isolated in almost 30% of the samples, but in lower numbers than in the land with manure (Kuhn et al., 2003). In a similar study in Germany, of 416 strains of enterococci isolated from 155 samples of food animal origin, 72% were *E. faecalis* and 13% *E. faecium* (Peters et al., 2003). In 2000, Aarestrup's research group reported isolation of *E. faecium* from 211 broilers and 55 pigs in Denmark; 55 poultry farms (turkeys and broilers) and four swine farms in Norway; and 52 poultry and 43 swine in Finland (Aarestrup et al., 2000b). Only a limited number of isolates were resistant to monensin or salinomycin. Isolates from broilers in Denmark were frequently resistant to avilamycin, while all isolates from Finland and Norway were susceptible. The same phenomenon could be observed for avoparcin, bacitracin, tylosin, and virginiamycin. In general, a correlation between the use of antimicrobial agents in each country and the occurrence of associated resistance was observed and indicated that use of antimicrobial agents as growth promoters selected for *E. faecium*-resistant strains in food animals most commonly in Denmark, while only to a limited extent in Finland and Norway (Aarestrup et al., 2000b). In another study, Aarestrup (2000) reported that glycopeptide-resistant enterococcal (GRE) isolates from poultry and pigs were separated into two distinct pulsed field gel electrophoresis (PFGE) types. The persistence of GRE in Danish food animals after the ban on glycopeptides may be due to continued use of macrolides in growth promotion and prophylaxis, which co-selected the *ermB* and *vanA* genes (Aarestrup, 2000). Multidrug-resistant (tetracyclines, chloramphenicol, ampicillin, ciprofloxacin, nitrofurantoin, vancomycin, aminoglycosides) *E. faecalis* and *E. faecium* were recently detected in six Portuguese swine farms (samples included feces, nostril swabs, feed, water, waste lagoons, walls, and air) (Novais et al., 2013).

Enterococci from pigs in some European countries (Denmark, Spain, and Sweden) were also examined for susceptibility to antimicrobial agents and copper and presence of selected resistance genes. Isolates from Spain and Denmark displayed greatest levels of resistance to macrolides, aminoglycosides, tetracycline, and vancomycin when compared to those from Sweden, where the use of antimicrobials for growth promotion was banned in 1985. Similar genes were found to encode resistance in different countries, but *tetL* and *tetS* genes were more frequently found among isolates from Spain. Notably, the transferable copper resistance gene *tcrB* was present in all copper-resistant isolates from several countries (Aarestrup et al., 2002). Later, Hasman and Aarestrup (2005) found an association between *tcrB* (located closely upstream of Tn*1546*) and *ermB* in *E. faecium* strain from a pig. However, it was noted that the long-term use of copper sulfate did not result in maintenance of high levels of macrolide and glycopeptide resistances. In contrast, a recent study from Portugal (Silveira et al., 2014) investigating the possible cotransfer of copper tolerance and antibiotic resistance traits in enterococci from various environments showed that copper tolerance may contribute to maintenance of multidrug-resistant enterococci on animal farms and in hospitals.

After the discontinuation of avoparcin in New Zealand broiler production in 2000, a surveillance study was conducted (2002–2003) to determine the prevalence of VRE on farms. A total of 382 enterococci were isolated from 213 fecal samples collected from 147 individual poultry farms using the enrichment broth without antibiotics. Of the 382 isolates, 5.8% (22 isolates) were resistant to vancomycin and 64.7% were resistant to erythromycin while 98.7% and 14.9% of isolates showed very high MICs for bacitracin and avilamycin, respectively. However, none of the isolates showed resistance to ampicillin or gentamicin (Manson et al., 2004). In parallel, when the 213 fecal enrichment broths were plated on mEnterococcus agar-containing vancomycin, 86 VRE strains were recovered; 66% of these isolates were *E. faecium* and the remainder were *E. faecalis*. These data demonstrated that persistence of VRE on farms in New Zealand was irrespective of the current status of its use on farms (Manson et al., 2004).

Four years after the ban on avoparcin in Korean livestock, the prevalence of VRE in 342 chicken meat samples and 214 swine fecal samples was 17% and 2%, respectively; while all 110 bovine samples were negative for VRE. All 61 VRE isolates were VanA-type *E. faecium* (vancomycin-resistant *E. faecium*, VREF) that expressed high-level resistance to vancomycin and teicoplanin. Although the VREF isolates had heterogeneous PFGE patterns, identical or closely related profiles were observed among strains isolated from the same farm (Lim et al., 2006a). Similarly, after the ban of avoparcin as a feed additive in Taiwan in 2000, a nationwide surveillance study was conducted between 2000 and 2003 to determine the prevalence of VRE in chicken farms. The authors observed that among the *E. faecalis* (*n* = 1021) and *E. faecium* (*n* = 967) isolates studied, resistance to tetracycline, erythromycin, high-level aminoglycosides, ciprofloxacin, and chloramphenicol either increased or remained high. In contrast, during the study period, the proportion of VRE decreased by 10% (from 13.7% to 3.7%) for *E. faecalis*, and from 3.4% to 0% for *E. faecium*. Moreover, there was an overall decrease in the prevalence of VRE-positive chicken farms: from 25% (15/60) in 2000 to only 8.8% (7/80) in 2003 (Lauderdale et al., 2007). Another research group in China tested chicken and pig swab samples collected from various provinces and isolated a total of 453 enterococcal isolates that represented six different species. All isolates were sensitive to vancomycin; however, frequent resistance (50–93%) was detected to tetracycline, amikacin, erythromycin, rifampin, and high-level streptomycin; intermediate resistance rate (~30%) to phenicols (chloramphenicol and florfenicol), enrofloxacin, and high-level gentamicin and low resistance rate to penicillins (8% to penicillin and 4% to ampicillin). It was evident that resistance of enterococci to most antimicrobials was more prevalent in China than in European or other Asian countries (Liu et al., 2013).

Resistance to aminoglycosides, lincosamides, macrolides, nitrofurans, penicillins, quinolones, streptogramins, and tetracyclines has been reported from food animals in the United States (Frye and Jackson, 2013). With the exception of daptomycin, resistance to the newer antimicrobials (linezolid and tigecycline) has not been detected in food animals (Jackson et al., 2011). Debnam et al. (2005) collected samples and cultured enterococci from the poultry houses on US farms with and without use of antibiotics such as flavomycin, virginiamycin, and bacitracin. Nine different species of enterococci were identified with *E. faecalis* more frequently obtained from chick boxliners (*n* = 176; 92%) and carcass rinses (*n* = 491; 69%), whereas *E. faecium* was found more frequently in litter (*n* = 361; 77%) and feed (*n* = 67; 64%). *E. faecalis* (*n* = 763; 52%) and *E. faecium* (*n* = 578; 39%) were isolated most often from the farm and houses, regardless of antimicrobial treatment indicating the antimicrobial use on this farm did not alter the resident population of enterococci (Debnam et al., 2005). Until recently, vancomycin resistance in food animals had not been observed in the United States. The first report was published in 2010 in which VREF strains (MIC = 256 µg/mL) were isolated from swine in three Michigan counties (Donabedian et al., 2010). The origin of the detected *vanA* gene located on Tn*1546* in these samples remains undetermined. So far, no other vancomycin resistance genes have been reported in the US food production system.

FARM ENVIRONMENT AND ANTIBIOTIC-RESISTANT STRAINS

The farm environment itself and incoming (animal feed) and outgoing (manure/ slurry) material may act as a source and a reservoir of antibiotic-resistant strains. Previously, Manero et al. (2006) showed that VREF isolated from pig feces, pig slurry, and urban sewage in Spain had a high population similarity index based on biochemical fingerprints and predicted that certain strains circulate through the food chain from farms to humans. Nilsson et al. (2009) collected samples from environmental compartments including air inlets and outlets, feed and water lines, and the floor from three Swedish broiler farms. They found that the temporal changes in environmental contamination were similar among different farms during the sampling period starting from the arrival of the birds till their slaughter. All identified isolates ($n = 85$) were VREF with a MIC $\geq 128\,\mu g/mL$ and belonged to ST310, the clone previously described to dominate among Swedish broilers. The two farms that were positive for VRE in the environmental samples also had 55% and 70% slaughtered birds colonized with VRE, respectively. Although the number of VRE-positive samples differed among the farms, VRE persisted in the compartments even after cleaning (Nilsson et al., 2009). Further evidence on inter-batch contamination among Swedish poultry farms was provided by Jansson et al. (2012) where 9 out of 12 farms (75%) were culture-positive prior to and 31% after cleaning/disinfection. Interestingly, they found five out of six samples from forklift tires positive for VREF and suggested that these vehicles contribute to cross-contamination among farms. Another new study showed that about 88% (150/171) of dust samples collected in Portuguese swine facilities were positive for enterococci with *E. faecalis* as the most prevalent species (Braga et al., 2013). Moreover, using selective plating, enterococcal isolates were recovered from 24% (41/171) swine facilities and ten of the isolates were VRE with MIC $\geq 256\,\mu g/mL$. The *vanA* gene was detected in all *E. faecium* (VREF), *vanB* was present in two of the six vancomycin-resistant *E. hirae* and in one *E. gallinarum*. The MLST types of selected VREF strains were closely related to pig and human isolates from European countries and Brazil (Braga et al., 2013).

Frequent isolation of potential foodborne pathogens such as *Salmonella* spp., *E. coli* O157:H7 as well as antibiotic-resistant strains of *E. faecium* and *Campylobacter jejuni* from animal feed and its raw components (reviewed in Sapkota et al., 2007) suggests that feed represents one potential route by which foodborne pathogens and/or antibiotic-resistant strains are introduced into the food animal. Frequent enterococcal contamination of raw feed ingredients comprising of animal and plant byproducts has been reported (Ge et al., 2013; Kinley et al., 2010). Strains of VRE were detected in chicken feed in the

United States but the origin of VRE was not determined (Schwalbe et al., 1999). A Portuguese group reported that enterococci isolated from the commercial broiler feed and raw feeding ingredients were frequently resistant to rifampicin, erythromycin, tetracycline, and nitrofurantoin whereas a lower percentage of resistance was observed to chloramphenicol, ciprofloxacin, vancomycin, and ampicillin (de Costa et al., 2007). In another finding by Channaiah (2009), over half (51.6%) of the samples ($n = 89$) from the feed mills and swine farms in the United States were positive for enterococci. It was noted that the feed mill samples were less commonly contaminated compared to the farm samples (46% vs. 59%) indicating that some feed becomes contaminated on farms. The proper conditions for feed storage as well as pest management on farms greatly impact the likelihood of feed contamination with bacteria and fungi. Overall, there is a need for improved quality standards in feed mills and animal farms to prevent contamination of feed before it is consumed by animals and potentially enters the food chain.

STATUS AS FOOD CONTAMINANTS

Transfer of the bacterial pathogens of animal origin can occur via the consumption of contaminated animal or vegetable food products. Meat products can become contaminated by fecal material at the slaughterhouses, whereas vegetables may be contaminated in the field with manure or sewage water used for fertilization and irrigation (Nilsson, 2012; WHO, 2011). While the role of enterococci as an opportunistic nosocomial pathogen is well established, their ability to cause foodborne illnesses remains largely unknown. Enterococci have been found as a contaminant in raw meat in concentrations of 10^2 to 10^4 CFU/g; in fermented food products, such as salami (10^2–10^5 CFU/g) and mozzarella cheese (10^5–10^7 CFU/g) (Klein, 2003). Although enterococci have been reported to cause diarrhea and other infections in animals, this has not been demonstrated in humans. In some cases, vomiting and headaches indicative of food intoxication are believed to be caused by the ingestion of fermented food-containing enterococci that have produced biogenic amines (Giraffa, 2002). Importantly, due to their ability to acquire and transfer antibiotic-resistant traits that could establish in the resident microbiome of the human digestive tract, enterococci represent an important foodborne risk. Hence, enterococci have been included in the "Bad Bug Book" published by the Center for Food Safety and Applied Nutrition of the Food and Drug Administration (FDA, 2012), that provides the most current information on the major known agents that cause foodborne illness.

Enterococci can be detected not only in raw materials but also in ready-to-eat food, of animal (e.g., milk, cheeses, meat, and fish products) and plant (e.g., vegetable and fruit) origins (Abriouel et al., 2008). Huys et al. (2004) reported that tetracycline resistance in food enterococci mainly originating from European cheeses, was conferred by the same types of *tet* genes (*tetM*, *tetL*, *tetS*) as those previously found among veterinary or clinical enterococcal isolates. Moreover, co-resistance to erythromycin and/or chloramphenicol reinforced the notion that *tet* genes located on mobile genetic elements may select for multiple resistance in food isolates. Macovei and Zurek (2007) showed that ready-to-eat food from fast-food restaurants was commonly contaminated with seven different species of *Enterococcus*. Although there was seasonal variation (summer greater than winter) in the prevalence of enterococcal contamination of various food types, the overall concentration throughout the year averaged ~10^3 CFU/g. Enterococci from summer samples were resistant to tetracycline (22.8% of isolates), erythromycin (22.1%), and kanamycin (13.0%). The *tetM* gene was detected among the summer (12%) and winter (35%) isolates; however, occurrence of putative virulence factors such as *gelE*, *asa1*, *cylA*, and *esp* was low (Macovei and Zurek, 2007). McGowan et al. (2006) reported the presence of antibiotic-resistant strains of enterococci in retail fruits, vegetables, and meat, and the majority was represented by *E. casseliflavus*. Later, the authors also found that isolates commonly carried the virulence factors *cob* and *gelE* but rarely contained *cylM*, *cylB*, *cylA*, and *efaAfm* (McGowan-Spicer et al., 2008). Recently, an Italian research group analyzed a total of 1,315 food samples purchased at 150 supermarkets and found *E. faecalis* as the dominant species in raw meat (beef, poultry, pork) and *E. faecium* in retail products (cheese, ready-to-eat salads, ham). In addition, *E. faecalis* was commonly resistant to tetracycline (60.6%) and gentamicin (21.9%) (Pesavento et al., 2014).

Tanimoto et al. (2005) reported the isolation of VRE from Japanese chicken meat that was imported from China; but in 2012, VanN-type clonal VREF strains with low MIC (12 μg/mL) were isolated from a sample of domestic meat as well (Nomura et al., 2012). VRE with acquired mechanisms of resistance (*vanA* and *vanB2*) were detected in 9 of 229 samples (3.9%, obtained from chicken, veal, and rabbit) tested after a decade of avoparcin ban in Spanish animal husbandry (Lopez et al., 2009). The two *vanB2* isolates displayed multiresistance to ampicillin, erythromycin, tetracycline, streptomycin, kanamycin, ciprofloxacin, chloramphenicol, and trimethoprim–sulfamethoxazole and harbored the *ermB*, *tetM*, *ant(6)*, and *aph(3′)-III* genes. Most of *vanA* enterococci showed erythromycin and tetracycline resistance and contained the *ermB* and *tetM* genes. One *vanB2* ST17 *E. faecium* was obtained from each of chicken meat and veal while one *vanA* ST78 *E. faecium* was isolated from rabbit meat. Importantly, cotransfer of the *ermB* and *vanA* genes by conjugation was demonstrated in one *vanA*-positive *E. faecium* isolate (Lopez et al., 2009).

The combination of the streptogramins, quinupristin, and dalfopristin, was approved in the United States in late 1999 for the treatment of VREF infections. Another streptogramin, virginiamycin, has been used at subtherapeutic concentrations to promote growth of farm animals, including chickens since 1974. In 2001, McDonald and others showed that the enterococcal isolates (11 out of 407 samples, 3%) from chickens sold in US supermarkets had comparatively high levels of resistance (MICs: 4–32 μg/mL) to streptogramins. The low prevalence

(3/334, 1%) and low level of resistance of *E. faecium* strains in tested outpatient stool specimens suggest that the use of virginiamycin in animals did not substantially affect human enterococcal microbiota (McDonald et al., 2001).

SHARING OF ENTEROCOCCAL STRAINS AMONG FOOD ANIMALS, FOOD, AND HUMANS

Several studies from different parts of the world have demonstrated that enterococci of food–animal origin, particularly strains that are vancomycin resistant, have a capacity to colonize the human digestive tract (Table 9.1). In Europe, where avoparcin was used for years as an animal growth promoter, carriage of VRE in the community was recorded as high as 28%. This is considerably higher than that in the United States, where detection of VRE is uncommon outside the hospital environment and where glycopeptides have never been approved as growth promoters in animal agriculture (Coque et al., 1996). Although VRE was not the primary cause of nosocomial infections in Europe, *vanA*-positive enterococci were readily detected outside hospitals in several European countries (Simjee et al., 2006). Ingesting VREF associated with chickens or virginiamycin-resistant *E. faecium* associated with pigs, resulted in resistant strains appearing in stools of volunteers for up to 14 days, suggesting multiplication during intestinal transit (Sorensen et al., 2001). In another experiment involving human volunteers, after the ingestion of a VREF isolate of chicken origin, together with a vancomycin-susceptible *E. faecium* recipient of human origin, transconjugants were recovered in three of six volunteers (Lester et al., 2006). This study showed the occurrence of intraspecies transfer of the *vanA* gene (residing on mobile genetic elements) between isolates of two different origins. This suggested that transient intestinal colonization represents a risk for spread of resistance traits to other commensal enterococci with potentially serious consequences, especially for immunocompromised patients and patients with internal wounds (Lester et al., 2006). Other studies demonstrating transfer of resistance genes of enterococci between animal and human origin in mice models have been reviewed by Hammerum et al. (2004).

In the late 1990s, van den Bogaard and others collected fecal samples from turkeys at 47 farms and from 47 turkey farmers to test the hypothesis that VRE from animals can colonize humans. Fifty percent of turkeys and 39% of the samples from the turkey farmers were positive for VRE containing the identical *vanA* gene cluster. PFGE analysis showed the indistinguishable pattern (identical pattern of 17 bands) between samples from one farmer and his turkey flock. Furthermore, VRE were found in 20% of fecal specimens from turkey slaughterers and 14% of specimens from area residents (van den Bogaard et al., 1997). Subsequent study showed that turkey flocks receiving avoparcin in feed had a higher prevalence of VRE (60%) than flocks without the glycopeptide (8%) (Stobberingh et al., 1999). Although the PFGE patterns of VRE isolated from different treatment groups were heterogeneous, the same PFGE pattern was

TABLE 9.1 Clonal Spread of VRE and HLGR Enterococci among Healthy Humans, Animals, Food, and Patients Based on PFGE ± Multilocus Sequence Typing (MLST)

	Host	Organism	Resistance Genes	PFGE ± MLST	References
VRE	Turkey, turkey farmer	E. faecium	vanA gene cluster	Identical clone	van den Bogaard et al. (1997)
	Turkeys, turkey farmers	E. faecium	vanA gene cluster	Closely related clone	Stobberingh et al. (1999)
	Broiler, broiler farmer	E. hirae	vanA gene cluster	Identical clone	van den Bogaard et al. (2002)
	Broilers, patients	E. faecalis	vanA gene cluster	Identical clone	Manson et al. (2003)
		E. faecium		Closely related clone	
	Human, pigs	E. faecium	vanA gene cluster	Closely related clone	Hammerum et al. (2004)
	Poultry, poultry farmers	E. faecium	vanA gene cluster	Identical clone	Johnsen et al. (2005)
	Humans, patients, animals[b], foods[a,b]	E. faecium	vanA gene cluster	Closely related clone, CC17	Biavasco et al. (2007)
	Turkey meat[a], human, patient	E. faecalis	vanA gene cluster	Closely related clone, ST116	Agerso et al. (2008)
	Pigs, humans	E. faecium	vanA gene cluster	Closely related clone, CC5	Freitas et al. (2011)
	Pigs, patients	E. faecium		Identical clone, CC17-ST132	
		E. faecalis		Identical clone, CC2-ST6	
HLGR	Patients, pork[a]	E. faecalis	aac(6′)Ie-aph(2″)Ia	Closely related clone	Donabedian et al. (2003)
	Human, chicken[a]	E. faecalis		Identical clone	
	Human, poultry	E. faecalis	aac(6′)-aph(2″)Ia	Closely related clone	Novais et al. (2006)
		E. faecium	aph(3′)IIIa	Closely related clone	
	Pigs, patients, pork[a], humans	E. faecalis	aac(6)Ie-aph(2″)Ia	Closely related clone, ST16	Larsen et al. (2010)

VRE, vancomycin resistant enterococci; HLGR, high level gentamicin resistant enterococci.
[a]Indicates food samples; the other samples are feces/fecal swabs originated from animals and humans.
[b]Animals, (poultry; pig); food, (poultry meat, pork meat, cheese).

found among the human and animal isolates, and similar *vanA*-containing transposons were found in VRE isolates from both groups (Stobberingh et al., 1999). Widespread resistance to chloramphenicol, macrolides, kanamycin, streptomycin, and tetracycline was found among *E. faecalis* and *E. faecium* isolated from the general public (98 and 65 isolates), broilers (126 and 122), and pigs (102 and 88) in 1998 (Aarestrup et al., 2000a). Avilamycin resistance was only found in 35% of isolates from broilers and none of the isolates from humans and pigs. All *E. faecium* isolates from humans were susceptible to vancomycin; however, one VREF was isolated from the human stool sample using selective enrichment. All other VRE were obtained from broilers (10%) and pigs (17%). Resistances to erythromycin, kanamycin, and tetracycline were common in both hosts. All VRE contained the *vanA* gene, all chloramphenicol-resistant isolates were positive for the *cat*(pIP501) gene, and all gentamicin-resistant isolates contained the *aac6-aph2* gene. In addition, resistant strains of human and animal origin frequently harbored *ermB* (erythromycin resistance), *aphA3* (kanamycin resistance), and *tetM* (tetracycline resistance). Similar resistance patterns and resistance genes were detected frequently, indicating that transmission of resistant enterococci or resistance genes likely takes place among humans, broilers, and pigs (Aarestrup et al., 2000a).

In a Norwegian poultry farm study (van den Bogaard et al., 2002), a total of 73 *vanA* VRE were recovered from broilers, laying hens, their respective farmers, and poultry slaughterers. Identical or closely related *E. hirae* isolates were obtained from broilers and broiler farmers. Widespread occurrence of similar *vanA* elements across various PFGE patterns among humans and chickens suggested frequent transfer of transposons (van den Bogaard et al., 2002). Another study focused on two Norwegian poultry farms with previous use of avoparcin and obtained 136 fecal glycopeptide-resistant *E. faecium* (GREF) and 86 glycopeptide-susceptible *E. faecium* (GSEF) isolates from farmers and poultry on three separate occasions in 1998 and 1999. PFGE and plasmid DNA analyses revealed 22 GREF and 32 GSEF PFGE types within shifting polyclonal animal and human *E. faecium* populations and indicated the presence of transferable plasmid-mediated *vanA* resistance. Examples of the dominant and persistent GREF PFGE types supported the notion that environmentally well-adapted GREF types may counteract the reversal of resistance and a "plasmid addiction system" possibly contributes to the persistence of GREF in nonselective environments (Johnsen et al., 2005). A subsequent study showed that the prevalence of GRE in poultry flocks declined gradually between 1998 and 2003 from 100% to 78.5%, while GRE among farmers did not change (Sorum et al., 2006). Since macrolides have never been used in Norwegian poultry farms, the co-selection does not explain this situation. Detection of the variety of STs of GRE among farmers suggested that the poultry GRE were reintroduced to the farmers. Stable distribution and long-term persistence of antimicrobial resistance in reservoirs without apparent antimicrobial selective pressure indicates an important role of the "post-segregation killing system" (Sorum et al., 2006). After 7 years of the ban of avoparcin, a VREF isolate was detected in a Danish healthy volunteer who consumed probiotics containing a high concentration of *E. faecium*. The PFGE profile of the VREF was not related to the probiotic strains rather it had close resemblance to a porcine VRE isolate, indicating a pig origin of the strain (Hammerum et al., 2004).

Biavasco et al. (2007) studied a total of 154 VanA-type VRE from humans ($n = 69$), animals ($n = 49$), and food ($n = 36$). Most of the strains ($n = 142$) were collected in Italy between 1997 and 2003, and the rest from Belgium ($n = 1$) and Norway ($n = 10$). PFGE revealed two small genetically related clusters representing *E. faecium* from animal and human feces and these belonged to the epidemic lineage of CC17. Human intestinal and animal *E. faecium* isolates carried large (>150 kb) plasmids with *vanA* and 80% of strains contained the Tn*1546*. Virulence determinants were detected in all reservoirs but were significantly more frequent ($P < 0.02$) among clinical strains. Multiple virulence determinants were found in clinical and meat *E. faecalis* isolates. The presence of indistinguishable *vanA* elements (mostly plasmid-borne) and virulence determinants in different enterococcal species and with PFGE-diverse patterns suggested that all VRE might be potential reservoirs of resistance determinants and virulence traits transferable to human-adapted clusters (Biavasco et al., 2007).

In Portugal, analysis of 247 fecal enterococcal isolates from 99 healthy Portuguese individuals revealed the presence of VRE (5%) that were highly resistant to streptomycin (52% of total VRE), kanamycin (40%), or gentamicin (11%). Most isolates were also resistant to tetracycline, erythromycin, ciprofloxacin, and quinupristin–dalfopristin. A number of resistance traits including the *vanA* (two Tn*1546* types), *vanC1*, *ermB*, *aac(6')-aph(2")-Ia*, *aph(3')-IIIa*, *vatE*, and *vatD* genes were detected. *E. faecalis* and *E. faecium* isolates with high-level resistance to gentamicin were related to previously described Portuguese poultry isolates based on PFGE analysis (Novais et al., 2006). These authors further characterized the molecular diversity of *vanA* VRE (176 isolates/87 PFGE types) from different sources and cities in Portugal (1996–2004): food animals ($n = 38$ isolates out of 31 samples), hospitalized humans ($n = 101/101$), healthy human volunteers ($n = 7/4$), and environmental sources ($n = 30/10$) (Novais et al., 2008). Twenty-four Tn*1546* variants were identified, all located on plasmids (30–250 kb) and some were source-specific. This reflected a complex epidemiology of VRE in Portugal involving both clonal spread and plasmid dissemination. Apparent Tn*1546* heterogeneity among enterococci from a variety of sources may reflect frequent genetic exchange events and evolution of particular widely disseminated genetic elements (Novais et al., 2008).

Extended sampling of poultry meat in the Danish Integrated Antimicrobial Resistance Monitoring and Research Program (DANMAP) during 2005–2006 detected no *vanA*-positive *E. faecalis* (VREFs) in chicken meat while three VREFs were isolated from imported turkey meat without selective enrichment (Agerso et al., 2008). Two VREF were isolated out of 525 fecal samples from healthy human volunteers using a VRE-selective approach. All VREF isolates harbored *vanA*, *tetM*, and *ermB* conferring resistance to vancomycin/teicoplanin, tetracycline, and erythromycin, respectively. They all belonged to ST116 and showed similar PFGE restriction patterns. These findings suggested that imported turkey meat containing VREFs may be a source of VREFs in the gut of healthy human subjects (Agerso et al., 2008).

A wide surveillance study based in Portugal, Denmark, Spain, Switzerland, and the United States (1995–2008) compared VRE isolates from pigs ($n = 29$) and healthy persons ($n = 12$) with clinical strains of VRE ($n = 190$) collected from various countries over a period of two decades (Freitas et al., 2011). Thirty clonally related *E. faecium* CC5 isolates were obtained from feces of swine and healthy humans. The PFGE subtype of one *E. faecalis* isolate (CC2-ST6) obtained from a pig also corresponded to a nosocomial clone widely disseminated in European countries. Some CC5 and CC17 PFGE subtype strains from swine were indistinguishable from VRE of the clinical origin, confirming the relevance of reverse and alternative routes for dissemination of commensal and opportunistic bacteria. Plasmid profiling and gene sequencing also confirmed that these VRE strains of both human and swine origin harbor Tn*1546* on indistinguishable plasmids, indicating a current intra- and international spread of *E. faecium* and *E. faecalis* clones and their plasmids among swine and humans (Freitas et al., 2011).

Outside of Europe, several countries also reported the persistence of VRE in animals and humans after the ban of specific dietary supplements. In New Zealand, 82 VRE isolates were recovered from poultry fecal samples only in broilers that had previously used avoparcin (Manson et al., 2003). They all contained *vanA* and *ermB* that co-localized on the same plasmid. Most of these showed very high MICs to avilamycin and bacitracin and harbored the *tetM* gene. Of these VRE isolates, 73 (89%) were VanA-type *E. faecalis* that had an identical or closely related PFGE pattern and were susceptible to both ampicillin and gentamicin and most of them were closely related to VRE from humans in New Zealand. In contrast, nine (11%) VRE isolates were VanA-type *E. faecium* with heterogeneous PFGE patterns. Therefore, a clonal lineage of VanA-type *E. faecalis* containing identical Tn*1546*-like elements dominated the VRE population from poultry and humans in New Zealand (Manson et al., 2003). A Korean research group profiled the drug resistances and plasmid contents of a total of 85 VRE strains (54 from chicken feces, 31 from hospital patients). The data provided evidence for an exchange of genetic material between humans and poultry and suggested the potential for horizontal transmission of multiple drug resistance, including vancomycin resistance, via a pheromone-responsive conjugative plasmid. This indicates that VRE ingested in food can temporarily colonize the human intestine and can transfer plasmids to human resident *E. faecalis* (Lim et al., 2006b).

Donabedian et al. (2003) studied the prevalence of gentamicin-resistant enterococcal isolates (MIC: $\geq 128\,\mu g/mL$) in humans, retail food, and farm animals in the United States. They found 72% (259/360) of the gentamicin-resistant isolates possessed the *aac(6)-Ie-aph (2")-Ia* gene, 18% (66/360) contained the *aph(2")-Ic* gene, and 9% (33/360) the *aph(2")-Id* gene. Moreover, genotyping revealed that one human and 18 pork *E. faecalis* from Michigan with the *aac(6')-Ie-aph(2")-Ia* gene had related PFGE patterns and two *E. faecalis* from Oregon (one human and one grocery store chicken isolate) were identical clones. The *aph(2")-Ib* gene was not detected in any of the isolates. Notably, in spite of genetic diversity among the resistant isolates, the same gentamicin resistance gene was identified from an animal and its product from geographically diverse areas; thus emphasizing the argument that the gentamicin-resistant enterococci spread from animals to humans through the food supply (Donabedian et al., 2003).

High-level gentamicin-resistant (HLGR) *E. faecalis* isolates carrying the *aac(6')Ie-aph(2")Ia* gene were obtained from two endocarditis patients, pigs, pork samples, and community people in Denmark. These isolates clustered into one major clonal group with a minimum of 86% relatedness based on PFGE patterns and belonged to HLGR ST16. Detection of the same clone in pork supports foodborne transmission while the possibility of direct transmission from animals to humans cannot be ruled out (Larsen et al., 2010). In a following study, Larsen et al. (2011) expanded their study design and obtained a total of 20 gentamicin-susceptible *E. faecalis* isolates from endocarditis patients and compared them with several *E. faecalis* STs available via DANMAP. Based on STs and PFGE subtypes, the authors suggested that the clinical isolates originate from the community reservoir and also showed that the porcine-origin clonal types ST97:A and ST40:D of gentamicin-susceptible *E. faecalis* were linked to patients with infective endocarditis (Larsen et al., 2011).

OTHER IMPORTANT LINKS

Another potential source of transfer of the antibiotic-resistant strains from farms to the urban community is livestock insects because of their ability to acquire, multiply, and disseminate bacteria originating from animal feces/manure. The confined type of production system and organic waste on animal farms provide an excellent habitat for the development of flies and cockroaches (reviewed in Zurek and Ghosh, 2014). In a study from poultry farms in the United States, *E. faecalis* represented the majority of the enterococcal species isolated from 70% of poultry litter samples and 87% of house flies collected at and near confined chicken operations. The antibiotic resistance profile of enterococci recovered from both sources matched genotypically and phenotypically (Graham et al., 2009). In another study, Ahmad et al. (2011) compared enterococci from house flies, German cockroaches, and pig feces from two commercial swine operations in the United States and found that the majority (>89%) of all samples were positive for enterococci. Based on PFGE analysis, they detected identical clonal matches among enterococci from all three sources. This indicated that insects likely acquired enterococci from swine manure (Ahmad et al., 2011).

Macovei and Zurek (2006) screened the digestive tract of house flies collected at five fast-food restaurants and reported that 97% of fly samples were positive for enterococci with a mean CFU of 10^3 per fly. *E. faecalis* was the most abundant species (88.2%) harboring resistance to tetracycline (66.3% of isolates), erythromycin (23.8%), streptomycin (11.6%), ciprofloxacin (9.9%), and kanamycin (8.3%). In addition, members of the Tn*916*/Tn*1545* family of transposons were detected in 35% of the identified isolates. Further sampling of the ready-to-eat food from the same restaurants showed that these were commonly contaminated with antibiotic-resistant enterococci (Macovei and Zurek, 2007). This suggested that house flies play a role in this contamination. Consequently, insects likely represent a link between food and food animals for fecal bacteria, including multidrug-resistant strains of enterococci (Zurek and Ghosh, 2014).

Although not in direct contact with food production or the clinical environment, wildlife animals can acquire multidrug-resistant strains, including VRE, from wastewater, manure, human and clinical waste, and other sources. Reports on the carriage of VRE by wild mammals, birds, and fish as well as their potential dissemination in the environment have been reported (reviewed by Radhouani et al., 2014). For example, Silva et al. (2011) analyzed 220 cloacal samples from passerine and game birds in Azores Archipelago islands and detected VRE in 6% (13/220) of the samples. Four of ten VRE isolates were *vanA*-containing *E. faecium* and had high-level resistance to vancomycin (MIC $\geq$ 128 µg/mL) and teicoplanin (MIC = 64 µg/mL). Some of the VREF possessed *ermB* and the virulence gene *ace*. Furthermore, studies in several European countries reported prevalence of VRE including VanA-type *E. faecium* in the song thrush (prevalence: 1.3%, 2/154) and in migratory rooks (prevalence: 6%, 62/1,073) (Oravcova et al., 2013; Silva et al., 2012). In a recent study, fecal samples from American crows collected in Massachusetts, New York, and Kansas were positive for VRE at a prevalence of 2.5% (15/590). Based on the MLST analysis, it was shown that avian VREF STs (ST18 and ST555) were closely related to the hospital-adapted clonal complex CC-17 (Oravcova et al., 2014). These studies indicate that wild birds can be colonized by resistant bacteria, including VRE, although the sources of this contamination are unknown. The migrating nature of these birds may contribute to dissemination of VRE over large distances. In addition, marine fish (Araujo et al., 2011; Barros et al., 2012), wild rodents (Figueiredo et al., 2009; Mallon et al., 2002), wild boars (Poeta et al., 2007), Iberian wolves and Iberian lynxes (Goncalves et al., 2011), and red foxes (Radhouani et al., 2011) have been described as carriers of VRE.

CONCLUDING REMARKS

Due to an extensive use of antimicrobials in the food animal industry in many parts of the world, food animals and the food animal environment have become a reservoir of multi-drug-resistant pathogenic and nonpathogenic bacterial strains. Consequently, food of animal origin can be a source of antibiotic-resistant bacteria, including enterococci. While enterococci are not major food-borne pathogens, due to their ability to acquire and transfer antibiotic-resistant traits that could establish in the resident microbiome of the human digestive tract, this bacterial group represents an important foodborne risk. Reduction in the use of antibiotics in animal agriculture that are "critically important" in human medicine, is an important step toward preserving the benefits of antimicrobials for the treatment of human and animal infections. Clinicians, regulatory agencies, policy makers, and other stakeholders are recommended to use the ranking of antimicrobials based on their relative importance in human medicine when developing risk management strategies for the use of antimicrobials in food production animals (Collignon et al., 2009).

REFERENCES

Aarestrup, F.M., 2000. Characterization of glycopeptide-resistant *Enterococcus faecium* (GRE) from broilers and pigs in Denmark: genetic evidence that persistence of GRE in pig herds is associated with coselection by resistance to macrolides. J. Clin. Microbiol. 38, 2774–2777.

Aarestrup, F.M., Agerso, Y., Gerner-Smidt, P., Madsen, M., Jensen, L.B., 2000a. Comparison of antimicrobial resistance phenotypes and resistance genes in *Enterococcus faecalis* and *Enterococcus faecium* from humans in the community, broilers, and pigs in Denmark. Diagn. Microbiol. Infect. Dis. 37, 127–137.

Aarestrup, F.M., Kruse, H., Tast, E., Hammerum, A.M., Jensen, L.B., 2000b. Associations between the use of antimicrobial agents for growth promotion and the occurrence of resistance among *Enterococcus faecium* from broilers and pigs in Denmark, Finland, and Norway. Microb. Drug Resist. 6, 63–70.

Aarestrup, F.M., Hasman, H., Jensen, L.B., Moreno, M., Herrero, I.A., Dominguez, L., et al., 2002. Antimicrobial resistance among enterococci from pigs in three European countries. Appl. Environ. Microbiol. 68, 4127–4129.

Abe, Y., Nakamura, K., Yamada, M., Yamamoto, Y., 2006. Encephalomalacia with *Enterococcus durans* infection in the brain stem and cerebral hemisphere in chicks in Japan. Avian Dis. 50, 139–141.

Abriouel, H., Omar, N.B., Molinos, A.C., Lopez, R.L., Grande, M.J., Martinez-Viedma, P., et al., 2008. Comparative analysis of genetic diversity and incidence of virulence factors and antibiotic resistance among enterococcal populations from raw fruit and vegetable foods, water and soil, and clinical samples. Int. J. Food Microbiol. 123, 38–49.

Agerso, Y., Lester, C.H., Porsbo, L.J., Orsted, I., Emborg, H.D., Olsen, K.E., et al., 2008. Vancomycin-resistant *Enterococcus faecalis* isolates from a Danish patient and two healthy human volunteers are possibly related to isolates from imported turkey meat. J. Antimicrob. Chemother. 62, 844–845.

Ahmad, A., Ghosh, A., Schal, C., Zurek, L., 2011. Insects in confined swine operations carry a large antibiotic resistant and potentially virulent enterococcal community. BMC Microbiol. 11, 23.

Araujo, C., Torres, C., Goncalves, A., Carneiro, C., Lopez, M., Radhouani, H., et al., 2011. Genetic detection and multilocus sequence typing of *vanA*-containing *Enterococcus* strains from mullets fish (*Liza ramada*). Microb. Drug Resist. 17, 357–361.

Arias, C.A., Murray, B.E., 2008. Emergence and management of drug-resistant enterococcal infections. Expert. Rev. Anti. Infect. Ther. 6, 637–655.

Arias, C.A., Murray, B.E., 2012. The rise of the *Enterococcus*: beyond vancomycin resistance. Nat. Rev. Microbiol. 10, 266–278.

Barros, J., Andrade, M., Radhouani, H., Lopez, M., Igrejas, G., Poeta, P., et al., 2012. Detection of *vanA*-containing *Enterococcus* species in faecal microbiota of gilthead seabream (*Sparus aurata*). Microbes Environ. 27, 509–511.

Biavasco, F., Foglia, G., Paoletti, C., Zandri, G., Magi, G., Guaglianone, E., et al., 2007. VanA-type enterococci from humans, animals, and food: species distribution, population structure, Tn*1546* typing and location, and virulence determinants. Appl. Environ. Microbiol. 73, 3307–3319.

Bradley, C.R., Fraise, A.P., 1996. Heat and chemical resistance of enterococci. J. Hosp. Infect. 34, 191–196.

Braga, T.M., Pomba, C., Lopes, M.F., 2013. High-level vancomycin resistant *Enterococcus faecium* related to humans and pigs found in dust from pig breeding facilities. Vet. Microbiol. 161, 344–349.

Byappanahalli, M.N., Nevers, M.B., Korajkic, A., Staley, Z.R., Harwood, V.J., 2012. Enterococci in the environment. Microbiol. Mol. Biol. Rev. 76, 685–706.

Cao, G.T., Zeng, X.F., Chen, A.G., Zhou, L., Zhang, L., Xiao, Y.P., et al., 2013. Effects of a probiotic, *Enterococcus faecium*, on growth performance, intestinal morphology, immune response, and cecal microflora in broiler chickens challenged with *Escherichia coli* K88. Poult. Sci. 92, 2949–2955.

Cardona, C.J., Bickford, A.A., Charlton, B.R., Cooper, G.L., 1993. *Enterococcus durans* infection in young chickens associated with bacteremia and encephalomalacia. Avian Dis. 37, 234–239.

Chadfield, M.S., Christensen, J.P., Juhl-Hansen, J., Christensen, H., Bisgaard, M., 2005. Characterization of *Enterococcus hirae* outbreaks in broiler flocks demonstrating increased mortality because of septicemia and endocarditis and/or altered production parameters. Avian Dis. 49, 16–23.

Channaiah, L.H., 2009. Polyphasic characterization of antibiotic resistant and virulent enterococci isolated from animal feed and stored-product insects (Doctoral Thesis), Grain Science and Industry. Kansas State University.

Christoffersen, T.E., Jensen, H., Kleiveland, C.R., Dorum, G., Jacobsen, M., Lea, T., 2012. *In vitro* comparison of commensal, probiotic and pathogenic strains of *Enterococcus faecalis*. Br. J. Nutr. 108, 2043–2053.

Collignon, P., Powers, J.H., Chiller, T.M., Aidara-Kane, A., Aarestrup, F.M., 2009. World Health Organization ranking of antimicrobials according to their importance in human medicine: a critical step for developing risk management strategies for the use of antimicrobials in food production animals. Clin. Infect. Dis. 49, 132–141.

Coque, T.M., Tomayko, J.F., Ricke, S.C., Okhyusen, P.C., Murray, B.E., 1996. Vancomycin-resistant enterococci from nosocomial, community, and animal sources in the United States. Antimicrob. Agents Chemother. 40, 2605–2609.

de Costa, P.M., Oliveira, M., Bica, A., Vaz-Pires, P., Bernardo, F., 2007. Antimicrobial resistance in *Enterococcus* spp. and *Escherichia coli* isolated from poultry feed and feed ingredients. Vet. Microbiol. 120, 122–131.

de Perio, M.A., Yarnold, P.R., Warren, J., Noskin, G.A., 2006. Risk factors and outcomes associated with non-*Enterococcus faecalis*, non-*Enterococcus faecium* enterococcal bacteremia. Infect. Control Hosp. Epidemiol. 27, 28–33.

Debnam, A.L., Jackson, C.R., Avellaneda, G.E., Barrett, J.B., Hofacre, C.L., 2005. Effect of growth promotant usage on enterococci species on a poultry farm. Avian Dis. 49, 361–365.

Donabedian, S.M., Thal, L.A., Hershberger, E., Perri, M.B., Chow, J.W., Bartlett, P., et al., 2003. Molecular characterization of gentamicin-resistant enterococci in the United States: evidence of spread from animals to humans through food. J. Clin. Microbiol. 41, 1109–1113.

Donabedian, S.M., Perri, M.B., Abdujamilova, N., Gordoncillo, M.J., Naqvi, A., Reyes, K.C., et al., 2010. Characterization of vancomycin-resistant *Enterococcus faecium* isolated from swine in three Michigan counties. J. Clin. Microbiol. 48, 4156–4160.

EFSA, 2007. Opinion of the scientific committee on a request from EFSA on the introduction of a Qualified Presumption of Safety (QPS) approach for assessment of selected microorganisms referred to EFSA. EFSA J. 587, 1–16.

Emmanuel, D.G., Jafari, A., Beauchemin, K.A., Leedle, J.A., Ametaj, B.N., 2007. Feeding live cultures of *Enterococcus faecium* and *Saccharomyces cerevisiae* induces an inflammatory response in feedlot steers. J. Anim. Sci. 85, 233–239.

Facklam, R.R., Collins, M.D., 1989. Identification of *Enterococcus* species isolated from human infections by a conventional test scheme. J. Clin. Microbiol. 27, 731–734.

FDA, Food and Drug Administration, 2012. *Enterococcus*, Bad Bug Book, Foodborne Pathogenic Microorganisms and Natural Toxins, second ed. Center for Food Safety and Applied Nutrition, College Park, MD, pp. 113–115.

Figueiredo, N., Radhouani, H., Goncalves, A., Rodrigues, J., Carvalho, C., Igrejas, G., et al., 2009. Genetic characterization of vancomycin-resistant enterococci isolates from wild rabbits. J. Basic Microbiol. 49, 491–494.

Fisher, K., Phillips, C., 2009. The ecology, epidemiology and virulence of *Enterococcus*. Microbiology 155, 1749–1757.

Flannagan, S.E., Chow, J.W., Donabedian, S.M., Brown, W.J., Perri, M.B., Zervos, M.J., et al., 2003. Plasmid content of a vancomycin-resistant *Enterococcus faecalis* isolate from a patient also colonized by *Staphylococcus aureus* with a VanA phenotype. Antimicrob. Agents Chemother. 47, 3954–3959.

Foulquie Moreno, M.R., Sarantinopoulos, P., Tsakalidou, E., De Vuyst, L., 2006. The role and application of enterococci in food and health. Int. J. Food Microbiol. 106, 1–24.

Franz, C.M., Huch, M., Abriouel, H., Holzapfel, W., Galvez, A., 2011. Enterococci as probiotics and their implications in food safety. Int. J. Food Microbiol. 151, 125–140.

Franz, C.M.A.P., Holzapfel, W.H., Stiles, M.E., 1999. Enterococci at the crossroads of food safety? Int. J. Food Microbiol. 47, 1–24.

Freitas, A.R., Coque, T.M., Novais, C., Hammerum, A.M., Lester, C.H., Zervos, M.J., et al., 2011. Human and swine hosts share vancomycin-resistant *Enterococcus faecium* CC17 and CC5 and *Enterococcus faecalis* CC2 clonal clusters harboring Tn*1546* on indistinguishable plasmids. J. Clin. Microbiol. 49, 925–931.

Frolkova, P., Ghosh, A., Svec, P., Zurek, L., Literak, I., 2012. Use of the manganese-dependent superoxide dismutase gene *sodA* for rapid identification of recently described enterococcal species. Folia Microbiol. (Praha) 57, 439–442.

Frye, J.G., Jackson, C.R., 2013. Genetic mechanisms of antimicrobial resistance identified in *Salmonella enterica*, *Escherichia coli*, and *Enteroccocus* spp. isolated from U.S. food animals. Front Microbiol. 4, 135.

Ge, B., LaFon, P.C., Carter, P.J., McDermott, S.D., Abbott, J., Glenn, A., et al., 2013. Retrospective analysis of *Salmonella*, *Campylobacter*, *Escherichia coli*, and *Enterococcus* in animal feed ingredients. Foodborne Pathog. Dis. 10, 684–691.

Giraffa, G., 2002. Enterococci from foods. FEMS Microbiol. Rev. 26, 163–171.

Goncalves, A., Igrejas, G., Radhouani, H., Lopez, M., Guerra, A., Petrucci-Fonseca, F., et al., 2011. Detection of vancomycin-resistant enterococci from faecal samples of Iberian wolf and Iberian lynx, including *Enterococcus faecium* strains of CC17 and the new singleton ST573. Sci. Total Environ. 410–411, 266–268.

Graham, J.P., Price, L.B., Evans, S.L., Graczyk, T.K., Silbergeld, E.K., 2009. Antibiotic resistant enterococci and staphylococci isolated from flies collected near confined poultry feeding operations. Sci. Total Environ. 407, 2701–2710.

Hammerum, A.M., Lester, C.H., Neimann, J., Porsbo, L.J., Olsen, K.E., Jensen, L.B., et al., 2004. A vancomycin-resistant *Enterococcus faecium* isolate from a Danish healthy volunteer, detected 7 years after the ban of avoparcin, is possibly related to pig isolates. J. Antimicrob. Chemother. 53, 547–549.

Hasman, H., Aarestrup, F.M., 2005. Relationship between copper, glycopeptide, and macrolide resistance among *Enterococcus faecium* strains isolated from pigs in Denmark between 1997 and 2003. Antimicrob. Agents Chemother. 49, 454–456.

Hidron, A.I., Edwards, J.R., Patel, J., Horan, T.C., Sievert, D.M., Pollock, D.A., et al., 2008. NHSN annual update: antimicrobial-resistant pathogens associated with healthcare-associated infections: annual summary of data reported to the National Healthcare Safety Network at the Centers for Disease Control and Prevention, 2006–2007. Infect. Control Hosp. Epidemiol. 29, 996–1011.

Huys, G., D'Haene, K., Collard, J.M., Swings, J., 2004. Prevalence and molecular characterization of tetracycline resistance in *Enterococcus* isolates from food. Appl. Environ. Microbiol. 70, 1555–1562.

Jackson, C.R., Lombard, J.E., Dargatz, D.A., Fedorka-Cray, P.J., 2011. Prevalence, species distribution and antimicrobial resistance of enterococci isolated from US dairy cattle. Lett. Appl. Microbiol. 52, 41–48.

Jansson, D.S., Nilsson, O., Lindblad, J., Greko, C., Bengtsson, B., 2012. Inter-batch contamination and potential sources of vancomycin-resistant *Enterococcus faecium* on broiler farms. Br. Poult. Sci. 53, 790–799.

Johnsen, P.J., Osterhus, J.I., Sletvold, H., Sorum, M., Kruse, H., Nielsen, K., et al., 2005. Persistence of animal and human glycopeptide-resistant enterococci on two Norwegian poultry farms formerly exposed to avoparcin is associated with a widespread plasmid-mediated *vanA* element within a polyclonal *Enterococcus faecium* population. Appl. Environ. Microbiol. 71, 159–168.

Karchmer, A.W., 2000. Nosocomial bloodstream infections: organisms, risk factors, and implications. Clin. Infect. Dis. 31, S139–S143.

Kinley, B., Rieck, J., Dawson, P., Jiang, X., 2010. Analysis of *Salmonella* and enterococci isolated from rendered animal products. Can. J. Microbiol. 56, 65–73.

Kirschner, C., Maquelin, K., Pina, P., Ngo Thi, N.A., Choo-Smith, L.P., Sockalingum, G.D., et al., 2001. Classification and identification of enterococci: a comparative phenotypic, genotypic, and vibrational spectroscopic study. J. Clin. Microbiol. 39, 1763–1770.

Klein, G., 2003. Taxonomy, ecology and antibiotic resistance of enterococci from food and the gastro-intestinal tract. Int. J. Food Microbiol. 88, 123–131.

Kolbjornsen, O., David, B., Gilhuus, M., 2011. Bacterial osteomyelitis in a 3-week-old broiler chicken associated with *Enterococcus hirae*. Vet. Pathol. 48, 1134–1137.

Kuhn, I., Iversen, A., Burman, L.G., Olsson-Liljequist, B., Franklin, A., Finn, M., et al., 2003. Comparison of enterococcal populations in animals, humans, and the environment—a European study. Int. J. Food Microbiol. 88, 133–145.

Lang, M.M., Ingham, S.C., Ingham, B.H., 2001. Differentiation of *Enterococcus* spp. by cell membrane fatty acid methyl ester profiling, biotyping and ribotyping. Lett. Appl. Microbiol. 33, 65–70.

Larsen, J., Schonheyder, H.C., Lester, C.H., Olsen, S.S., Porsbo, L.J., Garcia-Migura, L., et al., 2010. Porcine-origin gentamicin-resistant *Enterococcus faecalis* in humans, Denmark. Emerg. Infect. Dis. 16, 682–684.

Larsen, J., Schonheyder, H.C., Singh, K.V., Lester, C.H., Olsen, S.S., Porsbo, L.J., et al., 2011. Porcine and human community reservoirs of *Enterococcus faecalis*, Denmark. Emerg. Infect. Dis. 17, 2395–2397.

Lauderdale, T.L., Shiau, Y.R., Wang, H.Y., Lai, J.F., Huang, I.W., Chen, P.C., et al., 2007. Effect of banning vancomycin analogue avoparcin on vancomycin-resistant enterococci in chicken farms in Taiwan. Environ. Microbiol. 9, 819–823.

Lester, C.H., Frimodt-Moller, N., Sorensen, T.L., Monnet, D.L., Hammerum, A.M., 2006. *In vivo* transfer of the *vanA* resistance gene from an *Enterococcus faecium* isolate of animal origin to an *E. faecium* isolate of human origin in the intestines of human volunteers. Antimicrob. Agents Chemother. 50, 596–599.

Lim, S.K., Kim, T.S., Lee, H.S., Nam, H.M., Joo, Y.S., Koh, H.B., 2006a. Persistence of VanA-type *Enterococcus faecium* in Korean livestock after ban on avoparcin. Microb. Drug Resist. 12, 136–139.

Lim, S.K., Tanimoto, K., Tomita, H., Ike, Y., 2006b. Pheromone-responsive conjugative vancomycin resistance plasmids in *Enterococcus faecalis* isolates from humans and chicken feces. Appl. Environ. Microbiol. 72, 6544–6553.

Limbago, B.M., Kallen, A.J., Zhu, W., Eggers, P., McDougal, L.K., Albrecht, V.S., 2014. Report of the 13th vancomycin-resistant *Staphylococcus aureus* isolate from the United States. J. Clin. Microbiol. 52, 998–1002.

Liu, Y., Liu, K., Lai, J., Wu, C., Shen, J., Wang, Y., 2013. Prevalence and antimicrobial resistance of *Enterococcus* species of food animal origin from Beijing and Shandong Province, China. J. Appl. Microbiol. 114, 555–563.

Lopez, M., Saenz, Y., Rojo-Bezares, B., Martinez, S., del Campo, R., Ruiz-Larrea, F., et al., 2009. Detection of *vanA* and *vanB2*-containing enterococci from food samples in Spain, including *Enterococcus faecium* strains of CC17 and the new singleton ST425. Int. J. Food Microbiol. 133, 172–178.

Love, R.M., 2001. *Enterococcus faecalis*—a mechanism for its role in endodontic failure. Int. Endod. J. 34, 399–405.

Macovei, L., Zurek, L., 2006. Ecology of antibiotic resistance genes: characterization of enterococci from houseflies collected in food settings. Appl. Environ. Microbiol. 72, 4028–4035.

Macovei, L., Zurek, L., 2007. Influx of enterococci and associated antibiotic resistance and virulence genes from ready-to-eat food to the human digestive tract. Appl. Environ. Microbiol. 73, 6740–6747.

Mallon, D.J., Corkill, J.E., Hazel, S.M., Wilson, J.S., French, N.P., Bennett, M., et al., 2002. Excretion of vancomycin-resistant enterococci by wild mammals. Emerg. Infect. Dis. 8, 636–638.

Manero, A., Vilanova, X., Cerda-Cuellar, M., Blanch, A.R., 2006. Vancomycin- and erythromycin-resistant enterococci in a pig farm and its environment. Environ. Microbiol. 8, 667–674.

Manson, J.M., Keis, S., Smith, J.M., Cook, G.M., 2003. A clonal lineage of VanA-type *Enterococcus faecalis* predominates in vancomycin-resistant enterococci isolated in New Zealand. Antimicrob. Agents Chemother. 47, 204–210.

Manson, J.M., Smith, J.M.B., Cook, G.M., 2004. Persistence of vancomycin-resistant enterococci in New Zealand broilers after discontinuation of avoparcin use. Appl. Environ. Microbiol. 70, 5764–5768.

Martone, W.J., 1998. Spread of vancomycin-resistant enterococci: why did it happen in the United States? Infect. Control Hosp. Epidemiol. 19, 539–545.

McBride, S.M., Coburn, P.S., Baghdayan, A.S., Willems, R.J., Grande, M.J., Shankar, N., et al., 2009. Genetic variation and evolution of the pathogenicity island of *Enterococcus faecalis*. J. Bacteriol. 191, 3392–3402.

McDonald, L.C., Rossiter, S., Mackinson, C., Wang, Y.Y., Johnson, S., Sullivan, M., et al., 2001. Quinupristin-dalfopristin-resistant *Enterococcus faecium* on chicken and in human stool specimens. N. Engl. J. Med. 345, 1155–1160.

McGowan, L.L., Jackson, C.R., Barrett, J.B., Hiott, L.M., Fedorka-Cray, P.J., 2006. Prevalence and antimicrobial resistance of enterococci isolated from retail fruits, vegetables, and meats. J. Food Prot. 69, 2976–2982.

McGowan-Spicer, L.L., Fedorka-Cray, P.J., Frye, J.G., Meinersmann, R.J., Barrett, J.B., Jackson, C.R., 2008. Antimicrobial resistance and virulence of *Enterococcus faecalis* isolated from retail food. J. Food Prot. 71, 760–769.

Murray, B.E., 1990. The life and times of the *Enterococcus*. Clin. Microbiol. Rev. 3, 46–65.

Nilsson, O., 2012. Vancomycin resistant enterococci in farm animals—occurrence and importance. Infect. Ecol. Epidemiol. 2.

Nilsson, O., Greko, C., Bengtsson, B., 2009. Environmental contamination by vancomycin resistant enterococci (VRE) in Swedish broiler production. Acta. Vet. Scand. 51, 49.

Nocek, J.E., Kautz, W.P., 2006. Direct-fed microbial supplementation on ruminal digestion, health, and performance of pre- and postpartum dairy cattle. J. Dairy Sci. 89, 260–266.

Nomura, T., Tanimoto, K., Shibayama, K., Arakawa, Y., Fujimoto, S., Ike, Y., et al., 2012. Identification of VanN-type vancomycin resistance in an *Enterococcus faecium* isolate from chicken meat in Japan. Antimicrob. Agents Chemother. 56, 6389–6392.

Novais, C., Coque, T.M., Sousa, J.C., Peixe, L.V., 2006. Antimicrobial resistance among faecal enterococci from healthy individuals in Portugal. Clin. Microbiol. Infect. 12, 1131–1134.

Novais, C., Freitas, A.R., Sousa, J.C., Baquero, F., Coque, T.M., Peixe, L.V., 2008. Diversity of Tn*1546* and its role in the dissemination of vancomycin-resistant enterococci in Portugal. Antimicrob. Agents Chemother. 52, 1001–1008.

Novais, C., Freitas, A.R., Silveira, E., Antunes, P., Silva, R., Coque, T.M., et al., 2013. Spread of multidrug-resistant *Enterococcus* to animals and humans: an underestimated role for the pig farm environment. J. Antimicrob. Chemother. 68, 2746–2754.

Oetzel, G.R., Emery, K.M., Kautz, W.P., Nocek, J.E., 2007. Direct-fed microbial supplementation and health and performance of pre- and postpartum dairy cattle: a field trial. J. Dairy Sci. 90, 2058–2068.

Oravcova, V., Ghosh, A., Zurek, L., Bardon, J., Guenther, S., Cizek, A., et al., 2013. Vancomycin-resistant enterococci in rooks (*Corvus frugilegus*) wintering throughout Europe. Environ. Microbiol. 15, 548–556.

Oravcova, V., Zurek, L., Townsend, A., Clark, A.B., Ellis, J.C., Cizek, A., et al., 2014. American crows as carriers of vancomycin-resistant enterococci with *vanA* gene. Environ. Microbiol. 16, 939–949.

Palmer, K.L., Kos, V.N., Gilmore, M.S., 2010. Horizontal gene transfer and the genomics of enterococcal antibiotic resistance. Curr. Opin. Microbiol. 13, 632–639.

Pesavento, G., Calonico, C., Ducci, B., Magnanini, A., Lo Nostro, A., 2014. Prevalence and antibiotic resistance of *Enterococcus* spp. isolated from retail cheese, ready-to-eat salads, ham, and raw meat. Food Microbiol. 41, 1–7.

Peters, J., Mac, K., Wichmann-Schauer, H., Klein, G., Ellerbroek, L., 2003. Species distribution and antibiotic resistance patterns of enterococci isolated from food of animal origin in Germany. Int. J. Food Microbiol. 88, 311–314.

Poeta, P., Costa, D., Igrejas, G., Rojo-Bezares, B., Saenz, Y., Zarazaga, M., et al., 2007. Characterization of *vanA*-containing *Enterococcus faecium* isolates carrying Tn*5397*-like and Tn*916*/Tn*1545*-like transposons in wild boars (*Sus Scrofa*). Microb. Drug Resist. 13, 151–156.

Pollmann, M., Nordhoff, M., Pospischil, A., Tedin, K., Wieler, L.H., 2005. Effects of a probiotic strain of *Enterococcus faecium* on the rate of natural chlamydia infection in swine. Infect. Immun. 73, 4346–4353.

Poyart, C., Quesnes, G., Trieu-Cuot, P., 2000. Sequencing the gene encoding manganese-dependent superoxide dismutase for rapid species identification of enterococci. J. Clin. Microbiol. 38, 415–418.

Radhouani, H., Igrejas, G., Carvalho, C., Pinto, L., Goncalves, A., Lopez, M., et al., 2011. Clonal lineages, antibiotic resistance and virulence factors in vancomycin-resistant enterococci isolated from fecal samples of red foxes (*Vulpes vulpes*). J. Wildl. Dis. 47, 769–773.

Radhouani, H., Silva, N., Poeta, P., Torres, C., Correia, S., Igrejas, G., 2014. Potential impact of antimicrobial resistance in wildlife, environment and human health. Front Microbiol. 5, 23.

Rogers, D.G., Zeman, D.H., Erickson, E.D., 1992. Diarrhea associated with *Enterococcus durans* in calves. J. Vet. Diagn. Invest. 4, 471–472.

Sapkota, A.R., Lefferts, L.Y., McKenzie, S., Walker, P., 2007. What do we feed to food-production animals? A review of animal feed ingredients and their potential impacts on human health. Environ. Health. Perspect. 115, 663–670.

Scharek, L., Guth, J., Reiter, K., Weyrauch, K.D., Taras, D., Schwerk, P., et al., 2005. Influence of a probiotic *Enterococcus faecium* strain on development of the immune system of sows and piglets. Vet. Immunol. Immunopathol. 105, 151–161.

Schwalbe, R.S., McIntosh, A.C., Qaiyumi, S., Johnson, J.A., Morris Jr., J.G., 1999. Isolation of vancomycin-resistant enterococci from animal feed in USA. Lancet 353, 722.

Sghir, A., Gramet, G., Suau, A., Rochet, V., Pochart, P., Dore, J., 2000. Quantification of bacterial groups within human fecal flora by oligonucleotide probe hybridization. Appl. Environ. Microbiol. 66, 2263–2266.

Silva, N., Igrejas, G., Rodrigues, P., Rodrigues, T., Goncalves, A., Felgar, A.C., et al., 2011. Molecular characterization of vancomycin-resistant enterococci and extended-spectrum beta-lactamase-containing *Escherichia coli* isolates in wild birds from the Azores Archipelago. Avian. Pathol. 40, 473–479.

Silva, N., Igrejas, G., Felgar, A., Goncalves, A., Pacheco, R., Poeta, P., 2012. Molecular characterization of *vanA*-containing *Enterococcus* from migratory birds: song thrush (*Turdus philomelos*). Braz. J. Microbiol. 43, 1026–1029.

Silveira, E., Freitas, A.R., Antunes, P., Barros, M., Campos, J., Coque, T.M., et al., 2014. Co-transfer of resistance to high concentrations of copper and first-line antibiotics among *Enterococcus* from different origins (humans, animals, the environment and foods) and clonal lineages. J. Antimicrob. Chemother. 69, 899–906.

Simjee, S., Jensen, L.B., Donabedian, S.M., Zervos, M.J., 2006. Enterococcus. In: Aarestrup, F.M. (Ed.), Antimicrobial Resistance in Bacteria of Animal Origin ASM Press, Washington, DC, pp. 315–328.

Sorensen, T.L., Blom, M., Monnet, D.L., Frimodt-Moller, N., Poulsen, R.L., Espersen, F., 2001. Transient intestinal carriage after ingestion of antibiotic-resistant *Enterococcus faecium* from chicken and pork. N. Engl. J. Med. 345, 1161–1166.

Sorum, M., Johnsen, P.J., Aasnes, B., Rosvoll, T., Kruse, H., Sundsfjord, A., et al., 2006. Prevalence, persistence, and molecular characterization of glycopeptide-resistant enterococci in Norwegian poultry and poultry farmers 3 to 8 years after the ban on avoparcin. Appl. Environ. Microbiol. 72, 516–521.

Steentjes, A., Veldman, K.T., Mevius, D.J., Landman, W.J., 2002. Molecular epidemiology of unilateral amyloid arthropathy in broiler breeders associated with *Enterococcus faecalis*. Avian. Pathol. 31, 31–39.

Stobberingh, E., van den Bogaard, A., London, N., Driessen, C., Top, J., Willems, R., 1999. Enterococci with glycopeptide resistance in turkeys, turkey farmers, turkey slaughterers, and (sub)urban residents in the south of the Netherlands: evidence for transmission of vancomycin resistance from animals to humans? Antimicrob. Agents Chemother. 43, 2215–2221.

Szabo, I., Wieler, L.H., Tedin, K., Scharek-Tedin, L., Taras, D., Hensel, A., et al., 2009. Influence of a probiotic strain of *Enterococcus faecium* on *Salmonella enterica* serovar Typhimurium DT104 infection in a porcine animal infection model. Appl. Environ. Microbiol. 75, 2621–2628.

Tanimoto, K., Nomura, T., Hamatani, H., Xiao, Y.H., Ike, Y., 2005. A vancomycin-dependent VanA-type *Enterococcus faecalis* strain isolated in Japan from chicken imported from China. Lett. Appl. Microbiol. 41, 157–162.

Tankson, J.D., Thaxton, J.P., Vizzier-Thaxton, Y., 2001. Pulmonary hypertension syndrome in broilers caused by *Enterococcus faecalis*. Infect. Immun. 69, 6318–6322.

Taras, D., Vahjen, W., Macha, M., Simon, O., 2006. Performance, diarrhea incidence, and occurrence of *Escherichia coli* virulence genes during long-term administration of a probiotic *Enterococcus faecium* strain to sows and piglets. J. Anim. Sci. 84, 608–617.

Top, J., Willems, R., Bonten, M., 2008. Emergence of CC17 *Enterococcus faecium*: from commensal to hospital-adapted pathogen. FEMS Immunol. Med. Microbiol. 52, 297–308.

van den Bogaard, A.E., Jensen, L.B., Stobberingh, E.E., 1997. Vancomycin-resistant enterococci in turkeys and farmers. N. Engl. J. Med. 337, 1558–1559.

van den Bogaard, A.E., Willems, R., London, N., Top, J., Stobberingh, E.E., 2002. Antibiotic resistance of faecal enterococci in poultry, poultry farmers and poultry slaughterers. J. Antimicrob. Chemother. 49, 497–505.

Vancanneyt, M., Snauwaert, C., Cleenwerck, I., Baele, M., Descheemaeker, P., Goossens, H., et al., 2001. *Enterococcus villorum* sp. nov., an enteroadherent bacterium associated with diarrhoea in piglets. Int. J. Syst. Evol. Microbiol. 51, 393–400.

WHO, 2011. World Health Organization. Programmes and projects, Zoonoses and veterinary public health. <http://www.who.int/zoonoses/en/>.

Willems, R.J., Top, J., van Santen, M., Robinson, D.A., Coque, T.M., Baquero, F., et al., 2005. Global spread of vancomycin-resistant *Enterococcus faecium* from distinct nosocomial genetic complex. Emerg. Infect. Dis. 11, 821–828.

Wyder, A.B., Boss, R., Naskova, J., Kaufmann, T., Steiner, A., Graber, H.U., 2011. *Streptococcus* spp. and related bacteria: their identification and their pathogenic potential for chronic mastitis—a molecular approach. Res. Vet. Sci. 91, 349–357.

Yamaguchi, T., Miura, Y., Matsumoto, T., 2013. Antimicrobial susceptibility of *Enterococcus* strains used in clinical practice as probiotics. J. Infect. Chemother. 19, 1109–1115.

Ziakas, P.D., Thapa, R., Rice, L.B., Mylonakis, E., 2013. Trends and significance of VRE colonization in the ICU: a meta-analysis of published studies. PLoS ONE 8, e75658.

Zurek, L., Ghosh, A., 2014. Insects represent a link between food animal farms and the urban environment for antibiotic resistance traits. Appl. Environ. Microbiol. 80, 3562–3567.

Chapter 10

Clostridium difficile: A Food Safety Concern?

Jane W. Marsh and Lee H. Harrison

Department of Medicine, Division of Infectious Diseases, University of Pittsburgh, Pittsburgh, PA, USA

Chapter Outline

INTRODUCTION

Clostridium difficile is a Gram-positive, spore-forming anaerobe and the primary cause of antibiotic-associated, hospital-acquired diarrhea. Disease pathogenesis requires antibiotic exposure and fecal–oral transmission of *C. difficile*—events that commonly occur in the healthcare environment. As a result, most *C. difficile* infections (CDI) are hospital-associated. However, the epidemiology of *C. difficile* has changed dramatically in the new millennium coinciding with the rapid, global spread of a new emergent clone and increased incidence of community-associated disease. The increased incidence of CDI in the community suggests that sources of *C. difficile* outside of the hospital environment are responsible for disease transmission. These results and the finding of *C. difficile* in food animals, meat, and retail foods of the same genetic lineages as those causing human infection have led to speculation that *C. difficile* may be a foodborne pathogen. The goal of this chapter is to provide an overview of *C. difficile* pathogenesis; epidemiology in humans, food animals, and foods; antibiotic resistance; and potential food safety risk.

Antimicrobial Resistance and Food Safety. DOI: http://dx.doi.org/10.1016/B978-0-12-801214-7.00010-7

PATHOGENESIS

Toxins

The major virulence factors of *C. difficile* are the large clostridial toxins—TcdA and TcdB, which are encoded on a 19.6 kb pathogenicity locus (PaLoc). The PaLoc encodes three additional proteins that are involved in regulation of toxin gene expression and secretion. TcdA and TcdB are glucosyltransferases which modify key signaling molecules in intestinal epithelial cells resulting in inflammation, cell death, and diarrhea (Voth and Ballard, 2005). The toxins belong to the family of large clostridial toxins and show high sequence similarity to *Clostridium sordillii* toxin. The inflammatory response can result in severe abdominal pain, fever, elevated white blood cell count, low serum albumin, and pseudomembranous colitis. In severe cases, toxic megacolon and death may occur even after drastic measures such as colectomy. The toxins may act systemically in severe cases, resulting in multiorgan dysfunction including cardiopulmonary arrest and renal failure (Hunt and Ballard, 2013). Some lineages produce an ADP-ribosylating binary toxin that is encoded separately from the toxins on the PaLoc. Pro-inflammatory responses and strong humoral immunity to TcdA has been demonstrated to significantly reduce the risk of mortality in patients infected with *C. difficile* (Solomon et al., 2013). Thus, recent *C. difficile* vaccine strategies target TcdA and TcdB.

The Spore

Another important determinant of *C. difficile* pathogenesis is the bacterial endospore. As a strict anaerobe, survival and dissemination of *C. difficile* is dependent upon the spore which consists of a complex, multilayered outer structure surrounding an inner core of nucleic acid and enzymes. Unique features of each structural layer contribute to the spore's resistance to disinfectants, heat, and drying (Paredes-Sabja et al., 2014). Under the appropriate environmental conditions, specific small molecules induce spore germination and the production of vegetative cells resulting in toxin production and infection (Paredes-Sabja et al., 2014). Thus, *C. difficile* can persist indefinitely in a variety of environments. This characteristic and the ability of the spore to survive recommended cooking temperatures has led to speculation that community acquisition of *C. difficile* could occur through ingestion of contaminated foods (Hoover and Rodriguez-Palacios, 2013; Rodriguez-Palacios and Lejeune, 2011).

Risk Factors

The major risk factor for development of CDI is antibiotic exposure. Historically, clindamycin therapy has been associated with the highest risk of CDI. However, many antibiotic classes are associated with CDI, including cephalosporins and fluoroquinolones. Other risk factors for CDI include advanced age, immune compromise, and the presence of comorbid conditions including kidney and inflammatory bowel disease, malignancy and solid organ transplantation and use of proton pump inhibitors (PPIs) (Janarthanan et al., 2012; Kwok et al., 2012). Presumably, a reduction in stomach acid permits survival of vegetative *C. difficile* which then colonize the gut and, under favorable conditions, cause disease (Jump et al., 2007). The US Food and Drug Administration (FDA, 2012) issued a public health warning in February of 2012 regarding the association of stomach acid drugs and *C. difficile* disease.

Another potential risk factor which has not been proven is the ingestion of foods contaminated with *C. difficile*. The finding of *C. difficile* in retail foods of the same types that cause human disease suggests that food may be a source of *C. difficile*. Recent unexplained increases in community-associated CDI suggest a possible association but, to date, no evidence of food-related disease has been conclusively demonstrated.

The Microbiome

The normal gut microflora provides an effective barrier to infection—a biologic phenomenon termed colonization resistance whereby the diverse microbial community of the intestine prevents enteric infection (Freter, 1955). Normal, healthy individuals with an intact gut microbiome who ingest foods contaminated with *C. difficile* are unlikely to develop disease. In fact, 7–15% of healthy individuals (non-healthcare workers) have been shown to be asymptomatically colonized with *C. difficile* (Galdys et al., 2014; Ozaki et al., 2004). Antibiotic exposure disrupts the normal gut microflora permitting *C. difficile* colonization and outgrowth. Thus, patients undergoing antibiotic treatment who are either colonized or have ingested *C. difficile* are susceptible to infection. When colonization resistance is destroyed, *C. difficile* grows unchecked and produces toxins that lead to the development of pseudomembranous colitis characteristic of CDI.

Therapy

Removal of the inciting antibiotic can resolve disease in less severe CDI cases. Paradoxically however, antibiotic treatment is required in most cases. Metronidazole is prescribed for cases of mild to moderate disease. Oral vancomycin is effective in treating severe disease. In clinical trials, vancomycin was shown to be more effective than metronidazole in treating severe infections (Zar et al., 2007). Fidaxomicin, like vancomycin, achieves high concentrations in the intestine and is non-inferior to vancomycin for treatment of CDI. In clinical trials, fidaxomicin demonstrated similar effectiveness as vancomycin but also showed reduced incidence of recurrent disease 25 days after therapy (Cornely et al., 2012a,b). The clinical relevance of this difference is uncertain, as many recurrences occur far beyond this endpoint (Marsh et al., 2012). Because antibiotics may cause collateral deleterious changes to the intestinal microflora, alternative, non-antimicrobial therapies have been explored. Early studies demonstrating that antibodies to toxins are protective against disease have led to the development of several immunization strategies (Kyne et al., 2000). Passive immunization with humanized monoclonal antibodies directed against TcdA and TcdB reduced recurrence in patients with CDI (Lowy et al., 2010). A phase I clinical trial of a toxoid vaccine has demonstrated safety, dosing, and immunogenicity in healthy human volunteers of various age groups (Greenberg et al., 2012). Another novel approach under investigation is the use of non-toxigenic *C. difficile*. Studies in the hamster model of disease demonstrated that certain nontoxigenic *C. difficile* strains prevented disease (Sambol et al., 2002). A phase I clinical trial demonstrated safety and colonization in healthy human volunteers pretreated with vancomycin (Villano et al., 2012). A potential drawback of this approach is acquisition of the PaLoc through horizontal gene transfer from a toxigenic to a non-toxigenic strain, which has been demonstrated to occur at very low frequencies *in vitro* (Brouwer et al., 2013).

Recurrent *C. difficile* occurs in ~25% of patients, causing substantial morbidity, mortality, and expense. The diversity of bacteria inhabiting the intestine of patients who experience recurrent CDI has been demonstrated to be significantly reduced (Chang et al., 2008). Given the lack of effective therapies for relapsing CDI, fecal microbiota transplant (FMT) has been explored and used successfully as a strategy to restore colonization resistance in patients with recurrent disease. A randomized, open-label, controlled pilot study demonstrated a 90% cure rate among patients with recurrent disease who received frozen, healthy donor stool (Youngster et al., 2014). In a randomized, control clinical trial comparing vancomycin to FMT, 94% of patients with recurrent disease had resolution of disease overall compared to 31% of patients receiving vancomycin (van Nood et al., 2013). Moreover, this study demonstrated restoration of bacterial diversity in patients receiving FMT compared to vancomycin. In a March 2014 guidance document, the US FDA exercised enforcement discretion regarding investigational new drug requirements for FMT (FDA, 2014). Thus, FMT is becoming widely implemented as an effective therapy for recurrent CDI, although the long-term safety of FMT remains a significant unknown.

EPIDEMIOLOGY

Molecular Genotyping

A variety of molecular epidemiologic genotyping methods with varying degrees of discriminatory power are currently in use to describe *C. difficile* population structure and track transmission. PCR-ribotyping, the most widely used genotyping method in Europe establishes genetic lineage based upon variation in ribosomal RNA spacer regions (Stubbs et al., 1999). Toxinotyping uses restriction fragment polymorphism (RFLP) analysis of the toxin genes and is commonly used to establish genetic lineage. Multi-locus sequence typing

(MLST) is an alternative, objective method to establish genetic lineage but is not sufficiently discriminatory to track transmission events in an outbreak situation (Griffiths et al., 2010). Restriction enzyme analysis (REA) and pulsed-field gel electrophoresis (PFGE) are more discriminatory methods for outbreak investigations but are subjective, labor-intensive, and rely on the use of reference strains. Multi-locus variable number tandem repeat analysis (MLVA) is a cost-effective, highly discriminatory method for outbreak investigations and has been used to determine the genetic relationships among *C. difficile* isolates of food, animal, and human origin (Curry et al., 2012; Marsh et al., 2006, 2011). As costs of whole-genome sequencing decline, genetic relationships based upon phylogenies from single nucleotide polymorphisms (SNPs) will likely replace the more traditional genotyping methods (Eyre et al., 2013). Whole-genome sequencing (WGS) is currently being used to understand *C. difficile* transmission dynamics in the hospital environment and to investigate clonal emergence.

Emergence of a Drug-Resistant Epidemic Clone

In the early 2000s, a dramatic increase in the incidence of CDI occurred and was associated with a single clone known as BI/NAP1/027—so-named according to REA, PFGE, and PCR ribotype, respectively. Comparative WGS analysis of a global collection of BI/NAP1/027 isolates demonstrated the origins and spread of two fluoroquinolone-resistant sublineages (He et al., 2010). The non-judicious use of antibiotics, fluoroquinolones in particular, likely contributed to the rapid emergence of this clone. In a retrospective, case–control study investigating a *C. difficile* outbreak, an association of CDI with fluoroquinolone use was observed (Muto et al., 2005). This BI/NAP1/027 outbreak was brought under control through implementation of a number of infection control measures including antibiotic stewardship (Muto et al., 2007). Most BI/NAP1/027 isolates collected after 2005 harbor mutations in *gyrA* that confer resistance to the fluoroquinolone class of antimicrobials including ciprofloxacin, moxifloxicin, and levofloxacin (Drudy et al., 2007). The rapid emergence and spread of the BI/NAP1/027 clone illustrate the dynamic evolutionary capability of *C. difficile* in response to the environment.

Asymptomatic Carriage

Studies in the community indicate that up to 13% of healthy human subjects are asymptomatically colonized with *C. difficile* (Galdys et al., 2014; Miyajima et al., 2011; Ozaki et al., 2004). A US study of 106 individuals found 6.6% toxigenic *C. difficile* carriage (Galdys et al., 2014). Healthy human subjects represent a potential reservoir for *C. difficile* both in the community as well as the hospital. In a US hospital surveillance study, 29% of hospital-associated *C. difficile* cases were attributed to asymptomatic carriers using MLVA to define genetic relatedness (Curry et al., 2013). In the United Kingdom, WGS demonstrated that only 35% (333/957) of CDI cases were genetically related to another CDI case and 45% of CDI case isolates were genetically distinct from all other hospital-associated cases; of genetically related cases, only 38% (126/333) could be attributed to a close hospital contact (Eyre et al., 2013). These data challenge the central dogma of *C. difficile* transmission in the hospital—that infected patients are responsible for ongoing transmission. While asymptomatic carriers are an important reservoir of *C. difficile* in the hospital, other sources must contribute to the observed genetic diversity in the healthcare environment and are likely to be community-based.

Community-Associated Disease

Reports of CDI in patients without traditional risk factors have increased with surveillance studies demonstrating 20–40% of cases as community-associated (Dumyati et al., 2012; Khanna et al., 2012; Noren et al., 2004). This change in the epidemiology of CDI coincided with the emergence of the BI/NAP1/027 epidemic clone and may in part reflect increased CDI awareness and testing. Community-associated CDI occurs in younger patients without recent hospitalization (<3 months) and with few underlying health conditions (CDC, 2005; Khanna et al., 2012; Kutty et al., 2010). Surveillance studies indicate that previous antimicrobial use is a risk factor for community-associated disease but to a lesser extent than hospital-associated disease (Chitnis et al., 2013; Dumyati et al., 2012). In a study of 984 patients with community-associated disease, 82% had been exposed to an outpatient healthcare facility within the previous 12 weeks of diagnosis (Chitnis et al., 2013). Outpatient clinic and emergency rooms have been shown to be contaminated with *C. difficile* suggesting these environments may contribute to *C. difficile* transmission in the community (Curry et al., 2011; Jury et al., 2013). Infant care may also be associated with community acquisition of *C. difficile* as children <2 years of age are often asymptomatic carriers of *C. difficile* (Rousseau et al., 2011; Stoesser et al., 2011).

Another potential reservoir of *C. difficile* in the community is household pets. An early study of common household pets including dogs, cats, and birds demonstrated 23% (19/82) *C. difficile* carriage (Borriello et al., 1983). While the majority of the isolates were non-toxigenic, several toxigenic strains were identified. In a more recent study, 58% of hospital-visiting dogs in Ontario hospitals were shown to be colonized with *C. difficile* and 71% of the isolates were toxigenic (Lefebvre et al., 2006). Thus, multiple potential sources of *C. difficile* exist which may contribute to community-associated disease.

The true incidence of community-associated CDI may be an overestimate. Patients with other common causes of diarrhea may only be colonized with *C. difficile*. In a community-based prospective study of acute outpatient diarrheal illness, CDI without traditional risk factors or other coinfections for common causes of gastroenteritis was reported in only three of 1,091 (0.3%) outpatients with diarrhea (Hirshon et al., 2011). Thus, CDI may be over reported in the community due to the widespread availability of *C. difficile* testing.

FOOD ANIMALS AND FOOD

Consumption of contaminated foods and/or exposure to food animals on farms represent potential sources of *C. difficile*. Numerous reports of *C. difficile* in retail meats and vegetables have led to speculation that *C. difficile* could be a foodborne pathogen (Gould and Limbago, 2010). Genetic lineages that predominate in food animals including PCR ribotype 078 (see below) also cause human disease (Jhung et al., 2008; Rodriguez-Palacios et al., 2006). Several studies suggest that transmission from swine to humans may occur on farms or in large integrated swine operations (Keessen et al., 2013a; Norman et al., 2011). Aerial dissemination of *C. difficile* from pig farms has been demonstrated, raising the possibility that living near swine production facilities may be a risk factor for community-associated CDI (Keessen et al., 2011a).

PCR Ribotype 078

Early studies in the Netherlands describe an increased incidence of severe disease in the community associated with PCR ribotype 078 (Goorhuis et al., 2008a,b). In Europe, PCR ribotype 078 is the third most common lineage responsible for CDI (Bauer et al., 2011). In the United States, the incidence of PCR ribotype 078 CDI is increasing and may be more common in the community (Black et al., 2011; Jhung et al., 2008). Reports from the United Kingdom indicate that PCR ribotype 078 is the most common lineage among patients with community-associated disease and the risk of mortality is reduced compared to PCR ribotype 027 (Patterson et al., 2012; Taori et al., 2014). PCR ribotype 078 is commonly isolated from swine and other food animals (Bakker et al., 2010; Keel et al., 2007; Koene et al., 2012). Molecular genotyping suggest that PCR ribotype 078 isolates of human and swine origin are highly related and may therefore represent potential zoonotic transmission (Bakker et al., 2010; Debast et al., 2009; Jhung et al., 2008). A characteristic of many PCR ribotype 078 strains is the presence of the Tn*916* mobile genetic element which confers tetracycline resistance. Chlortetracycline and oxytetracycline are commonly used in food animal production to prevent infection. Thus, antibiotic practices in animal production may contribute to the emergence of drug-resistant *C. difficile* in food animals.

Swine

Many studies investigating *C. difficile* prevalence have focused on swine and pork production (Avbersek et al., 2009; Baker et al., 2010; Fry et al., 2012; Hawken et al., 2013; Hopman et al., 2011; Keessen et al., 2011b; Norman et al., 2009; Songer and Anderson, 2006; Weese et al., 2010b). *C. difficile* can cause significant mortality in piglets and is therefore a major concern to the pork industry (Songer and Anderson, 2006). Numerous worldwide surveillance studies have demonstrated a high prevalence of *C. difficile* in piglets and PCR ribotype 078 has been shown to be the predominant genetic lineage identified (Table 10.1). While clinical disease is often limited to piglets, colonization is common throughout

TABLE 10.1 Summary of *C. difficile* Prevalence Studies in Swine

Animal	Prevalence % (Proportion)	Predominant Ribotype	Country	Time Period	References
Piglets	45/67 (67%)	046	Sweden	5–11/2012	Noren et al. (2014)
Piglets	114/185 (62%)	237	Australia	7–11/2009	Squire et al. (2013)
Piglets	147/201 (73%)	078	Germany	4–11/2012	Schneeberg et al. (2013)
Piglets	28/30 (93%)	078	Canada	na	Hawken et al. (2013)
Piglets	140/541 (26%)	078	Spain	na	Alvarez-Perez et al. (2009)
Piglets	183/251 (73%)	078	Ohio/NC	na	Fry et al. (2012)
Sows	32/68 (47%)	078	Ohio/NC	na	Fry et al. (2012)
Piglets	133/257 (52%)	078	Slovenia	na	Pirs et al. (2008)
Finisher	14/50 (28%)	015	Netherlands	na	Hopman et al. (2011)
Finisher	30/436 (7%)	078	Canada	na	Weese et al. (2011)
Finisher	58/677 (9%)	078	Netherlands	na	Keessen et al. (2011b)
Piglets	241/513 (47%)	078	Wisconsin	1–9/2009	Baker et al. (2010)
Piglets	247/485 (51%)	066	Slovenia	na	Avbersek et al. (2009)
Finisher	55/345 (16%)	ND	United States	11/06–9/08	Thitaram et al. (2011)
Piglets	61/122 (50%)	078	Texas	2004–2005	Norman et al. (2009)
Finishers	15/382 (4%)	078	Texas	2004–2005	Norman et al. (2009)

swine production from farrow to finisher (Fry et al., 2012; Norman et al., 2011; Thitaram et al., 2011). Reports from the United States and Europe describe *C. difficile* prevalence rates of up to 83% in piglets (Avbersek et al., 2009; Keel et al., 2007; Pirs et al., 2008). However, a recent longitudinal investigation in Canada demonstrated a significant reduction in *C. difficile* prevalence in swine from farrow (74%) to nursery (4%) (Weese et al., 2010b). In the same study, 40% of sows were colonized with *C. difficile* and a recent US survey demonstrated 16% (55/345) *C. difficile* prevalence in grower–finisher populations (Thitaram et al., 2011). Data from composite fecal samples obtained on an integrated swine operation in Texas demonstrated 50% (61/122) *C. difficile* prevalence in piglets, 24% (34/143) in lactating sows and 4% (15/382) prevalence for grower/finishers (Norman et al., 2009). In the Netherlands, a prevalence of 9% (58/677) was detected in slaughter-age pigs from 52 farms (Keessen et al., 2011b). In Belgium, 1% (1/100) of pig intestinal contents and 7% (7/100) of carcasses were positive for *C. difficile* at slaughter (Rodriguez et al., 2013). In a study of 436 slaughter-age pigs in Canada, 3.4% *C. difficile* prevalence was observed (Weese et al., 2011). These studies demonstrate that rates of colonization in slaughter-aged swine are reduced relative to piglets. However, even low prevalence of *C. difficile* at slaughter could represent a potential source of food contamination during processing. Moreover, high rates of colonization in breeders and lactating sows suggest that *C. difficile* may become endemic in some swine operations through the continued reintroduction of colonized sows.

Indeed, farm management practices may affect the prevalence of *C. difficile*. For instance the size of the operation may affect prevalence. In a recent study, the prevalence of *C. difficile* on large, integrated swine production facilities was 58% versus 27% on smaller, independently owned and operated farms in the US Midwest (Baker et al., 2010). Molecular genotyping in this study indicated greater genetic diversity on the smaller, regional farms compared to the large integrated operations (Baker et al., 2010). On the other hand, in a study performed on 52 farms in the Netherlands, the size and type of farm (organic or conventional) had no significant effect on *C. difficile* prevalence in swine (Keessen et al., 2011b). Further investigation of farm management practices is required to determine factors specific to *C. difficile* transmission between animals.

The recent emergence of PCR ribotype 078 as a cause of significant human disease coincided with multiple studies reporting high prevalence of *C. difficile* in animals and, in particular, pigs (Goorhuis et al., 2008b; Keel et al., 2007; Rupnik et al., 2008). Several early studies found PCR ribotype 078 of human and animal origin to be indistinguishable by PFGE and MLVA (Debast et al., 2009; Jhung et al., 2008). In a study from four different European countries, PCR ribotype 078 isolates of human and swine origin were found to be highly related by MLVA (Bakker et al., 2010). These studies suggested that food animals are a common reservoir for zoonotic transmission (Bakker et al., 2010). A *C. difficile* prevalence study in a closed, integrated swine operation in Texas found 12% (272/2,292) of human wastewater samples and 9% (252/2,936) of swine fecal samples positive for *C. difficile*, respectively (Norman et al., 2011). The majority of the human wastewater and swine isolates had molecular genotypes consistent with PCR ribotype 078. This finding and the closed environment of the operation suggest a potential common source of *C. difficile*, although significant differences in antimicrobial susceptibilities between human and swine isolates were observed (Norman et al., 2011, 2014b). In a recent study of swine workers with daily pig contact in the Netherlands, *C. difficile* from healthy workers was indistinguishable genetically from pig isolates (Keessen et al., 2013a). These data support zoonotic transmission from animals to humans. Alternatively, PCR ribotype 078 may lack genetic diversity and methods such as whole-genome sequencing may be required to better discriminate highly related clonal populations such as PCR ribotype 078.

While PCR ribotype 078 is predominant in most studies of swine, a recent report from Sweden found only PCR ribotype 046 positive piglets sampled from three farms (Noren et al., 2014). This finding is of interest as nosocomial spread of ribotype 046 in humans is a current problem in the south of Sweden (Noren et al., 2014). While antimicrobial susceptibility profiles between the swine and human isolate populations vary, the finding of identical ribotypes in swine and human disease has led to speculation regarding temporal/spatial distribution, bacterial evolution, and possible zoonotic transmission. Further molecular epidemiologic investigation with highly discriminatory genotyping methods such as comparative whole-genome sequencing is necessary to address genetic relationships.

Cattle

Enteritis in calves is associated with CDI and carriage is also found (Hopman et al., 2011). These data and recent studies suggest that beef, veal, and dairy cattle are potential reservoirs of *C. difficile* (Table 10.2). In a Canadian study, 11% (31/278) of dairy calves were shown to be positive for toxigenic *C. difficile* belonging to PCR ribotypes commonly associated with human disease (Rodriguez-Palacios et al., 2006). Calf isolates belonged to the PCR ribotype 027 epidemic clone as well as PCR ribotype 017, a lineage associated with recent epidemic disease in Asia (Baker et al., 2010; Rodriguez-Palacios et al., 2006; Weese et al., 2011). In Ohio, 13% of beef cattle arriving at feedlot were colonized with *C. difficile* and prevalence decreased to 1.2% prior to slaughter (Rodriguez-Palacios et al., 2011). The majority of isolates were non-toxigenic in this study but at least one animal harbored PCR ribotype 078 (Rodriguez-Palacios et al., 2011). A recent study in Belgium found 10% (10/101) and 8% (8/101) of cattle intestinal and carcass samples positive for *C. difficile* at slaughter respectively (Knight et al., 2013). More recent studies have found a lower prevalence of *C. difficile* in cattle at slaughter (Rodriguez-Palacios et al., 2011; Rupnik et al., 2008; Thitaram et al., 2011). In a Canadian study, 3.7% of cattle arriving at feedlot were positive for *C. difficile* and prevalence increased slightly to 6.2% at mid-feeding with PCR ribotype 078 the major lineage recovered (Costa et al., 2012).

TABLE 10.2 Summary of C. *difficile* Prevalence Studies in Cattle

Animal	Prevalence	Predominant Ribotype	Country	Time Period	References
Beef cattle	18/539 (3%)	078	Canada	1–9/2009	Costa et al. (2012)
Beef cattle	24/186 (13%)	mix	Ohio	10/2007–4/2008	Rodriguez-Palacios et al. (2011)
Dairy calves	31/278 (11%)	Mix	Canada	5–9/2004	Rodriguez-Palacios et al. (2006)
Calves	1/56 (2%)	NTCD	Slovenia	na	Pirs et al. (2008)
Calves	4/42 (10%)	Mix	Slovenia	na	Avbersek et al. (2009)
Dairy cattle	32/1325 (2%)	ND	United States	11/06–9/08	Thitaram et al. (2011)
Beef cattle	188/2965 (6%)	ND	United States	11/06–9/08	Thitaram et al. (2011)
Beef cattle	4/944 (0.4%)	Mix	United States	7–9/2008	Rupnik et al. (2008)
Beef cattle	2/168 (1.2%)	078	United States		Rodriguez-Palacios et al. (2011)
Beef cattle	3/67 (4.5%)	Mix	Austria	3–7/2008	Indra et al. (2009)
Veal calves	56/200 (28%)	ND	PA	na	Houser et al. (2012)
Beef cattle	10/101(10%)	Mix	Belgium	9/11–5/12	Rodriguez et al. (2013)
Veal calves	203/360(56%)	Mix	Australia	11/07–1/08	Knight et al. (2013)
Calves	90/150 (60%)	ND	Iran	2013	Rahimi et al. (2014)

Studies in veal calves have documented high *C. difficile* prevalence. In a longitudinal study from the US, 28% (56/200) of calf fecal samples were positive at least once for *C. difficile* by multiplex PCR detection of toxin genes and 12% were *C. difficile* positive at slaughter (Houser et al., 2012). Similarly, a Canadian study found high rates of *C. difficile* in young animals (51%) but the prevalence dropped to 2% in animals at slaughter (Rodriguez et al., 2013). In a recent survey from Australia, *C. difficile* was recovered from 56% (203/360) of calves <7 days of age and 1.8% (5/280) of adult cattle (Knight et al., 2013). The majority of *C. difficile* from veal calves in this study belong to PCR ribotypes 127, 033, and 126, which are genetically related to PCR ribotype 078 (Knight et al., 2013). PCR ribotype 078 has not been found in Australian food animals leading to speculation that PCR ribotype 127 occupies a similar niche as 078 in other countries and may develop human pathogenic potential (Knight et al., 2013). Thus, like swine, *C. difficile* is commonly present in young animals but diminishes with age and while a variety of PCR ribotypes are present, PCR ribotype 078 predominates.

Poultry

A survey of fecal samples taken from broiler chickens at marketplaces in an urban area of Zimbabwe found 29% *C. difficile* prevalence and the majority of isolates (90%) were toxigenic (Simango and Mwakurudza, 2008). In contrast, much higher rates were found on a poultry farm in Slovenia where 62% (38/61) of fecal samples were culture positive for *C. difficile* (Zidaric et al., 2008). This study described a diversity of PCR ribotypes among *C. difficile* isolates and notably no PCR ribotype 078 was observed (Zidaric et al., 2008). In Austria, only 5% (3/59) of chickens were positive for *C. difficile* (Indra et al., 2009). Similarly, the Netherlands reported a lower prevalence of *C. difficile* in poultry with only 6% (7/121) of fecal samples positive (Koene et al., 2012). In both studies, several PCR ribotypes associated with human disease were detected including PCR ribotypes 001 and 014 (Indra et al., 2009; Koene et al., 2012). In a Texas study of market-age broiler chickens, 2.3% (7/300) *C. difficile* prevalence was observed and all isolates had molecular genotypes consistent with PCR ribotype 078 (toxinotype V, NAP7) (Harvey et al., 2011a).

Meat Processing Facilities

Contamination of retail meats with *C. difficile* spores likely occurs during processing. *C. difficile* carriage in animals at the time of slaughter has been documented. A US study of beef cattle found 1.2% *C. difficile* prevalence at slaughter (Rodriguez-Palacios et al., 2011). In a study from Belgium, *C. difficile* was isolated from 1% of swine and 10% of cattle at slaughter (Rodriguez et al., 2013). Thus, *C. difficile* in animals at slaughter is present at variable rates. Contamination of surfaces and processing equipment could lead to widespread dissemination of infectious spores. In a US study of ground pork products, Curry et al. found 38% of Brand A pork products contaminated with *C. difficile* (Curry et al., 2012). Further sampling identified 57% of products from a single processing facility were contaminated with five distinct *C. difficile* MLVA genotypes, all belonging to PCR ribotype 078 (Curry et al., 2012). These data demonstrate that *C. difficile* carried by animals entering meat processing facilities can result in widespread and persistent contamination of the facility and the food that is produced (Table 10.2).

Retail Foods

Reports worldwide have documented the isolation of *C. difficile* from retail foods including raw vegetables, meats, and seafood. The majority of reports documenting the isolation of *C. difficile* from retail foods are from ground meats including beef and pork (Table 10.3). However, *C. difficile* has been isolated from a variety of meats including veal, chicken, turkey, goat, sheep, and buffalo (Harvey et al., 2011a; Houser et al., 2012; Rahimi et al., 2014; Weese et al., 2010a). A few reports of *C. difficile* in raw vegetables and ready-to-eat salads indicate that these food items are a potential source of *C. difficile* with prevalence ranging from 2.4–7.5% (al Saif and Brazier, 1996; Bakri et al., 2009; Eckert et al., 2013; Metcalf et al., 2010b). A few studies have looked for *C. difficile* in seafood. A single study in the United States isolated *C. difficile* from 4.5% of shellfish and finfish samples (Norman et al., 2014a). A similar prevalence in seafood and fish was observed in Canada where 5/119 (4.8%) samples were found to be contaminated with *C. difficile* (Metcalf et al., 2011). In a study from Naples, Italy, 49% (26/53) of bivalve molluscs were found to be positive for *C. difficile* (Pasquale et al., 2012). Fifteen of the isolates were toxigenic and belonged to PCR ribotypes commonly associated with human disease (Pasquale et al., 2012). The authors suggest that concentration of sewage contaminants may occur in filter feeders such as molluscs (Pasquale et al., 2012). Recovery of toxigenic *C. difficile* ribotypes associated with human disease from the effluent of wastewater treatment facilities has been demonstrated (Romano et al., 2012). Taken together, these data suggest that sewage treatment plants may contribute to contamination of *C. difficile* in fish and shellfish and human transmission.

In Europe, the prevalence of *C. difficile* in retail meats is generally reported to be <3% (Table 10.3). This observation is in contrast to North America where a higher prevalence of *C. difficile* in retail meats is observed (Table 10.3). While the majority of studies from North America demonstrate <20% of meats contaminated with *C. difficile* (Table 10.3), these rates are still significantly higher than those in Europe. Studies in Canada and Texas have found 13% of poultry samples positive for *C. difficile* (Harvey et al., 2011a; Weese et al., 2010a). Studies on ground meats including pork, beef, and veal have shown a wide range of *C. difficile* prevalence from 42% to 0% (Table 10.3). In contrast, two large US studies found no *C. difficile* in over 2,000 ground meat samples

TABLE 10.3 Summary of C. *difficile* Prevalence in Retail Foods

Source	Prevalence	Country	Ribotype	References
Vegetables—potato, mushroom, onion, radish, cucumber	7/300 (2.4%)	Wales	ND	al Saif and Brazier (1996)
Ground beef, veal	12/60 (20%)	Ontario	027-like	Rodríguez-Palacios et al. (2007)
Ground meats—beef, pork, sheep, reindeer, moose, poultry, sausage	2/82 (2.4%)	Sweden	ND	Von Abercron et al. (2009)
Ground beef and pork, chicken	0/84 (0%)	Austria		Indra et al. (2009)
Ground beef and pork	28/230 (12%)	Canada	078	Weese et al. (2009)
Ready-to-eat salads	3/40 (7.5%)	Scotland	017/001	Bakri et al. (2009)
Ground beef and pork, cat foods	2/105 (1.9%)	France	012	Bouttier et al. (2010)
Lamb, chicken	8/500 (1.6%)	The Netherlands	001	de Boer et al. (2011)
Vegetables, ginger, carrot, and eddoes from United States/China	5/111 (4.5%)	Ontario	078	Metcalf et al. (2010b)
Ground meats	3/100 (3%)	Austria	NTCD, common Austrian ribotype	Jobstl et al. (2010)
Ground meats	0/46 (0%)	Switzerland		Hoffer et al. (2010)

Food	Prevalence	Location	Ribotype	Reference
Ground pork and pork chops	7/393 (1.8%)	Canada	027	Metcalf et al. (2010a)
Poultry	4/32 (12.5%)	United States—TX	078	Harvey et al. (2011a)
Poultry	26/203 (12.8%)	Ontario	078	Weese et al. (2010a)
Ground pork, chorizo, turkey, and environmental	23/245 (9.5%)	United States—TX	078	Harvey et al. (2011b)
Ground veal	4/50 (8%)	United States—PA	ND	Houser et al. (2012)
Ground beef, buffalo, veal, chicken, lamb, pork, turkey	2/102 (1.9%)	United States—PA	078	Curry et al. (2012)
Molluscs	26/53 (49%)	Italy	NTCD, multiple	Pasquale et al. (2012)
Ground beef	0/956 (0%)	United States		Kalchayanand et al. (2013)
Ground meats and pork chops, chicken breasts	0/1755 (0%)	United States		Limbago et al. (2012)
Ready-to-eat salads, vegetables	3/104 (2.9%)	France	001/014/015	Eckert et al. (2013)
Seafood	3/67 (4.5%)	United States—TX	078	Norman et al. (2014a)
Buffalo, beef, sheep goat	13/660 (2%)	Iran	078	Rahimi et al. (2014)

(Kalchayanand et al., 2013; Limbago et al., 2012). In the US Department of Agriculture (USDA) study, 956 ground beef samples from seven microbiological monitoring regions throughout the United States were examined (Kalchayanand et al., 2013). In the Centers for Disease Control and Prevention (CDC) study, 1,755 ground and whole meat samples from retail stores in eight states were investigated (Limbago et al., 2012). These negative results from large sample sizes are unexpected based on previous studies. A consensus method for culture of *C. difficile* from meat proposed in the CDC study is based upon a spore inoculum of 100 spores/g (Limbago et al., 2012). This method may not detect low-level spore contamination. Moreover, broth enrichment without antibiotic selection may allow organisms with shorter doubling times to grow and outcompete *C. difficile*. In a study using selective broth enrichment in the presence of taurocholate and lysozyme without subsequent ethanol shock, <0.18 to <0.45 spores per gram were detected in retail ground pork (Curry et al., 2012). In a study comparing several *C. difficile* culture methods from animal fecal samples, considerable variation in recovery was observed indicating that some methodologies generate a high proportion of false-negative cultures (Blanco et al., 2013). Thus, differences in culture methods likely explain some of the discrepancies in studies of *C. difficile* prevalence from retail foods. Other factors including meat composition as well as strain or spore properties may impact recovery rates. Lineage-specific differences in germination efficiencies have been documented and may explain variable prevalence of *C. difficile* in food studies (Moore et al., 2013). These variables represent a few of the challenges in studying the relationship of *C. difficile* and food safety.

Farm Environment and Waste Management Practices

Farm workers who come in direct contact with animals or the farm environment may be at risk for acquiring *C. difficile*. A recent study in the Netherlands provided evidence supporting on-farm transmission of *C. difficile* from pigs to humans. In this study, *C. difficile* carriage was detected in 25% (12/48) of workers who had daily contact with pigs (Keessen et al., 2013a). Pig and human isolates from the majority of the farms surveyed were predominantly PCR ribotype 078 and either genetically related or indistinguishable by MLVA genotyping (Keessen et al., 2013a). This finding and similar antibiotic susceptibility results for human and pig isolates suggested that direct transmission from pigs to humans had occurred on-farm (Keessen et al., 2013a).

The current inventory of swine in the United States is ~67 million head across 69,000 farm operations according to the USDA National Agricultural Statistics Service (USDA-NASS). Over the past 30 years, a shift in swine production has occurred from many small, regional niche farms to fewer, large confined animal feeding operations of ≥1,000 head (USDA-NASS) (Davies, 2011). As a result, the volume of swine manure produced and stored on these operations is large, highly concentrated and potentially contaminated with many different strains of *C. difficile* (Cook et al., 2010). On-farm waste management occurs in either closed anaerobic systems or in open lagoons. These systems function to reduce and prevent off-farm transport of pathogenic manure (Ziemer et al., 2010). *Clostridium* spp. are commonly found in swine waste lagoons and contribute to the anaerobic degradation of materials present in the manure (Cook et al., 2010; Cotta et al., 2003; Snell-Castro et al., 2005). Several studies have demonstrated that clostridia are endemic in swine waste management systems (Cook et al., 2010; McLaughlin et al., 2012). These data suggest that *C. difficile* may survive, persist, and even replicate in swine waste lagoons.

Swine manure is used as fertilizer for agricultural purposes and this practice may introduce zoonotic pathogens such as *C. difficile* to the environment. Regulation of microbial levels in animal wastewater destined for crop production varies widely (US EPA) and is based on fecal coliform counts which may not be an appropriate surrogate for spore-forming organisms such as *C. difficile* (McLaughlin et al., 2012). The prevalence of *C. difficile* in composted manure destined for use as feed crop fertilizer is unknown. Highly resistant *C. difficile* spores may be transferred to agricultural soils and animal feed crops as well as fruits and vegetables for human consumption. Studies investigating the prevalence of *C. difficile* in farm waste management systems and composted material used in food production are required.

Antibiotic Resistance

Multidrug-resistant strains belonging to a variety of genetic lineages have been described for *C. difficile* of both animal and human origin (Pirs et al., 2013; Spigaglia et al., 2011; Tenover et al., 2012). In a 2005 European study, 55% (88/148) of drug-resistant clinical isolates were resistant to three or more antimicrobials (Spigaglia et al., 2011). *C. difficile* is characterized by a highly mobile, mosaic genome (Sebaihia et al., 2006). Genetic exchange and recombination contribute to antibiotic resistance and the presence of multiple conjugative transposons is an important determinant in the emergence and persistence of new *C. difficile* clones (Spigaglia et al., 2011). A recent study in the United States documented multidrug resistance in multiple genetic lineages commonly associated with human disease and reported reduced susceptibility to vancomycin in 39.1% of PCR ribotype 027 isolates (Tickler et al., 2014). In addition this study described increased prevalence of PCR ribotype 014 (Tickler et al., 2014). This lineage is characterized by low-level antibiotic resistance but has a broad host range—having been isolated from swine, horse, cattle, poultry, cats, dogs, and goats (Janezic et al., 2014). These observations support a model whereby intermingling and genetic recombination among *C. difficile* of animal and human origin may contribute to the emergence of new drug-resistant lineages.

The use of antibiotics to promote growth, prevent, and treat infections in food animal production is common and is associated with the development of resistance (Garcia-Migura et al., 2014). The frequent use of tetracycline in pork production may explain the prevalence of PCR ribotype 078 in swine. Tetracycline resistance in this lineage is conferred by the *tetM* gene carried on Tn*916* and may have been selected for by the widespread use of tetracycline in pork production. Tn*916* belongs to the family of conjugative transposons. These mobile genetic elements are prevalent in the *C. difficile* genome and are capable of acquiring multiple clinically relevant resistance genes (Roberts and Mullany, 2011). Of particular concern is the possible transfer of vancomycin resistance from *Enterococcus faecium* to *C. difficile* in co-colonized patients (Roberts and Mullany, 2011). This potential event would have severe implications for the clinical management of CDI.

Indeed, antibiotic practices in both animal production and the hospital setting impact the emergence of antibiotic-resistant *C. difficile* lineages. For instance, the use of fluoroquinolones is associated with *C. difficile* mutations in *gyrA* and *gyrB* and the emergence of the epidemic clone, BI/NAP1/027 (He et al., 2012; Muto et al., 2005). Similarly, the use of fluoroquinolones in the pork industry may have contributed to the emergence of the multidrug-resistant PCR ribotype 078. In the Netherlands, 37% (37/99) of human and swine PCR ribotype 078 isolates were demonstrated to have resistance to four or more antimicrobials (Keessen et al., 2013b). This study also found a significant association of fluoroquinolone use with the isolation of moxifloxicin-resistant isolates from both animals and humans (Keessen et al., 2013b). Thus, antibiotic management practices drive the selection and evolution of drug-resistant *C. difficile.*

Public health concern regarding nontherapeutic use of antibiotics in the agricultural setting has resulted in new FDA guidelines included in "The Judicious Use of Medically Important Antimicrobial Drugs in Food-Producing Animals," and a call for voluntary limitations on the use of antimicrobials in food animal production with veterinary oversight of such drugs (FDA, 2013). A ban on animal feeds containing antimicrobials for growth promotion in Europe which began over 30 years ago has been largely successful in reducing drug resistance (Marshall and Levy, 2011). The goals of the current FDA regulations are similar—to limit the use of important therapeutics to prevent and reduce development of drug resistance. The recent emergence of the "double-host adapted" MRSA CC398 exemplifies bacterial evolution in response to selective pressures and crowded farm environments (Price et al., 2012). The zoonotic potential of *C. difficile* remains to be proven. However, the epidemiology of *C. difficile* continues to evolve. Prudent antibiotic use and multidisciplinary surveillance studies in veterinary, food science, and clinical medicine are necessary to understand the emergence of drug-resistant *C. difficile.*

CONCLUSIONS

Despite evidence that *C. difficile* could be a foodborne pathogen; transmission of *C. difficile* through food has not been documented. Demonstrating foodborne transmission of *C. difficile* is challenging as many contributing factors including host susceptibility, infectious dose, and incubation period are poorly defined. Carefully designed studies are required to address the role of food and food animals in *C. difficile* carriage and disease. A large, prospective study of sufficient size to identify risk factors for *C. difficile* carriage in healthy human volunteers would ideally include a detailed food diary and samples of ingested foods. Longitudinal sampling of an integrated cohort of healthy individuals would establish duration of carriage, and identify risk factors for asymptomatic carriage and subsequent community-associated disease. Questions regarding the role of food and food animals in CDI and transmission will remain until these types of studies are performed.

REFERENCES

al Saif, N., Brazier, J.S., 1996. The distribution of *Clostridium difficile* in the environment of South Wales. J. Med. Microbiol. 45, 133–137.

Alvarez-Perez, S., Blanco, J.L., Bouza, E., Alba, P., Gibert, X., Maldonado, J., et al., 2009. Prevalence of *Clostridium difficile* in diarrhoeic and non-diarrhoeic piglets. Vet. Microbiol. 137, 302–305.

Avbersek, J., Janezic, S., Pate, M., Rupnik, M., Zidaric, V., Logar, K., et al., 2009. Diversity of *Clostridium difficile* in pigs and other animals in Slovenia. Anaerobe 15, 252–255.

Baker, A.A., Davis, E., Rehberger, T., Rosener, D., 2010. Prevalence and diversity of toxigenic *Clostridium perfringens* and *Clostridium difficile* among swine herds in the Midwest. Appl. Environ. Microbiol. 76, 2961–2967.

Bakker, D., Corver, J., Harmanus, C., Goorhuis, A., Keessen, E.C., Fawley, W.N., et al., 2010. Relatedness of human and animal *Clostridium difficile* PCR ribotype 078 isolates determined on the basis of multilocus variable-number tandem-repeat analysis and tetracycline resistance. J. Clin. Microbiol. 48, 3744–3749.

Bakri, M.M., Brown, D.J., Butcher, J.P., Sutherland, A.D., 2009. *Clostridium difficile* in ready-to-eat salads, Scotland. Emerg. Infect. Dis. 15, 817–818.

Bauer, M.P., Notermans, D.W., van Benthem, B.H., Brazier, J.S., Wilcox, M.H., Rupnik, M., et al., 2011. *Clostridium difficile* infection in Europe: a hospital-based survey. Lancet 377, 63–73.

Black, S.R., Weaver, K.N., Jones, R.C., Ritger, K.A., Petrella, L.A., Sambol, S.P., et al., 2011. *Clostridium difficile* outbreak strain BI is highly endemic in Chicago area hospitals. Infect. Control. Hosp. Epidemiol. 32, 897–902.

Blanco, J.L., Alvarez-Perez, S., Garcia, M.E., 2013. Is the prevalence of *Clostridium difficile* in animals underestimated? Vet. J. 197, 694–698.

Borriello, S.P., Honour, P., Turner, T., Barclay, F., 1983. Household pets as a potential reservoir for *Clostridium difficile* infection. J. Clin. Pathol. 36, 84–87.

Bouttier, S., Barc, M.C., Felix, B., Lambert, S., Collignon, A., Barbut, F., 2010. *Clostridium difficile* in ground meat, France. Emerg. Infect. Dis. 16, 733–735.

Brouwer, M.S., Roberts, A.P., Hussain, H., Williams, R.J., Allan, E., Mullany, P., 2013. Horizontal gene transfer converts non-toxigenic *Clostridium difficile* strains into toxin producers. Nat. Commun. 4, 2601.

CDC, 2005. Severe *Clostridium difficile*-associated disease in populations previously at low risk—four states, 2005. MMWR Morb. Mortal. Wkly. Rep. 54, 1201–1205.

Chang, J.Y., Antonopoulos, D.A., Kalra, A., Tonelli, A., Khalife, W.T., Schmidt, T.M., et al., 2008. Decreased diversity of the fecal Microbiome in recurrent *Clostridium difficile*-associated diarrhea. J. Infect. Dis. 197, 435–438.

Chitnis, A.S., Holzbauer, S.M., Belflower, R.M., Winston, L.G., Bamberg, W.M., Lyons, C., et al., 2013. Epidemiology of community-associated *Clostridium difficile* infection, 2009 through 2011. JAMA Intern. Med. 173, 1359–1367.

Cook, K.L., Rothrock Jr., M.J., Lovanh, N., Sorrell, J.K., Loughrin, J.H., 2010. Spatial and temporal changes in the microbial community in an anaerobic swine waste treatment lagoon. Anaerobe 16, 74–82.

Cornely, O.A., Crook, D.W., Esposito, R., Poirier, A., Somero, M.S., Weiss, K., et al., 2012a. Fidaxomicin versus vancomycin for infection with *Clostridium difficile* in Europe, Canada, and the USA: a double-blind, non-inferiority, randomised controlled trial. Lancet Infect. Dis. 12, 281–289.

Cornely, O.A., Miller, M.A., Louie, T.J., Crook, D.W., Gorbach, S.L., 2012b. Treatment of first recurrence of *Clostridium difficile* infection: fidaxomicin versus vancomycin. Clin. Infect. Dis. 55 (Suppl. 2), S154–161.

Costa, M.C., Reid-Smith, R., Gow, S., Hannon, S.J., Booker, C., Rousseau, J., et al., 2012. Prevalence and molecular characterization of *Clostridium difficile* isolated from feedlot beef cattle upon arrival and mid-feeding period. BMC Vet. Res. 8, 38.

Cotta, M.A., Whitehead, T.R., Zeltwanger, R.L., 2003. Isolation, characterization and comparison of bacteria from swine faeces and manure storage pits. Environ. Microbiol. 5, 737–745.

Curry, S., Brown, N., Marsh, J., Muto, C., Harrison, L.H., Binion, D., 2011. A Point Prevalence Study of Hospital-Based GI Clinic Environmental Contamination with Toxigenic *Clostridium difficile*. Society for Healthcare Epidemiology of America, Inc., Dallas, TX.

Curry, S.R., Marsh, J.W., Schlackman, J.L., Harrison, L.H., 2012. Prevalence of *Clostridium difficile* in uncooked ground meat products from Pittsburgh, Pennsylvania. Appl. Environ. Microbiol. 78, 4183–4186.

Curry, S.R., Muto, C.A., Schlackman, J.L., Pasculle, A.W., Shutt, K.A., Marsh, J.W., et al., 2013. Use of multilocus variable number of tandem repeats analysis genotyping to determine the role of asymptomatic carriers in *Clostridium difficile* transmission. Clin. Infect. Dis. 57, 1094–1102.

Davies, P.R., 2011. Intensive swine production and pork safety. Foodborne Pathog. Dis. 8, 189–201.

de Boer, E., Zwartkruis-Nahuis, A., Heuvelink, A.E., Harmanus, C., Kuijper, E.J., 2011. Prevalence of *Clostridium difficile* in retailed meat in the Netherlands. Int. J. Food Microbiol. 144, 561–564.

Debast, S.B., van Leengoed, L.A., Goorhuis, A., Harmanus, C., Kuijper, E.J., Bergwerff, A.A., 2009. *Clostridium difficile* PCR ribotype 078 toxinotype V found in diarrhoeal pigs identical to isolates from affected humans. Environ. Microbiol. 11, 505–511.

Drudy, D., Kyne, L., O'Mahony, R., Fanning, S., 2007. gyrA mutations in fluoroquinolone-resistant *Clostridium difficile* PCR-027. Emerg. Infect. Dis. 13, 504–505.

Dumyati, G., Stevens, V., Hannett, G.E., Thompson, A.D., Long, C., Maccannell, D., et al., 2012. Community-associated *Clostridium difficile* infections, Monroe County, New York, USA. Emerg. Infect. Dis. 18, 392–400.

Eckert, C., Burghoffer, B., Barbut, F., 2013. Contamination of ready-to-eat raw vegetables with *Clostridium difficile* in France. J. Med. Microbiol. 62, 1435–1438.

Eyre, D.W., Cule, M.L., Wilson, D.J., Griffiths, D., Vaughan, A., O'Connor, L., et al., 2013. Diverse sources of *C. difficile* infection identified on whole-genome sequencing. N. Engl. J. Med. 369, 1195–1205.

FDA, 2012. *Clostridium difficile*-associated diarrhea can be associated with stomach acid drugs known as proton pump inhibitors (PPIs), FDA Drug Safety Communication.

FDA, 2013. Judicious use of medically important antimicrobial drugs in food producing animals.

FDA, 2014. Guidance to industry: enforcement policy regarding investigational new drug requirements for use of fecal microbiota for transplantation to treat *Clostridium difficile* infection not responsive to standard therapies.

Freter, R., 1955. The fatal enteric cholera infection in the guinea pig, achieved by inhibition of normal enteric flora. J. Infect. Dis. 97, 57–65.

Fry, P.R., Thakur, S., Abley, M., Gebreyes, W.A., 2012. Antimicrobial resistance, toxinotype, and genotypic profiling of *Clostridium difficile* isolates of swine origin. J. Clin. Microbiol. 50, 2366–2372.

Galdys, A.L., Nelson, J.S., Shutt, K.A., Schlackman, J.L., Pakstis, D.L., Pasculle, A.W., et al., 2014. Prevalence and duration of asymptomatic *Clostridium difficile* carriage among healthy subjects in Pittsburgh, Pennsylvania. J. Clin. Microbiol. 52, 2406–2409.

Garcia-Migura, L., Hendriksen, R.S., Fraile, L., Aarestrup, F.M., 2014. Antimicrobial resistance of zoonotic and commensal bacteria in Europe: the missing link between consumption and resistance in veterinary medicine. Vet. Microbiol. 170, 1–9.

Goorhuis, A., Bakker, D., Corver, J., Debast, S.B., Harmanus, C., Notermans, D.W., et al., 2008a. Emergence of *Clostridium difficile* infection due to a new hypervirulent strain, polymerase chain reaction ribotype 078. Clin. Infect. Dis. 47, 1162–1170.

Goorhuis, A., Debast, S.B., van Leengoed, L.A., Harmanus, C., Notermans, D.W., Bergwerff, A.A., et al., 2008b. *Clostridium difficile* PCR ribotype 078: an emerging strain in humans and in pigs? J. Clin. Microbiol. 46, 1157 (author reply 1158).

Gould, L.H., Limbago, B., 2010. *Clostridium difficile* in food and domestic animals: a new foodborne pathogen? Clin. Infect. Dis. 51, 577–582.

Greenberg, R.N., Marbury, T.C., Foglia, G., Warny, M., 2012. Phase I dose finding studies of an adjuvanted *Clostridium difficile* toxoid vaccine. Vaccine 30, 2245–2249.

Griffiths, D., Fawley, W., Kachrimanidou, M., Bowden, R., Crook, D.W., Fung, R., et al., 2010. Multilocus sequence typing of *Clostridium difficile*. J. Clin. Microbiol. 48, 770–778.

Harvey, R.B., Norman, K.N., Andrews, K., Hume, M.E., Scanlan, C.M., Callaway, T.R., et al., 2011a. *Clostridium difficile* in poultry and poultry meat. Foodborne Pathog. Dis. 8, 1321–1323.

Harvey, R.B., Norman, K.N., Andrews, K., Norby, B., Hume, M.E., Scanlan, C.M., et al., 2011b. *Clostridium difficile* in retail meat and processing plants in Texas. J. Vet. Diagn. Invest. 23, 807–811.

Hawken, P., Weese, J.S., Friendship, R., Warriner, K., 2013. Longitudinal study of *Clostridium difficile* and methicillin-resistant *Staphylococcus aureus* associated with pigs from weaning through to the end of processing. J. Food Prot. 76, 624–630.

He, M., Sebaihia, M., Lawley, T.D., Stabler, R.A., Dawson, L.F., Martin, M.J., et al., 2010. Evolutionary dynamics of *Clostridium difficile* over short and long time scales. Proc. Natl. Acad. Sci. USA 107, 7527–7532.

He, M., Miyajima, F., Roberts, P., Ellison, L., Pickard, D.J., Martin, M.J., et al., 2012. Emergence and global spread of epidemic healthcare-associated *Clostridium difficile*. Nat. Genet. 45 (1), 109–113.

Hirshon, J.M., Thompson, A.D., Limbago, B., McDonald, L.C., Bonkosky, M., Heimer, R., et al., 2011. *Clostridium difficile* infection in outpatients, Maryland and Connecticut, USA, 2002–2007. Emerg. Infect. Dis. 17, 1946–1949.

Hoffer, E., Haechler, H., Frei, R., Stephan, R., 2010. Low occurrence of *Clostridium difficile* in fecal samples of healthy calves and pigs at slaughter and in minced meat in Switzerland. J. Food Prot. 73, 973–975.

Hoover, D.G., Rodriguez-Palacios, A., 2013. Transmission of *Clostridium difficile* in foods. Infect. Dis. Clin. North Am. 27, 675–685.

Hopman, N.E., Oorburg, D., Sanders, I., Kuijper, E.J., Lipman, L.J., 2011. High occurrence of various *Clostridium difficile* PCR ribotypes in pigs arriving at the slaughterhouse. Vet. Q. 31, 179–181.

Houser, B.A., Soehnlen, M.K., Wolfgang, D.R., Lysczek, H.R., Burns, C.M., Jayarao, B.M., 2012. Prevalence of *Clostridium difficile* toxin genes in the feces of veal calves and incidence of ground veal contamination. Foodborne Pathog. Dis. 9, 32–36.

Hunt, J.J., Ballard, J.D., 2013. Variations in virulence and molecular biology among emerging strains of *Clostridium difficile*. Microbiol. Mol. Biol. Rev. 77, 567–581.

Indra, A., Lassnig, H., Baliko, N., Much, P., Fiedler, A., Huhulescu, S., et al., 2009. *Clostridium difficile*: a new zoonotic agent? Wien. Klin. Wochenschr. 121, 91–95.

Janarthanan, S., Ditah, I., Adler, D.G., Ehrinpreis, M.N., 2012. *Clostridium difficile*-associated diarrhea and proton pump inhibitor therapy: a meta-analysis. Am. J. Gastroenterol. 107, 1001–1010.

Janezic, S., Zidaric, V., Pardon, B., Indra, A., Kokotovic, B., Blanco, J.L., et al., 2014. International *Clostridium difficile* animal strain collection and large diversity of animal associated strains. BMC Microbiol. 14, 173.

Jhung, M.A., Thompson, A.D., Killgore, G.E., Zukowski, W.E., Songer, G., Warny, M., et al., 2008. Toxinotype V *Clostridium difficile* in humans and food animals. Emerg. Infect. Dis. 14, 1039–1045.

Jobstl, M., Heuberger, S., Indra, A., Nepf, R., Kofer, J., Wagner, M., 2010. *Clostridium difficile* in raw products of animal origin. Int. J. Food Microbiol. 138, 172–175.

Jump, R.L., Pultz, M.J., Donskey, C.J., 2007. Vegetative *Clostridium difficile* survives in room air on moist surfaces and in gastric contents with reduced acidity: a potential mechanism to explain the association between proton pump inhibitors and *C. difficile*-associated diarrhea? Antimicrob. Agents Chemother. 51, 2883–2887.

Jury, L.A., Sitzlar, B., Kundrapu, S., Cadnum, J.L., Summers, K.M., Muganda, C.P., et al., 2013. Outpatient healthcare settings and transmission of *Clostridium difficile*. PLoS One 8, e70175.

Kalchayanand, N., Arthur, T.M., Bosilevac, J.M., Brichta-Harhay, D.M., Shackelford, S.D., Wells, J.E., et al., 2013. Isolation and characterization of *Clostridium difficile* associated with beef cattle and commercially produced ground beef. J. Food Prot. 76, 256–264.

Keel, K., Brazier, J.S., Post, K.W., Weese, S., Songer, J.G., 2007. Prevalence of PCR ribotypes among *Clostridium difficile* isolates from pigs, calves, and other species. J. Clin. Microbiol. 45, 1963–1964.

Keessen, E.C., Donswijk, C.J., Hol, S.P., Hermanus, C., Kuijper, E.J., Lipman, L.J., 2011a. Aerial dissemination of *Clostridium difficile* on a pig farm and its environment. Environ. Res. 111, 1027–1032.

Keessen, E.C., van den Berkt, A.J., Haasjes, N.H., Hermanus, C., Kuijper, E.J., Lipman, L.J., 2011b. The relation between farm specific factors and prevalence of *Clostridium difficile* in slaughter pigs. Vet. Microbiol. 154, 130–134.

Keessen, E.C., Harmanus, C., Dohmen, W., Kuijper, E.J., Lipman, L.J., 2013a. *Clostridium difficile* infection associated with pig farms. Emerg. Infect. Dis. 19, 1032–1034.

Keessen, E.C., Hensgens, M.P., Spigaglia, P., Barbanti, F., Sanders, I.M., Kuijper, E.J., et al., 2013b. Antimicrobial susceptibility profiles of human and piglet *Clostridium difficile* PCR-ribotype 078. Antimicrob. Resist. Infect. Control 2, 14.

Khanna, S., Pardi, D.S., Aronson, S.L., Kammer, P.P., Orenstein, R., St Sauver, J.L., et al., 2012. The epidemiology of community-acquired *Clostridium difficile* infection: a population-based study. Am. J. Gastroenterol. 107, 89–95.

Knight, D.R., Thean, S., Putsathit, P., Fenwick, S., Riley, T.V., 2013. Cross-sectional study reveals high prevalence of *Clostridium difficile* non-PCR ribotype 078 strains in Australian veal calves at slaughter. Appl. Environ. Microbiol. 79, 2630–2635.

Koene, M.G., Mevius, D., Wagenaar, J.A., Harmanus, C., Hensgens, M.P., Meetsma, A.M., et al., 2012. *Clostridium difficile* in Dutch animals: their presence, characteristics and similarities with human isolates. Clin. Microbiol. Infect. 18, 778–784.

Kutty, P.K., Woods, C.W., Sena, A.C., Benoit, S.R., Naggie, S., Frederick, J., et al., 2010. Risk factors for and estimated incidence of community-associated *Clostridium difficile* infection, North Carolina, USA. Emerg. Infect. Dis. 16, 197–204.

Kwok, C.S., Arthur, A.K., Anibueze, C.I., Singh, S., Cavallazzi, R., Loke, Y.K., 2012. Risk of *Clostridium difficile* infection with acid suppressing drugs and antibiotics: meta-analysis. Am. J. Gastroenterol. 107, 1011–1019.

Kyne, L., Warny, M., Qamar, A., Kelly, C.P., 2000. Asymptomatic carriage of *Clostridium difficile* and serum levels of IgG antibody against toxin A. N. Engl. J. Med. 342, 390–397.

Lefebvre, S.L., Waltner-Toews, D., Peregrine, A.S., Reid-Smith, R., Hodge, L., Arroyo, L.G., et al., 2006. Prevalence of zoonotic agents in dogs visiting hospitalized people in Ontario: implications for infection control. J. Hosp. Infect. 62, 458–466.

Limbago, B., Thompson, A.D., Greene, S.A., MacCannell, D., MacGowan, C.E., Jolbitado, B., et al., 2012. Development of a consensus method for culture of *Clostridium difficile* from meat and its use in a survey of U.S. retail meats. Food Microbiol. 32, 448–451.

Lowy, I., Molrine, D.C., Leav, B.A., Blair, B.M., Baxter, R., Gerding, D.N., et al., 2010. Treatment with monoclonal antibodies against *Clostridium difficile* toxins. N. Engl. J. Med. 362, 197–205.

Marsh, J.W., O'Leary, M.M., Shutt, K.A., Pasculle, A.W., Johnson, S., Gerding, D.N., et al., 2006. Multilocus variable-number tandem-repeat analysis for investigation of *Clostridium difficile* transmission in Hospitals. J. Clin. Microbiol. 44, 2558–2566.

Marsh, J.W., Tulenko, M.M., Shutt, K.A., Thompson, A.D., Weese, J.S., Songer, J.G., et al., 2011. Multi-locus variable number tandem repeat analysis for investigation of the genetic association of *Clostridium difficile* isolates from food, food animals and humans. Anaerobe 17, 156–160.

Marsh, J.W., Arora, R., Schlackman, J.L., Shutt, K.A., Curry, S.R., Harrison, L.H., 2012. Association of relapse of *Clostridium difficile* Disease with BI/NAP1/027. J. Clin. Microbiol. 50, 4078–4082.

Marshall, B.M., Levy, S.B., 2011. Food animals and antimicrobials: impacts on human health. Clin. Microbiol. Rev. 24, 718–733.

McLaughlin, M.R., Brooks, J.P., Adeli, A., 2012. Temporal flux and spatial dynamics of nutrients, fecal indicators, and zoonotic pathogens in anaerobic swine manure lagoon water. Water Res. 46, 4949–4960.

Metcalf, D., Reid-Smith, R.J., Avery, B.P., Weese, J.S., 2010a. Prevalence of *Clostridium difficile* in retail pork. Can. Vet. J. 51, 873–876.

Metcalf, D.S., Costa, M.C., Dew, W.M., Weese, J.S., 2010b. *Clostridium difficile* in vegetables, Canada. Lett. Appl. Microbiol. 51, 600–602.

Metcalf, D., Avery, B.P., Janecko, N., Matic, N., Reid-Smith, R., Weese, J.S., 2011. *Clostridium difficile* in seafood and fish. Anaerobe 17, 85–86.

Miyajima, F., Roberts, P., Swale, A., Price, V., Jones, M., Horan, M., et al., 2011. Characterisation and carriage ratio of *Clostridium difficile* strains isolated from a community-dwelling elderly population in the United Kingdom. PLoS One 6, e22804.

Moore, P., Kyne, L., Martin, A., Solomon, K., 2013. Germination efficiency of clinical *Clostridium difficile* spores and correlation with ribotype, disease severity and therapy failure. J. Med. Microbiol. 62, 1405–1413.

Muto, C.A., Pokrywka, M., Shutt, K., Mendelsohn, A.B., Nouri, K., Posey, K., et al., 2005. A large outbreak of *Clostridium difficile*-associated disease with an unexpected proportion of deaths and colectomies at a teaching hospital following increased fluoroquinolone use. Infect. Control Hosp. Epidemiol. 26, 273–280.

Muto, C.A., Blank, M.K., Marsh, J.W., Vergis, E.N., O'Leary, M.M., Shutt, K.A., et al., 2007. Control of an outbreak of infection with the hypervirulent *Clostridium difficile* BI strain in a university hospital using a comprehensive "bundle" approach. Clin. Infect. Dis. 45, 1266–1273.

Noren, T., Akerlund, T., Back, E., Sjoberg, L., Persson, I., Alriksson, I., et al., 2004. Molecular epidemiology of hospital-associated and community-acquired *Clostridium difficile* infection in a Swedish county. J. Clin. Microbiol. 42, 3635–3643.

Noren, T., Johansson, K., Unemo, M., 2014. *Clostridium difficile* PCR ribotype 046 is common among neonatal pigs and humans in Sweden. Clin. Microbiol. Infect. 20, O2–6.

Norman, K.N., Harvey, R.B., Scott, H.M., Hume, M.E., Andrews, K., Brawley, A.D., 2009. Varied prevalence of *Clostridium difficile* in an integrated swine operation. Anaerobe 15, 256–260.

Norman, K.N., Scott, H.M., Harvey, R.B., Norby, B., Hume, M.E., Andrews, K., 2011. Prevalence and genotypic characteristics of *Clostridium difficile* in a closed and integrated human and swine population. Appl. Environ. Microbiol. 77, 5755–5760.

Norman, K.N., Harvey, R.B., Andrews, K., Hume, M.E., Callaway, T.R., Anderson, R.C., et al., 2014a. Survey of *Clostridium difficile* in retail seafood in College Station, Texas. Food Addit. Contam. Part A Chem. Anal. Control Expo. Risk. Assess. 31, 1127–1129.

Norman, K.N., Scott, H.M., Harvey, R.B., Norby, B., Hume, M.E., 2014b. Comparison of antimicrobial susceptibility among *Clostridium difficile* isolated from an integrated human and swine population in Texas. Foodborne Pathog. Dis. 11, 257–264.

Ozaki, E., Kato, H., Kita, H., Karasawa, T., Maegawa, T., Koino, Y., et al., 2004. *Clostridium difficile* colonization in healthy adults: transient colonization and correlation with enterococcal colonization. J. Med. Microbiol. 53, 167–172.

Paredes-Sabja, D., Shen, A., Sorg, J.A., 2014. *Clostridium difficile* spore biology: sporulation, germination, and spore structural proteins. Trends Microbiol. 22 (7), 406–416.

Pasquale, V., Romano, V., Rupnik, M., Capuano, F., Bove, D., Aliberti, F., et al., 2012. Occurrence of toxigenic *Clostridium difficile* in edible bivalve molluscs. Food Microbiol. 31, 309–312.

Patterson, L., Wilcox, M.H., Fawley, W.N., Verlander, N.Q., Geoghegan, L., Patel, B.C., et al., 2012. Morbidity and mortality associated with *Clostridium difficile* ribotype 078: a case–case study. J. Hosp. Infect. 82, 125–128.

Pirs, T., Ocepek, M., Rupnik, M., 2008. Isolation of *Clostridium difficile* from food animals in Slovenia. J. Med. Microbiol. 57, 790–792.

Pirs, T., Avbersek, J., Zdovc, I., Krt, B., Andlovic, A., Lejko-Zupanc, T., et al., 2013. Antimicrobial susceptibility of animal and human isolates of *Clostridium difficile* by broth microdilution. J. Med. Microbiol. 62, 1478–1485.

Price, L.B., Stegger, M., Hasman, H., Aziz, M., Larsen, J., Andersen, P.S., et al., 2012. *Staphylococcus aureus* CC398: host adaptation and emergence of methicillin resistance in livestock. MBio 3, e00305–e00311.

Rahimi, E., Jalali, M., Weese, J.S., 2014. Prevalence of *Clostridium difficile* in raw beef, cow, sheep, goat, camel and buffalo meat in Iran. BMC Public. Health. 14, 119.

Roberts, A.P., Mullany, P., 2011. Tn*916*-like genetic elements: a diverse group of modular mobile elements conferring antibiotic resistance. FEMS Microbiol. Rev. 35, 856–871.

Rodriguez-Palacios, A., Lejeune, J.T., 2011. Moist-heat resistance, spore aging, and superdormancy in *Clostridium difficile*. Appl. Environ. Microbiol. 77, 3085–3091.

Rodriguez-Palacios, A., Stampfli, H.R., Duffield, T., Peregrine, A.S., Trotz-Williams, L.A., Arroyo, L.G., et al., 2006. *Clostridium difficile* PCR ribotypes in calves, Canada. Emerg. Infect. Dis. 12, 1730–1736.

Rodriguez-Palacios, A., Staempfli, H.R., Duffield, T., Weese, J.S., 2007. *Clostridium difficile* in retail ground meat, Canada. Emerg. Infect. Dis. 13, 485–487.

Rodriguez-Palacios, A., Pickworth, C., Loerch, S., LeJeune, J.T., 2011. Transient fecal shedding and limited animal-to-animal transmission of *Clostridium difficile* by naturally infected finishing feedlot cattle. Appl. Environ. Microbiol. 77, 3391–3397.

Rodriguez, C., Avesani, V., Van Broeck, J., Taminiau, B., Delmee, M., Daube, G., 2013. Presence of *Clostridium difficile* in pigs and cattle intestinal contents and carcass contamination at the slaughterhouse in Belgium. Int. J. Food Microbiol. 166, 256–262.

Romano, V., Pasquale, V., Krovacek, K., Mauri, F., Demarta, A., Dumontet, S., 2012. Toxigenic *Clostridium difficile* PCR ribotypes from wastewater treatment plants in southern Switzerland. Appl. Environ. Microbiol. 78, 6643–6646.

Rousseau, C., Lemee, L., Le Monnier, A., Poilane, I., Pons, J.L., Collignon, A., 2011. Prevalence and diversity of *Clostridium difficile* strains in infants. J. Med. Microbiol. 60, 1112–1118.

Rupnik, M., Widmer, A., Zimmermann, O., Eckert, C., Barbut, F., 2008. *Clostridium difficile* toxinotype V, ribotype 078, in animals and humans. J. Clin. Microbiol. 46, 2146.

Sambol, S.P., Merrigan, M.M., Tang, J.K., Johnson, S., Gerding, D.N., 2002. Colonization for the prevention of *Clostridium difficile* disease in hamsters. J. Infect. Dis. 186, 1781–1789.

Schneeberg, A., Neubauer, H., Schmoock, G., Baier, S., Harlizius, J., Nienhoff, H., et al., 2013. *Clostridium difficile* genotypes in piglet populations in Germany. J. Clin. Microbiol. 51, 3796–3803.

Sebaihia, M., Wren, B.W., Mullany, P., Fairweather, N.F., Minton, N., Stabler, R., et al., 2006. The multidrug-resistant human pathogen *Clostridium difficile* has a highly mobile, mosaic genome. Nat. Genet. 38, 779–786.

Simango, C., Mwakurudza, S., 2008. *Clostridium difficile* in broiler chickens sold at market places in Zimbabwe and their antimicrobial susceptibility. Int. J. Food Microbiol. 124, 268–270.

Snell-Castro, R., Godon, J.J., Delgenes, J.P., Dabert, P., 2005. Characterisation of the microbial diversity in a pig manure storage pit using small subunit rDNA sequence analysis. FEMS Microbiol. Ecol. 52, 229–242.

Solomon, K., Martin, A.J., O'Donoghue, C., Chen, X., Fenelon, L., Fanning, S., et al., 2013. Mortality in patients with *Clostridium difficile* infection correlates with host pro-inflammatory and humoral immune responses. J. Med. Microbiol. 62, 1453–1460.

Songer, J.G., Anderson, M.A., 2006. *Clostridium difficile*: an important pathogen of food animals. Anaerobe 12, 1–4.

Spigaglia, P., Barbanti, F., Mastrantonio, P., 2011. Multidrug resistance in European *Clostridium difficile* clinical isolates. J. Antimicrob. Chemother. 66, 2227–2234.

Squire, M.M., Carter, G.P., Mackin, K.E., Chakravorty, A., Noren, T., Elliott, B., et al., 2013. Novel molecular type of *Clostridium difficile* in neonatal pigs, Western Australia. Emerg. Infect. Dis. 19, 790–792.

Stoesser, N., Crook, D.W., Fung, R., Griffiths, D., Harding, R.M., Kachrimanidou, M., et al., 2011. Molecular epidemiology of *Clostridium difficile* strains in children compared with that of strains circulating in adults with *Clostridium difficile*-associated infection. J. Clin. Microbiol. 49, 3994–3996.

Stubbs, S.L., Brazier, J.S., O'Neill, G.L., Duerden, B.I., 1999. PCR targeted to the 16S–23S rRNA gene intergenic spacer region of *Clostridium difficile* and construction of a library consisting of 116 different PCR ribotypes. J. Clin. Microbiol. 37, 461–463.

Taori, S.K., Wroe, A., Hardie, A., Gibb, A.P., Poxton, I.R., 2014. A prospective study of community-associated *Clostridium difficile* infections: the role of antibiotics and co-infections. J. Infect. 69 (2), 134–144.

Tenover, F.C., Tickler, I.A., Persing, D.H., 2012. Antimicrobial-resistant strains of *Clostridium difficile* from North America. Antimicrob. Agents Chemother. 56, 2929–2932.

Thitaram, S.N., Frank, J.F., Lyon, S.A., Siragusa, G.R., Bailey, J.S., Lombard, J.E., et al., 2011. *Clostridium difficile* from healthy food animals: optimized isolation and prevalence. J. Food. Prot. 74, 130–133.

Tickler, I.A., Goering, R.V., Whitmore, J.D., Lynn, A.N., Persing, D.H., Tenover, F.C., 2014. Strain types and antimicrobial resistance patterns of *Clostridium difficile* isolates from the United States, 2011 to 2013. Antimicrob. Agents Chemother. 58, 4214–4218.

van Nood, E., Vrieze, A., Nieuwdorp, M., Fuentes, S., Zoetendal, E.G., de Vos, W.M., et al., 2013. Duodenal infusion of donor feces for recurrent *Clostridium difficile*. N. Engl. J. Med. 368, 407–415.

Villano, S.A., Seiberling, M., Tatarowicz, W., Monnot-Chase, E., Gerding, D.N., 2012. Evaluation of an oral suspension of VP20621, spores of nontoxigenic *Clostridium difficile* strain M3, in healthy subjects. Antimicrob. Agents Chemother. 56, 5224–5229.

Von Abercron, S.M., Karlsson, F., Wigh, G.T., Wierup, M., Krovacek, K., 2009. Low occurrence of *Clostridium difficile* in retail ground meat in Sweden. J. Food Prot. 72, 1732–1734.

Voth, D.E., Ballard, J.D., 2005. *Clostridium difficile* toxins: mechanism of action and role in disease. Clin. Microbiol. Rev. 18, 247–263.

Weese, J.S., Avery, B.P., Rousseau, J., Reid-Smith, R.J., 2009. Detection and enumeration of *Clostridium difficile* spores in retail beef and pork. Appl. Environ. Microbiol. 75, 5009–5011.

Weese, J.S., Reid-Smith, R.J., Avery, B.P., Rousseau, J., 2010a. Detection and characterization of *Clostridium difficile* in retail chicken. Lett. Appl. Microbiol. 50, 362–365.

Weese, J.S., Wakeford, T., Reid-Smith, R., Rousseau, J., Friendship, R., 2010b. Longitudinal investigation of *Clostridium difficile* shedding in piglets. Anaerobe 16, 501–504.

Weese, J.S., Rousseau, J., Deckert, A., Gow, S., Reid-Smith, R.J., 2011. *Clostridium difficile* and methicillin-resistant *Staphylococcus aureus* shedding by slaughter-age pigs. BMC Vet. Res. 7, 41.

Youngster, I., Sauk, J., Pindar, C., Wilson, R.G., Kaplan, J.L., Smith, M.B., et al., 2014. Fecal microbiota transplant for relapsing *Clostridium difficile* infection using a frozen inoculum from unrelated donors: a randomized, open-label, controlled pilot study. Clin. Infect. Dis. 58, 1515–1522.

Zar, F.A., Bakkanagari, S.R., Moorthi, K.M., Davis, M.B., 2007. A comparison of vancomycin and metronidazole for the treatment of *Clostridium difficile*-associated diarrhea, stratified by disease severity. Clin. Infect. Dis. 45, 302–307.

Zidaric, V., Zemljic, M., Janezic, S., Kocuvan, A., Rupnik, M., 2008. High diversity of *Clostridium difficile* genotypes isolated from a single poultry farm producing replacement laying hens. Anaerobe 14, 325–327.

Ziemer, C.J., Bonner, J.M., Cole, D., Vinje, J., Constantini, V., Goyal, S., et al., 2010. Fate and transport of zoonotic, bacterial, viral, and parasitic pathogens during swine manure treatment, storage, and land application. J. Anim. Sci. 88, E84–94.

Chapter 11

Methods for the Detection of Antimicrobial Resistance and the Characterization of *Staphylococcus aureus* Isolates from Food-Producing Animals and Food of Animal Origin

Kristina Kadlec, Sarah Wendlandt, Andrea T. Feßler and
Stefan Schwarz
*Institute of Farm Animal Genetics, Friedrich-Loeffler-Institut (FLI),
Neustadt-Mariensee, Germany*

Chapter Outline

Antimicrobial Resistance and Food Safety. DOI: http://dx.doi.org/10.1016/B978-0-12-801214-7.00011-9
© 2015 Elsevier Inc. All rights reserved.

INTRODUCTION

Staphylococcus aureus is a commensal colonizer of the skin in humans and animals and can also be found on mucous membranes of the body, particularly in the nose and the throat. About 20–30% of the human population is considered to be colonized by *S. aureus* (van Belkum et al., 2009). However, *S. aureus* can also cause severe infections. In humans, predominantly skin and soft tissue infections, but also bone, joint, and implant infections, necrotizing pneumonia, and septicemia are associated with *S. aureus* (Monecke et al., 2011). Moreover, *S. aureus* can produce a number of toxins which may be involved in specific diseases, such as toxic shock syndrome or staphylococcal food poisoning (Argudín et al., 2010b, 2012; Chiang et al., 2008). In animals, *S. aureus* can also cause a wide variety of infections, including skin infections, wound infections, and mastitis (Schwarz et al., 2013).

S. aureus infections in humans and animals are commonly treated with antimicrobial agents. Due to the bacterium's ability to acquire resistance genes, *S. aureus* isolates from humans and animals have gained a considerable number of resistance genes over the last 60 years (Wendlandt et al., 2013a). This accumulation of resistance genes can be seen as the bacterium's ability to effectively cope with changed environmental conditions. Meticillin-resistant *S. aureus* (MRSA), which also carry numerous other resistance genes, are of particular interest (Monecke et al., 2011). The dissemination of virulent and multi-resistant *S. aureus* isolates poses a serious threat to public health.

As MRSA, but also meticillin-susceptible *S. aureus* (MSSA), can be transferred between humans and animals in both directions via direct and indirect contact, *S. aureus* isolates present in or on food animals can also be transferred to carcasses during slaughter and further processing of the carcasses (Wendlandt et al., 2013f; Lassok and Tenhagen, 2013). In this regard, it should be noted that food of animal origin, such as meat, milk, or cheese, and even ice cream, constitutes a good medium for *S. aureus* multiplication. To identify (multi-) resistant *S. aureus* in food animals and food of animal origin, it is important to properly conduct antimicrobial susceptibility testing (AST) and to evaluate the results obtained (Schwarz et al., 2010). Moreover, detailed strain identification and characterization is necessary to reliably trace back the dissemination of *S. aureus* found in food-producing animals and food of animal origin to identify sources of contamination along the food chain (Wendlandt et al., 2013f).

A summary of the most commonly used methods for AST, but also for typing and comprehensive characterization of *S. aureus* is presented in this chapter. Advantages and disadvantages of the different methods are listed and examples for the use of these methods in recent studies are provided.

DETECTION AND ANALYSIS OF ANTIMICROBIAL RESISTANCE IN *S. AUREUS*

Various methods have been described to detect antimicrobial resistance in *S. aureus* and other staphylococci. Among the phenotypic methods, agar disk diffusion and broth microdilution are most common in veterinary and food routine diagnostics. Genotypic tests, which focus on the detection of specific resistance genes either by specific PCR assays or DNA microarray, are also available.

AST of *S. aureus*

There are a few basic rules for the correct performance of AST: (1) Regardless of the method, the performance of AST must follow an approved AST guideline. For bacteria of animal origin, the Clinical and Laboratory Standards Institute (CLSI) has recently published two documents, which summarize the latest information on how to correctly perform AST of staphylococci and other bacteria from animals (CLSI, 2013a) and provide the largest selection of veterinary-specific breakpoints (CLSI, 2013b). (2) For reliable AST, the use of pure bacterial cultures is indispensable. (3) Since different bacteria may need different AST conditions, it is necessary to identify the bacterial isolates to genus or species level prior to AST. (4) For any AST procedure, it is a fundamental requirement to run defined and approved quality control (QC) strains, such as *S. aureus* ATCC®25923 for disk diffusion and *S. aureus* ATCC®29213 for broth dilution, side-by-side with the test strains. The CLSI documents VET01-S2 (CLSI, 2013b) and M100-S24 (CLSI, 2014) list acceptable zone diameter and minimal inhibitory concentration (MIC) ranges of these and other QC strains for the different antimicrobial agents.

   With regard to the interpretation of AST results, two different types of interpretive criteria for zone diameters and MIC values are available: clinical breakpoints and epidemiological cutoff values (ECOFFs). Both types of interpretive criteria differ in their informative values and in the way in which they are determined (CLSI, 2011; Buß et al., 2012; Schwarz et al., 2013). *Clinical breakpoints* take into account (i) pharmacological parameters including the concentration of the antimicrobial agent at the site of infection achievable after regular dosing, (ii) clinical efficacy, and (iii) microbiological parameters such as MIC values of the causative bacterial pathogens (CLSI, 2011; Schwarz et al., 2010). Clinical breakpoints are often specific for a combination of an antimicrobial agent, a bacterial pathogen/group of bacterial pathogens, and a specific disease condition in a defined host animal species, for example, pirlimycin + *S. aureus*, *Streptococcus uberis*, *Streptococcus agalactiae*, *Streptococcus dysgalactiae* + bovine mastitis. Clinical breakpoints are intended to guide antimicrobial chemotherapy by facilitating the choice of the antimicrobial agent that is likely to lead to therapeutic success. When using clinical breakpoints, the bacteria are classified as resistant, susceptible, or—if this category is available—intermediate (CLSI, 2011; Schwarz et al., 2010). *ECOFFs* enable a description of a bacterial population based on their MIC or zone diameter values (http://eucast.org, last accessed 31.07.2014). ECOFFs do not take into account the aforementioned aspects and therefore should be used outside a clinical context. By using ECOFFs, bacterial populations are subdivided into a wild-type population—which is believed to lack resistance mechanisms—and a non-wild-type population that usually exhibits higher MIC values or smaller zone diameters and is believed to have acquired resistance mechanisms. When using ECOFFs, the bacteria should be classified as wild-type or non-wild-type (CLSI, 2011; Schwarz et al., 2010).

Diffusion Methods

Agar disk diffusion is a method which involves the determination of diameters of growth-inhibition zones around a paper disk that is impregnated with a defined amount of an antimicrobial agent. For this test, a defined amount of the bacteria to be tested—the so-called inoculum—is evenly spread on an agar plate, for example, Mueller–Hinton agar (MHA) for staphylococci. Subsequently, the disks are applied and the agar plate is incubated for a defined time under defined conditions, for example, for staphylococci 16–18 h (except 24 h when testing for oxacillin, meticillin, nafcillin, and vancomycin susceptibility) at $35 \pm 2°C$. During this time, the antimicrobial agent diffuses from the disk into the agar and produces a concentration gradient of the agent around the disk with the highest concentration near the disk which then decreases in the periphery. Growth of the bacteria is suppressed depending on their level of susceptibility and as a consequence differently sized zones of growth inhibition are formed around the disks (Figure 11.1). After the incubation period, the zone diameters around the disks are measured and compared with the zone diameter breakpoints given in the respective AST documents. Based on this comparison, the bacteria are classified as either susceptible, intermediate, or resistant to the antimicrobial agent tested (Buß et al., 2012; Schwarz et al., 2013).

Advantages and disadvantages: Advantages of the agar disk diffusion test are that it is a cost-efficient test that is easy to conduct and easy to evaluate, several antimicrobial agents can be tested simultaneously on the same plate, and the antimicrobial agents to be tested can be varied by simply exchanging a disk cartouche in the disk dispenser. Disadvantages are that only qualitative test results (susceptible–intermediate–resistant) can be obtained, no determination of MIC values is possible, and standardization in inter-laboratory trials can be problematic.

The *E-test* is an agar diffusion method allowing the determination of MIC values. A standardized inoculum is spread onto the surface of an agar plate. Then a strip containing a concentration gradient of the antimicrobial agent to be tested is applied. While the antimicrobial agent diffuses from the strip into the agar, an elliptical zone of growth inhibition forms around the strip (Figure 11.1) (Schwarz et al., 2013). The MIC is read where the inhibitory zone intersects the E-test strip and the MIC value is obtained from the scale printed on the E-test strip.

Advantages and disadvantages: The advantages of the E-test are that it provides a quantitative test result and allows MIC determination by a diffusion method. The disadvantages are that it is a comparatively expensive method,

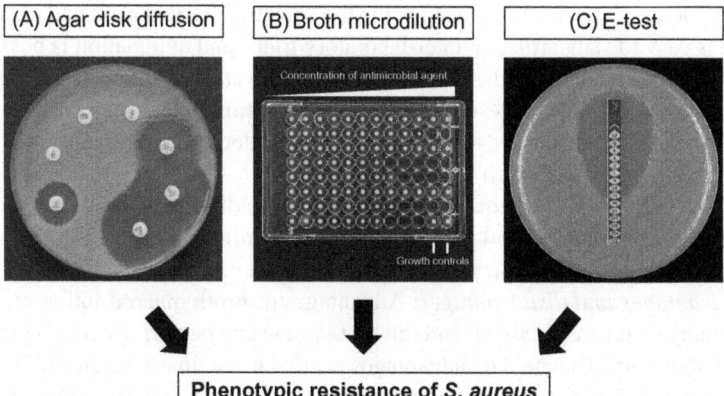

FIGURE 11.1   Presentation of three different methods used for phenotypic AST of *S. aureus*: (A)—agar disk diffusion, (B)—broth microdilution, and (C)—E-test.

a maximum of two E-test strips per regular agar plate can be used, and MIC reading can be difficult when bacteria grow mucoid or are hemolytic. No E-test strips are available for several antimicrobial agents solely approved for veterinary medicine.

Dilution Methods

Broth microdilution is a commonly used AST method to determine MICs. Usually, custom-made or commercially available microtiter plates are used in which the antimicrobial agents are present freeze-dried in twofold dilution series. Moreover, two or three antibiotic-free wells—which serve as growth controls—must be included in each microtiter plate. For susceptibility testing, defined inocula (for staphylococci: ca. 5×10^5 CFU/mL) prepared in the test medium (for staphylococci: cation-adjusted Mueller–Hinton broth (CAMHB); CAMHB + 2% NaCl when testing for oxacillin, meticillin, and nafcillin susceptibility) are added. Then the microtiter plate is sealed to avoid drying-off and is incubated for a defined time (for staphylococci: 16–18 h; 24 h when testing for oxacillin, meticillin, nafcillin, and vancomycin susceptibility) at a defined temperature. After the incubation period, bacterial growth in the wells is examined either by visual inspection or by semi- or full-automated systems (Figure 11.1) (Buß et al., 2012; Schwarz et al., 2013). The lowest concentration of an antimicrobial agent that completely inhibits visible growth of the bacteria is defined as the MIC. MIC values are compared with the MIC breakpoints given in the respective AST documents. Based on this comparison, the bacteria are classified as either susceptible, intermediate, or resistant to the antimicrobial agent tested.

Advantages and disadvantages: Advantages of broth microdilution are that quantitative data are obtained which tell how susceptible or resistant an isolate is, microtiter plate formats are commercially available, which supports standardized and harmonized susceptibility testing, the test system is easy to learn and to evaluate, it is easy to standardize in inter-laboratory trials, and automation is possible. The main disadvantage is that it is not possible to vary the antimicrobial agents from sample to sample. To test other or additional antimicrobial agents, either new microtiter plate formats have to be designed and produced or additional tests (e.g., E-test or broth macrodilution) must be applied.

Broth macrodilution requires the same test conditions as broth microdilution, but uses test tubes and larger volumes (according to CLSI 2–5 mL). The twofold dilution series is prepared freshly and not commercially available.

Advantages and disadvantages: Advantages of broth macrodilution are that quantitative data are obtained, and this test system can be performed in virtually every laboratory. The main disadvantages are that it is a time-consuming and not very cost-efficient test system, it requires the availability of the antimicrobial agents to be tested as pure substance, individual mistakes in the preparation of stock concentrations or dilution series can occur and no automation is possible.

Agar dilution typically investigates multiple bacterial strains on a single agar plate. The antimicrobial agent to be tested is incorporated in the respective concentration(s)—usually a twofold dilution series—into the agar plates. A standardized inoculum of the bacteria is prepared and commonly inoculated onto the agar plates using a multi-point inoculator. After incubation for a defined time, the agar plates are examined for growth (Schwarz et al., 2013). The result is also a MIC value, but in contrast to broth micro- or macrodilution, solid media (for *S. aureus* MHA or MHA + 2% NaCl for testing oxacillin, meticillin, and nafcillin) are used.

Advantages and disadvantages: Agar dilution is a suitable method when testing large numbers of bacterial isolates against a limited number of antimicrobial agents in a limited number of concentrations. Especially in the case of lower concentrations, an even distribution within the MHA has to be assured. Other disadvantages are the same as for broth macrodilution, but automation is at least in part possible.

Specific Phenotypic Tests for S. aureus

The CLSI document VET01-S2 (CLSI, 2013b) contains a number of specific tests for *S. aureus*. These include (1) an agar disk diffusion test for the prediction of *mecA*-mediated resistance, which is performed using 30 μg cefoxitin disks for *S. aureus*, (2) screening tests for β-lactamase production in *S. aureus*, and (3) screening tests for inducible resistance to clindamycin. In addition, the CLSI document M100-S24 (CLSI, 2014) also provides screening tests for vancomycin MICs ≥8 mg/L and high-level mupirocin resistance in *S. aureus*.

Genotypic Detection of Antimicrobial Resistance

Genotypic tests provide information about which resistance genes are present in a bacterial isolate. In staphylococci, it is known that different resistance genes account for the same resistance phenotype, for example, the resistance genes *erm*(A), *erm*(B), *erm*(C), and *erm*(T) confer macrolide/lincosamide/strepto-gramin B resistance, and these genes can occur alone or in different combinations per isolate (Wendlandt et al., 2013a; Feßler et al., 2010, 2011). To find out which resistance genes are present, basically two main approaches are commonly used: specific PCR assays or DNA microarray analysis. One general *disadvantage* is that only known nucleotide sequences can be detected. Thus, even if the presence of all known staphylococcal resistance genes is excluded on the basis of PCR and/or microarray results, the option remains that the strain in question carries a completely novel resistance gene/resistance mediating mutation or an unusual resistance gene acquired from other bacteria, for example, enterococci (Kadlec et al., 2012; Wendlandt et al., 2013d,e). Moreover, genotypic methods cannot provide information on whether a detected gene is intact and functionally active and therefore, they cannot replace phenotypic susceptibility testing.

PCR Assays for Resistance Genes

As for any gene-specific PCR assay, the primers should be located within the coding sequence of the resistance gene and enable the amplification of a sufficiently large segment. This is important to be able to confirm the amplicon by restriction analysis and/or sequence analysis. For resistance genes found in *S. aureus*, a larger number of PCR assays have been published, for example, by Schnellmann et al. (2006).

PCR detection of resistance genes can be performed as monoplex PCR for a single resistance gene (Schnellmann et al., 2006) or as multiplex PCR for two or more resistance genes. Strommenger et al. (2003) described a multiplex PCR assay for the simultaneous detection of nine clinically relevant antibiotic resistance genes of *S. aureus*. Multiplex PCRs developed for the simultaneous detection of antimicrobial resistance and virulence genes in *S. aureus* have also been described (Jones et al., 2001; Mehrotra et al., 2000). One variation is performing specific PCR assays in a real-time cycler—including the use of multiple color fluorophores to detect multiple genes simultaneously—and thereby saving time and avoiding subsequent gel electrophoresis.

The detection of specific resistance genes can also be coupled with the detection of species-specific markers. As such, a multiplex real-time PCR assay capable of the simultaneous detection of *mecA* and *S. aureus* has recently been described (Wang et al., 2014). Another multiplex real-time PCR targeted the *nuc* gene for identification of *S. aureus*, the *mecA* gene for meticillin resistance, and the Panton–Valentine leukocidin (PVL) virulence gene in animals and retail meat (Velasco et al., 2014).

Since some resistance genes of staphylococci are often physically linked with each other, PCR assays have also been developed that specifically confirm the linkage between resistance genes, such as *tet*(L)-*dfrK* or *spc-erm*(A)

(Feßler et al., 2010). A set of ten linkage PCRs has recently been developed and applied to analyze multi-resistance gene clusters in *S. aureus* isolates from pigs, humans, and food of animal origin (Li et al., 2013; Wendlandt et al., 2014).

In the case of inducible clindamycin resistance in staphylococci, the genes *erm*(A), *erm*(C), and *erm*(T) are most frequently involved. Since the switch from inducible to constitutive *erm* gene expression is often due to deletions or tandem duplications in the *erm* gene-associated regulatory region, PCR assays have also been described that amplify the entire regulatory region plus the 5′ end of the respective *erm* gene (Werckenthin et al., 1999; Werckenthin and Schwarz, 2000; Schmitz et al., 2002).

Advantages and disadvantages: PCR assays are very specific, fast, and in general easy to perform. PCR products can be used for digests or sequencing when needed. However, positive controls are needed. Another disadvantage— especially for simplex PCRs—is that PCR assays are time-consuming with a long hands-on time.

DNA Microarray for the Detection of Resistance Genes

The commonly used *S. aureus*-specific diagnostic DNA microarray StaphyType (Alere Technologies, Jena, Germany) is described as an example for this technique. This DNA microarray recognizes a total of 330 different genes and alleles thereof, including antimicrobial resistance genes. Arrays are mounted in tubes or strips and processed according to the manufacturer's protocols. In principle, all targets are amplified and labeled in a primer elongation reaction. Since a single primer per target is used, this reaction is linear rather than exponential and facilitates to cover all targets simultaneously in a single multiplex reaction. This reaction is also used to incorporate biotin-16-dUTP into the amplicons which are then hybridized to the array and detected. The hybridization pattern of the isolate is recorded (Figure 11.2) and analyzed using a designated reader and software (Arraymate, Iconoclust, both by Alere Technologies) (Kadlec et al., 2009; Monecke et al., 2011).

Advantages and disadvantages: The microarray-based detection of antimicrobial resistance genes has the huge advantage that a large number of resistance genes and other genes can be detected simultaneously. However, it requires the microarrays and equipment to record and evaluate the results.

TYPING OF *S. AUREUS*

Typing of bacterial isolates requires a more or less complex characterization which allows distinguishing particular isolates from others of the same species. Along the food chain, typing of *S. aureus* is relevant to identify outbreak strains of foodborne infections and intoxications, but also to trace back *S. aureus* in food of animal origin and to identify the most likely sources of contamination. Several typing methods are available for MSSA and MRSA isolates (Figure 11.2) and can be divided into genotypic and phenotypic methods.

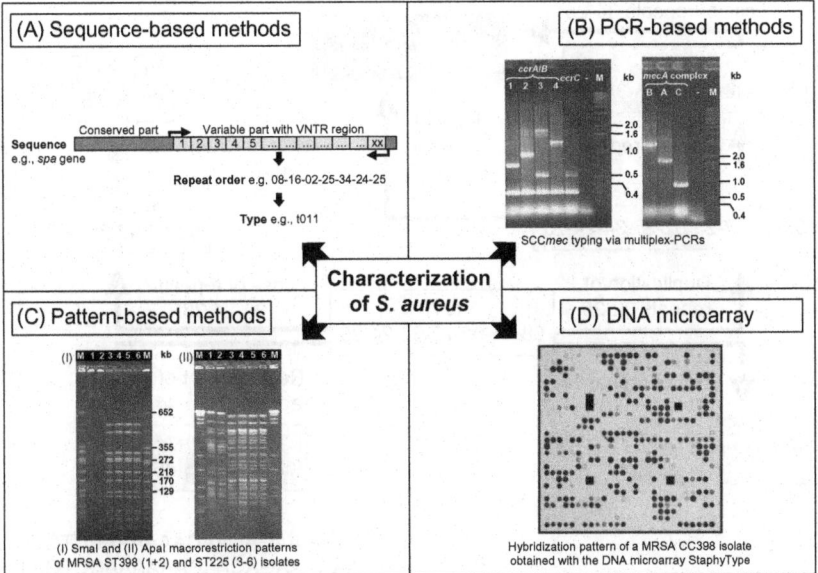

FIGURE 11.2 Presentation of sequence-based methods (A, *spa* typing), PCR-based methods (B, SCC*mec* typing), pattern-based methods (C, PFGE), and a DNA microarray (D) used for a complex characterization of *S. aureus* (MSSA and/or MRSA).

Genotypic Typing Methods

Targeting the chromosomal DNA can be done by sequence-based, PCR-based, or pattern-based methods.

Sequence-Based Typing Methods

The sequence-based methods are very frequently used. Some of them target a single locus, while others target multiple loci in the *S. aureus* genome. The most frequently used typing method for *S. aureus* is the so-called *spa* typing (Cuny et al., 2010; Feßler et al., 2011; Köck et al., 2013). This method targets the staphylococcal protein A, which consists of conserved regions and a variable region in the X domain, which is composed of a variable number of tandem repeats (VNTR) with sizes of 21–30 bp (Figure 11.2). The X region is amplified by PCR and the PCR product is sequenced. A common database for *spa* repeats and *spa* types is used for the analysis of the sequence data (Harmsen et al., 2003). By comparison with this database, the different *spa* repeats are identified (Figure 11.2). It should be noted that a single nucleotide polymorphism leads to a different repeat number (Figure 11.3). In a second step, the *spa* type is deduced from the repeat order (Figure 11.2). Again, the change of a single repeat also results in a different *spa* type (Figure 11.3). The numbers/names for the repeats and the *spa* types are given chronologically and similar numbers do

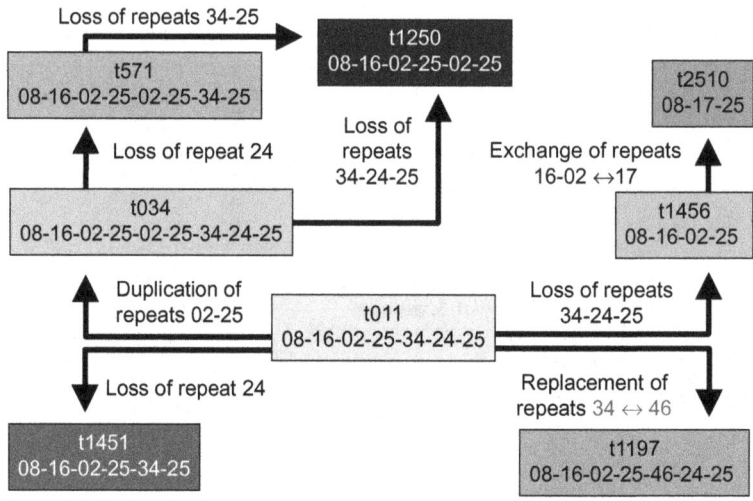

r34: AAAGAAGACAACAAAAAACCTGGT
r46: AAACAAGACAACAAAAAACCTGGT

FIGURE 11.3 Interrelationships between different *spa* types detected among MRSA CC398 from pigs (Kadlec et al., 2009). The structural variations seen are based on the loss, the duplication or the replacement of repeats. The replacement of single repeats is often based on the exchange of only a single nucleotide as shown in the case of r34 ↔ r46.

not necessarily indicate genetic relationship and vice versa the genetic unrelatedness (Figure 11.3). So far, 653 different repeats and 14,104 different *spa* types have been described (http://spaserver.ridom.de; last accessed 14.09.2014).

Another single locus sequence-based method has been established for MRSA, the so-called *dru* typing (Bartels et al., 2013; Feßler et al., 2011; Goering et al., 2008; Shore et al., 2010). The direct repeat unit (*dru*) region is a noncoding region within the SCC*mec* element that also consists of a VNTR region. The *dru* repeats have a length of 29–40 bp. The freely available, continuously updated *dru* typing database contains 95 *dru* repeats and 477 *dru* types with one to 23 repeats (http://dru-typing.org; last accessed 14.09.2014). The *dru* type designations consist of a number followed by one or two letters in chronological order. As such, *dru* type (dt) 9a consists of nine repeats and was the first one identified with nine repeats. As for *spa* typing, similar designations neither indicate similar sequences nor a relatedness of the isolates that carry the respective *dru* types.

The widely used *multi-locus sequence typing* (*MLST*) is also applicable to *S. aureus*. Internal parts of seven housekeeping genes are amplified by PCR and subsequently sequenced (Enright et al., 2000). In the database, various different alleles of these housekeeping genes are listed by numbers and the MLST sequence type (ST)—currently 2,770 STs are known—is deduced from these allele numbers in the correct order (http://saureus.mlst.net; last accessed 31.07.2014). Related MLST types can be summarized into a clonal complex (CC), in which each MLST type shares at least five alleles with all other members of the same CC. For the differentiation of staphylococcal isolates that share the same ST or CC, for example, MRSA isolates of ST398 or CC398, additional molecular typing methods need to be employed. To facilitate the identification of CC398, two PCR assays specific for CC398 isolates have been developed (van Wamel et al., 2010).

Advantages and disadvantages: Sequence-based typing methods are highly reproducible, and the results can be very easily exchanged between laboratories and/or entered into common databases. While *spa* and *dru* typing are good tools for short-term epidemiology and outbreak investigation, MLST is a suitable method for long-term epidemiological and phylogenetic studies, but is unsuitable for short-term epidemiological studies or outbreak investigations. A further disadvantage is that all PCR products have to be sequenced. In rare cases, MRSA nontypable by *spa* typing (Ghaznavi-Rad et al., 2011) or *dru* typing have been observed (Feßler et al., 2014).

PCR-Based Typing Methods

SCCmec typing enables the identification of the *mec* gene complex and the *ccr* gene complex. Actually, at least 11 different main variants of SCC*mec*, which differ in their *mec* gene complex, their *ccr* genes and the so-called junkyard regions, are currently known in MRSA (IWG-SCC, 2009). For *mecA*-containing SCC*mec* elements, several multiplex PCR assays have been described (e.g., Chongtrakool et al., 2006; Kondo et al., 2007). With the detection of the *mecC* gene and its allotypes on novel SCC*mec* cassettes (García-Álvarez et al., 2011; Shore et al., 2011), additional PCR assays have been developed which reliably detect the *mecC* gene (e.g., Becker et al., 2013; Pichon et al., 2012).

Advantages and disadvantages: The SCC*mec* typing is a good method to confirm and characterize MRSA isolates. Since there are also recombinations between different SCC*mec* possible elements, it may become problematic to identify the SCC*mec* type correctly by PCR only.

The *Multiple Locus VNTR Analysis* (*MLVA*) is based on the fact that bacteria harbor different VNTR regions and all such regions can be used as a target. In contrast to the sequence-based typing methods, the MLVA typing targets several VNTR regions by multiplex PCRs and subsequent comparison of the amplicon patterns (Rasschaert et al., 2009; Sabat et al., 2003). In *S. aureus*, a MLVA approach, using 16 VNTR loci amplified in two multiplex PCRs, was described and applied to MRSA and MSSA isolates from various human and animal sources (Sobral et al., 2012). Since VNTR regions are used instead of highly conserved regions of housekeeping genes as in MLST, MLVA typing might be more appropriate for short-term epidemiological investigations (Pourcel et al., 2009). Moreover, an automated sequencing of fluorescence-labeled PCR products was described as an alternative to the use of agarose gel electrophoresis for the detection of MLVA results, enhancing the accuracy and portability of this method (Schouls et al., 2009). The MLVA typing usually showed a good correlation with macrorestriction analysis, which is used as the gold standard because of the high discriminatory power, and proved to be superior to *spa* typing and MLST (Rasschaert et al., 2009; Schouls et al., 2009).

Advantages and disadvantages: MLVA is an appropriate method for typing large numbers of isolates and short-term epidemiological investigations. The availability of several different MLVA approaches can hamper the comparability of MLVA data from different studies.

Pattern-Based Typing Methods

Macrorestriction analysis is still considered as the gold standard. Whole-cell DNA is prepared by embedding bacteria into agarose plugs followed by careful lysis of the bacteria within these agarose plugs. This ensures that the whole-cell DNA remains largely intact for subsequent restriction analysis with rare-cutting endonucleases (Wendlandt et al., 2013f). For *S. aureus*, a standard protocol, the so-called HARMONY protocol, using the enzyme SmaI has been developed (Murchan et al., 2003). However, a restriction–modification system of MRSA CC398 modifies the cleavage site of SmaI and thereby renders MRSA CC398 not digestible by SmaI (Bens et al., 2006). For MRSA CC398, other restriction endonucleases, such as ApaI (Figure 11.2), Cfr9I, or EagI have been used successfully (Argudín et al., 2010a; Huber et al., 2011; Kadlec et al., 2009). The fragments obtained with these rare-cutting enzymes are separated by pulsed-field gel electrophoresis (PFGE) and the resulting fragment patterns are compared for similarity either by visual inspection or by using specific software (Wendlandt et al., 2013f).

Advantages and disadvantages: Macrorestriction analysis is considered as the most discriminative method, especially when performed with more than one restriction endonuclease in parallel. It is also the most suitable method for short-term epidemiological analyses and outbreak investigations (Wendlandt et al., 2013f). Despite the harmonized protocol, fragment patterns from different gels or from gels run on different PFGE equipments are often difficult to compare. Moreover, the method is quite expensive, needs some practice, and requires considerable hands-on time.

Phenotypic Typing Methods

Biochemical assays are commonly used to identify staphylococci to species or even subspecies level. Although certain species show variable metabolic abilities or differences in the production of specific enzymes, such as utilization of turanose or production of urease, arginine dihydrolase, or pyrrolidonyl arylamidase among *S. aureus*, these properties are not used for typing purposes.

MALDI-TOF mass spectrometry is a frequently used technique to identify bacteria to genus and species level (Clark et al., 2013; Dubois et al., 2010).

The difficulty in typing bacteria to a strain level is mainly due to the fact that members of the same species often exhibit remarkably similar MALDI-TOF MS profiles (Sandrin et al., 2013). Although strain-specific peaks have been reported, they commonly represent only a very small portion of the MS profile. Sandrin et al. (2013) stated that poor and/or not-quantified profile reproducibility can hinder the identification of reliable peaks as strain-specific biomarkers. Wolters et al. (2011) described reproducible differences in the spectra of 25 MRSA isolates of the clonal complexes CC5, CC8, CC22, CC30, and CC45. Lasch et al. (2014) recently investigated 59 diverse and mostly MRSA isolates of the aforementioned clonal complexes plus CC398, but were unable to identify biomarkers allowing a consistent and reliable identification of phylogenetic lineages, CCs, or ST.

Advantages and disadvantages: The MALDI-TOF MS is comparably easy to perform and suitable for species identification. However, further typing of *S. aureus* is currently not possible. This relatively new technology requires the use of expensive equipment. Moreover, MALDI-TOF MS is only as good as is the underlying database. This can be a problem when rarely detected staphylococcal species of veterinary and food origin shall be reliably identified.

Typing by DNA Microarray

The previously described DNA microarray StaphyType (see section "DNA Microarray for the Detection of Resistance Genes") can also be used for the characterization of *S. aureus* isolates and allowed the assignment of the MRSA isolates to 34 distinct lineages which could be clearly defined based on nonmobile genes. The results were in accordance with MLST data (Monecke et al., 2011). Moreover, the SCC*mec* type can also be detected by this microarray (Monecke et al., 2011).

Advantages and disadvantages see section "DNA Microarray for the Detection of Resistance Genes".

DETECTION OF VIRULENCE GENES IN *S. AUREUS*

As mentioned in sections "PCR Assays for Resistance Genes" and "DNA Microarray for the Detection of Resistance Genes", virulence genes in *S. aureus* can also be detected by either PCR or DNA microarray. Among the various virulence genes carried by *S. aureus* isolates, those coding for the PVL (*lukP/V*), the toxic shock syndrome toxin 1 (*tst1*), exfoliative toxins (*etA*, *etB*), and enterotoxins are the most relevant. Besides the molecular detection systems mentioned below, specific toxins can also be detected by other methods, for example, the detection of PVL via MALDI-TOF MS (Bittar et al., 2009) or by a lateral flow assay using monoclonal antibodies (Monecke et al., 2013a), respectively, or the detection of enterotoxins using a chromogenic macroarray (Lin et al., 2009).

PCR Assays for Virulence Genes

The PVL genes are nowadays commonly detected in multiplex PCR approaches, which also recognize other genes, including *S. aureus*-specific markers and the *mecA* gene (Renwick et al., 2013; Velasco et al., 2014), *S. aureus*-specific markers and the *tst1* gene for TSST-1 (Fosheim et al., 2011) as well as antimicrobial resistance genes (Granger et al., 2010). The genes for the exfoliative toxins A and B are also commonly detected in multiplex PCR assays, which target *tst1*, enterotoxin genes (Becker et al., 1998; Johnson et al., 1991) and additional resistance genes (Mehrotra et al., 2000). From a food safety perspective, the detection of enterotoxins plays an important role. Numerous staphylococcal enterotoxins (SE) and related staphylococcal enterotoxin-like proteins (SE*l*) have been identified although an emetic activity has not been confirmed for all of them (Argudín et al., 2010b). These toxins are resistant to heat treatment, low pH, and proteolytic enzymes and, thus, retain their activity in the digestive tract after ingestion. Moreover, they are not inactivated by cooking/heating of *S. aureus*-contaminated food (Argudín et al., 2010b). A number of examples of food-poisoning outbreaks due to enterotoxigenic *S. aureus* and the enterotoxins identified in the respective strains are given in a review by Argudín et al. (2010b). Detection of enterotoxin genes usually also occurs by multiplex PCRs (Chiang et al., 2008, 2006; Hwang et al., 2007).

For the *advantages and disadvantages* see section "PCR Assays for Resistance Genes".

Detection of Virulence Genes by DNA Microarray

Various DNA microarrays have been described, some of which are specific for the detection of enterotoxin genes (Sergeev et al., 2004) or virulence genes in general (Vautor et al., 2009), while others are suitable for the detection of a wide variety of genes including virulence genes (El Garch et al., 2009; McCarthy et al., 2012; Monecke et al., 2011). All these DNA microarrays differ in the number of target genes and consequently in their informative values.

The *advantages and disadvantages* of this method are described in detail in section "DNA Microarray for the Detection of Resistance Genes".

EXAMPLES FOR THE COMPLEX CHARACTERIZATION OF *S. AUREUS* FROM FOOD-PRODUCING ANIMALS AND FOOD OF ANIMAL SOURCES

Numerous studies have been conducted in which *S. aureus* isolates, including MRSA, from different food samples have been investigated for their virulence and resistance properties, but also for their genotypic relationships (Wendlandt et al., 2013f). The following examples have been chosen to illustrate the application of the aforementioned methods for a comprehensive characterization of MRSA and MSSA from a diverse range of food samples of animal origin. More examples are listed in Table 11.1.

TABLE 11.1 Examples for the Application of Methods for a Complex Characterization of MRSA from Food and Food Animal Origin

Study	No. of Samples	No. of MRSA	Origin of MRSA	AST	Testing for Resistance Genes	spa Typing	dru Typing	MLST	SCCmec Typing	MLVA	PFGE	Testing for Virulence Genes	DNA Micro-array
Jackson et al. (2013)	300	57	Human (n = 50), beef (n = 4), pork (n = 3)	Broth microdilution	✓	✓	–	✓	✓	–	✓	✓	–
Vester-gaard et al. (2012)	25	11	Pig (n = 6), pork (n = 5)	Broth microdilution	✓	✓	–	✓	✓	–	✓	✓	Staphy Type
Haran et al. (2012)	150	2	Milk (n = 2)	Broth microdilution	–	✓	–	(✓)[a]	✓	–	✓	✓	–
Argudín et al. (2012)	n.a.	2	Hamburger, (n = 2)	Agar disk diffusion, agar dilution	✓	✓	–	(✓)[a]	✓	–	–	✓	–
Feßler et al. (2011)	86	32	Turkey (n = 22), chicken (n = 10)	Broth microdilution	✓	✓	✓	✓	✓	–	✓	✓	Staphy Type
Kraushaar and Fetsch (2014)	351	28	Wild boar (n = 28)	Broth microdilution	✓	✓	–	(✓)[a]	✓	–	(✓)[a]	✓	Staphy Type

(Continued)

TABLE 11.1 (Continued)

Study	No. of Samples	No. of MRSA	Origin of MRSA	AST	Testing for Resistance Genes	spa Typing	dru Typing	MLST	SCCmec Typing	MLVA	PFGE	Testing for Virulence Genes	DNA Micro-array
Tenhagen et al. (2014)	3819	632	Cattle (n = 288), veal (n = 103), calf carcass (n = 90), dust at farm (n = 39), beef (n = 83), milk (n = 29)	Broth microdilution	–	✓	–	(✓)[a]	✓	–	–	–	–
van Duijkeren et al. (2014)	411	16	Cow (n = 16)	Broth microdilution	–	✓	–	–	–	✓	–	✓	–
Boost et al. (2013)	1400	126	Pork (n = 78), chicken (n = 31), beef (n = 17)	n.a.	–	✓	–	n.a.	✓	–	–	✓	–
Hammad et al. (2012)	200	5	fish (n = 5)	Agar disk diffusion, test for inducible clindamycin resistance	✓	✓	–	✓	✓	–	–	✓	–

O'Brien et al. (2012)	395	26	Pork (n = 26)	Broth microdilution	–	✓	–	(✓)[a]	–	–	✓	–
Hanson et al. (2011)	165	2	Pork (n = 2)	Broth dilution	–	✓	–	✓	–	–	✓	–
Rhee and Woo (2010)	165	4	Beef (n = 2), sea bass (n = 1), rockfish (n = 1)	Agar disk diffusion, agar dilution	–	–	–	✓	–	–	✓	–
Weese et al. (2010)	402	31	Pork (n = 31)	–	–	✓	–	–	–	✓	✓	–
de Boer et al. (2009)	2217	267	Chicken (n = 83), beef (n = 42), turkey (n = 41), veal (n = 39), pork (n = 33), lamb or mutton (n = 20), fowl (n = 4), game–bird (n = 4)	–	–	✓	–	–	–	–	–	–
Lozano et al. (2009)	318	5	Pork, chicken, veal, rabbit, wild boar (each n = 1)	Agar disk diffusion, agar dilution	✓	✓	–	✓	–	–	✓	–

n.a., not available.
[a]Selected isolates.

Jackson et al. (2013) identified 45 *S. aureus* from 100 pork and 63 *S. aureus* from 100 beef products obtained from 14 retail shops in Athens, Georgia, United States, and compared them to 100 human *S. aureus* from a local hospital. Among them, three from pork, four from beef, and 50 from humans were MRSA and further characterized. The three MRSA isolates from pork were ST5 (MLST)/t002 (*spa*)/IV (SCC*mec*), ST9/t337/IV, and ST30/t012/IV, while the four MRSA from beef were ST8/t008/IV (*n* = 1) and ST5/t002/IV (*n* = 3). One of the pork MRSA was only resistant to penicillins and gentamicin while all other MRSA from pork and beef showed more expanded resistance patterns. One MRSA from beef was PVL-positive. Three retail beef MRSA isolates were indistinguishable in their PFGE pattern, ST, and *spa* type to two human clonal MRSA isolates (USA100 and USA300) while the remaining retail beef MRSA isolate had a PFGE pattern similar to that of a human MRSA isolate. In contrast, none of the retail pork MRSA isolates had PFGE patterns similar to those of human MRSA isolates. The authors concluded that based on their comprehensive characterization the retail beef samples were contaminated by a human source, possibly during processing of the meat, and may present a source of MRSA for consumers and others who handle raw meat (Jackson et al., 2013).

Vestergaard et al. (2012) identified five MRSA from ten pork samples and six MRSA from 15 nasal swabs of pigs in the Samuth Songkhram province in Thailand. All 11 MRSA belonged to CC9 and were identified as ST2136/t337/IX (*n* = 8), ST9/t337/IX (*n* = 2), and ST2278/t337/IX (*n* = 1). All pig isolates and most of the pork isolates were positive for enterotoxin genes (*entG*, *entI*, *entM*, *entN*, and *entO*). All 11 MRSA were multi-resistant with the pork isolates showing even more expanded resistance properties than the pig isolates. PFGE revealed a high diversity among the 11 isolates; the pork and pig isolates formed two clusters with a similarity of 75%. Based on the genetic diversity observed, the authors concluded that these isolates have been in the agriculture sector for some time and that their carriage of resistance and enterotoxin genes is of great concern (Vestergaard et al., 2012).

Haran et al. (2012) screened a total of 150 pooled bulk tank milk samples from 50 dairy farms in Minnesota, United States, for the presence of MSSA and MRSA. They found 93 MSSA and two MRSA isolates. Susceptibility testing of *S. aureus* isolates showed pansusceptibility in 54 isolates, resistance to a single antibiotic class in 21 isolates, resistance to two antibiotic classes in 13 isolates, and resistance to more than three antibiotic classes and thus multidrug resistance in five isolates (including the two MRSA isolates). The *spa* types t529 and t034 were most frequent among the MSSA. One of the two MRSA was identified as ST8/t121/IVa and resembled the USA300 clone in its PFGE pattern. This isolate had the SE genes C, D, and E and was PVL-positive. The other MRSA was an ST5/II isolate with an unknown *spa* type and was closest related to the USA100 clone in its PFGE pattern. This isolate had the enterotoxins B, C, and D genes. Based on their results, the authors concluded that MRSA genotypes associated with hospitals and community can be isolated from milk at very low rates (Haran et al., 2012).

Argudín et al. (2012) investigated 64 *S. aureus* isolates from food and food handlers, associated or not associated with food-poisoning outbreaks in Spain. Based on the same *spa* type, ST, *agr* type, virulence profile, and resistance profile, 31 strains were identified, which belonged to ten CCs: CC5 (29.0%), CC30 (25.8%), CC45 (16.1%), CC8, CC15 (two strains each), CC1, CC22, CC25, CC59, and CC121 (one strain each). Exfoliatin genes, *tst1*, and numerous enterotoxin or enterotoxin-like genes were detected. Two MRSA isolates from hamburgers were identified as ST5/t002/IVd and ST5/t2173/IVd. Six isolates (19.4%) were mupirocin-resistant, and one MSSA ST15/t120 isolate from a food handler carried the gene *mupA*. Resistance to antimicrobial agents was observed at different frequencies: ampicillin (*blaZ*) (61.3%), erythromycin [*erm*(A)-*erm*(C) or *erm*(C)] (25.8%), tetracycline [*tet*(K)] (3.2%), and amikacin–gentamicin–kanamycin–tobramycin (*aphA* with *aacA/aphD* or *aadD*) (6.5%). The authors concluded that the presence of *S. aureus* strains with a considerable number of virulence and resistance genes in the food chain represents a potential health hazard for consumers (Argudín et al., 2012).

Feßler et al. (2011) investigated 32 MRSA from poultry meat and poultry meat products obtained in retail shops in the south-western part of Germany. Twenty-eight isolates were identified as CC398/t011/IV/dt10q (dru), CC398/t011/IV/dt10at, CC398/t899/IV/dt10as, or CC398/V/exhibiting *spa* types t011, t034, t2346, or t6574 and *dru* types dt2b, dt6j, dt6m, dt10a, dt11a, dt11v, or dt11ab. Moreover, two MRSA from chicken products were ST9/t1430/IV/dt10a and another two MRSA from fresh turkey meat were ST5/III/t002/dt9v or ST1791/III/t002/dt9v. The ST9, ST5, and ST1791 isolates carried the enterotoxin gene cluster which comprises the genes for the enterotoxins G, I, M, N, O, and U. None of the isolates was PVL-positive. Two MRSA isolates were only resistant to β-lactams and fluoroquinolones or β-lactams and tetracyclines, whereas the other 30 MRSA showed multi-resistance pheno- and genotypes. This study confirmed that MRSA are present in/on chicken and turkey meat and meat products (Feßler et al., 2011). The MRSA types detected corresponded closely to those found in healthy chickens on farms (Wendlandt et al., 2013b), chickens at slaughter (Wendlandt et al., 2013c), or in diseased chickens and turkeys (Monecke et al., 2013b).

CONCLUSIONS

Numerous studies have reported the presence of *S. aureus*, MSSA, and MRSA, in/on food and food products of animal origin. This can be due to a spillover from the food-animal's residential microbiota, which is transferred from the animal to its carcass during the killing, dehairing/defeathering, or evisceration processes. Another option—especially when dealing with MSSA/MRSA in food products—is contamination by humans or by contaminated equipment in the further processing of the carcasses (Doyle et al., 2012; Wendlandt et al., 2013f). To trace back the *S. aureus* strains and to identify potential sources of contamination, a detailed characterization—mainly by using molecular tools—is a suitable approach. When the risk for the consumer should be evaluated, several other aspects need to be taken into account. On the one hand, the ingested dose of *S. aureus* is relevant. On the other hand, the type of MRSA (and also MSSA) and its pathogenic potential for humans, for example, its ability to produce enterotoxins and other virulence factors and to carry antimicrobial resistance genes (Wendlandt et al., 2013f) need to be determined. For this, numerous methods, including PCR assays and DNA microarrays, exist and a comprehensive phenotypic and genotypic characterization of the *S. aureus* isolates in question, as described in this chapter, is an excellent basis. When infections due to *S. aureus* occur and must be treated with antimicrobial agents, it is indispensable to perform AST accurately. AST of bacteria and subsequent interpretation of the results are complex topics which can be sources of multiple errors. The most common mistakes in AST of bacteria of animal origin have been described, including hints on how to avoid them (CLSI, 2011; Schwarz et al., 2010).

Although comparatively strict hygiene regulations have been implemented into different areas of life, such as hospitals, canteen kitchens, restaurants, but also in the food-producing industry. It is the consumers' own responsibility to avoid cross-contamination in their own kitchen or staphylococcal multiplication in/on food due to improper storage. In terms of tracing back staphylococcal contaminations or intoxications, suitable methods are already available, some of which, however, are laborious and require specific equipment. More investment into the development of fast and accurate test systems—especially for (i) the reliable identification, (ii) the differentiation, and (iii) the antimicrobial susceptibility of causative *S. aureus* is necessary.

ACKNOWLEDGMENTS

We apologize in advance to all the investigators whose research could not be appropriately cited owing to space limitations. We thank all cooperation partners in the studies on *S. aureus*. Studies on MRSA and MSSA by the authors (2010–2016) are supported by the German Federal Ministry of Education and Research (BMBF) through the German Aerospace Center (DLR), grant numbers 01KI1014D (MedVet-Staph), and 01KI1301D (MedVet-Staph II).

REFERENCES

Argudín, M.A., Fetsch, A., Tenhagen, B.-A., Hammerl, J.A., Hertwig, S., Kowall, J., et al., 2010a. High heterogeneity within methicillin-resistant *Staphylococcus aureus* ST398 isolates, defined by Cfr9I macrorestriction-pulsed-field gel electrophoresis profiles and *spa* and SCC*mec* types. Appl. Environ. Microbiol. 76, 652–658.

Argudín, M.Á., Mendoza, M.C., Rodicio, M.R., 2010b. Food poisoning and *Staphylococcus aureus* enterotoxins. Toxins (Basel) 2, 1751–1773.

Argudín, M.A., Mendoza, M.C., González-Hevia, M.A., Bances, M., Guerra, B., Rodicio, M.R., 2012. Genotypes, exotoxin gene content, and antimicrobial resistance of *Staphylococcus aureus* strains recovered from foods and food handlers. Appl. Environ. Microbiol. 78, 2930–2935.

Bartels, M.D., Boye, K., Oliveira, D.C., Worning, P., Goering, R., Westh, H., 2013. Associations between *dru* Types and SCC*mec* cassettes. PLoS ONE 8, e61860.

Becker, K., Roth, R., Peters, G., 1998. Rapid and specific detection of toxigenic *Staphylococcus aureus*: use of two multiplex PCR enzyme immunoassays for amplification and hybridization of staphylococcal enterotoxin genes, exfoliative toxin genes, and toxic shock syndrome toxin 1 gene. J. Clin. Microbiol. 36, 2548–2553.

Becker, K., Larsen, A.R., Skov, R.L., Paterson, G.K., Holmes, M.A., Sabat, A.J., et al., 2013. Evaluation of a modular multiplex-PCR methicillin-resistant *Staphylococcus aureus* detection assay adapted for *mecC* detection. J. Clin. Microbiol. 51, 1917–1919.

Bens, C.C., Voss, A., Klaassen, C.H., 2006. Presence of a novel DNA methylation enzyme in methicillin-resistant *Staphylococcus aureus* isolates associated with pig farming leads to uninterpretable results in standard pulsed-field gel electrophoresis analysis. J. Clin. Microbiol. 44, 1875–1876.

Bittar, F., Ouchenane, Z., Smati, F., Raoult, D., Rolain, J.M., 2009. MALDI-TOF-MS for rapid detection of staphylococcal Panton–Valentine leukocidin. Int. J. Antimicrob. Agents 34, 467–470.

Boost, M.V., Wong, A., Ho, J., O'Donoghue, M., 2013. Isolation of methicillin-resistant *Staphylococcus aureus* (MRSA) from retail meats in Hong Kong. Foodborne Pathog. Dis. 10, 705–710.

Buß, M., Feßler, A.T., Kadlec, K., Peters, T., Schwarz, S., 2012. Antimicrobial resistance: mechanisms, dissemination, diagnostics. M^2-Magazine 2, 8–14.

Chiang, Y.C., Chang, L.T., Lin, C.W., Yang, C.Y., Tsen, H.Y., 2006. PCR primers for the detection of staphylococcal enterotoxins K, L, and M and survey of staphylococcal enterotoxin types in *Staphylococcus aureus* isolates from food poisoning cases in Taiwan. J. Food. Prot. 69, 1072–1079.

Chiang, Y.C., Liao, W.W., Fan, C.M., Pai, W.Y., Chiou, C.S., Tsen, H.Y., 2008. PCR detection of Staphylococcal enterotoxins (SEs) N, O, P, Q, R, U, and survey of SE types in *Staphylococcus aureus* isolates from food-poisoning cases in Taiwan. Int. J. Food Microbiol. 121, 66–73.

Chongtrakool, P., Ito, T., Ma, X.X., Kondo, Y., Trakulsomboon, S., Tiensasitorn, C., et al., 2006. Staphylococcal cassette chromosome *mec* (SCC*mec*) typing of methicillin-resistant *Staphylococcus aureus* strains isolated in 11 Asian countries: a proposal for a new nomenclature for SCC*mec* elements. Antimicrob. Agents Chemother. 50, 1001–1012.

Clark, A.E., Kaleta, E.J., Arora, A., Wolk, D.M., 2013. Matrix-assisted laser desorption ionization-time of flight mass spectrometry: a fundamental shift in the routine practice of clinical microbiology. Clin. Microbiol. Rev. 26, 547–603.

Clinical and Laboratory Standards Institute (CLSI), 2011. CLSI Document VET05-R: Generation, Presentation, and Application of Antimicrobial Susceptibility Test Data for Bacteria of Animal Origin: A Report. Clinical and Laboratory Standards Institute, Wayne, PA.

Clinical and Laboratory Standards Institute (CLSI), 2013a. CLSI Document VET01-A4: Performance Standards for Antimicrobial Disk and Dilution Susceptibility Tests for Bacteria Isolated from Animals; Approved Standard—Fourth Edition. Clinical and Laboratory Standards Institute, Wayne.

Clinical and Laboratory Standards Institute (CLSI), 2013b. CLSI Document VET01-S2: Performance Standards for Antimicrobial Disk and Dilution Susceptibility Tests for Bacteria Isolated from Animals; Second Informational Supplement. Clinical and Laboratory Standards Institute, Wayne, PA.

Clinical and Laboratory Standards Institute (CLSI), 2014. CLSI Document M100-S24: Performance Standards for Antimicrobial Susceptibility Testing; Twenty-Fourth Informational Supplement. Clinical and Laboratory Standards Institute, Wayne, PA.

Cuny, C., Friedrich, A., Kozytska, S., Layer, F., Nübel, U., Ohlsen, K., et al., 2010. Emergence of methicillin-resistant *Staphylococcus aureus* (MRSA) in different animal species. Int. J. Med. Microbiol. 300, 109–117.

de Boer, E., Zwartkruis-Nahuis, J.T., Wit, B., Huijsdens, X.W., de Neeling, A.J., Bosch, T., et al., 2009. Prevalence of methicillin-resistant *Staphylococcus aureus* in meat. Int. J. Food Microbiol. 134, 52–56.

Doyle, M.E., Hartmann, F.A., Lee Wong, A.C., 2012. Methicillin-resistant staphylococci: implications for our food supply? Anim. Health Res. Rev. 13, 157–180.

Dubois, D., Leyssene, D., Chacornac, J.P., Kostrzewa, M., Schmit, P.O., Talon, R., et al., 2010. Identification of a variety of *Staphylococcus* species by matrix-assisted laser desorption ionization-time of flight mass spectrometry. J. Clin. Microbiol. 48, 941–945.

El Garch, F., Hallin, M., De Mendonça, R., Denis, O., Lefort, A., Struelens, M.J., 2009. StaphVar-DNA microarray analysis of accessory genome elements of community-acquired methicillin-resistant *Staphylococcus aureus*. J. Antimicrob. Chemother. 63, 877–885.

Enright, M.C., Day, N.P., Davies, C.E., Peacock, S.J., Spratt, B.G., 2000. Multilocus sequence typing for characterization of methicillin-resistant and methicillin-susceptible clones of *Staphylococcus aureus*. J. Clin. Microbiol. 38, 1008–1015.

Feßler, A., Scott, C., Kadlec, K., Ehricht, R., Monecke, S., Schwarz, S., 2010. Characterization of methicillin-resistant *Staphylococcus aureus* ST398 from cases of bovine mastitis. J. Antimicrob. Chemother. 65, 619–625.

Feßler, A.T., Kadlec, K., Hassel, M., Hauschild, T., Eidam, C., Ehricht, R., et al., 2011. Characterization of methicillin-resistant *Staphylococcus aureus* isolates from food and food products of poultry origin in Germany. Appl. Environ. Microbiol. 77, 7151–7157.

Feßler, A.T., Calvo, N., Gutiérrez, N., Muñoz Bellido, J.L., Fajardo, M., Garduño, E., et al., 2014. Cfr-mediated linezolid resistance in methicillin-resistant *Staphylococcus aureus* and *Staphylococcus haemolyticus* associated with clinical infections in humans: two case reports. J. Antimicrob. Chemother. 69, 268–270.

Fosheim, G.E., Nicholson, A.C., Albrecht, V.S., Limbago, B.M., 2011. Multiplex real-time PCR assay for detection of methicillin-resistant *Staphylococcus aureus* and associated toxin genes. J. Clin. Microbiol. 49, 3071–3073.

García-Álvarez, L., Holden, M.T., Lindsay, H., Webb, C.R., Brown, D.F., Curran, M.D., et al., 2011. Methicillin-resistant *Staphylococcus aureus* with a novel *mecA* homologue in human and bovine populations in the UK and Denmark: a descriptive study. Lancet Infect. Dis. 11, 595–603.

Ghaznavi-Rad, E., Goering, R.V., Nor Shamsudin, M., Weng, P.L., Sekawi, Z., Tavakol, M., et al., 2011. *mec*-associated *dru* typing in the epidemiologic analysis of ST239 MRSA in Malaysia. Eur. J. Clin. Microbiol. Infect. Dis. 30, 1365–1369.

Goering, R.V., Morrison, D., Al-Doori, Z., Edwards, G.F., Gemmell, C.G., 2008. Usefulness of *mec*-associated direct repeat unit (*dru*) typing in the epidemiological analysis of highly clonal methicillin-resistant *Staphylococcus aureus* in Scotland. Clin. Microbiol. Infect. 14, 964–969.

Granger, K., Rundell, M.S., Pingle, M.R., Shatsky, R., Larone, D.H., Golightly, L.M., et al., 2010. Multiplex PCR-ligation detection reaction assay for simultaneous detection of drug resistance and toxin genes from *Staphylococcus aureus*, *Enterococcus faecalis*, and *Enterococcus faecium*. J. Clin. Microbiol. 48, 277–280.

Hammad, A.M., Watanabe, W., Fujii, T., Shimamoto, T., 2012. Occurrence and characteristics of methicillin-resistant and -susceptible *Staphylococcus aureus* and methicillin-resistant coagulase-negative staphylococci from Japanese retail ready-to-eat raw fish. Int. J. Food Microbiol. 156, 286–289.

Hanson, B.M., Dressler, A.E., Harper, A.L., Scheibel, R.P., Wardyn, S.E., Roberts, L.K., et al., 2011. Prevalence of *Staphylococcus aureus* and methicillin-resistant *Staphylococcus aureus* (MRSA) on retail meat in Iowa. J. Infect. Public Health 4, 169–174.

Haran, K.P., Godden, S.M., Boxrud, D., Jawahir, S., Bender, J.B., Sreevatsan, S., 2012. Prevalence and characterization of *Staphylococcus aureus*, including methicillin-resistant *Staphylococcus aureus*, isolated from bulk tank milk from Minnesota dairy farms. J. Clin. Microbiol. 50, 688–695.

Harmsen, D., Claus, H., Witte, W., Rothgänger, J., Claus, H., Turnwald, D., et al., 2003. Typing of methicillin-resistant *Staphylococcus aureus* in a university hospital setting by using novel software for *spa* repeat determination and database management. J. Clin. Microbiol. 41, 5442–5448.

Huber, H., Giezendanner, N., Stephan, R., Zweifel, C., 2011. Genotypes, antibiotic resistance profiles and microarray-based characterization of methicillin-resistant *Staphylococcus aureus* strains isolated from livestock and veterinarians in Switzerland. Zoonoses Public Health 58, 343–349.

Hwang, S.Y., Kim, S.H., Jang, E.J., Kwon, N.H., Park, Y.K., Koo, H.C., et al., 2007. Novel multiplex PCR for the detection of the *Staphylococcus aureus* superantigen and its application to raw meat isolates in Korea. Int. J. Food Microbiol. 117, 99–105.

International Working Group on the Classification of Staphylococcal Cassette Chromosome Elements (IWG-SCC), 2009. Classification of staphylococcal cassette chromosome *mec* (SCC*mec*): guidelines for reporting novel SCC*mec* elements. Antimicrob. Agents Chemother. 53, 4961–4967.

Jackson, C.R., Davis, J.A., Barrett, J.B., 2013. Prevalence and characterization of methicillin-resistant *Staphylococcus aureus* isolates from retail meat and humans in Georgia. J. Clin. Microbiol. 51, 1199–1207.

Johnson, W.M., Tyler, S.D., Ewan, E.P., Ashton, F.E., Pollard, D.R., Rozee, K.R., 1991. Detection of genes for enterotoxins, exfoliative toxins, and toxic shock syndrome toxin 1 in *Staphylococcus aureus* by the polymerase chain reaction. J. Clin. Microbiol. 29, 426–430.

Jones, M.E., Köhrer, K., Schmitz, F.J., 2001. Multiplex PCR for the rapid simultaneous speciation and detection of methicillin-resistance and genes encoding toxin production in *Staphylococcus aureus*. Methods Mol. Med. 48, 103–112.

Kadlec, K., Ehricht, R., Monecke, S., Steinacker, U., Kaspar, H., Mankertz, J., et al., 2009. Diversity of antimicrobial resistance pheno- and genotypes of methicillin-resistant *Staphylococcus aureus* ST398 from diseased swine. J. Antimicrob. Chemother. 64, 1156–1164.

Kadlec, K., Feßler, A.T., Hauschild, T., Schwarz, S., 2012. Novel and uncommon antimicrobial resistance genes in livestock-associated methicillin-resistant *Staphylococcus aureus*. Clin. Microbiol. Infect. 18, 745–755.

Köck, R., Schaumburg, F., Mellmann, A., Köksal, M., Jurke, A., Becker, K., et al., 2013. Livestock-associated methicillin-resistant *Staphylococcus aureus* (MRSA) as causes of human infection and colonization in Germany. PLoS ONE 8, e55040.

Kondo, Y., Ito, T., Ma, X.X., Watanabe, S., Kreiswirth, B.N., Etienne, J., et al., 2007. Combination of multiplex PCRs for staphylococcal cassette chromosome *mec* type assignment: rapid identification system for *mec*, *ccr*, and major differences in junkyard regions. Antimicrob. Agents Chemother. 51, 264–274.

Kraushaar, B., Fetsch, A., 2014. First description of PVL-positive methicillin-resistant *Staphylococcus aureus* (MRSA) in wild boar meat. Int. J. Food Microbiol. 186, 68–73.

Lasch, P., Fleige, C., Stämmler, M., Layer, F., Nübel, U., Witte, W., et al., 2014. Insufficient discriminatory power of MALDI-TOF mass spectrometry for typing of *Enterococcus faecium* and *Staphylococcus aureus* isolates. J. Microbiol. Methods 100, 58–69.

Lassok, B., Tenhagen, B.-A., 2013. From pig to pork: methicillin-resistant *Staphylococcus aureus* in the pork production chain. J. Food Prot. 76, 1095–1108.

Li, B., Wendlandt, S., Yao, J., Liu, Y., Zhang, Q., Shi, Z., et al., 2013. Detection and new genetic environment of the pleuromutilin-lincosamide-streptogramin A resistance gene *lsa*(E) in methicillin-resistant *Staphylococcus aureus* of swine origin. J. Antimicrob. Chemother. 68, 1251–1255.

Lin, C.M., Chiang, Y.C., Tsen, H.Y., 2009. Development and use of a chromogenic macroarray system for the detection of *Staphylococcus aureus* with enterotoxin A, B, C, D, E, and G genes in food and milk samples. Foodborne Pathog. Dis. 6, 445–452.

Lozano, C., López, M., Gómez-Sanz, E., Ruiz-Larrea, F., Torres, C., Zarazaga, M., 2009. Detection of methicillin-resistant *Staphylococcus aureus* ST398 in food samples of animal origin in Spain. J. Antimicrob. Chemother. 64, 1325–1326.

McCarthy, A.J., Breathnach, A.S., Lindsay, J.A., 2012. Detection of mobile-genetic-element variation between colonizing and infecting hospital-associated methicillin-resistant *Staphylococcus aureus* isolates. J. Clin. Microbiol. 50, 1073–1075.

Mehrotra, M., Wang, G., Johnson, W.M., 2000. Multiplex PCR for detection of genes for *Staphylococcus aureus* enterotoxins, exfoliative toxins, toxic shock syndrome toxin 1, and methicillin resistance. J. Clin. Microbiol. 38, 1032–1035.

Monecke, S., Coombs, G., Shore, A.C., Coleman, D.C., Akpaka, P., Borg, M., et al., 2011. A field guide to pandemic, epidemic and sporadic clones of methicillin-resistant *Staphylococcus aureus*. PLoS ONE 6, e17936.

Monecke, S., Müller, E., Buechler, J., Rejman, J., Stieber, B., Akpaka, P.E., et al., 2013a. Rapid detection of Panton–Valentine leukocidin in *Staphylococcus aureus* cultures by use of a lateral flow assay based on monoclonal antibodies. J. Clin. Microbiol. 51, 487–495.

Monecke, S., Ruppelt, A., Wendlandt, S., Schwarz, S., Slickers, P., Ehricht, R., et al., 2013b. Genotyping of *Staphylococcus aureus* isolates from diseased poultry. Vet. Microbiol. 162, 806–812.

Murchan, S., Kaufmann, M.E., Deplano, A., de Ryck, R., Struelens, M., Zinn, C.E., et al., 2003. Harmonization of pulsed-field gel electrophoresis protocols for epidemiological typing of strains of methicillin-resistant *Staphylococcus aureus*: a single approach developed by consensus in 10 European laboratories and its application for tracing the spread of related strains. J. Clin. Microbiol. 41, 1574–1585.

O'Brien, A.M., Hanson, B.M., Farina, S.A., Wu, J.Y., Simmering, J.E., Wardyn, S.E., et al., 2012. MRSA in conventional and alternative retail pork products. PLoS ONE 7, e30092.

Pichon, B., Hill, R., Laurent, F., Larsen, A.R., Skov, R.L., Holmes, M., et al., 2012. Development of a real-time quadruplex PCR assay for simultaneous detection of *nuc*, Panton-Valentine leucocidin (PVL), *mecA* and homologue *mecA*$_{LGA251}$. J. Antimicrob. Chemother. 67, 2338–2341.

Pourcel, C., Hormigos, K., Onteniente, L., Sakwinska, O., Deurenberg, R.H., Vergnaud, G., 2009. Improved multiple-locus variable-number tandem-repeat assay for *Staphylococcus aureus* genotyping, providing a highly informative technique together with strong phylogenetic value. J. Clin. Microbiol. 47, 3121–3128.

Rasschaert, G., Vanderhaeghen, W., Dewaele, I., Janez, N., Huijsdens, X., Butaye, P., et al., 2009. Comparison of fingerprinting methods for typing methicillin-resistant *Staphylococcus aureus* sequence type 398. J. Clin. Microbiol. 47, 3313–3322.

Renwick, L., Holmes, A., Templeton, K., 2013. Multiplex real-time PCR assay for the detection of methicillin-resistant *Staphylococcus aureus* and Panton-Valentine leukocidin from clinical samples. Methods Mol. Biol. 943, 105–113.

Rhee, C.H., Woo, G.J., 2010. Emergence and characterization of foodborne methicillin-resistant *Staphylococcus aureus* in Korea. J. Food Prot. 73, 2285–2290.

Sabat, A., Krzyszton-Russjan, J., Strzalka, W., Filipek, R., Kosowska, K., Hryniewicz, W., et al., 2003. New method for typing *Staphylococcus aureus* strains: multiple-locus variable-number tandem repeat analysis of polymorphism and genetic relationships of clinical isolates. J. Clin. Microbiol. 41, 1801–1804.

Sandrin, T.R., Goldstein, J.E., Schumaker, S., 2013. MALDI TOF MS profiling of bacteria at the strain level: a review. Mass Spectrom. Rev. 32, 188–217.

Schmitz, F.J., Petridou, J., Jagusch, H., Astfalk, N., Scheuring, S., Schwarz, S., 2002. Molecular characterization of ketolide-resistant *erm*(A)-carrying *Staphylococcus aureus* isolates selected *in vitro* by telithromycin, ABT-773, quinupristin and clindamycin. J. Antimicrob. Chemother. 49, 611–617.

Schnellmann, C., Gerber, V., Rossano, A., Jaquier, V., Panchaud, Y., Doherr, M.G., et al., 2006. Presence of new *mecA* and *mph*(C) variants conferring antibiotic resistance in *Staphylococcus* spp. isolated from the skin of horses before and after clinic admission. J. Clin. Microbiol. 44, 4444–4454.

Schouls, L.M., Spalburg, E.C., van Luit, M., Huijsdens, X.W., Pluister, G.N., van Santen-Verheuvel, M.G., et al., 2009. Multiple-locus variable number tandem repeat analysis of *Staphylococcus aureus*: comparison with pulsed-field gel electrophoresis and *spa*-typing. PLoS ONE 4, e5082.

Schwarz, S., Silley, P., Simjee, S., Woodford, N., van Duijkeren, E., Johnson, A.P., et al., 2010. Editorial: assessing the antimicrobial susceptibility of bacteria obtained from animals. J. Antimicrob. Chemother. 65, 601–604.

Schwarz, S., Kadlec, K., Silley, P., 2013. Antimicrobial Resistance in Bacteria of Animal Origin, first ed. Zett-Verlag, Steinen.

Sergeev, N., Volokhov, D., Chizhikov, V., Rasooly, A., 2004. Simultaneous analysis of multiple staphylococcal enterotoxin genes by an oligonucleotide microarray assay. J. Clin. Microbiol. 42, 2134–2143.

Shore, A.C., Rossney, A.S., Kinnevey, P.M., Brennan, O.M., Creamer, E., Sherlock, O., et al., 2010. Enhanced discrimination of highly clonal ST22-methicillin-resistant *Staphylococcus aureus* IV isolates achieved by combining *spa*, *dru*, and pulsed-field gel electrophoresis typing data. J. Clin. Microbiol. 48, 1839–1852.

Shore, A.C., Deasy, E.C., Slickers, P., Brennan, G., O'Connell, B., Monecke, S., et al., 2011. Detection of staphylococcal cassette chromosome *mec* type XI carrying highly divergent *mecA*, *mecI*, *mecR1*, *blaZ*, and *ccr* genes in human clinical isolates of clonal complex 130 methicillin-resistant *Staphylococcus aureus*. Antimicrob. Agents Chemother. 55, 3765–3773.

Sobral, D., Schwarz, S., Bergonier, D., Brisabois, A., Feßler, A.T., Gilbert, F.B., et al., 2012. High throughput multiple locus variable number of tandem repeat analysis (MLVA) of *Staphylococcus aureus* from human, animal and food sources. PLoS ONE 7, e33967.

Strommenger, B., Kettlitz, C., Werner, G., Witte, W., 2003. Multiplex PCR assay for simultaneous detection of nine clinically relevant antibiotic resistance genes in *Staphylococcus aureus*. J. Clin. Microbiol. 41, 4089–4094.

Tenhagen, B.A., Vossenkuhl, B., Käsbohrer, A., Alt, K., Kraushaar, B., Guerra, B., et al., 2014. Methicillin-resistant *Staphylococcus aureus* in cattle food chains—prevalence, diversity, and antimicrobial resistance in Germany. J. Anim. Sci. 92, 2741–2751.

van Belkum, A., Melles, D.C., Nouwen, J., van Leeuwen, W.B., van Wamel, W., Vos, M.C., et al., 2009. Co-evolutionary aspects of human colonisation and infection by *Staphylococcus aureus*. Infect. Genet. Evol. 9, 32–47.

van Duijkeren, E., Hengeveld, P.D., Albers, M., Pluister, G., Jacobs, P., Heres, L., et al., 2014. Prevalence of methicillin-resistant *Staphylococcus aureus* carrying *mecA* or *mecC* in dairy cattle. Vet. Microbiol. 171, 364–367.

van Wamel, W.J., Hansenová Manásková, S., Fluit, A.C., Verbrugh, H., de Neeling, A.J., van Duijkeren, E., et al., 2010. Short term micro-evolution and PCR-detection of methicillin-resistant and -susceptible *Staphylococcus aureus* sequence type 398. Eur. J. Clin. Microbiol. Infect. Dis. 29, 119–122.

Vautor, E., Magnone, V., Rios, G., Le Brigand, K., Bergonier, D., Lina, G., et al., 2009. Genetic differences among *Staphylococcus aureus* isolates from dairy ruminant species: a single-dye DNA microarray approach. Vet. Microbiol. 133, 105–114.

Velasco, V., Sherwood, J.S., Rojas-García, P.P., Logue, C.M., 2014. Multiplex real-time PCR for detection of *Staphylococcus aureus*, *mecA* and Panton–Valentine Leukocidin (PVL) genes from selective enrichments from animals and retail meat. PLoS ONE 9, e97617.

Vestergaard, M., Cavaco, L.M., Sirichote, P., Unahalekhaka, A., Dangsakul, W., Svendsen, C.A., et al., 2012. SCC*mec* type IX element in methicillin resistant *Staphylococcus aureus* *spa* type t337 (CC9) isolated from pigs and pork in Thailand. Front. Microbiol. Available from: http://dx.doi.org/10.3389/fmicb.2012.00103

Wang, H.Y., Kim, S., Kim, J., Park, S.D., Uh, Y., Lee, H., 2014. Multiplex real-time PCR assay for rapid detection of methicillin-resistant staphylococci directly from positive blood cultures. J. Clin. Microbiol. 52, 1911–1920.

Weese, J.S., Reid-Smith, R., Rousseau, J., Avery, B., 2010. Methicillin-resistant *Staphylococcus aureus* (MRSA) contamination of retail pork. Can. Vet. J. 51, 749–752.

Wendlandt, S., Feßler, A.T., Monecke, S., Ehricht, R., Schwarz, S., Kadlec, K., 2013a. The diversity of antimicrobial resistance genes among staphylococci of animal origin. Int. J. Med. Microbiol. 303, 338–349.

Wendlandt, S., Kadlec, K., Feßler, A.T., Mevius, D., van Essen-Zandbergen, A., Hengeveld, P.D., et al., 2013b. Transmission of methicillin-resistant *Staphylococcus aureus* isolates on broiler farms. Vet. Microbiol. 167, 632–637.

Wendlandt, S., Kadlec, K., Feßler, A.T., Monecke, S., Ehricht, R., van de Giessen, A.W., et al., 2013c. Resistance phenotypes and genotypes of methicillin-resistant *Staphylococcus aureus* isolates from broiler chickens at slaughter and abattoir workers. J. Antimicrob. Chemother. 68, 2458–2463.

Wendlandt, S., Li, B., Lozano, C., Ma, Z., Torres, C., Schwarz, S., 2013d. Identification of the novel spectinomycin resistance gene *spw* in methicillin-resistant and methicillin-susceptible *Staphylococcus aureus* of human and animal origin. J. Antimicrob. Chemother. 68, 1679–1680.

Wendlandt, S., Lozano, C., Kadlec, K., Gómez-Sanz, E., Zarazaga, M., Torres, C., et al., 2013e. The enterococcal ABC transporter gene *lsa*(E) confers combined resistance to lincosamides, pleuromutilins and streptogramin A antibiotics in methicillin-susceptible and methicillin-resistant *Staphylococcus aureus*. J. Antimicrob. Chemother. 68, 473–475.

Wendlandt, S., Schwarz, S., Silley, P., 2013f. Methicillin-resistant *Staphylococcus aureus*: a foodborne pathogen? Annu. Rev. Food Sci. Technol. 4, 117–139.

Wendlandt, S., Li, J., Ho, J., Porta, M.A., Feßler, A.T., Wang, Y., et al., 2014. Enterococcal multiresistance gene cluster in methicillin-resistant *Staphylococcus aureus* from various origins and geographical locations. J. Antimicrob. Chemother. 69, 2573–2575.

Werckenthin, C., Schwarz, S., 2000. Molecular analysis of the translational attenuator of a constitutively expressed *erm*(A) gene from *Staphylococcus intermedius*. J. Antimicrob. Chemother. 46, 785–788.

Werckenthin, C., Schwarz, S., Westh, H., 1999. Structural alterations in the translational attenuator of constitutively expressed *ermC* genes. Antimicrob. Agents Chemother. 43, 1681–1685.

Wolters, M., Rohde, H., Maier, T., Belmar-Campos, C., Franke, G., Scherpe, S., et al., 2011. MALDI-TOF MS fingerprinting allows for discrimination of major methicillin-resistant *Staphylococcus aureus* lineages. Int. J. Med. Microbiol. 301, 64–68.

Chapter 12

Non-Phenotypic Tests to Detect and Characterize Antibiotic Resistance Mechanisms in *Enterobacteriaceae*

Agnese Lupo[1], Krisztina M. Papp-Wallace[2,3], Robert A. Bonomo[2,3,4] and Andrea Endimiani[1]

[1]*Institute for Infectious Diseases, University of Bern, Bern, Switzerland,* [2]*Research Service, Louis Stokes Cleveland Department of Veteran Affairs Medical Center, Cleveland, OH, USA,* [3]*Department of Medicine, Case Western Reserve University, Cleveland, OH, USA,* [4]*Department of Pharmacology, Molecular Biology and Microbiology, Case Western Reserve University, Cleveland, OH, USA*

Chapter Outline

The spread of *Enterobacteriaceae* (*Ent*) resistant to extended-spectrum cephalosporins (ESCs) is challenging our therapeutic armamentarium (Coque et al., 2008; Endimiani and Paterson, 2007; Grayson, 2010; Meyer et al., 2010; Michalopoulos and Falagas, 2010). These multidrug-resistant pathogens are frequently co-resistant to quinolones and aminoglycosides (Giamarellou and Poulakou, 2009; Hawser et al., 2010), and may also develop resistance to carbapenems (Walsh, 2010).

Antimicrobial Resistance and Food Safety. DOI: http://dx.doi.org/10.1016/B978-0-12-801214-7.00012-0

More troublesome is the fact that many food products are found contaminated with these life-threatening pathogens (Seiffert et al., 2013b).

Resistance to ESCs is usually due to the production of extended-spectrum β-lactamases (ESBLs; e.g., CTX-Ms, TEMs, and SHVs), plasmid-mediated (pAmpCs; e.g., CMYs), or chromosomal AmpC (cAmpCs) (Bush and Jacoby, 2009; Harris and Ferguson, 2012; Hilty et al., 2013; Jacoby, 2009; Seiffert et al., 2013a), whereas resistance to carbapenems is due to the production of carbapenemase enzymes (e.g., KPCs, NDMs, VIMs, IMPs, OXA-48-types) (Lascols et al., 2013; Mathers et al., 2013; Papp-Wallace et al., 2011; Poirel et al., 2012).

Quinolone resistance is typically mediated by chromosomal mutations in the quinolone resistance determining region (QRDR) that encodes DNA gyrase (*gyrA* and *parC*) genes (Hooper, 2001; Jacoby, 2005). Low-level resistance can also arise from the expression of plasmid-mediated quinolone resistance (PMQR) determinants (e.g., *Qnr*, *aac(6')-Ib-cr*) (Strahilevitz et al., 2009). Aminoglycoside resistance is caused by enzymatic inactivation mediated by aminoglycoside-modifying enzymes (AMEs) or by production of 16S rRNA methylases (e.g., ArmA, Rmt) (Bueno et al., 2013; Doi and Arakawa, 2007; Hidalgo et al., 2013; Magnet and Blanchard, 2005).

Ideally, a rapid diagnostic system should investigate all of these important resistance traits in one reaction and should provide easy-to-interpret results (Table 12.1). Moreover, the methodology should be easy to perform, relatively cheap, accurate, and versatile enough to be regularly updated.

ENDPOINT PCRs

The single PCR allows identification of one gene and requires the design of specific primers (Mullis and Faloona, 1987; Predari et al., 1991; Zhou et al., 1994). This approach is advantageous if the detection of a single gene at the generic level is sufficient (Table 12.1) (Boyd et al., 2004). For specific gene characterization, DNA sequencing is necessary.

The multiplex endpoint PCRs are based on the same principle of the single PCRs but use multiple primer sets allowing the amplification of different targets in the same reaction (Markoulatos et al., 2002). Several multiplex PCRs able to detect resistance traits have been proposed (Table 12.2). One example is a multiplex designed to detect the bla_{pAmpCs}. The DNA template was obtained by boiling the colonies and the total time from primary incubation to test results (TTR) was estimated at around 3 h (Perez-Perez and Hanson, 2002). Doyle et al. designed a multiplex for the simultaneous detection of bla_{KPC}, bla_{NDM}, bla_{VIM}, bla_{IMP}, and $bla_{OXA-48-like}$. The methodology showed 100% sensitivity and specificity, much higher than the standard phenotypic tests (Doyle et al., 2012). Cattoir et al. (2007) designed a multiplex for the detection of several *qnr* genes. Bercot et al. (2011) developed a protocol for the detection of 16S rRNA methylase genes (Williamson et al., 2012).

TABLE 12.1 Most Important Antibiotic Resistance Traits that Molecular Methods Should Be Able to Detect in *Enterobacteriaceae*

Antibiotics	Mechanism of Resistance	Most Frequent Resistance Traits	Importance	Ideal Level of Detection/Characterization
ESCs	ESBLs	CTX-M	++++	Generic
		TEM and SHV	++++	Distinguishing ESBLs from non-ESBLs[a]
		PER	++	Generic
		GES	++	Distinguishing ESBLs (e.g., GES-1) from carbapenemases (e.g., GES-5)[b]
		SFO	+	Generic
	pAmpCs	CMY	++++	Generic
		DHA	+++	Generic
		FOX	++	Generic
		LAT, MIR, ACT, MOX	+	Generic
	cAmpCs	Mutations in the promoter (*E. coli*)	++	Identifying substitutions (e.g., at −32/−42) and insertions (between −10 and −35)
Carbapenems	Carbapenemases			
	Class A	KPC	++++	Generic
		GES	++	Distinguishing ESBLs (e.g., GES-1) from carbapenemases (e.g., GES-5)[b]
	Class B	NDM	++++	Generic
		SME	+	Generic
		VIM	++++	Generic
		IMP	+++	Generic
		SPM and GIM	++	Generic
		AIM, SIM	+	Generic

(Continued)

TABLE 12.1 (Continued)

Antibiotics	Mechanism of Resistance	Most Frequent Resistance Traits	Importance	Ideal Level of Detection/Characterization
	Class D	OXA-48	++++	Specific
		OXA-181, -162, -163	+++	Specific
Quinolones	Amino acid substitutions in the QRDR	GyrA	++++	Mutation at position −248, −259/260[c]
		ParC	++++	Mutation at position 238/239 and 250/251[d]
	PMQR determinants	QnrA/B/S/D	+	Generic
		Aac(6)-Ib-cr	++	Distinguishing the aac(6′)-Ib-cr from the AME aac(6′)-Ib
		QepA	++	Generic
		OqxAB	+	Generic
Aminoglycosides	AMEs	Aac(6)-I-like, Ant(3)-like, Aph(3)-like	+++	Specific
	16S rRNA methylases	ArmA, RmtA/B/C/D/E, NpmA	++++	Generic

Note: QRDR, quinolone resistance determining region; AMEs, aminoglycoside-modifying enzymes. The list is based on: (i) the frequency of observation during epidemiological studies conducted worldwide; (ii) the clinical importance of the antimicrobials affected by the resistance mechanism(s).
[a]Amino acidic substitutions E104K, R164S/H, G238S, and E240K for TEM; D179N, G238S, and E240K for SHV (Gniadkowski, 2008; Perilli et al., 2008; Randegger et al., 2000).
[b]Amino acidic substitution G170A/S (Girlich et al., 2012).
[c]Lead to modification of S83 and D87 (Jacoby, 2005; Qiang et al., 2002).
[d]Lead to modification of S80 and E84 (Qiang et al., 2002; Jacoby, 2005).

TABLE 12.2 Single and Multiplex Endpoint PCRs Used for the Detection of the Main Antibiotic Resistance Genes in *Enterobacteriaceae*

Approach	Detected Resistance Target	No. Primers Sets	Cycling Conditions	TTR (h)[d]	Comments	References
Single PCRs[a]	CTX-M	4	1[b]	2	Identification of CTX-M at group level	Pitout et al. (2004)
	CTX-M	1	1	2	Generic detection of CTX-M	Boyd et al. (2004)
	TEM, SHV, PER	1, 1, 1	1, 1, 1	2	Generic detection	De Champs et al. (2004)
	VEB, GES, SFO	1, 1, 1	1, 1, 1	2	Generic detection	Poirel et al. (2000)
	KPC	1	1	2	Generic detection	Bradford et al. (2004)
	IMP and VIM	1, 1	1, 1	2	Generic detection	Pitout et al. (2005)
	NDM	1	1	2	Generic detection	Nordmann et al. (2011)
	GyrA	1	1	2	Sequencing always required	Weigel et al. (1998)
	ParC	1	1	2	Sequencing always required	McDonald et al. (2001)
	QnrA, QnrB	1	1	2	Generic detection	Wang et al. (2003)
	QnrS	1	1	2	Generic detection	Robicsek et al. (2006)
	Aac(6)-Ib-cr	1	1	2	Sequencing always required	Park et al. (2006)
	AMEs	3	1	2	Specific detection	Miro et al. (2012)
Multiplex PCRs	TEM, SHV, CTX-M	NA	1	2.5	Commercial kit (Hyplex)	NA
	CMY, DHA, ACC, FOX, MOX, LAT, ACT, MIR	6	1	2–4	Generic detection. DNA sequencing is required for CMY. Cross reaction with chromosomal genes	Perez-Perez and Hanson (2002)

(Continued)

TABLE 12.2 (Continued)

Approach	Detected Resistance Target	No. Primers Sets	Cycling Conditions	TTR (h)[d]	Comments	References
	TEM, SHV, CTX-M, GES, VEB, PER KPC, IMP, VIM CMY, DHA, ACC, FOX, MOX, LAT, ACT, MIR OXA (including OXA-48-like)	21	3[c]	3	NDM-types not included Cross reaction with chromosomal genes	Dallenne et al. (2010)
	TEM, SHV, CTX-M, GES, VEB, PER KPC, SME, IMI, NMC-A, IMP, VIM, GIM, SIM, NDM, SPM CMY, DHA, MOX, ACC, ACT, MIR, FOX OXA (including OXA-48-like)	20	1	3	Partial cross reaction with chromosomal genes	Voets et al. (2011)
	KPC, BIC SPM, VIM, GIM, AIM, IMP, NDM, SIM, DIM OXA-48, OXA-163, OXA-181	11	1	<4	Limited to carbapenemases detection	Poirel et al. (2011)
	KPC, IMP, VIM, NDM, OXA-48-like	NA	1	2.5	Commercial kit (Hyplex)	Kaase et al. (2012)
	SME, NMC-A, KPC, GES	4	1	<3	No detection of NDM	Hong et al. (2012)
	KPC, NDM, VIM, IMP, OXA-48-like	5	1	<3	It uses previously reported primers	Doyle et al. (2012)
	QnrA, QnrB, QnrS	3	1	<3	No detection of other PMQR determinants	Cattoir et al. (2007)
	ArmA, RmtB, RmtC, RmtD, NpmA	6	1	<4	Generic detection	Bercot et al. (2011)

Note: NA, not available.

[a]Numerous single PCRs have been designed to detect the listed antibiotic resistance genes. In this table, we present the most frequently implemented PCR schemes.

[b]Each couple of primers operates with one cycling condition (i.e., duration and temperature of: initial DNA denaturation; intermediate DNA denaturation; primers annealing; elongation; number of cycles; final elongation).

[c]The primers designed for this assay operated with three different annealing temperatures (55°C, 57°C, and 60°C), thus three cycling conditions are necessary.

[d]The TTR is calculated starting from DNA template.

Commercially Available Multiplex PCRs

Amplex Diagnostics (Germany) has designed three commercial PCR-ELISA kits to rapidly detect (2.5–4 h) *bla* genes. In particular, hyplex ESBL ID, hyplex SuperBug ID, and hyplex CarbOXA ID detect respectively: (i) bla_{TEM}, bla_{SHV}, bla_{CTX-M}, and bla_{OXA}; (ii) bla_{KPC}, bla_{IMP}, bla_{VIM}, bla_{NDM}, $bla_{OXA-48-like}$; and (iii) bla_{OXA-23}, bla_{OXA-24}, and $bla_{OXA-51-like}$. Hyplex SuperBug ID showed a 97% agreement with the standard PCR/sequencing (Avlami et al., 2010; Kaase et al., 2012).

Advantage and Disadvantages of PCRs

PCRs require a high copy number of the intended target to produce a detectable amplification product; thus, a good DNA extraction is essential (Schrader et al., 2012). Results can be obtained in <4 h (simple amplification) to ≥24 h (determination of the specific DNA sequence). Typically, multiplex PCRs provide a generic identification of genes. However, such PCRs detect numerous genes simultaneously, the execution is simple, and costs are relatively low.

REAL-TIME PCRs

The real-time PCR consists of an amplification reaction coupled with the simultaneous detection of the exponentially amplified target (Heid et al., 1996; Higuchi et al., 1993). Monitoring of fluorescence emission occurring during the amplification (e.g., with SYBR Green) is the mechanism on which the dynamics of real-time PCR is based (Arya et al., 2005; Higuchi et al., 1992). The real-time apparatus can also analyze the denaturing temperature of synthesized DNA fragments giving information on small variations in the sequence (high-resolution melting (HRM) analysis) (Erali et al., 2008; Tong and Giffard, 2012).

One of the first applications of real-time PCR for detecting antibiotic resistance in *Ent* was performed by Hammond et al. where a single nucleotide polymorphism (SNP) of bla_{SHV} was analyzed to distinguish ESBL from non-ESBL variants (Table 12.3) (Hammond et al., 2005). Using TaqMan probes specific for CTX-Ms, Birkett et al. (2007) implemented a method that avoids sequencing costs and provides a faster alternative to conventional PCR. Raghunathan et al. (2011) designed a real-time PCR using the SYBR Green for the detection of bla_{KPC}, whereas Kruttgen et al. (2011) designed a real time PCR for the detection of bla_{NDM}. Oxacelay et al. (2009) engineered a HYB probe for the detection of CTX-Ms in urine. Naas et al. (2011b) engineered a TaqMan probe for the detection of bla_{NDM-1} from stools (detection limit of 10^1 CFU/100 mg). A real-time TaqMan probe possessing 100% sensitivity was also implemented for the detection of bla_{OXA-48} in stools (detection limit of 10–50 CFU/100 mg) (Naas et al., 2013).

Bisiklis et al. (2007) used HYB probe to amplify and distinguish bla_{VIM} and bla_{IMP}. Guillard et al. (2011) elaborated the first real-time PCR for the detection of PMQR genes using the ResoLight dye in <2 h. These two assays exploited HRM analysis for discriminating amplified products (Table 12.3).

TABLE 12.3 Single and Multiplex Real-Time PCRs Used for the Detection of the Main Antibiotic Resistance Genes in *Enterobacteriaceae*

Approach	Detected Resistance Target	No. Primers Sets/ Probes	Cycling Conditions	TTR (h)	Comments	References
Single real-time PCR	SHV (ESBLs variants)	2	1[a]	2	SNPs are detected by the specificity of the primers	Hammond et al. (2005)
	TEM (ESBLs variants)	2/2	1	2	SNPs are detected by the probes	Mroczkowska and Barlow (2008)
	CTX-M	1/1	1	2	Specific characterization	Oxacelay et al. (2009)
	CTX-M	1/4	1	2	Specific detection	Birkett et al. (2007)
	KPC	NA	1	2	Commercial kit (EasyQ KPC test, bioMérieux)	Spanu et al. (2012)
	KPC	1/1	1	<4	Clinical implementation using enriched rectal swabs	Singh et al. (2012)
	KPC	1/SG	1	2	Used previously published primers	Raghunathan et al. (2011)
	NDM-1	2/3	1	3	Synthesized a positive control	Kruttgen et al. (2011)
	NDM-1	1/1	1	2	Good performance of the TaqMan probe for copies number determination	Manchanda et al. (2011)
	NDM-1	1/1	1	<4	From spiked stools	Naas et al. (2011b)
	OXA-48-like	1/1	1	<4	From spiked stools	Naas et al. (2013)
	CMY-2	1/SG	1	2	Quantification purpose	Kurpiel and Hanson (2011)

Method	Targets				Comments	References
Multiplex real-time PCR	TEM and SHV (ESBL variants), CTX-M	NA	1	3.5	Commercial kit (Check-Points)	Nijhuis et al. (2012)
	CIT, MOX, FOX, DHA, ACC, EBC	6/SG	1	3	Performed with previous published primers	Brolund et al. (2010)
	CMY, ACT, DHA, MOX, ACC, FOX, 16S rRNA	7/7	1	3	No cross reaction with cAmpC of E. coli	Geyer et al. (2012)
	KPC, 16S rRNA	2/2	1	<4	From overnight cultures	Singh et al. (2012)
	IMP, VIM	2/4	1	<1	Specific detection	Bisiklis et al. (2007)
	KPC, RNaseP	2/2	1	3	Clinical implementation from blood cultures	Hindiyeh et al. (2011)
	KPC, TonB, GapA	4/4	1	4	Identifies peculiar Klebsiella pneumoniae sequence types	Chen et al. (2012)
	GES, IMI/NMC, KPC, SME, OXA-48	5/5	1	3	It does not discriminate the carbapenemase GES variants	Swayne et al. (2011)
	KPC, NDM, GES, OXA-48, IMP, VIM	6/EG	1	3	Specific identification by Tm analysis	Monteiro et al. (2012)
	KPC, NDM, VIM, IMP, OXA-48	NA	1	3.5	Commercial kit (Check-Points). Not yet evaluated	Cuzon et al. (2013)
	QnrA/B/S/C/D, QepA	6	2	<2	Specific identification	Guillard et al. (2011)
	QnrA/B/S	3	1	<2	Coupled with HRM curve analysis	Guillard et al. (2012)
	Aac(6)-Ib-cr	1/1	1	<1	Coupled with HRM curve analysis, specific detection	Bell et al. (2010)

Note: NA, not available; SG, SYBR Green; EG, EvaGreen dye.
[a] Each couple of primers operates with one cycling condition (i.e., duration and temperature: initial DNA denaturation; intermediate DNA denaturation; primers annealing; elongation; final elongation).

Geyer et al. developed a TaqMan assay for pAmpCs showing 100% specificity and sensitivity (Geyer et al., 2012). Interestingly, Chen et al. combined multilocus sequencing typing (MLST) and bla_{KPC} rapid detection (1 h) using molecular beacons in a unique multiplex (Chen et al., 2012).

The loop-mediated isothermal amplification of DNA (LAMP), based on the detection of fluorescence and a peculiar engineering of the primer set, allows DNA target amplification at a constant temperature (Notomi et al., 2000). This method, being cheaper, could represent an alternative to PCR and to real-time PCR. Anjum et al. (2013) has used this approach to correctly amplify bla_{CTX-M} and $bla_{OXA-10-like}$ in meat samples.

Commercially Available Real-Time PCRs

The New NucliSENS EasyQ KPC (bioMérieux) is a very rapid (<2 h), highly sensitive, and specific commercial real-time PCR system for the detection of bla_{KPC} (Spanu et al., 2012). Check-Points Health BV (the Netherlands) has developed another rapid (3.5 h) multiplex real-time PCR (Check-MDR ESBL kit) to detect SHV/TEM variants and CTX-Ms (Nijhuis et al., 2012). Similarly, the Check-MDR Carba kit rapidly detects bla_{KPC}, bla_{NDM}, bla_{VIM}, bla_{IMP}, and bla_{OXA-48} (Cuzon et al., 2013).

Implementation of Real-Time PCRs

A TaqMan probe design enabled bla_{KPC} detection directly from rectal/perianal swabs. The method was rapid and had a superior sensitivity compared to selective cultures (Hindiyeh et al., 2008). Similarly, Singh et al. designed two TaqMan probes for rectal screening of KPC carriers (both sensitivity and specificity of 97%) (Singh et al., 2012). Hindiyeh et al. (2011) adapted the above protocol for detecting bla_{KPC} within 4 h in blood culture bottles.

Advantages and Disadvantages of Real-Time PCR

Although the setup is complicated, the real-time PCR avoids time consuming steps, is more sensitive, and usually does not require DNA sequencing (Bustin et al., 2009). Overall, this methodology is a reliable and cost-effective choice compared to the other molecular approaches.

MICROARRAY

Microarray allows simultaneous identification of a very large number of genes (even >2 million). It consists of oligonucleotide probes that are immobilized on the solid surface of an array. If the targeted allele is present, it is labeled and subsequently hybridized to the probe; such reactions are then measured with dedicated scanners (Miller and Tang, 2009; Sibley et al., 2012).

Glenn et al. (2013) used a microarray to detect many resistance genes in *Salmonella enterica* found in retail meats. Braun et al. (2012) developed an array dedicated to the O- and H-serotyping and resistance genes. Huehn et al. (2009) designed a microarray for the characterization of virulence, O-antigen, phase variation, insertion sequences, plasmids, and resistance genes. The same platform was used for characterizing *Salmonella* from pork (Hauser et al., 2010, 2011). van Hoek et al. (2005) included 28 probes to detect resistance genes, genes for identification, and integrons and integrase genes in *Salmonella*. A microarray developed by Bruant et al. (2006) is able to detect specific markers allowing phylogenetic *Escherichia coli* group tracking.

Although the above in-house microarrays demonstrated excellent ability to detect numerous genes, their systematic implementation appears difficult, especially because of problems related to the standardization of the procedures.

Commercially Available Microarrays

Check-Points (the Netherlands) has developed several DNA microarrays to detect *bla* genes. Results are automatically interpreted by software and the overall TTR is rapid (~6 h) (Endimiani et al., 2010a). Cohen Stuart et al. evaluated a prototype platform to detect bla_{ESBLs}. Specific probes were designed to distinguish non-ESBL and ESBL variants (e.g., SNPs determining substitutions G238S for TEMs and G238S/A, E240K for SHVs) and the CTX-Ms; sensitivity and specificity were 95% and 100%, respectively (Cohen Stuart et al., 2010). Then, we evaluated an updated platform (Check KPC/ESBL) able to further detect the bla_{KPC}. Overall, there was a 91.5% agreement with the DNA sequencing (Endimiani et al., 2010a; Lascols et al., 2012; Naas et al., 2010; Platteel et al., 2011). More recently, new Check-Points kits have been designed. The Check-MDR CT101 assay detects bla_{ESBLs} and bla_{KPC}, but also bla_{pAmpCs} and bla_{NDM}; it has 100% agreement with the PCR/DNA sequencing (Bogaerts et al., 2011). The Check-MDR CT102 detects bla_{ESBLs}, but also more carbapenemase genes (bla_{VIM}, bla_{IMP}, and the $bla_{OXA-48-like}$) rather than only bla_{KPC} and bla_{NDM}. This assay has an agreement with standard sequencing of 99% (Naas et al., 2011a). Anjum et al. (2013) and Veldman et al. (2014) implemented this microarray for the characterization of *Ent* from meat samples and from culinary herbs, respectively. So far, the Check-MDR CT103 is the most complete platform because it merges the characteristics of CT101 and CT102 (Cuzon et al., 2012).

Batchelor et al. (2008) developed a microarray for the detection of 47 resistance genes. An updated assay (Identibac, Alere, Germany) containing probes to detect numerous virulence factors of *E. coli* and further important resistance alleles has been implemented for the characterization of foodborne *E. coli* (Szmolka et al., 2012; Vogt et al., 2014). Verigene Nucleic Acid Test (Nanosphere, USA) is the most recently developed automated microarray for detection of bacterial species and resistance genes within 2 h. Its clinical implementation has shown a very promising future (Mancini et al., 2014; Sullivan et al., 2014).

Clinical Implementation of the Microarrays

The Check KPC/ESBL was used to detect *bla* genes directly in positive blood cultures. This approach could reduce the notification time of ESBL and/or KPC production by 18–20 h (Fishbain et al., 2012; Wintermans et al., 2012). Peter et al. evaluated an in-house DNA microarray to directly detect all variants of bla_{KPC} from urine within 5–6 h (Peter et al., 2012). The Verigene Gram-negative Blood Culture Nucleic Acid Test is able to identify the species of Gram-negatives along with bla_{CTX-M}, bla_{KPC}, bla_{NDM}, bla_{VIM}, bla_{IMP}, and bla_{OXA}. Several authors evaluated its performance with positive blood cultures obtaining sensitivities and specificities of 87–90% and 100%, respectively (Mancini et al., 2014; Sullivan et al., 2014).

Advantages and Disadvantages of Microarrays

Microarrays possess greater analytical capacity compared to most of the other molecular methods because they can simultaneously analyze a larger number of target genes (Cuzon et al., 2012; Endimiani et al., 2010a). Moreover, the commercial platforms are easy to perform and update. On the other hand, the TTR is moderately high (6–8 h) (Batchelor et al., 2008; Bogaerts et al., 2011). Devices and commercial kits are also quite expensive but more affordable prices are expected in the near future (Table 12.4).

NEXT-GENERATION DNA-SEQUENCING METHODOLOGIES

The Sanger methodology is still the most used for the sequencing of genes. However, since this method is laborious, recent efforts have been made to overcome these limitations giving rise to the so-called next-generation sequencing (NGS) methodologies. Mostly, these methods are based on the "real-time" fluorescence/light detection occurring during the sequencing process (Liu et al., 2012; Maxam and Gilbert, 1977; Sanger et al., 1977).

The pyrosequencing (available as "454"; Roche) is nowadays a benchtop tool (Ronaghi et al., 1996). The sequencing by synthesis, commercially referred to as "Illumina," consists of the neo-synthesis of DNA which is detected in real-time (Liu et al., 2012). The sequencing by ligation, implemented by Life-technology with the commercial name "SOLiD," is based on the action of a DNA ligase which adds fluorescent labeled dNTPs to an anchor probe hybridizing a single strand DNA (Liu et al., 2012). The Ion torrent sequencing is another powerful NGS approach based on the detection of the hydrogen ions released at each addition of a nucleic base (Liu et al., 2012). This technology was applied to the investigation of an *E. coli* outbreak, allowing not only the detection of CTX-Ms, but providing the genome sequences of these strains (Sherry et al., 2013).

The advent of NGS methodologies and their commercialization promise to be revolutionary for the improvement of pathogen detection and

TABLE 12.4 Main Characteristics of the Rapid Molecular Tools So Far Available for the Detection and Characterization of Antibiotic Resistance Genes in *Enterobacteriaceae*

Molecular Tool	Time	Costs[c]	Main Setting(s) of Application	Comments
Single PCR coupled with standard DNA sequencing	2–3 h (PCR)[a]	+	Research/ Epidemiology	Detect all genes but labor intensive, especially for multiple genes
	1–3 or more days (sequencing)	+++		
Multiplex endpoint PCR	3–4 h[a]	+	Epidemiology	PCR reaction should be optimized. Moderately labor intensive
Hyplex platforms (commercial)	2.5–4 h[a]	+++	Clinical	Helpful for treatment and screening of carriers
Real-time PCR	2–3 h[a]	++	Clinical	Helpful for treatment and screening of carriers. However, only one target gene is tested
NucliSENS EasyQ KPC platform (commercial)	2 h[a]	+++	Clinical	It detects only *bla*$_{KPC}$
Multiplex real-time PCR	2–3 h[a]	++	Clinical	Reactions should be optimized. Moderately labor-intensive
Check-Points (commercial)	4.5 h[b]	+++	Clinical	Helpful for treatment and screening of carriers of ESBL-*Ent* and/or carbapenemase producers
LAMP	1–1.5 h[a]	+	Clinical	Very rapid and cheap. Good for low-income countries
Microarray	6–8 h or more[a]	++	Epidemiology	In-house platforms are difficult to be implemented for clinical applications

(Continued)

TABLE 12.4 (Continued)

Molecular Tool	Time	Costs[c]	Main Setting(s) of Application	Comments
Check-Points (commercial kits)	5–6 h[a]	+++	Epidemiology/ Clinical	Very versatile and easy to use
Identibac (commercial)	5–6 h[a]	+++	Epidemiology/ Research	Capable to detect many genes (including those for non-β-lactams) but not yet validated testing large and well-characterized collections of isolates
Verigene (commercial)	2 h	+++	Clinical	Very rapid. Implementation on positive blood cultures
NGS methodologies	3 h[a]	+++	Clinical	Much more rapid than standard sequencing, especially if coupled to real-time PCR. However, they are still unavailable in most laboratories
MALDI-TOF MS (for detection of catalytic activity)	3 h[b]	+	Clinical/ Research	It detects only the spectrum of β-lactamases without distinguishing the specific kind of enzymes
MALDI-TOF MS (for detection of atomic mass)	3 h[b]	+	Clinical/ Research	It can detect the class of β-lactamases. The specific kind of enzymes can be distinguished if the accuracy and the databases are improved
MALDI-TOF MS (for mini-sequencing)	3 h[a]	++	Research	Difficult to be implemented when compared to the other easier and more rapid systems
PCR/ESI-MS (PLEX-ID, commercial)	4–6 h (starting from sample)	++++	(Clinical)/ Research	Excellent for clinical applications. However, so far available only for research

[a]TTR calculated starting from DNA template.
[b]TTR calculated starting from colonies.
[c] +, lowest cost; + + + +, highest cost.

antimicrobial resistance determination. Undoubtedly, these methodologies are fast (e.g., 5 days from a positive culture to completion of sequencing), costs are becoming reasonably cheap (e.g., ~US$300 per strain, reagents only), and the development of bioinformatics is simplifying data interpretation.

MASS SPECTROMETRY-BASED METHODOLOGIES

MALDI-TOF MS

Matrix-assisted laser desorption ionization-time of flight mass spectroscopy (MALDI-TOF MS) allows clinical microbiologists to discern unique protein signatures (e.g., ribosomal proteins) in order to identify the pathogen (Wieser et al., 2012); in addition, there is potential for the detection of resistance mechanisms (Hrabak et al., 2013). This last application includes identification of the antibiotic itself and its modified/degraded derivatives (Burckhardt and Zimmermann, 2011; Hooff et al., 2012; Hrabak et al., 2011, 2012; Sparbier et al., 2012), detection of the resistance proteins within the cell (Cai et al., 2012; Camara and Hays, 2007; Schaumann et al., 2012), and discovery of mutations within resistance genes through mini-sequencing (Ikryannikova et al., 2008).

Several studies have assessed the utility of MALDI-TOF MS for the identification of β-lactams and β-lactam degradation products (Burckhardt and Zimmermann, 2011; Hooff et al., 2012; Hoyos-Mallecot et al., 2014; Hrabak et al., 2011, 2012; Jung et al., 2014; Sparbier et al., 2012; Wang et al., 2013). Typically, β-lactams are incubated with the enzyme(s) extract and then are analyzed for β-lactam degradation or modified products by MALDI-TOF MS. The time required to do this assay is ~1–4 h. For instance, this method is effective for identifying carbapenemase-producing *Ent* (Burckhardt and Zimmermann, 2011; Hrabak et al., 2011, 2012; Wang et al., 2013). Recently, MALDI-TOF MS was used to identify hydrolyzed ampicillin and cefotaxime directly from blood cultures (Jung et al., 2014). However, this method can only detect the presence of β-lactamases as a resistance mechanism.

Detection of the actual proteins conferring resistance in *Ent* by MALDI-TOF MS is more problematic because the protein signatures of bacteria are very complex. Camara and Hays (2007) established a "proof of principle" for this method in which they dissected the differences between a wild-type *E. coli* and an ampicillin-resistant strain. They identified a 29,000 atomic mass unit peak in resistant *E. coli* confirming that this peak represented a β-lactamase.

Mini-sequencing using MALDI-TOF MS may detect SNPs within resistance genes. This method requires PCR amplification of the resistance gene followed by MALDI-TOF MS analysis. As a "proof of concept," Ikryannikova et al. (2008) focused on three nucleotide positions that correspond to the amino acid positions 104, 164, and 238, as mutations conferring ESBL phenotype in TEMs. The system identified polymorphisms accurately at these three sites.

Advantages and Disadvantages of MALDI-TOF MS

The unique advantage of MALDI-TOF MS is that it is fast and inexpensive (Emonet et al., 2010; Ho and Reddy, 2011; Lavigne et al., 2012). However, its implementation to detect β-lactam degradation is just in addition to the second-line phenotypic tests (e.g., synergy with clavulanate). In fact, this approach defines only if an enzyme with ESBL or carbapenemase hydrolytic spectrum is expressed by the tested pathogens (Burckhardt and Zimmermann, 2011; Hrabak et al., 2011). We note that this information can be easily obtained by implementing rapid (<2 h) and cost-effective phenotypic tests (Nordmann et al., 2012a,b). Moreover, the use of MALDI-TOF MS to perform mini-sequencing is, so far, too difficult and time-consuming to replace the standard methods. On the other hand, implementation of MALDI-TOF MS for detecting the specific atomic mass of β-lactamases may be a future application (Table 12.4).

PCR/ESI MS

The PCR/electrospray ionization mass spectrometry (PCR/ESI MS) is a methodology that performs mini-sequencing of small PCR products (100–450 bp) measuring their exact molecular mass and interpreting such data with an advanced software. This methodology has been integrated into a fully automated system (PLEX-ID; Abbott Biosciences) with results available within 4–6 h starting from clinical samples (Emonet et al., 2010; Ho and Reddy, 2011; Lavigne et al., 2012; Wolk et al., 2012). An important study revealed the utility of PCR/ESI MS for detecting bla_{KPC} expressing *Ent.* PCR/ESI MS correctly detected 100% of the bla_{KPC}-producing isolates (Endimiani et al., 2010b).

Advantages and Disadvantages of PCR/ESI MS

PCR/ESI MS represents a very promising technology. The rapid and fully automated ability to detect multiple genes (not only those conferring antibiotic resistance) in one reaction may represent a revolution for the diagnosis of infectious diseases (Ecker et al., 2008). This platform also offers the ability to detect pathogens at very low copy numbers in "culture-negative" specimens and determine genetic relatedness based upon an MLST-like approach (Ecker et al., 2006), thus providing "real-time" data essential to prevent outbreaks. Unfortunately, PCR/ESI MS is still under development and currently only available for research applications (Table 12.4).

CONCLUSIONS

It is clear that we are at the threshold of a "new dawn" in clinical microbiology. Each method described above promises early and more sensitive detection.

The field of microbiology has long awaited this promise. It is true that we will need a considerable amount of expertise to work with these advanced approaches. However, it is also imperative for us to accept these challenges and to enact wise choices in the selection of diagnostics. We are not certain which method will best suit our needs, but, perhaps, *"chance will favor a prepared mind."*

REFERENCES

Anjum, M.F., Lemma, F., Cork, D.J., Meunier, D., Murphy, N., North, S.E., et al., 2013. Isolation and detection of extended spectrum β-lactamase (ESBL)-producing *Enterobacteriaceae* from meat using chromogenic agars and isothermal loop-mediated amplification (LAMP) assays. J. Food Sci. 78, M1892–M1898.

Arya, M., Shergill, I.S., Williamson, M., Gommersall, L., Arya, N., Patel, H.R., 2005. Basic principles of real-time quantitative PCR. Expert Rev. Mol. Diagn. 5, 209–219.

Avlami, A., Bekris, S., Ganteris, G., Kraniotaki, E., Malamou-Lada, E., Orfanidou, M., et al., 2010. Detection of metallo-β-lactamase genes in clinical specimens by a commercial multiplex PCR system. J. Microbiol. Methods 83, 185–187.

Batchelor, M., Hopkins, K.L., Liebana, E., Slickers, P., Ehricht, R., Mafura, M., et al., 2008. Development of a miniaturised microarray-based assay for the rapid identification of antimicrobial resistance genes in Gram-negative bacteria. Int. J. Antimicrob. Agents 31, 440–451.

Bell, J.M., Turnidge, J.D., Andersson, P., 2010. *aac(6')-Ib-cr* genotyping by simultaneous high-resolution melting analyses of an unlabeled probe and full-length amplicon. Antimicrob. Agents Chemother. 54, 1378–1380.

Bercot, B., Poirel, L., Nordmann, P., 2011. Updated multiplex polymerase chain reaction for detection of 16S rRNA methylases: high prevalence among NDM-1 producers. Diagn. Microbiol. Infect. Dis. 71, 442–445.

Birkett, C.I., Ludlam, H.A., Woodford, N., Brown, D.F., Brown, N.M., Roberts, M.T., et al., 2007. Real-time TaqMan PCR for rapid detection and typing of genes encoding CTX-M extended-spectrum β-lactamases. J. Med. Microbiol. 56, 52–55.

Bisiklis, A., Papageorgiou, F., Frantzidou, F., Alexiou-Daniel, S., 2007. Specific detection of bla_{VIM} and bla_{IMP} metallo-β-lactamase genes in a single real-time PCR. Clin. Microbiol. Infect. 13, 1201–1203.

Bogaerts, P., Hujer, A.M., Naas, T., de Castro, R.R., Endimiani, A., Nordmann, P., et al., 2011. Multicenter evaluation of a new DNA microarray for rapid detection of clinically relevant *bla* genes from β-lactam-resistant gram-negative bacteria. Antimicrob. Agents Chemother. 55, 4457–4460.

Boyd, D.A., Tyler, S., Christianson, S., McGeer, A., Muller, M.P., Willey, B.M., et al., 2004. Complete nucleotide sequence of a 92-kilobase plasmid harboring the CTX-M-15 extended-spectrum β-lactamase involved in an outbreak in long-term-care facilities in Toronto, Canada. Antimicrob. Agents Chemother. 48, 3758–3764.

Bradford, P.A., Bratu, S., Urban, C., Visalli, M., Mariano, N., Landman, D., et al., 2004. Emergence of carbapenem-resistant *Klebsiella* species possessing the class A carbapenem-hydrolyzing KPC-2 and inhibitor-resistant TEM-30 β-lactamases in New York City. Clin. Infect. Dis. 39, 55–60.

Braun, S.D., Ziegler, A., Methner, U., Slickers, P., Keiling, S., Monecke, S., et al., 2012. Fast DNA serotyping and antimicrobial resistance gene determination of *Salmonella enterica* with an oligonucleotide microarray-based assay. PLoS ONE 7, e46489.

Brolund, A., Wisell, K.T., Edquist, P.J., Elfstrom, L., Walder, M., Giske, C.G., 2010. Development of a real-time SYBR Green PCR assay for rapid detection of acquired AmpC in *Enterobacteriaceae*. J. Microbiol. Methods 82, 229–233.

Bruant, G., Maynard, C., Bekal, S., Gaucher, I., Masson, L., Brousseau, R., et al., 2006. Development and validation of an oligonucleotide microarray for detection of multiple virulence and antimicrobial resistance genes in *Escherichia coli*. Appl. Environ. Microbiol. 72, 3780–3784.

Bueno, M.F., Francisco, G.R., O'Hara, J.A., de Oliveira Garcia, D., Doi, Y., 2013. Coproduction of 16S rRNA methyltransferase RmtD or RmtG with KPC-2 and CTX-M group extended-spectrum β-lactamases in *Klebsiella pneumoniae*. Antimicrob. Agents Chemother. 57, 2397–2400.

Burckhardt, I., Zimmermann, S., 2011. Using matrix-assisted laser desorption ionization-time of flight mass spectrometry to detect carbapenem resistance within 1 to 2.5 hours. J. Clin. Microbiol. 49, 3321–3324.

Bush, K., Jacoby, G.A., 2009. Updated functional classification of β-lactamases. Antimicrob. Agents Chemother. 54, 969–976.

Bustin, S.A., Benes, V., Garson, J.A., Hellemans, J., Huggett, J., Kubista, M., et al., 2009. The MIQE guidelines: minimum information for publication of quantitative real-time PCR experiments. Clin. Chem. 55, 611–622.

Cai, J.C., Hu, Y.Y., Zhang, R., Zhou, H.W., Chen, G.X., 2012. Detection of OmpK36 porin loss in *Klebsiella* spp. by matrix-assisted laser desorption ionization-time of flight mass spectrometry. J. Clin. Microbiol. 50, 2179–2182.

Camara, J.E., Hays, F.A., 2007. Discrimination between wild-type and ampicillin-resistant *Escherichia coli* by matrix-assisted laser desorption/ionization time-of-flight mass spectrometry. Anal. Bioanal. Chem. 389, 1633–1638.

Cattoir, V., Poirel, L., Rotimi, V., Soussy, C.J., Nordmann, P., 2007. Multiplex PCR for detection of plasmid-mediated quinolone resistance qnr genes in ESBL-producing enterobacterial isolates. J. Antimicrob. Chemother. 60, 394–397.

Chen, L., Chavda, K.D., Mediavilla, J.R., Zhao, Y., Fraimow, H.S., Jenkins, S.G., et al., 2012. Multiplex real-time PCR for detection of an epidemic KPC-producing *Klebsiella pneumoniae* ST258 clone. Antimicrob. Agents Chemother. 56, 3444–3447.

Cohen Stuart, J., Dierikx, C., Al Naiemi, N., Karczmarek, A., Van Hoek, A.H., Vos, P., et al., 2010. Rapid detection of TEM, SHV and CTX-M extended-spectrum β-lactamases in *Enterobacteriaceae* using ligation-mediated amplification with microarray analysis. J. Antimicrob. Chemother. 65, 1377–1381.

Coque, T.M., Baquero, F., Canton, R., 2008. Increasing prevalence of ESBL-producing *Enterobacteriaceae* in Europe. Euro. Surveill. 13.

Cuzon, G., Naas, T., Bogaerts, P., Glupczynski, Y., Nordmann, P., 2012. Evaluation of a DNA microarray for the rapid detection of extended-spectrum β-lactamases (TEM, SHV and CTX-M), plasmid-mediated cephalosporinases (CMY-2-like, DHA, FOX, ACC-1, ACT/MIR and CMY-1-like/MOX) and carbapenemases (KPC, OXA-48, VIM, IMP and NDM). J. Antimicrob. Chemother. 67, 1865–1869.

Cuzon, G., Naas, T., Bogaerts, P., Glupczynski, Y., Nordmann, P., 2013. Probe ligation and real-time detection of KPC, OXA-48, VIM, IMP, and NDM carbapenemase genes. Diagn. Microbiol. Infect. Dis. 76, 502–505.

Dallenne, C., Da Costa, A., Decre, D., Favier, C., Arlet, G., 2010. Development of a set of multiplex PCR assays for the detection of genes encoding important β-lactamases in *Enterobacteriaceae*. J. Antimicrob. Chemother. 65, 490–495.

De Champs, C., Chanal, C., Sirot, D., Baraduc, R., Romaszko, J.P., Bonnet, R., et al., 2004. Frequency and diversity of class A extended-spectrum β-lactamases in hospitals of the Auvergne, France: a 2 year prospective study. J. Antimicrob. Chemother. 54, 634–639.

Doi, Y., Arakawa, Y., 2007. 16S ribosomal RNA methylation: emerging resistance mechanism against aminoglycosides. Clin. Infect. Dis. 45, 88–94.

Doyle, D., Peirano, G., Lascols, C., Lloyd, T., Church, D.L., Pitout, J.D., 2012. Laboratory detection of *Enterobacteriaceae* that produce carbapenemases. J. Clin. Microbiol. 50, 3877–3880.

Ecker, D.J., Sampath, R., Massire, C., Blyn, L.B., Hall, T.A., Eshoo, M.W., et al., 2008. Ibis T5000: a universal biosensor approach for microbiology. Nat. Rev. Microbiol. 6, 553–558.

Ecker, J.A., Massire, C., Hall, T.A., Ranken, R., Pennella, T.T., Agasino Ivy, C., et al., 2006. Identification of *Acinetobacter* species and genotyping of *Acinetobacter baumannii* by multilocus PCR and mass spectrometry. J. Clin. Microbiol. 44, 2921–2932.

Emonet, S., Shah, H.N., Cherkaoui, A., Schrenzel, J., 2010. Application and use of various mass spectrometry methods in clinical microbiology. Clin. Microbiol. Infect. 16, 1604–1613.

Endimiani, A., Paterson, D.L., 2007. Optimizing therapy for infections caused by *Enterobacteriaceae* producing extended-spectrum β-lactamases. Semin. Respir. Crit. Care Med. 28, 646–655.

Endimiani, A., Hujer, A.M., Hujer, K.M., Gatta, J.A., Schriver, A.C., Jacobs, M.R., et al., 2010a. Evaluation of a commercial microarray system for detection of SHV-, TEM-, CTX-M-, and KPC-type β-lactamase genes in Gram-negative isolates. J. Clin. Microbiol. 48, 2618–2622.

Endimiani, A., Hujer, K.M., Hujer, A.M., Sampath, R., Ecker, D.J., Bonomo, R.A., 2010b. Rapid identification of *bla*$_{KPC}$-possessing *Enterobacteriaceae* by PCR/electrospray ionization-mass spectrometry. J. Antimicrob. Chemother. 65, 1833–1834.

Erali, M., Voelkerding, K.V., Wittwer, C.T., 2008. High resolution melting applications for clinical laboratory medicine. Exp. Mol. Pathol. 85, 50–58.

Fishbain, J.T., Sinyavskiy, O., Riederer, K., Hujer, A.M., Bonomo, R.A., 2012. Detection of extended-spectrum β-lactamase and *Klebsiella pneumoniae* Carbapenemase genes directly from blood cultures by use of a nucleic acid microarray. J. Clin. Microbiol. 50, 2901–2904.

Geyer, C.N., Reisbig, M.D., Hanson, N.D., 2012. Development of a TaqMan multiplex PCR assay for detection of plasmid-mediated ampC β-lactamase genes. J. Clin. Microbiol. 50, 3722–3725.

Giamarellou, H., Poulakou, G., 2009. Multidrug-resistant Gram-negative infections: what are the treatment options? Drugs 69, 1879–1901.

Girlich, D., Poirel, L., Szczepanowski, R., Schluter, A., Nordmann, P., 2012. Carbapenem-hydrolyzing GES-5-encoding gene on different plasmid types recovered from a bacterial community in a sewage treatment plant. Appl. Environ. Microbiol. 78, 1292–1295.

Glenn, L.M., Lindsey, R.L., Folster, J.P., Pecic, G., Boerlin, P., Gilmour, M.W., et al., 2013. Antimicrobial resistance genes in multidrug-resistant *Salmonella enterica* isolated from animals, retail meats, and humans in the United States and Canada. Microb. Drug Resist. 19, 175–184.

Gniadkowski, M., 2008. Evolution of extended-spectrum β-lactamases by mutation. Clin. Microbiol. Infect. 14 (Suppl. 1), 11–32.

Grayson, M.L., 2010. Cephalosporins and related drugs Kocers' the Use of Antibiotics, sixth ed. ASM Press, London.

Guillard, T., Moret, H., Brasme, L., Carlier, A., Vernet-Garnier, V., Cambau, E., et al., 2011. Rapid detection of qnr and *qepA* plasmid-mediated quinolone resistance genes using real-time PCR. Diagn. Microbiol. Infect. Dis. 70, 253–259.

Guillard, T., de Champs, C., Moret, H., Bertrand, X., Scheftel, J.M., Cambau, E., 2012. High-resolution melting analysis for rapid characterization of qnr alleles in clinical isolates and detection of two novel alleles, *qnrB25* and *qnrB42*. J. Antimicrob. Chemother. 67, 2635–2639.

Hammond, D.S., Schooneveldt, J.M., Nimmo, G.R., Huygens, F., Giffard, P.M., 2005. bla(SHV) Genes in *Klebsiella pneumoniae*: different allele distributions are associated with different promoters within individual isolates. Antimicrob. Agents Chemother. 49, 256–263.

252 Antimicrobial Resistance and Food Safety

Harris, P.N., Ferguson, J.K., 2012. Antibiotic therapy for inducible AmpC β-lactamase-producing Gram-negative bacilli: what are the alternatives to carbapenems, quinolones and aminoglycosides? Int. J. Antimicrob. Agents 40, 297–305.

Hauser, E., Tietze, E., Helmuth, R., Junker, E., Blank, K., Prager, R., et al., 2010. Pork contaminated with *Salmonella enterica* serovar 4,[5],12:i:-, an emerging health risk for humans. Appl. Environ. Microbiol. 76, 4601–4610.

Hauser, E., Hebner, F., Tietze, E., Helmuth, R., Junker, E., Prager, R., et al., 2011. Diversity of *Salmonella enterica* serovar Derby isolated from pig, pork and humans in Germany. Int. J. Food Microbiol. 151, 141–149.

Hawser, S.P., Bouchillon, S.K., Hoban, D.J., Badal, R.E., Canton, R., Baquero, F., 2010. Incidence and antimicrobial susceptibility of *Escherichia coli* and *Klebsiella pneumoniae* with extended-spectrum β-lactamases in community- and hospital-associated intra-abdominal infections in Europe: results of the 2008 Study for Monitoring Antimicrobial Resistance Trends (SMART). Antimicrob. Agents Chemother. 54, 3043–3046.

Heid, C.A., Stevens, J., Livak, K.J., Williams, P.M., 1996. Real time quantitative PCR. Genome Res. 6, 986–994.

Hidalgo, L., Hopkins, K.L., Gutierrez, B., Ovejero, C.M., Shukla, S., Douthwaite, S., et al., 2013. Association of the novel aminoglycoside resistance determinant RmtF with NDM carbapenemase in *Enterobacteriaceae* isolated in India and the UK. J. Antimicrob. Chemother. 68 (7), 1543–1550.

Higuchi, R., Dollinger, G., Walsh, P.S., Griffith, R., 1992. Simultaneous amplification and detection of specific DNA sequences. Biotechnology 10, 413–417.

Higuchi, R., Fockler, C., Dollinger, G., Watson, R., 1993. Kinetic PCR analysis: real-time monitoring of DNA amplification reactions. Biotechnology 11, 1026–1030.

Hilty, M., Sendi, P., Seiffert, S.N., Droz, S., Perreten, V., Hujer, A.M., et al., 2013. Characterisation and clinical features of *Enterobacter cloacae* bloodstream infections occurring at a tertiary care university hospital in Switzerland: is cefepime adequate therapy? Int. J. Antimicrob. Agents 41, 236–249.

Hindiyeh, M., Smollen, G., Grossman, Z., Ram, D., Davidson, Y., Mileguir, F., et al., 2008. Rapid detection of *bla*$_{KPC}$ carbapenemase genes by real-time PCR. J. Clin. Microbiol. 46, 2879–2883.

Hindiyeh, M., Smollan, G., Grossman, Z., Ram, D., Robinov, J., Belausov, N., et al., 2011. Rapid detection of *bla*$_{KPC}$ carbapenemase genes by internally controlled real-time PCR assay using BACTEC blood culture bottles. J. Clin. Microbiol. 49, 2480–2484.

Ho, Y.P., Reddy, P.M., 2011. Advances in mass spectrometry for the identification of pathogens. Mass Spectrom. Rev. 30, 1203–1224.

Hong, S.S., Kim, K., Huh, J.Y., Jung, B., Kang, M.S., Hong, S.G., 2012. Multiplex PCR for rapid detection of genes encoding class A carbapenemases. Ann. Lab. Med. 32, 359–361.

Hooff, G.P., van Kampen, J.J., Meesters, R.J., van Belkum, A., Goessens, W.H., Luider, T.M., 2012. Characterization of β-lactamase enzyme activity in bacterial lysates using MALDI-mass spectrometry. J. Proteome Res. 11, 79–84.

Hooper, D.C., 2001. Mechanisms of action of antimicrobials: focus on fluoroquinolones. Clin. Infect. Dis. 32 (Suppl. 1), S9–S15.

Hoyos-Mallecot, Y., Cabrera-Alvargonzalez, J.J., Miranda-Casas, C., Rojo-Martin, M.D., Liebana-Martos, C., Navarro-Mari, J.M., 2014. MALDI-TOF MS, a useful instrument for differentiating metallo-beta-lactamases in *Enterobacteriaceae* and *Pseudomonas* spp. Lett. Appl. Microbiol. 58, 325–329.

Hrabak, J., Walkova, R., Studentova, V., Chudackova, E., Bergerova, T., 2011. Carbapenemase activity detection by matrix-assisted laser desorption ionization-time of flight mass spectrometry. J. Clin. Microbiol. 49, 3222–3227.

Hrabak, J., Studentova, V., Walkova, R., Zemlickova, H., Jakubu, V., Chudackova, E., et al., 2012. Detection of NDM-1, VIM-1, KPC, OXA-48, and OXA-162 carbapenemases by matrix-assisted laser desorption ionization-time of flight mass spectrometry. J. Clin. Microbiol. 50, 2441–2443.

Hrabak, J., Chudackova, E., Walkova, R., 2013. Matrix-assisted laser desorption ionization-time of flight (MALDI-TOF) mass spectrometry for detection of antibiotic resistance mechanisms: from research to routine diagnosis. Clin. Microbiol. Rev. 26, 103–114.

Huehn, S., Bunge, C., Junker, E., Helmuth, R., Malorny, B., 2009. Poultry-associated *Salmonella enterica* subsp. enterica serovar 4,12:d:- reveals high clonality and a distinct pathogenicity gene repertoire. Appl. Environ. Microbiol. 75, 1011–1020.

Ikryannikova, L.N., Shitikov, E.A., Zhivankova, D.G., Il'ina, E.N., Edelstein, M.V., Govorun, V.M., 2008. A MALDI TOF MS-based minisequencing method for rapid detection of TEM-type extended-spectrum β-lactamases in clinical strains of *Enterobacteriaceae*. J. Microbiol. Methods 75, 385–391.

Jacoby, G.A., 2005. Mechanisms of resistance to quinolones. Clin. Infect. Dis. 41 (Suppl. 2), S120–S126.

Jacoby, G.A., 2009. AmpC β-lactamases. Clin. Microbiol. Rev. 22, 161–182.

Jung, J.S., Popp, C., Sparbier, K., Lange, C., Kostrzewa, M., Schubert, S., 2014. Evaluation of matrix-assisted laser desorption ionization-time of flight mass spectrometry for rapid detection of beta-lactam resistance in Enterobacteriaceae derived from blood cultures. J. Clin. Microbiol. 52, 924–930.

Kaase, M., Szabados, F., Wassill, L., Gatermann, S.G., 2012. Detection of carbapenemases in *Enterobacteriaceae* by a commercial multiplex PCR. J. Clin. Microbiol. 50, 3115–3118.

Kruttgen, A., Razavi, S., Imohl, M., Ritter, K., 2011. Real-time PCR assay and a synthetic positive control for the rapid and sensitive detection of the emerging resistance gene New Delhi metallo-β-lactamase-1 (bla_{NDM-1}). Med. Microbiol. Immunol. 200, 137–141.

Kurpiel, P.M., Hanson, N.D., 2011. Association of IS5 with divergent tandem bla_{CMY-2} genes in clinical isolates of *Escherichia coli*. J. Antimicrob. Chemother. 66, 1734–1738.

Lascols, C., Hackel, M., Hujer, A.M., Marshall, S.H., Bouchillon, S.K., Hoban, D.J., et al., 2012. Using nucleic acid microarrays to perform molecular epidemiology and detect novel β-lactamases: a snapshot of extended-spectrum β-lactamases throughout the world. J. Clin. Microbiol. 50, 1632–1639.

Lascols, C., Peirano, G., Hackel, M., Laupland, K.B., Pitout, J.D., 2013. Surveillance and molecular epidemiology of *Klebsiella pneumoniae* isolates that produce Carbapenemases: first report of OXA-48-like enzymes in North America. Antimicrob. Agents Chemother. 57, 130–136.

Lavigne, J.P., Espinal, P., Dunyach-Remy, C., Messad, N., Pantel, A., Sotto, A., 2012. Mass spectrometry: a revolution in clinical microbiology? Clin. Chem. Lab. Med. CCLM/FESCC, 1–14.

Liu, L., Li, Y., Li, S., Hu, N., He, Y., Pong, R., et al., 2012. Comparison of next-generation sequencing systems. J. Biomed. Biotechnol. 2012, 251364.

Magnet, S., Blanchard, J.S., 2005. Molecular insights into aminoglycoside action and resistance. Chem. Rev. 105, 477–498.

Manchanda, V., Rai, S., Gupta, S., Rautela, R.S., Chopra, R., Rawat, D.S., et al., 2011. Development of TaqMan real-time polymerase chain reaction for the detection of the newly emerging form of carbapenem resistance gene in clinical isolates of *Escherichia coli, Klebsiella pneumoniae*, and *Acinetobacter baumannii*. Indian J. Med. Microbiol. 29, 249–253.

Mancini, N., Infurnari, L., Ghidoli, N., Valzano, G., Clementi, N., Burioni, R., et al., 2014. Potential impact of a microarray-based nucleic acid assay for rapid detection of Gram-negative bacteria and resistance markers in positive blood cultures. J. Clin. Microbiol. 52, 1242–1245.

Markoulatos, P., Siafakas, N., Moncany, M., 2002. Multiplex polymerase chain reaction: a practical approach. J. Clin. Lab. Anal. 16, 47–51.

Mathers, A.J., Hazen, K.C., Carroll, J., Yeh, A.J., Cox, H.L., Bonomo, R.A., et al., 2013. First clinical cases of OXA-48-producing carbapenem-resistant *Klebsiella pneumoniae* in the United States: the "menace" arrives in the new world. J. Clin. Microbiol. 51, 680–683.

Maxam, A.M., Gilbert, W., 1977. A new method for sequencing DNA. Proc. Natl. Acad. Sci. USA. 74, 560–564.

McDonald, L.C., Chen, F.J., Lo, H.J., Yin, H.C., Lu, P.L., Huang, C.H., et al., 2001. Emergence of reduced susceptibility and resistance to fluoroquinolones in *Escherichia coli* in Taiwan and contributions of distinct selective pressures. Antimicrob. Agents Chemother. 45, 3084–3091.

Meyer, E., Schwab, F., Schroeren-Boersch, B., Gastmeier, P., 2010. Dramatic increase of third-generation cephalosporin-resistant *E. coli* in German intensive care units: secular trends in antibiotic drug use and bacterial resistance, 2001 to 2008. Crit. Care 14, R113.

Michalopoulos, A., Falagas, M.E., 2010. Treatment of *Acinetobacter* infections. Expert Opin. Pharmacother. 11, 779–788.

Miller, M.B., Tang, Y.W., 2009. Basic concepts of microarrays and potential applications in clinical microbiology. Clin. Microbiol. Rev. 22, 611–633.

Miro, E., Grunbaum, F., Gomez, L., Rivera, A., Mirelis, B., Coll, P., et al., 2012. Characterization of aminoglycoside-modifying enzymes in *Enterobacteriaceae* clinical strains and characterization of the plasmids implicated in their diffusion. Microb. Drug Resist.

Monteiro, J., Widen, R.H., Pignatari, A.C., Kubasek, C., Silbert, S., 2012. Rapid detection of carbapenemase genes by multiplex real-time PCR. J. Antimicrob. Chemother. 67, 906–909.

Mroczkowska, J.E., Barlow, M., 2008. Fitness trade-offs in *bla*$_{TEM}$ evolution. Antimicrob. Agents Chemother. 52, 2340–2345.

Mullis, K.B., Faloona, F.A., 1987. Specific synthesis of DNA *in vitro* via a polymerase-catalyzed chain reaction. Methods Enzymol. 155, 335–350.

Naas, T., Cuzon, G., Truong, H., Bernabeu, S., Nordmann, P., 2010. Evaluation of a DNA microarray, the Check-Points ESBL/KPC array, for rapid detection of TEM, SHV, and CTX-M extended-spectrum β-lactamases and KPC carbapenemases. Antimicrob. Agents Chemother. 54, 3086–3092.

Naas, T., Cuzon, G., Bogaerts, P., Glupczynski, Y., Nordmann, P., 2011a. Evaluation of a DNA microarray (Check-MDR CT102) for rapid detection of TEM, SHV, and CTX-M extended-spectrum β-lactamases and of KPC, OXA-48, VIM, IMP, and NDM-1 carbapenemases. J. Clin. Microbiol. 49, 1608–1613.

Naas, T., Ergani, A., Carrer, A., Nordmann, P., 2011b. Real-time PCR for detection of NDM-1 carbapenemase genes from spiked stool samples. Antimicrob. Agents Chemother. 55, 4038–4043.

Naas, T., Cotellon, G., Ergani, A., Nordmann, P., 2013. Real-time PCR for detection of *bla*$_{OXA-48}$ genes from stools. J. Antimicrob. Chemother. 68, 101–104.

Nijhuis, R., van Zwet, A., Stuart, J.C., Weijers, T., Savelkoul, P., 2012. Rapid molecular detection of extended-spectrum β-lactamase gene variants with a novel ligation-mediated real-time PCR. J. Med. Microbiol. 61, 1563–1567.

Nordmann, P., Poirel, L., Carrer, A., Toleman, M.A., Walsh, T.R., 2011. How to detect NDM-1 producers. J. Clin. Microbiol. 49, 718–721.

Nordmann, P., Dortet, L., Poirel, L., 2012a. Rapid detection of extended-spectrum-β-lactamase-producing *Enterobacteriaceae*. J. Clin. Microbiol. 50, 3016–3022.

Nordmann, P., Poirel, L., Dortet, L., 2012b. Rapid detection of carbapenemase-producing *Enterobacteriaceae*. Emerg. Infect. Dis. 18, 1503–1507.

Notomi, T., Okayama, H., Masubuchi, H., Yonekawa, T., Watanabe, K., Amino, N., et al., 2000. Loop-mediated isothermal amplification of DNA. Nucleic Acids Res. 28, E63.

Oxacelay, C., Ergani, A., Naas, T., Nordmann, P., 2009. Rapid detection of CTX-M-producing *Enterobacteriaceae* in urine samples. J. Antimicrob. Chemother. 64, 986–989.

Papp-Wallace, K.M., Endimiani, A., Taracila, M.A., Bonomo, R.A., 2011. Carbapenems: past, present, and future. Antimicrob. Agents Chemother. 55, 4943–4960.

Park, C.H., Robicsek, A., Jacoby, G.A., Sahm, D., Hooper, D.C., 2006. Prevalence in the United States of *aac(6')-Ib-cr* encoding a ciprofloxacin-modifying enzyme. Antimicrob. Agents Chemother. 50, 3953–3955.

Perez-Perez, F.J., Hanson, N.D., 2002. Detection of plasmid-mediated AmpC β-lactamase genes in clinical isolates by using multiplex PCR. J. Clin. Microbiol. 40, 2153–2162.

Perilli, M., Celenza, G., De Santis, F., Pellegrini, C., Forcella, C., Rossolini, G.M., et al., 2008. E240V substitution increases catalytic efficiency toward ceftazidime in a new natural TEM-type extended-spectrum β-lactamase, TEM-149, from *Enterobacter aerogenes* and *Serratia marcescens* clinical isolates. Antimicrob. Agents Chemother. 52, 915–919.

Peter, H., Berggrav, K., Thomas, P., Pfeifer, Y., Witte, W., Templeton, K., et al., 2012. Direct detection and genotyping of *Klebsiella pneumoniae* carbapenemases from urine by use of a new DNA microarray test. J. Clin. Microbiol. 50, 3990–3998.

Pitout, J.D., Hossain, A., Hanson, N.D., 2004. Phenotypic and molecular detection of CTX-M-β-lactamases produced by *Escherichia coli* and *Klebsiella* spp. J. Clin. Microbiol. 42, 5715–5721.

Pitout, J.D., Gregson, D.B., Poirel, L., McClure, J.A., Le, P., Church, D.L., 2005. Detection of *Pseudomonas aeruginosa* producing metallo-β-lactamases in a large centralized laboratory. J. Clin. Microbiol. 43, 3129–3135.

Platteel, T.N., Stuart, J.W., Voets, G.M., Scharringa, J., van de Sande, N., Fluit, A.C., et al., 2011. Evaluation of a commercial microarray as a confirmation test for the presence of extended-spectrum β-lactamases in isolates from the routine clinical setting. Clin. Microbiol. Infect. 17, 1435–1438.

Poirel, L., Le Thomas, I., Naas, T., Karim, A., Nordmann, P., 2000. Biochemical sequence analyses of GES-1, a novel class A extended-spectrum β-lactamase, and the class 1 integron In52 from *Klebsiella pneumoniae*. Antimicrob. Agents Chemother. 44, 622–632.

Poirel, L., Walsh, T.R., Cuvillier, V., Nordmann, P., 2011. Multiplex PCR for detection of acquired carbapenemase genes. Diagn. Microbiol. Infect. Dis. 70, 119–123.

Poirel, L., Potron, A., Nordmann, P., 2012. OXA-48-like carbapenemases: the phantom menace. J. Antimicrob. Chemother. 67, 1597–1606.

Predari, S.C., Ligozzi, M., Fontana, R., 1991. Genotypic identification of methicillin-resistant coagulase-negative staphylococci by polymerase chain reaction. Antimicrob. Agents Chemother. 35, 2568–2573.

Qiang, Y.Z., Qin, T., Fu, W., Cheng, W.P., Li, Y.S., Yi, G., 2002. Use of a rapid mismatch PCR method to detect *gyrA* and *parC* mutations in ciprofloxacin-resistant clinical isolates of *Escherichia coli*. J. Antimicrob. Chemother. 49, 549–552.

Raghunathan, A., Samuel, L., Tibbetts, R.J., 2011. Evaluation of a real-time PCR assay for the detection of the *Klebsiella pneumoniae* carbapenemase genes in microbiological samples in comparison with the modified Hodge test. Am. J. Clin. Pathol. 135, 566–571.

Randegger, C.C., Keller, A., Irla, M., Wada, A., Hachler, H., 2000. Contribution of natural amino acid substitutions in SHV extended-spectrum β-lactamases to resistance against various β-lactams. Antimicrob. Agents Chemother. 44, 2759–2763.

Robicsek, A., Strahilevitz, J., Sahm, D.F., Jacoby, G.A., Hooper, D.C., 2006. qnr prevalence in ceftazidime-resistant *Enterobacteriaceae* isolates from the United States. Antimicrob. Agents Chemother. 50, 2872–2874.

Ronaghi, M., Karamohamed, S., Pettersson, B., Uhlen, M., Nyren, P., 1996. Real-time DNA sequencing using detection of pyrophosphate release. Anal. Biochem. 242, 84–89.

Sanger, F., Nicklen, S., Coulson, A.R., 1977. DNA sequencing with chain-terminating inhibitors. Proc. Natl. Acad. Sci. U.S.A. 74, 5463–5467.

Schaumann, R., Knoop, N., Genzel, G.H., Losensky, K., Rosenkranz, C., Stingu, C.S., et al., 2012. A step towards the discrimination of β-lactamase-producing clinical isolates of Enterobacteriaceae and Pseudomonas aeruginosa by MALDI-TOF mass spectrometry. Med. Sci. Monit. 18, MT71–MT77.

Schrader, C., Schielke, A., Ellerbroek, L., Johne, R., 2012. PCR inhibitors—occurrence, properties and removal. J. Appl. Microbiol. 113, 1014–1026.

Seiffert, S.N., Hilty, M., Perreten, V., Endimiani, A., 2013a. Extended-spectrum cephalosporin-resistant gram-negative organisms in livestock: an emerging problem for human health? Drug Resist. 16 (1–2), 22–45.

Seiffert, S.N., Tinguely, R., Lupo, A., Neuwirth, C., Perreten, V., Endimiani, A., 2013b. High prevalence of extended-spectrum-cephalosporin-resistant Enterobacteriaceae in poultry meat in Switzerland: emergence of CMY-2- and VEB-6-possessing Proteus mirabilis. Antimicrob. Agents Chemother. 57, 6406–6408.

Sherry, N.L., Porter, J.L., Seemann, T., Watkins, A., Stinear, T.P., Howden, B.P., 2013. Outbreak investigation using high-throughput genome sequencing within a diagnostic microbiology laboratory. J. Clin. Microbiol. 51 (5), 1396–1401.

Sibley, C.D., Peirano, G., Church, D.L., 2012. Molecular methods for pathogen and microbial community detection and characterization: current and potential application in diagnostic microbiology. Infect. Genet. Evol. 12, 505–521.

Singh, K., Mangold, K.A., Wyant, K., Schora, D.M., Voss, B., Kaul, K.L., et al., 2012. Rectal screening for Klebsiella pneumoniae carbapenemases: comparison of real-time PCR and culture using two selective screening agar plates. J. Clin. Microbiol. 50, 2596–2600.

Spanu, T., Fiori, B., D'Inzeo, T., Canu, G., Campoli, S., Giani, T., et al., 2012. Evaluation of the new NucliSENS EasyQ KPC test for rapid detection of Klebsiella pneumoniae carbapenemase genes (bla$_{KPC}$). J. Clin. Microbiol. 50, 2783–2785.

Sparbier, K., Schubert, S., Weller, U., Boogen, C., Kostrzewa, M., 2012. Matrix-assisted laser desorption ionization-time of flight mass spectrometry-based functional assay for rapid detection of resistance against β-lactam antibiotics. J. Clin. Microbiol. 50, 927–937.

Strahilevitz, J., Jacoby, G.A., Hooper, D.C., Robicsek, A., 2009. Plasmid-mediated quinolone resistance: a multifaceted threat. Clin. Microbiol. Rev. 22, 664–689.

Sullivan, K.V., Deburger, B., Roundtree, S.S., Ventrola, C.A., Blecker-Shelly, D.L., Mortensen, J.E., 2014. Rapid detection of inpatient Gram-negative bacteremia; extended-spectrum beta-lactamases and carbapenemase resistance determinants with the verigene BC-GN Test: a multicenter evaluation. J. Clin. Microbiol. 51 (11), 3579–3584.

Swayne, R.L., Ludlam, H.A., Shet, V.G., Woodford, N., Curran, M.D., 2011. Real-time TaqMan PCR for rapid detection of genes encoding five types of non-metallo- (class A and D) carbapenemases in Enterobacteriaceae. Int. J. Antimicrob. Agents 38, 35–38.

Szmolka, A., Anjum, M.F., La Ragione, R.M., Kaszanyitzky, E.J., Nagy, B., 2012. Microarray based comparative genotyping of gentamicin resistant Escherichia coli strains from food animals and humans. Vet. Microbiol. 156, 110–118.

Tong, S.Y., Giffard, P.M., 2012. Microbiological applications of high-resolution melting analysis. J. Clin. Microbiol. 50, 3418–3421.

van Hoek, A.H., Scholtens, I.M., Cloeckaert, A., Aarts, H.J., 2005. Detection of antibiotic resistance genes in different Salmonella serovars by oligonucleotide microarray analysis. J. Microbiol. Methods 62, 13–23.

Veldman, K., Kant, A., Dierikx, C., van Essen-Zandbergen, A., Wit, B., Mevius, D., 2014. Enterobacteriaceae resistant to third-generation cephalosporins and quinolones in fresh culinary herbs imported from Southeast Asia. Int. J. Food Microbiol. 177, 72–77.

Voets, G.M., Fluit, A.C., Scharringa, J., Cohen Stuart, J., Leverstein-van Hall, M.A., 2011. A set of multiplex PCRs for genotypic detection of extended-spectrum β-lactamases, carbapenemases, plasmid-mediated AmpC β-lactamases and OXA β-lactamases. Int. J. Antimicrob. Agents 37, 356–359.

Vogt, D., Overesch, G., Endimiani, A., Collaud, A., Thomann, A., Perreten, V., 2014. Occurrence and genetic characteristics of third-generation cephalosporin-resistant *Escherichia coli* in Swiss retail meat. Microb. Drug Resist. 20 (5), 485–494.

Walsh, T.R., 2010. Emerging carbapenemases: a global perspective. Int. J. Antimicrob. Agents 36 (Suppl. 3), S8–S14.

Wang, L., Han, C., Sui, W., Wang, M., Lu, X., 2013. MALDI-TOF MS applied to indirect carbapenemase detection: a validated procedure to clearly distinguish between carbapenemase-positive and carbapenemase-negative bacterial strains. Anal. Bioanal. Chem. 405, 5259–5266.

Wang, M., Tran, J.H., Jacoby, G.A., Zhang, Y., Wang, F., Hooper, D.C., 2003. Plasmid-mediated quinolone resistance in clinical isolates of *Escherichia coli* from Shanghai, China. Antimicrob. Agents Chemother. 47, 2242–2248.

Weigel, L.M., Steward, C.D., Tenover, F.C., 1998. gyrA mutations associated with fluoroquinolone resistance in eight species of *Enterobacteriaceae*. Antimicrob. Agents Chemother. 42, 2661–2667.

Wieser, A., Schneider, L., Jung, J., Schubert, S., 2012. MALDI-TOF MS in microbiological diagnostics-identification of microorganisms and beyond (mini review). Appl. Microbiol. Biotechnol. 93, 965–974.

Williamson, D.A., Sidjabat, H.E., Freeman, J.T., Roberts, S.A., Silvey, A., Woodhouse, R., et al., 2012. Identification and molecular characterisation of New Delhi metallo-β-lactamase-1 (NDM-1)- and NDM-6-producing *Enterobacteriaceae* from New Zealand hospitals. Int. J. Antimicrob. Agents 39, 529–533.

Wintermans, B.B., Reuland, E.A., Wintermans, R.G., Bergmans, A.M., Kluytmans, J.A., 2012. The cost-effectiveness of ESBL detection: towards molecular detection methods? Clin. Microbiol. Infect. 19 (7), 662–665.

Wolk, D.M., Kaleta, E.J., Wysocki, V.H., 2012. PCR–electrospray ionization mass spectrometry: the potential to change infectious disease diagnostics in clinical and public health laboratories. J. Mol. Diagn. 14, 295–304.

Zhou, X.Y., Bordon, F., Sirot, D., Kitzis, M.D., Gutmann, L., 1994. Emergence of clinical isolates of *Escherichia coli* producing TEM-1 derivatives or an OXA-1 β-lactamase conferring resistance to β-lactamase inhibitors. Antimicrob. Agents Chemother. 38, 1085–1089.

Chapter 13
Monitoring and Surveillance: The National Antimicrobial Resistance Monitoring System

Emily Crarey, Claudine Kabera and Heather Tate

Food and Drug Administration, Center for Veterinary Medicine, Laurel, MD, USA

Chapter Outline

INTRODUCTION

The emergence, distribution, and persistence of antimicrobial-resistant foodborne pathogens is a global problem. In some cases, resistant organisms have been linked to worse health outcomes, including greater incidence of hospitalization and longer treatment times (Krueger et al., 2014; Nelson et al., 2004; Varma et al., 2005a,b). In the United States alone, drug-resistant *Campylobacter* and *Salmonella* infections are estimated to cause more than 410,000 illnesses and 50 deaths annually (CDC, 2013a). Expert scientific communities have expressed concerns regarding the national and global trends in antimicrobial resistance and the complex issues that are associated with its increase (American Society for Microbiology Public and Scientific Affairs Board, 1995; Institute of Medicine Committee on Emerging Microbial Threats to Health, 1992). The nature and magnitude of the public health hazard arising from the veterinary use of antimicrobial agents has been widely debated for decades. The problem was first systematically reviewed in the United Kingdom during the 1960s,

Antimicrobial Resistance and Food Safety. DOI: http://dx.doi.org/10.1016/B978-0-12-801214-7.00013-2

which led to a 1968 meeting of the Joint Committee on the Use of Antibiotics in Animal Husbandry and Veterinary Medicine and the subsequent publication of what later became referred to as the "Swann Report." Based on scientific findings that resistant *Salmonella* could pass their resistance genes on to other bacteria, the Swann report recommended that only antibiotics which have little or no application in human or veterinary medicine be used for growth promotion in animals (Swann et al., 1969).

A number of interventions have been implemented in various countries to reduce resistance in foodborne bacteria, mainly through restricted antimicrobial use policies. An essential element of these efforts is an integrated antimicrobial resistance monitoring system that can be used both to establish baseline resistance levels, and to measure the impact of various tactics to control or reduce resistance arising from settings in which antibiotics are used. Integrated surveillance programs involve sampling, testing, and reporting of antimicrobial resistance data on bacteria from humans, foods, and food animals and coordinated farm-to-fork analysis of resistance trends at a national level. In 1995, following a government ban on avoparcin based on new evidence that this growth promoter was a potential threat to human health, Denmark launched the Danish Integrated Antimicrobial Resistance Monitoring and Research Program (DANMAP) (Hammerum et al., 2007), becoming the first country to establish a national program of continuous monitoring of antimicrobial use and resistance. Since then, a number of national integrated surveillance programs have been launched. These programs are intended to provide data for risk analysis of foodborne antimicrobial resistance hazards. Towards this end, national monitoring programs:

- Provide baseline resistance levels to new antimicrobial compounds in bacteria from different reservoirs
- Identify temporal and spatial trends in resistance
- Describe how resistance genes spread among strains of bacteria
- Generate hypotheses about sources and reservoirs of resistant bacteria
- Increase understanding of the association between use practices and resistance
- Identify risk factors and clinical outcomes of infections caused by resistant bacteria
- Educate the public on current and emerging hazards
- Guide evidence-based policy decisions related to food safety, drug safety, and animal health
- Identify public health intervention strategies to contain resistance.

In 2013, the World Health Organization (WHO) Advisory Group on Integrated Surveillance of Antimicrobial Resistance (AGISAR) released guidelines for creating and maintaining an integrated surveillance program (WHO, 2013). AGISAR asserts that the main features to be considered when developing an integrated system are: the sources to be sampled, the microorganisms to collect and test, the design of the sampling scheme, the susceptibility testing methodology, and the analysis and reporting of data. The US National Antimicrobial Resistance Monitoring System-Enteric Bacteria (NARMS) addresses all of these considerations, and as such was cited by WHO as one of a few model antimicrobial resistance monitoring programs that could be used as a reference for countries working to establish their own national programs. NARMS data are used by public health agencies, drug industry, consumer safety groups, food animal health and production stakeholders, and academicians to assess the risks and public health impact of antimicrobial use in food animals. This chapter focuses on the history and operation of NARMS and its application for both regulatory and nonregulatory purposes.

NARMS OVERVIEW

On August 18, 1995, the US Food and Drug Administration (FDA) approved sarafloxacin, the first fluoroquinolone antimicrobial agent for use in animals intended for food, to control illness caused by *Escherichia coli* (*E. coli*) in poultry. In 1996, the FDA also approved the fluoroquinolone, enrofloxacin, for use in poultry (CDC, 1996). Because fluoroquinolones are commonly used to treat many infectious conditions in humans, the approval of these drugs for use in food animals raised serious public health concerns over the potential risk of transfer of resistant bacteria and resistance genes from animals to humans. Due to these public health concerns, a Veterinary Medical Advisory Committee was formed to review the information associated with these approvals. This group recommended that the FDA establish a monitoring program for antimicrobial resistance to help ensure the continued safety and effectiveness of fluoroquinolones (Tollefson et al., 1999) and other antimicrobial products. Thus, NARMS was originally intended to function as a post-approval safety monitoring system for FDA-regulated products. It began in 1996 as a partnership among the FDA's Center for Veterinary Medicine (CVM), the Centers for Disease Control and Prevention (CDC), and the US Department of Agriculture (USDA). It was launched as part of an overall strategy to assess the potential impact of antimicrobial use in animal agriculture on the evolution of antimicrobial resistance in human clinical isolates.

The primary objectives of NARMS are to:

- Monitor trends in antimicrobial resistance among enteric bacteria from humans, retail meats, and food animals
- Disseminate timely information on antimicrobial resistance in pathogenic and commensal organisms to stakeholders in the Unites States and abroad to promote interventions that reduce resistance among foodborne bacteria
- Conduct research to better understand the emergence, persistence, and spread of antimicrobial resistance and
- Provide data that assist the FDA in making decisions relative to the approval of safe and effective antimicrobial drugs for animals.

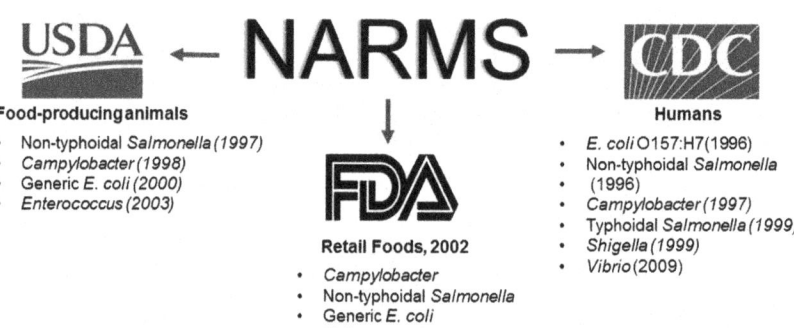

FIGURE 13.1   Participating NARMS Agencies, Study Populations, and Target Organisms.

NARMS is comprised of three components that collect data on enteric bacteria from retail meats, human clinical cases, and food animals (Figure 13.1). Pathogens included in NARMS surveillance were selected because they are among the most common bacterial causes of zoonotic foodborne illness. In the retail and animal components of NARMS, commensal bacteria, *E. coli* and *Enterococcus*, are used as Gram-negative and Gram-positive indicator organisms, respectively, to detect both emerging resistance patterns and specific resistance genes that could potentially be transferred to other pathogenic bacteria.

Strains from human clinical cases of foodborne illness are tested at the CDC laboratories in the National Center for Emerging and Zoonotic Infectious Diseases in Atlanta, Georgia. The CDC NARMS component conducts nationwide surveillance of typhoidal and nontyphoidal *Salmonella, Shigella, E. coli* O157:H7, and *Vibrio* species other than *V. cholerae* from humans. Testing for *Campylobacter* is conducted through the ten Foodborne Diseases Active Surveillance Network (FoodNet) sites, including California, Colorado, Connecticut, Georgia, Maryland, Minnesota, New Mexico, New York, Oregon, and Tennessee. Typhoidal *Salmonella* and *Shigella* are predominately human pathogens and are not used to assess the risk of veterinary drug use to human health. Additionally, *Vibrio* testing is not currently conducted in samples from foods or food animals. Therefore, those three organisms will not be addressed in detail in this chapter. More information on these organisms can be found in CDC NARMS reports (Centers for Disease Control and Prevention, 2013c).

Retail meat surveillance is conducted in collaboration with the ten FoodNet sites plus four State departments of public health in Louisiana, Missouri, Pennsylvania, and Washington. Participating state laboratories purchase chicken, ground turkey, ground beef, and pork chops from a stratified random selection of retail stores that are located at a convenient distance from the state laboratories. These particular types of retail meat are sampled because they are commonly consumed and are derived from major food animal species. State laboratory personnel collect ten each of these retail meat samples on a monthly basis totaling 40 samples per month. While all states culture each retail meat commodity for *Salmonella*, only retail poultry is cultured for *Campylobacter*. In addition, Georgia, Maryland, Oregon, and Tennessee culture retail meat for *E. coli* and *Enterococcus*. Isolates are sent to CVM's Office of Research in Laurel, Maryland for species and serotype confirmation, antimicrobial susceptibility testing and genetic analysis.

In addition to human and food isolates, NARMS conducts nationwide surveillance of *Salmonella, Campylobacter, Enterococcus*, and *E. coli* from animals at slaughter. The sampling protocol has evolved over the years. Currently, public health veterinarians within federally inspected slaughter establishments collect samples of cecal contents from large intestines of beef cow, dairy cow, heifer, steer, young chicken, market swine, sow, and young turkey. Sampling is based on a weighted proportional random sampling where larger plants with more production volume are sampled more frequently and smaller plants less frequently. Samples are shipped to the Eastern Laboratory of USDA's Food Safety Inspection Service (FSIS), located in Athens, Georgia, for culture and susceptibility testing.

DRUG SELECTION AND BREAKPOINTS USED

Because NARMS is designed to help measure the movement of resistance from the farm, and to provide data for assessing the public health consequences of this phenomenon, the antimicrobials that NARMS includes in testing are selected based on (i) their importance for treating human foodborne infections, (ii) their use in both human and veterinary medicine, and (iii) their value as epidemiological markers or as indicators of emerging resistance phenotypes. Currently, NARMS monitors *Salmonella* and *E. coli* for susceptibility to 14 antimicrobial compounds (Table 13.1), *Campylobacter* is tested against nine (Table 13.2), and *Enterococcus* is monitored for susceptibility to 16 compounds (Table 13.3). The array of compounds tested is evaluated periodically and amended if new resistances arise or when new classes of compounds are introduced into medical practice.

The interpretation of susceptibility testing results for *Salmonella, E. coli*, and *Enterococcus* is done in accordance with standards published by the Clinical and Laboratory Standards Institute (CLSI) (Clinical and Laboratory Standards Institute, 2008, 2010, 2014) (Tables 13.1 and 13.3). The clinical breakpoints take into account clinical efficacy as well as the pharmacokinetic and pharmacodynamic characteristics of each drug.

For some bacteria–antimicrobial combinations, there are no CLSI breakpoints available (Tables 13.1–13.3). In these cases, NARMS uses FDA-established criteria based on antimicrobial minimum inhibitory concentration (MIC) distribution data. NARMS maintains and reports MIC values as well as

TABLE 13.1 Interpretive Criteria Used for Antimicrobial Susceptibility Testing of *Salmonella* and *E. coli* in 2014

Antimicrobial Class	Antimicrobial Agent	Concentration Range (µg/mL)	Breakpoints (µg/mL)		
			Susceptible	Intermediate[a]	Resistant
Aminoglycosides	Gentamicin	0.25–16	≤4	8	≥16
	Streptomycin[b]	32–64	≤32	N/A	≥64
β-Lactam/β-Lactamase Inhibitor Combinations	Amoxicillin–Clavulanic acid	1/0.5–32/16	≤8/4	16/8	≥32/64
Cephems	Cefoxitin	0.5–32	≤8	16	≥32
	Ceftiofur	0.12–8	≤2	4	≥8
	Ceftriaxone	0.25–64	≤1	2	≥4
Folate Pathway Inhibitors	Sulfisoxazole	16–256	≤256	N/A	≥512
	Trimethoprim–sulfamethoxazole	0.12/2.4–4/76	≤2/38	N/A	≥4/76
Macrolides	Azithromycin[b]	0.12–16	≤16	N/A	≥32
Penicillins	Ampicillin	1–32	≤8	16	≥32
Phenicols	Chloramphenicol	2–32	≤8	16	≥32
Quinolones	Ciprofloxacin	0.015–4	≤0.06	0.12–0.5	≥1
	Nalidixic acid	0.5–32	≤16	N/A	≥32
Tetracyclines	Tetracyclines	4–32	≤4	8	≥16

[a]N/A indicates that no MIC range of intermediate susceptibility exists.
[b]No CLSI interpretative criteria for this bacterium/antimicrobial combination currently available.

TABLE 13.2 Interpretive Criteria Used for Antimicrobial Susceptibility Testing of *Campylobacter* in 2014

Antimicrobial Class	Antimicrobial Agent	Concentration Range (µg/mL)	Breakpoints (µg/mL)			
			jejuni		*coli*	
			Susceptible	Resistant	Susceptible	Resistant
Aminoglycosides	Gentamicin	0.12–32	≤2	≥4	≤2	≥4
Ketolides	Telithromycin	0.015–8	≤4	≥8	≤4	≥8
Lincosamides	Clindamycin	0.03–16	≤0.5	≥1	≤1	≥2
Macrolides	Azithromycin	0.015–64	≤0.25	≥0.5	≤0.5	≥1
	Erythromycin	0.03–64	≤4	≥8	≤8	≥16
Phenicols	Florfenicol	0.03–64	≤4	≥8	≤4	≥8
Quinolones	Ciprofloxacin	0.015–64	≤0.5	≥1	≤0.5	≥1
	Nalidixic acid	4–64	≤16	≥32	≤16	≥32
Tetracyclines	Tetracyclines	0.06–64	≤1	≥2	≤2	≥4

TABLE 13.3 Interpretive Criteria used for Antimicrobial Susceptibility Testing of *Enterococcus* in 2014

Antimicrobial Class	Antimicrobial Agent	Concentration Range (µg/mL)	Breakpoints (µg/mL)		
			Susceptible	Intermediate[a]	Resistant
Aminoglycosides	Gentamicin	128–1024	≤500	N/A	>500
	Kanamycin[b]	128–1024	≤512	N/A	≥1024
	Streptomycin	512–2048	≤1000	N/A	>1000
Glycopeptides	Vancomycin	0.25–32	≤4	8–16	≥32
Glycylcyclines	Tigecycline[b,c]	0.015–0.5	≤0.25	N/A	N/A
Lincosamides	Lincomycin[b]	1–8	≤2	4	≥8
Lipopeptides	Daptomycin[c]	0.25–16	≤4	N/A	N/A
Macrolides	Erythromycin	0.25–8	≤0.5	1–4	≥8
	Tylosin[b]	0.25–32	≤8	16	≥32
Nitrofurans	Nitrofurantoin	2–64	≤32	64	≥128
Oxazolidinones	Linezolid	0.5–8	≤2	4	≥8
Penicillins	Penicillin	0.25–16	≤8	N/A	≥16
Phenicols	Chloramphenicol	2–32	≤8	16	≥32
Quinolones	Ciprofloxacin	0.12–4	≤1	2	≥4
Streptogramins	Quinupristin/dalfopristin	0.5–32	≤1	2	≥4
Tetracyclines	Tetracycline	1–32	≤4	8	≥16

[a]N/A indicates that no MIC range of intermediate susceptibility exists.
[b]No CLSI interpretive criteria for this bacterium/antimicrobial combination currently available.
[c]Only a susceptible breakpoint has been established.

the susceptible, intermediate, and resistant interpretations. The advantage of maintaining numerical data is that when CLSI breakpoints are changed, they can be applied to historical data and reports can be easily updated with new interpretations.

Some surveillance systems use epidemiological cut-off values (ECOFFs) to interpret laboratory susceptibility testing results. These interpretive criteria are provided by the European Committee on Antimicrobial Susceptibility Testing (EUCAST) and are used across the European Union. (The CLSI guidelines for determining ECOFFs, and for reporting surveillance data based on ECOFFs, are in development.) The ECOFF is defined as the highest MIC value of the susceptible population. In NARMS, ECOFFS are used for reporting *Campylobacter* data, where formally established clinical breakpoints are not available for any antimicrobial agent. ECOFFs distinguish bacteria without resistance mechanisms ["wild type"; (WT)] from those with any acquired resistance mechanism ["non-wild type"; (NWT)], even if the acquired trait does not confer MICs above the clinical breakpoint. ECOFFs are based on the testing of large numbers of strains from different institutions to determine the MIC range of WT populations. ECOFFs are aimed at optimizing the detection of isolates with acquired resistance; they do not take into consideration any data on dosages or clinical efficacy. Because ECOFFs are adjacent to the wild-type population and tend to be lower than clinical breakpoints, they are more valuable than clinical breakpoints for detecting emerging resistance. Furthermore, because ECOFFs are based on biological mechanisms only, they are less likely than clinical breakpoints to change over time. Because many other national surveillance programs apply ECOFFs to their own data, NARMS' use of EUCAST ECOFFs is a step toward globally harmonized reporting methods for *Campylobacter* surveillance.

CURRENT TESTING METHODS

NARMS uses the same susceptibility testing methods across all three arms. A range of different acceptable methods for species identification and serotyping may be used by state public health laboratories responsible for isolating bacteria from human clinical samples and retail meats. For human samples, state laboratories use methods accepted by laboratory accreditation organizations, in which different media and incubation conditions may be permitted. State public health laboratories follow modified procedures in the FDA Bacteriological Analytical Manual (FDA, 2014) to culture bacteria from food matrices; however, protocols may be modified further if needed. All cecal samples collected at slaughter are tested at FSIS laboratories using USDA-established methods. Because of differences in laboratory methods, and to maintain proficiency in laboratory procedures, NARMS conducts cross-agency quality assurance programs every 2 years to compare methods across sites and assure different methods provide comparable and acceptable results.

Antimicrobial MICs are determined by automated broth microdilution according to CLSI standards (Clinical and Laboratory Standards Institute, 2004, 2008, 2014) using the Trek Diagnostic's Sensititre® platform, which employs a 96-well plate with dehydrated serial twofold dilutions of the test compounds. Because the CLSI testing method has been standardized and validated, reliability is ensured. Custom testing panels are developed for all organisms tested, using quality control (QC) organism dilution ranges for each drug. Custom panels are also created to test Gram-negative isolates for the presence of extended-spectrum beta lactamases (ESBLs).

NARMS isolates are also subjected to DNA fingerprinting analysis using pulsed-field gel electrophoresis (PFGE) to assess genetic relatedness, help with source attribution, and track the dispersion of subtypes. PFGE patterns are submitted to CDC's National Molecular Subtyping Network for Foodborne Disease Surveillance (PulseNet). PFGE data from NARMS isolates are used to help identify possible food sources in outbreak investigations (Hoffmann et al., 2012).

NARMS performs additional testing on select strains. *Salmonella* and *E. coli* isolates displaying resistance to either ceftriaxone or ceftiofur (both third-generation cephalosporins) are subsequently screened for the presence of ESBLs and carbapenemases using a supplemental testing panel of beta-lactam drugs (Table 13.4). ESBLs, which confer resistance to virtually all penicillins and last-generation cephalosporins, have been documented in humans for some time (Turner, 2005) and in recent years, they have also emerged in Gram-negative

TABLE 13.4 Interpretive Criteria Used for Antimicrobial Susceptibility Testing of ESBL Producing Isolates in 2014[a]

Antimicrobial Class	Antimicrobial Agent	Concentration Range (µg/mL)	Breakpoints (µg/mL)		
			Susceptible	Intermediate	Resistant
β-Lactam/ β-Lactamase Inhibitor Combinations	Piperacillin-tazobactam	0.5–128	≤16	32–64	≥128
Penems	Imipenem	0.125–16	≤1	2	≥4
Cephems	Cefepime	0.125–32	≤2	4–8	≥16
	Cefotaxime	0.125–128	≤1	2	≥4
	Ceftazidime	0.125–128	≤4	8	≥16
Monobactams	Aztreonam	0.125–32	≤4	8	≥16

[a]*Cefepime MICs above the susceptible range and below the resistant range are Susceptible Dose Dependent according to CLSI guidelines (Clinical Laboratory Standards Institute, 2014).*

bacteria isolated from animals (Trott, 2013). Co-expression of ESBL and carbapenamase genes mediates bacterial resistance against almost all available β-lactam agents, leaving physicians with limited therapeutic options. If susceptibly testing data indicate the possible presence of an ESBL or carbapenamase enzyme, isolates undergo additional genotypic characterization using PCR to detect ESBL (bla_{CTX-M}, bla_{TEM}, and bla_{SHV}) and carbapenamase (bla_{OXA}, bla_{VIM}, bla_{IMP}, and bla_{NDM-1}) genes. While a majority of ceftriaxone- and ceftiofur-resistant isolates express phenotypes indicative of plasmid-encoded AmpC β-lactamase (bla_{CMY}) production, NARMS has found *Salmonella* isolates of both animal and human origin containing ESBL genes, particularly those belonging to the TEM family (Sjolund-Karlsson et al., 2013). Human isolates have also been found harboring CTX-M genes (Sjolund-Karlsson et al., 2010, 2013). In 2011, CDC detected the first New Delhi metallo-β-lactamase (NDM) carbapenamase-positive *Salmonella* in the United States. This isolate came from a perirectal culture of an Indian patient who had been transferred to a US hospital by air ambulance (Savard et al., 2011).

THE EVOLUTION OF NARMS

Sampling

Since 1996, NARMS sampling and laboratory methods have evolved to meet changing needs.

Human Sampling

Before NARMS was established, CDC monitored antimicrobial resistance in *Salmonella*, *Shigella*, and *Campylobacter* using periodic surveys of isolates from sentinel county studies from 1979 to 1995 (CDC, 2006; Gupta et al., 2004; Lee et al., 1994; MacDonald et al., 1987; Riley et al., 1984). Data from these periodic studies showed that the epidemiology of resistance differed substantially between organisms, and such resistances were often associated with preceding exposure to an antimicrobial agent for an unrelated reason (Tauxe et al., 1990). Results from these prospective studies provided baseline data for the human sampling component of NARMS (Tollefson et al., 1999).

NARMS human sampling initially included antimicrobial resistance testing of nontyphoidal *Salmonella* and *E. coli* O157 isolates from 14 state and local health departments. In 1997, five of these states began antimicrobial resistance monitoring in human *Campylobacter* isolates. By 1999, 18 states were contributing to *Salmonella* surveillance, which brought the NARMS population to more than 100 million persons covering seven of the nine CDC regions (CDC, 2000). In 1999, testing of *Salmonella* Typhi and *Shigella* was also added to the human component.

By 2003, NARMS human sampling expanded to include nationwide surveillance with 50 states systematically selecting and forwarding representative isolates from laboratory-confirmed cases of nontyphoidal *Salmonella*, *Salmonella typhi*, *Shigella*, and *E. coli* O157 to CDC. During this time, *Campylobacter* surveillance also expanded from five to ten state laboratories. In 2009, NARMS began susceptibility testing of human isolates of *Vibrio* species other than *V. cholerae* from all 50 states. According to 2011 estimates published in the 2012 US Census Bureau report (US Census Bureau, 2012), NARMS nationwide human surveillance covers approximately 312 million persons, while *Campylobacter* surveillance in ten states represents approximately 48 million persons.

From 1996 to 2002, participating NARMS public health laboratories systematically forwarded every tenth nontyphoidal *Salmonella*, every *Salmonella Typhi*, every tenth *Shigella*, and every fifth *E. coli* O157 isolate received at their laboratory to CDC for antimicrobial susceptibility testing. Although every *Salmonella Typhi* isolate was sent, analysis was restricted to one isolate per patient. In 2003, when the human component became nationwide, the sampling was updated to systematically select every twentieth non-Typhi *Salmonella*, *Shigella*, and *E. coli* O157, and every *Salmonella Typhi* isolate received at their laboratory for antimicrobial susceptibility testing.

From 1997 to 2004, the first *Campylobacter* isolate received each week at either the state health department or a designated sentinel clinical laboratory in each participating state was forwarded to CDC for susceptibility testing. Unlike nontyphoidal *Salmonella*, where every twentieth isolate is sent to CDC, the number of *Campylobacter* isolates forwarded as of 2005 is dependent on the state's burden of illness, where a state may submit every isolate, every other isolate, every third isolate, or every fifth isolate received.

Animal Sampling

The animal component of NARMS began in 1997 with monitoring of *Salmonella* and later expanded to include *Campylobacter* (1998), *E. coli* (2000), and *Enterococcus* (2003). Initially, the animal component of NARMS was conducted through USDA's Agricultural Research Service (ARS) testing of Hazard Analysis Critical Control Point (HACCP) isolates and non-diagnostic isolate submissions from farms participating in the National Animal Health Monitoring System (NAHMS) studies. *Salmonella* recovery was from carcass rinsates (chicken), carcass swabs (turkey, cattle, and swine), and ground products (chicken, turkey, and beef) collected from periodic on-farm studies (i.e., NAHMS) and federally inspected slaughter and processing facilities throughout the United States. *Salmonella* isolates from ill animals were also received by ARS from veterinary diagnostic laboratories and the National Veterinary Service Laboratory. Recovery of *Campylobacter*, *E. coli*, and *Enterococcus* was through culture of chicken carcass rinsates received by the FSIS Eastern laboratory. Frequency-based sampling was not employed, and all food animal isolates collected from these sources were sent to ARS laboratories for antimicrobial susceptibility testing.

In 2006, USDA's HACCP sampling program became risk based, resulting in more frequent sampling of noncompliant plants. Sampling priority was also given to plants containing the greatest number of samples with serotypes most frequently associated with human infection. The inherent selection bias of the risk-based sampling posed the potential for biased antimicrobial susceptibility results. Additionally, culturing bacteria from post-processing samples increased the likelihood of plant contamination affecting antimicrobial resistance levels. Considering these factors, in 2013 NARMS began its current animal monitoring program, using unbiased sampling methods to collect cecal samples, which are more representative of antimicrobial status at slaughter. This enhancement is also helpful for determining differences in antimicrobial resistance among production subclasses that undergo different antimicrobial treatment regimens (i.e., beef vs. dairy cow).

Retail Meat Sampling

Before the retail meat component of NARMS was established, FDA conducted a pilot study from March 2001 to June 2002 in order to determine the prevalence and antimicrobial resistance of enteric bacteria isolated from retail meats from Iowa (Hayes et al., 2003). Samples of retail turkey, chicken, pork, and beef were collected weekly from 263 grocery stores in Iowa totaling 981 packages at the study's conclusion. Samples were transported to the FDA CVM laboratory for identification and antimicrobial susceptibility testing. Lessons learned from the study on methodology, workflow, meat availability, and bacterial load were later applied to NARMS retail meat surveillance. The results of this study established the baseline data for the retail meat component of NARMS.

The retail meat surveillance program became the third component of NARMS when it was launched by FDA in January 2002 in collaboration with five FoodNet sites. Initially retail meat sampling began with five states and has since expanded to 14. Two FoodNet sites joined in 2003 and the two remaining FoodNet sites joined in 2004. Since 2008, four additional state public health laboratories have joined the NARMS retail meat surveillance program.

Laboratory Methods

Since it was first established, NARMS has used broth microdilution as the susceptibility testing method for *Salmonella, E. coli*, and *Enterococcus*. When *Campylobacter* surveillance from humans and food animals was added to NARMS, there were no CLSI sanctioned *in vitro* antimicrobial susceptibility testing methods available so the AB Biodisk Etest® system was used to determine the MICs for antimicrobial agents. Based on Etest® manufacturer recommendations, MIC results that fell between the twofold dilutions were rounded up to the next twofold dilution for interpretations. When retail meat surveillance began in 2002, the FDA laboratory used an agar dilution method for *Campylobacter* (McDermott et al., 2004). Recognizing the need for a standardized semi-automated method, FDA developed a broth microdilution method for *Campylobacter* that was approved by CLSI in 2006 (McDermott et al., 2005). The retail component began using the method in 2004 and the human and food animal components adopted the method in 2005.

The dilution schemes and antimicrobial content of the susceptibility testing panels used by NARMS are periodically evaluated and have undergone several changes. The content of the panels has changed to accommodate new antimicrobial agents, to omit those no longer available or used, or to adjust dilution ranges for QC and monitoring purposes. For example, in 2011, amikacin was removed from the *Salmonella/E. coli* panel and was replaced with azithromycin, a macrolide increasingly used for treatment of shigellosis and invasive salmonellosis in humans. Additionally, in 2014, kanamycin was removed from the *Salmonella/E. coli* panel to accommodate the expansion of the streptomycin testing range in order to determine the QC ranges of that drug.

NARMS has recently begun to utilize next-generation sequencing (NGS) as part of routine surveillance in order to discover previously uncharacterized resistance genes and elucidate phylogenetic relationships of emerging clones. For example, NGS technology was used in a 2013 NARMS study examining the whole-genome sequence of gentamicin-resistant *Campylobacter coli* isolated from retail chicken (Chen et al., 2013). Prior to 2007, gentamicin resistance appeared sporadically among *C. coli* isolated from human samples; however, in 2007, gentamicin resistance emerged in retail chicken meat and rose to 18.1% by 2011. Whole-genome sequencing data revealed that the gentamicin resistance was conferred by a novel *aph(2")* gene. Furthermore, this gene was carried on a pN29710-1 plasmid, previously known to mainly carry tetracycline resistance (Batchelor et al., 2004). This level of detail was not possible using PCR methods based on the known genetics of *Campylobacter*, and illustrates the power of this new technique in supporting surveillance.

DATA ACQUISITION AND REPORTING

To date, NARMS data are collected and maintained at each participating federal agency. However, recently NARMS developed and launched an integrated database that allows for data sharing among NARMS partners in a secure environment. There are future plans to develop a method for sharing data with stakeholders by using information technology tools that allow efficient exploration and analysis of data across sample sources.

Comprehensive annual reports have been published for each NARMS component, in addition to an Executive Report. The Executive Report contains integrated aggregate data on *Salmonella* and *Campylobacter* recovered from human clinical cases, retail meats and food animals, and data on *E. coli* recovered from retail meats and food animals. The NARMS Executive Report contains graphs, select tables, and a summary that focus on resistance to drugs that are considered clinically important to human medicine as well as multidrug resistance patterns and specific co-resistant phenotypes that have been linked to severe illness in humans. Because NARMS seeks to identify sources of resistant strains entering the food supply, resistance patterns in serotypes with a host preference for specific food animal species are highlighted.

INTERNATIONAL HARMONIZATION

Foodborne disease is an international health challenge that requires coordinated global solutions. The worldwide trade in food-producing animals and food products has greatly increased over the last decade and will continue to grow. Likewise, antimicrobial resistance in pathogens found in food-producing animals and food products is an international problem. Consumers may become infected with antimicrobial-resistant pathogens both when they travel abroad and when they consume contaminated imported food (Aarestrup et al., 2007).

Since its inception, NARMS has collaborated with other countries to help build laboratory capacity for foodborne disease surveillance, including surveillance in developing countries. Strengthening global surveillance for antimicrobial-resistant pathogens in the food chain is an important step in preventing and mitigating the emergence of antimicrobial-resistant bacteria. Currently NARMS scientists engage in the following international activities:

• Participate as a member of the World Health Organization Advisory Group on Integrated Surveillance of Antimicrobial Resistance (WHO-AGISAR) and the WHO Global Food Infections Network (GFN) to help build international capacity and cooperation for the surveillance of foodborne disease and antimicrobial resistance.
• Support the World Organisation for Animal Health (OIE) food safety and animal welfare which works closely with international partners to harmonize antimicrobial susceptibility testing and reporting methods and to facilitate data sharing.
• Foster international research collaborations, special regional studies, and national pilot projects to characterize unique and common elements in the epidemiology of antimicrobial-resistant foodborne pathogens in different countries.

NARMS IN ACTION—PUBLIC HEALTH IMPACT

NARMS surveillance and research activities supply data that are used to help ensure food safety in a number of ways including: (i) development of risk assessments and math models that characterize the distribution and human health impact of antimicrobial-resistant foodborne bacteria and in turn help the agency assess the safety and effectiveness of animal drugs; (ii) development of science-based policies that address the use of antimicrobials in agriculture, and; (iii) attribution of foodborne outbreaks.

Risk Assessment

Risk assessment modeling will be described in detail in the next chapter, but, in short, risk assessment is an analytical process in which the chance of a defined adverse human health effect is estimated given exposures to a hazardous agent.

In the case of antimicrobial resistance, risk assessment can be used to determine whether human anti-infective therapy has been or might be compromised by antimicrobial drug use in animals. NARMS is part of FDA's overall strategy on antimicrobial resistance and is the primary tool used in both pre- and post-approval animal drug monitoring to detect changes in resistance patterns and trends. NARMS data can be integrated into risk analyses to evaluate the potential human and animal health impact of a drug that is in development or the safety of a drug that is already in use.

The first post-approval FDA risk assessment using NARMS data was conducted in 2000 to assess human health risk concerning fluoroquinolone resistance in *Campylobacter* from poultry (FDA, 2001). Fluoroquinolone drugs are used as empiric therapies for gastrointestinal illness in humans, and, at the time, fluoroquinolones were approved for use in poultry to control mortality associated with *E. coli*. A growing body of data implied that fluoroquinolone use in poultry was leading to a rise in domestically acquired fluoroquinolone-resistant *C. jejuni* infections (Smith et al., 1999). A quantitative risk model was developed using data from a number of sources, including NARMS. USDA-generated NARMS data on the prevalence of fluoroquinolone-resistant *Campylobacter* (USDA, 2012) were used to estimate the quantity of fluoroquinolone-resistant *Campylobacter*-contaminated chicken meat that was consumed. CDC-generated NARMS (CDC, 2000) and FoodNet *Campylobacter* Case Control Study (Friedman et al., 2004) data were used to derive the number of fluoroquinolone-resistant *Campylobacter* cases attributable to chicken. Additionally, the FDA conducted targeted studies to measure the biological effect of ciprofloxacin administration in poultry on the development of resistant *C. jejuni* (McDermott et al., 2002). The model estimated that in 1999, 1.2 billion pounds of chicken containing fluoroquinolone-resistant *Campylobacter* were consumed in the United States and as a result 153,580 people were infected with fluoroquinolone-resistant *Campylobacter*, 9,261 of which were treated with a fluoroquinolone (FDA, 2001). These data prompted FDA to propose withdrawal of the approval of new animal drug applications for fluoroquinolone use in poultry in 2000 (Nelson et al., 2007). After a lengthy legal dispute between FDA and the Bayer Corporation, sponsor of one of the approved fluoroquinolones, enrofloxacin, the action became effective in September of 2005 (Nelson et al., 2007).

In order to minimize the effects that veterinary drugs may have on the emergence of antimicrobial-resistant bacteria in animals and subsequent spread to humans, in October 2003 FDA began to require a hazard identification and qualitative risk assessment as part of the new animal drug pre-approval process, through its publication of the Guidance for Industry (GFI) #152 (FDA, 2003). These qualitative risk assessments address the probability that a human might be exposed to antimicrobial-resistant microbes through the ingestion of animal-derived food. The antimicrobial resistance risk assessment process is derived from Office International des Epizooties (OIE) Ad Hoc Group on Antimicrobial Resistance and is comprised of a release assessment, followed by an exposure assessment, and finally a consequence assessment. NARMS surveillance and research data are used in this pre-approval risk assessment process.

The *release assessment* is the probability that a drug used in a food animal will result in the emergence of resistant bacteria. This sub-assessment requires a number of relevant factors including but not limited to: (i) baseline data on the prevalence of resistant microorganisms in food animals, and; (ii) information on resistance mechanisms including the location and rate of transfer of the resistance determinants. NARMS data collected at the slaughterhouse are a comprehensive source of baseline resistance data, and research conducted through NARMS can be used to characterize molecular mechanisms of resistance.

The *exposure assessment* is the probability of humans ingesting foodborne bacteria from animal-derived food. In addition to food commodity consumption data, risk assessors must incorporate food commodity contamination data. NARMS retail meat data are particularly helpful here, as they represent the point at which customers are exposed. The number of meat samples collected is known, as is the number of samples that test positive for *Salmonella, Campylobacter, E. coli*, and *Enterococcus*. Risk assessors are encouraged to use this and other sources, such as USDA baseline data. Ongoing NARMS research on risk factors for acquiring resistant infections in humans may also help to inform the exposure assessment.

The *consequence assessment* is the likelihood that human exposure to resistant bacteria results in an adverse health consequence. Drugs are ranked as *important, highly important,* or *critically important* based on a number of set criteria that can be found in GFI #152. As of this book's publication, FDA is reevaluating these criteria. In its reevaluation, FDA will rely on NARMS research studies on clinical outcomes of resistant infections. These studies correlate severity of illness with exposure to drug-resistant pathogens that are most likely of foodborne origin.

These components are all integrated into overall *risk estimation* which represents the potential for human health to be adversely impacted by the selection or emergence of antimicrobial-resistant foodborne bacteria associated with the use of drugs in food-producing animals and is defined in qualitative categories as *high, medium,* or *low.*

NARMS Data and the Development of Regulatory Policy

NARMS data are vital to the formulation of FDA policy regarding the appropriate use of antimicrobials in veterinary medicine. One such example involves the use of cephalosporins. The cephalosporin, ceftriaxone, is considered a critically important therapy for a number of infections in humans including pneumonia, pelvic inflammatory disease, intra-abdominal infections, and skin and soft tissue infections. The cephalosporin, ceftiofur, an analog of ceftriaxone, is not used in human medicine but is approved for use in animal agriculture to treat and control certain diseases in food-producing animals. For many years, ceftiofur was also used in an extra-label manner meaning it was used at unapproved dose levels, frequencies, durations, and routes of administration, and it was used in animal species that were not indicated on the drug label. NARMS surveillance data had shown that while there were no ceftiofur-resistant *Salmonella* isolates from cattle and swine in 1997 and only 0.5% and 3.7% of *Salmonella* isolates from chickens and turkeys, respectively, by 2009, the proportion of *Salmonella* isolates that were ceftiofur resistant rose to 14.5% for cattle, 4.2% for swine, 12.7% for chickens, and 12.4% for turkeys (FDA, 2011). *Salmonella* resistance to ceftiofur largely coincides with resistance to ceftriaxone when using the 2010 CLSI breakpoints. Ceftriaxone resistance among nontyphoidal *Salmonella* isolates from humans rose from 0.2% in 1996 to 3.4% in 2009 (FDA, 2011). The rise in ceftiofur and ceftriaxone resistance among specific *Salmonella* serotypes was even more striking. NARMS surveillance found that ceftiofur resistance among all *Salmonella* Typhimurium isolates from chickens was 0% in 1997 and 33.3% in 2009. Among *Salmonella* Typhimurium isolates from cattle, ceftiofur resistance was 3% in 1998 and 27.8% in 2009. Ceftiofur resistance rose from 12.5% in 1998 to 58.8% in 2009 among *Salmonella* Newport isolates from cattle. Furthermore, while there was no ceftiofur resistance among *Salmonella* Heidelberg isolates from poultry in 1997, resistance rose to 17.6% in chicken isolates and 33.3% in turkey isolates in 2009 (FDA, 2011).

In human *Salmonella* isolates, ceftriaxone resistance was 0%, 0%, and 2.7% for serotypes Typhimurium, Newport, and Heidelberg, respectively. However, by 2009, ceftriaxone resistance among isolates from these serotypes was 6.5%, 6.4%, and 20.9%, respectively. Additionally, NARMS research data demonstrated that these isolates contained the *cmy* gene that confers resistance to first-, second-, and third-generation cephalosporins and other beta-lactams (Zhao et al., 2001, 2009). Few isolates contained any genes from the CTX-M family that provide resistance to all cephalosporins (Sjolund-Karlsson et al., 2010, 2011); however, FDA became concerned that these genes, often found on large mobile plasmids, could spread and cause cephalosporin resistance to escalate. These and other published data led FDA to become concerned that ceftiofur might compromise the efficacy of ceftriaxone for the treatment of human infections. Therefore in 2012, FDA issued a final order prohibiting the extra-label use of certain cephalosporin drugs in cattle, swine, chickens, and turkeys (FDA, 2012). NARMS continues to monitor cephalosporin resistance in enteric bacteria from humans, animals, and retail meats since the order went into effect.

NARMS Data and Outbreak Detection and Response

NARMS data have informed numerous epidemiological studies on trends in resistance (Gupta et al., 2004; Medalla et al., 2013), risk factors associated with resistant infections (Kassenborg et al., 2004), and burden of resistant illness estimates (Krueger et al., 2014; Nelson et al., 2004; Varma et al., 2005a,b), but only in recent years has it emerged as a valuable tool in outbreak detection and response. Because NARMS retail isolates are subjected to DNA fingerprinting via PFGE, and the PFGE patterns are uploaded to the PulseNet database, they are part of the reference database for CDC's molecular outbreak detection network. As the only ongoing retail-level microbiological food monitoring program in the United States, NARMS supplies a wealth of retail meat PFGE data to PulseNet. The value of NARMS as an outbreak detection tool was highlighted in 2011 during a multistate outbreak of *Salmonella* Heidelberg infections. Analysts found that PFGE patterns of *Salmonella* Heidelberg from outbreak cases matched those from two ground turkey samples purchased during the time of the outbreak as part of routine NARMS surveillance. These matches led public health officials to target their investigation towards ground turkey as a potential source of the illnesses. In all, a total of 136 persons from 34 states were reported in the outbreak, which resulted in a recall of over 36 million pounds of ground turkey products (CDC, 2014). Because participating NARMS laboratories also store the brand name and plant information captured from the retail meat packages they collect, NARMS has been used to link outbreak strains with particular brands. This was the case in 2012, when CDC investigated another outbreak of *Salmonella* Heidelberg infections. NARMS data were used to show that isolates with PFGE patterns matching the outbreak strain were significantly associated with retail chicken originating from a single poultry producer (CDC, 2013b). NARMS has also begun using additional tools such as whole-genome sequencing to further dissect clusters of retail isolates associated with human illness (Hoffmann et al., 2012). Using these advanced technologies, NARMS has also been able to uncover novel resistance genes (Chen et al., 2013).

STRENGTHS AND LIMITATIONS

The NARMS program has a number of strengths. NARMS is the only ongoing antimicrobial susceptibility monitoring program for foodborne bacteria in the United States and currently the most extensive program for integrated laboratory-based surveillance of foodborne bacteria in the world. It utilizes a tripartite structure, collecting isolates from food animals, retail meats, and humans, which is not only ideal for characterizing movement of resistance determinants along the farm-to-fork pathway, but it also highlights the advantage of utilizing expertise across disciplines and government agencies. Because NARMS is a multi-agency program that leverages existing public health infrastructure, it can be easily complemented by surveillance programs and databases housed in the partner agencies. The successes of the merger of NARMS susceptibility data with PulseNet data is one example. Other examples include integration of NARMS susceptibility data with epidemiological data from FoodNet. Dual analyses of these datasets allow NARMS to identify risk factors that contribute to antimicrobial-resistant infections (such as travel) and clinical outcomes of resistant infections. In instances where NARMS data alone cannot help ascertain the prevalence of *Salmonella* serotypes that are less frequently found in humans but common to foods, other CDC databases can be used such as that of the National Outbreak Reporting System and the Foodborne Disease Outbreak Surveillance System, both of which capture reports of enteric disease outbreaks from state, local, and territorial public health agencies. Linkage to other FDA, CDC, and USDA surveillance systems not mentioned here allows NARMS to answer other important public health questions.

At the core of NARMS lies integration. As was mentioned earlier, susceptibility testing methods are harmonized and quality assurance standards are followed by all participating agencies. Data capture methods are similar among the agencies and reports are closely aligned. Frequent communication through interagency meetings ensures direct comparability of results and consistency in data interpretation. As a national repository of data and isolates, NARMS can operate as an *ad hoc* research tool. NARMS isolates have been used to develop subtyping methods, characterize resistance genes, and test against new chemical compounds that are under review.

NARMS surveillance successfully fulfills its main role to document resistance levels in different reservoirs, describe the spread of resistant bacteria and their genes, identify temporal and spatial trends in resistance, help identify risk factors and clinical outcomes of infections by resistant bacteria, and provide data on current and emerging hazards for pre- and post-approval regulatory decisions. However, there is currently relatively limited information regarding the way antimicrobials are used in food-producing animals in the United States. As a result, it is difficult to interpret the resistance data that NARMS reports in the context of use practices that are driving resistance in animal agriculture. In 2013, the FDA and other participating NARMS agencies began to explore potential mechanisms to collect antimicrobial use data in food animals, including broad national surveys and more detailed on-farm studies that would pair use questionnaires with isolate collection. This information would help refine policies and stewardship principles to limit or reverse the development of resistance in food-producing animals.

While significant changes have been made to the food animal sampling component of NARMS to overcome biases in source selection, there are still some existing challenges in retail sampling. Retail products are collected from a limited number of areas for a small number of products. While major trends may be observed, the small sample size and lack of national sampling strategy make interpretation of these data difficult.

Timely reporting also remains a challenge. Timely reports are necessary in order to ensure that public health action is implemented early enough to affect change. Furthermore, as the use of advanced detection technologies that sequence whole genomes of bacterial isolates become more commonplace in NARMS and other surveillance programs, isolate-level data will make their way into the public domain at a faster rate. Therefore, it may be necessary to produce more reports in even shorter time intervals. Currently, surveillance reports are printed 1–3 calendar years after the sampling year. Some of the lag is due to logistical bottlenecks, laboratory and data auditing procedures, and repeat testing. Data interpretation can also prolong the time to complete reports. The NARMS agencies are currently working on ways to overcome delays imposed by other steps such as data collation and analysis by utilizing new epidemiological tools that will enhance data visualization and speed analytics and reporting.

CONCLUSION

Antimicrobial resistance monitoring systems are invaluable components of public health. Monitoring programs like NARMS provide data needed to inform and prioritize science-based approaches to minimize foodborne hazards resulting from antimicrobial use in food-producing animals. NARMS findings have played a key role in regulatory activities, in identifying risk factors and investigating outbreaks. The NARMS program continues to expand and to identify areas to improve. Changes in surveillance brought on by the advent of next-generation DNA sequencing technologies, along with better antibiotic use information, will strengthen integrated surveillance in the future. As the genotypic and phenotypic database and strain collections grow, researchers will better understand the ecology of resistant microorganisms in the food supply. As a result, public health officials will be better equipped to prioritize interventions to limit or reverse resistance in zoonotic foodborne pathogens.

REFERENCES

Aarestrup, F.M., Hendriksen, R.S., Lockett, J., Gay, K., Teates, K., McDermott, P.F., et al., 2007. International spread of multidrug-resistant *Salmonella* Schwarzengrund in food products. Emerg. Infect. Dis. 13 (5), 726–731.

American Scoiety for Microbiology Public and Scientific Affairs Board, 1995. Report of the ASM Task Fornce on Anibiotic Resistance. Washington, DC.

Batchelor, R.A., Pearson, B.M., Friis, L.M., Guerry, P., Wells, J.M., 2004. Nucleotide sequences and comparison of two large conjugative plasmids from different *Campylobacter* species. Microbiology 150 (Pt 10), 3507–3517.

CDC, 1996. Establishment of a national surveillance program for antimicrobial resistance in *Salmonella*. Morb. Mortal. Wkly. Rep. 45 (5), 110–111.

CDC, 2000. National Antimicrobial Resistance Monitoring System for Enteric Bacteria (NARMS): 1999 Human Isolates Final Report. Department of Health and Humand Services, CDC, Atlanta, GA.

CDC, 2006. National Antimicrobial Resistance Monitoring System for Enteric Bacteria (NARMS): 2003 Human Isolates Final Report. US Department of Health and Human Services, CDC, Atlanta, GA.

CDC, 2013a. Antibiotic Resistance Threats in the United States, 2013. Centers for Disease Control and Prevention, Atlanta, GA.

CDC, 2013b. Outbreak of *Salmonella* Heidelberg infections linked to a single poultry producer—13 states, 2012–2013. Morb. Mortal. Wkly. Rep. 62 (27), 553–556.

CDC, 2013c. National Antimicrobial Resistance Monitoring System for Enteric Bacteria (NARMS): Human Isolates Final Report, 2011. US Department of Health and Human Services, CDC, Atlanta, GA.

CDC, 2014. Multistate Outbreak of Human *Salmonella* Heidelberg Infections Linked to Ground Turkey. Centers for Disease Control, Atlanta, GA.

Chen, Y., Mukherjee, S., Hoffmann, M., Kotewicz, M.L., Young, S., Abbott, J., et al., 2013. Whole-genome sequencing of gentamicin-resistant *Campylobacter coli* isolated from U.S. retail meats reveals novel plasmid-mediated aminoglycoside resistance genes. Antimicrob. Agents Chemother. 57 (11), 5398–5405.

Clinical and Laboratory Standards Institute, 2004. Performance Standards for Antimicrobial Disk and Dilution Susceptibility Tests for Bacteria Isolated from Animals; Informational Supplement (M31-S1). Wayne, PA.

Clinical and Laboratory Standards Institute, 2008. Perfomance Standards for Antimicrobial Dilution and Disk Susceptibility Tests for Bacteria Isolated from Animals; Approved Standard—Third Edition. CLSI document M31-A3, Wayne, PA.

Clinical and Laboratory Standards Institute, 2010. Methods for Antimicrobial Dilution and Disk Susceptibility Testing of Infrequently Isolated or Fastidious Bacteria; Approved Guideline. CLSI Document M45-A2, Wayne, PA.

Clinical and Laboratory Standards Institute. Performance Standards for Antimicrobial Susceptibility Testing: Twenty-Fourth Informational Supplement. CLSI document M100-S24. 2014. Wayne, PA.

FDA, 2001. Risk assessment on the Human Health Impact of Fluoroquinolone Resistant *Campylobacter* Attributed to the Consumption of Chicken. US FDA, Rockville, MD, Jan 5.

FDA, 2003. Guidance for Industry #152: Evaluating te Saftey of Antimicrobial New Animal Drugs with Regard to Their Microbiological Effects on Bacteria of Human Health Concern. US Food and Drug Administration, Rockville, MD, Oct 23. Report No.: Docket Number [98D-1146].

FDA, 2011. National Antimicrobial Reistance Monitoring System—Enteric Bacteria (NARMS): 2009 Executive Report. US Department of Health and Human Services, Food and Drug Administration, Rockville, MD.

FDA, 2012. Cephalosporin Order of Prohibition. Food and Drug Administration, Rockville, MD, Jan 6. Docket No. [FDA-2008-N-0326].

FDA 2014. FDA Bacteriological Analytical Manual Online. <http://www.fda.gov/Food/FoodScienceResearch/LaboratoryMethods/ucm2006949.htm> (accessed 07.02.14.).

Friedman, C.R., Hoekstra, R.M., Samuel, M., Marcus, R., Bender, J., Shiferaw, B., et al., 2004. Risk factors for sporadic *Campylobacter* infection in the United States: a case–control study in FoodNet sites. Clin. Infect. Dis. 38 (Suppl. 3), S285–S296.

Gupta, A., Nelson, J.M., Barrett, T.J., Tauxe, R.V., Rossiter, S.P., Friedman, C.R., et al., 2004. Antimicrobial resistance among *Campylobacter* strains, United States, 1997–2001. Emerg. Infect. Dis. 10 (6), 1102–1109.

Hammerum, A.M., Heuer, O.E., Emborg, H.D., Bagger-Skjot, L., Jensen, V.F., Rogues, A.M., et al., 2007. Danish integrated antimicrobial resistance monitoring and research program. Emerg. Infect. Dis. 13 (11), 1632–1639.

Hayes, J.R., English, L.L., Carter, P.J., Proescholdt, T., Lee, K.Y., Wagner, D.D., et al., 2003. Prevalence and antimicrobial resistance of enterococcus species isolated from retail meats. Appl. Environ. Microbiol. 69 (12), 7153–7160.

Hoffmann, M., Zhao, S., Luo, Y., Li, C., Folster, J.P., Whichard, J., et al., 2012. Genome sequences of five *Salmonella enterica* serovar Heidelberg isolates associated with a 2011 multistate outbreak in the United States. J. Bacteriol. 194 (12), 3274–3275.

Institute of Medicine Committee on Emerging Microbial Threats to Health, 1992. In: Lederberg, J., Shope, R.E., Oaks, S.E. (Eds.), Emerging Infections: Microbial Threats to Health in the United States National Academy Press, Washington, DC, pp. 159–160.

Kassenborg, H.D., Smith, K.E., Vugia, D.J., Rabatsky-Ehr, T., Bates, M.R., Carter, M.A., et al., 2004. Fluoroquinolone-resistant *Campylobacter* infections: eating poultry outside of the home and foreign travel are risk factors. Clin. Infect. Dis. 38 (Suppl. 3), S279–S284.

Krueger, A.L., Greene, S.A., Barzilay, E.J., Henao, O., Vugia, D., Hanna, S., et al., 2014. Clinical outcomes of nalidixic acid, ceftriaxone, and multidrug-resistant nontyphoidal salmonella infections compared with pansusceptible infections in FoodNet Sites, 2006–2008. Foodborne Pathog. Dis.

Lee, L.A., Puhr, N.D., Maloney, E.K., Bean, N.H., Tauxe, R.V., 1994. Increase in antimicrobial-resistant *Salmonella* infections in the United States, 1989–1990. J. Infect. Dis. 170 (1), 128–134.

MacDonald, K.L., Cohen, M.L., Hargrett-Bean, N.T., Wells, J.G., Puhr, N.D., Collin, S.F., et al., 1987. Changes in antimicrobial resistance of *Salmonella* isolated from humans in the United States. JAMA 258 (11), 1496–1499.

McDermott, P.F., Bodeis, S.M., English, L.L., White, D.G., Walker, R.D., Zhao, S., et al., 2002. Ciprofloxacin resistance in *Campylobacter jejuni* evolves rapidly in chickens treated with fluoroquinolones. J. Infect. Dis. 185 (6), 837–840.

McDermott, P.F., Bodeis, S.M., Aarestrup, F.M., Brown, S., Traczewski, M., Fedorka-Cray, P., et al., 2004. Development of a standardized susceptibility test for campylobacter with quality-control ranges for ciprofloxacin, doxycycline, erythromycin, gentamicin, and meropenem. Microb. Drug Resist. 10 (2), 124–131.

McDermott, P.F., Bodeis-Jones, S.M., Fritsche, T.R., Jones, R.N., Walker, R.D., 2005. Broth microdilution susceptibility testing of *Campylobacter jejuni* and the determination of quality control ranges for fourteen antimicrobial agents. J. Clin. Microbiol. 43 (12), 6136–6138.

Medalla, F., Hoekstra, R.M., Whichard, J.M., Barzilay, E.J., Chiller, T.M., Joyce, K., et al., 2013. Increase in resistance to ceftriaxone and nonsusceptibility to ciprofloxacin and decrease in multidrug resistance among *Salmonella* strains, United States, 1996–2009. Foodborne Pathog. Dis. 10 (4), 302–309.

Nelson, J.M., Smith, K.E., Vugia, D.J., Rabatsky-Ehr, T., Segler, S.D., Kassenborg, H.D., et al., 2004. Prolonged diarrhea due to ciprofloxacin-resistant campylobacter infection. J. Infect. Dis. 190 (6), 1150–1157.

Nelson, J.M., Chiller, T.M., Powers, J.H., Angulo, F.J., 2007. Fluoroquinolone-resistant *Campylobacter* species and the withdrawal of fluoroquinolones from use in poultry: a public health success story. Clin. Infect. Dis. 44 (7), 977–980.

Riley, L.W., Cohen, M.L., Seals, J.E., Blaser, M.J., Birkness, K.A., Hargrett, N.T., et al., 1984. Importance of host factors in human salmonellosis caused by multiresistant strains of *Salmonella*. J. Infect. Dis. 149 (6), 878–883.

Savard, P., Gopinath, R., Zhu, W., Kitchel, B., Rasheed, J.K., Tekle, T., et al., 2011. First NDM-positive *Salmonella* sp. strain identified in the United States. Antimicrob. Agents Chemother. 55 (12), 5957–5958.

Sjolund-Karlsson, M., Rickert, R., Matar, C., Pecic, G., Howie, R.L., Joyce, K., et al., 2010. *Salmonella* isolates with decreased susceptibility to extended-spectrum cephalosporins in the United States. Foodborne Pathog. Dis. 7 (12), 1503–1509.

Sjolund-Karlsson, M., Howie, R., Krueger, A., Rickert, R., Pecic, G., Lupoli, K., et al., 2011. CTX-M-producing non-Typhi *Salmonella* spp. isolated from humans, United States. Emerg. Infect. Dis. 17 (1), 97–99.

Sjolund-Karlsson, M., Howie, R.L., Blickenstaff, K., Boerlin, P., Ball, T., Chalmers, G., et al., 2013. Occurrence of beta-lactamase genes among non-Typhi *Salmonella enterica* isolated from humans, food animals, and retail meats in the United States and Canada. Microb. Drug Resist. 19 (3), 191–197.

Smith, K.E., Besser, J.M., Hedberg, C.W., Leano, F.T., Bender, J.B., Wicklund, J.H., et al., 1999. Quinolone-resistant *Campylobacter jejuni* infections in Minnesota, 1992–1998. Investigation team. N. Engl. J. Med. 340 (20), 1525–1532.

Swann, M.M., Baxter, K.L., Field, H.I., Howie, J.W., Lucas, I.A.M., Millar, E.L.M., et al., 1969. Report of the Joint Committee on the Use of Antibiotics in Animal Husbandry and Veterinary Medicine. Her Majesty's Stationary Office, London.

Tauxe, R.V., Puhr, N.D., Wells, J.G., Hargrett-Bean, N., Blake, P.A., 1990. Antimicrobial resistance of *Shigella* isolates in the USA: the importance of international travelers. J. Infect. Dis. 162 (5), 1107–1111.

Tollefson, L., Fedorka-Cray, P.J., Angulo, F.J., 1999. Public health aspects of antibiotic resistance monitoring in the USA. Acta. Vet. Scand. Suppl. 92, 67–75.

Trott, D., 2013. beta-lactam resistance in gram-negative pathogens isolated from animals. Curr. Pharm. Des. 19 (2), 239–249.

Turner, P.J., 2005. Extended-spectrum beta-lactamases. Clin. Infect. Dis. 41 (Suppl. 4), S273–S275.

US Census Bureau, 2012. 2012 U.S. Census Bureau Report.

USDA, 2012. National Antimicrobial Resistance Monitoring System-Enteric Bacteria, Animal Arm (NARMS): 2010 NARMS Animal Arm Annual Report. USDA, Athens, GA.

Varma, J.K., Greene, K.D., Ovitt, J., Barrett, T.J., Medalla, F., Angulo, F.J., 2005a. Hospitalization and antimicrobial resistance in *Salmonella* outbreaks, 1984–2002. Emerg. Infect. Dis. 11 (6), 943–946.

Varma, J.K., Molbak, K., Barrett, T.J., Beebe, J.L., Jones, T.F., Rabatsky-Ehr, T., et al., 2005b. Antimicrobial-resistant nontyphoidal *Salmonella* is associated with excess bloodstream infections and hospitalizations. J. Infect. Dis. 191 (4), 554–561.

WHO, 2013. Integrated Surveillance of Antimicrobial Resistance:Guidance from a WHO Advisory Group. World Health Organization, Geneva.

Zhao, S., White, D.G., McDermott, P.F., Friedman, S., English, L., Ayers, S., et al., 2001. Identification and expression of cephamycinase bla(CMY) genes in *Escherichia coli* and *Salmonella* isolates from food animals and ground meat. Antimicrob. Agents Chemother. 45 (12), 3647–3650.

Zhao, S., Blickenstaff, K., Glenn, A., Ayers, S.L., Friedman, S.L., Abbott, J.W., et al., 2009. Beta-Lactam resistance in *Salmonella* strains isolated from retail meats in the United States by the National Antimicrobial Resistance Monitoring System between 2002 and 2006. Appl. Environ. Microbiol. 75 (24), 7624–7630.

Chapter 14

Risk Assessment of Antimicrobial Resistance

H. Gregg Claycamp

Center for Veterinary Medicine, US Food and Drug Administration, Rockville, MD, USA

Chapter Outline

INTRODUCTION

It is generally recognized that most antimicrobial resistance in human medicine is due to human medical uses; however, the occurrence of antimicrobial resistance in food animals and bacteria that contaminate food products is widely believed to also contribute to the burden of human antimicrobial resistance. This is particularly true for animal drugs used in animal and plant agriculture that use identical chemical antibiotics or otherwise have close human drug analogs. The risks of adverse human health effects among antimicrobial–pathogen or antimicrobial–commensal combinations range from the theoretical to some that have substantial footing in the epidemiological and controlled study observations. Antimicrobial resistance risk assessments for research, policy development, or regulatory compliance have already contributed significantly to understanding the nature and scope of antimicrobial resistance and also in methods for evaluating risk management interventions. Given that use of antimicrobials in human and animal medicine is likely to continue for the foreseeable future, risk assessment will remain a useful tool for informing public health decisions.

This chapter looks at the some of the scientific foundations of risk assessment and the quantitative approaches that have been used for the risk assessment

Antimicrobial Resistance and Food Safety. DOI: http://dx.doi.org/10.1016/B978-0-12-801214-7.00014-4
283

of antimicrobial resistance in the food chain.[1] Given the diversity of bacteria and the antibiotics used against them, the varying patterns of use and the complexity of the food chain, it is not unexpected that public health agencies implement risk assessment principles in many ways. The chapter does not review or recommend US or international policies or guidelines nor does it recommend a unique process for the assessment of human risks from antimicrobial drugs in the food chain.

RISK ASSESSMENT

Risk assessment is the scientific and analytical process to estimate the chance of adverse outcomes from exposure to a hazard (biological, chemical, or physical agent) or hazardous event. "Risk assessment is widely recognized as a systematic way to prepare, organize, and analyze information to help make regulatory decisions, establish programs, and prioritize research and development efforts" (USEPA, 2014a, p. 6). US and international public health agencies have proposed or implemented a variety of frameworks for microbial risk assessments, yet few frameworks have been proposed for antimicrobial resistance risk assessment (Claycamp and Hooberman, 2004). There are several possible reasons for a relative absence of formal frameworks for antimicrobial resistance risk assessment. First, antimicrobial resistance risk assessment might be thought of as a subset of microbial risk assessments—that is, a proportion of the microbes are resistant or acquire resistance to the antimicrobial drug of interest. Second, features of antimicrobial resistance add significant complexity to the already complex array of features for microbial risk assessments, leaving risk assessors to develop unique models for each microbe–antimicrobial drug combination. Third, antimicrobial resistance is expressed as an adverse health consequence to individuals only after treatment of an infection with an antimicrobial drug complicating putative dose–response relationships. Taken together, these observations suggest that a unique, all-encompassing framework for antimicrobial resistance risk assessment is unlikely to emerge except as a "principles-driven" proposal.

This chapter presents an overview of principles for risk assessment for antimicrobial resistance occurring in the food pathway. Although "antimicrobial" is a general term referring to chemical or physical agents having activity against microscopic pathogens including bacteria, viruses, fungi, and, perhaps, prions, this chapter concerns pathogenic and commensal bacteria in the human food chain. Antimicrobial resistance risk assessment builds on microbial risk assessment with the overlaid concept of "antimicrobial resistance." Thus, the reader is referred to thorough existing work on microbial risk assessment in texts

---

[1] This chapter does not recommend any specific methodology other than referring to the considerable foundation of scientific risk assessment in the risk analysis and chemical risk assessment literature.

(Cox, 2002; Forsythe, 2002; Haas et al., 1999; Salisbury et al., 2002; Vose, 1998; Yoe, 2012) and guidelines (FAO and WHO, 2009; USEPA and USDA, 2012).

A convenient starting point for conceptualizing the antimicrobial resistance risk assessment process is concepts and principles for risk assessment discussed in *Risk Assessment in the Federal Government: Managing the Process* (National Research Council, 1983). The review principles and articulation of the 1983, "four-step paradigm" for risk assessment was clearly a watershed moment in human health risk analysis. Although alternative frameworks have been proposed during the intervening decades, the analytical principles embodied in the four process steps of hazard identification, dose–response assessment, exposure assessment, and risk characterization remain scientifically sound starting points for risk assessment (Table 14.1). The relatively few antimicrobial resistance risk assessment paradigms discussed in research and regulatory settings generally have origins in the 1983 paradigm (Claycamp and Hooberman, 2004; European Food Safety Authority, 2014; World Health Organization, 2001). The National Academy of Sciences (NAS) recommendations on risk assessment and food risk assessment frameworks for health risk assessment refer to "risk analysis" as the overall process encompassing risk assessment, risk management, and risk communication. In some EU reports and international standards; however, risk management is the overall process that includes a quantitative process of "risk analysis" as part of risk assessment (Claycamp, 2007; Claycamp and Hooberman, 2004).

TABLE 14.1 Generalized "Four-Step" Paradigm for Risk Assessment[a]

Hazard identification is the process of determining whether or not exposures to a hazardous biological, chemical, or physical agent or a hazardous event can cause an increase in an adverse health effect.

Dose–response assessment "is the process of characterizing the relation between the dose of an agent administered or received and the incidence of an adverse health effect in exposed populations and in estimating the incidence of effect as a function of exposure to the agent."

Exposure assessment "is the process of measuring or estimating the intensity, frequency, and duration of human exposures to an agent currently present in the environment or of estimating hypothetical exposures that might arise from the release of new [agents] into the environment."

Risk characterization "is the process of estimating the incidence of a health effect under the various conditions of human exposure described in exposure assessment. It is performed by combining the exposure and dose–response assessments. The summary effects of the uncertainties in the preceding steps are described in this step."

[a]*National Research Council (1983).*

The Initiation of a Risk Assessment

Public health risk assessments are called for by risk managers for research into emerging health hazards, research to support policy-making or as part of regulatory compliance. The policy making includes setting regulatory guidelines or proposing the conditions for "bright line" safety assessments. The bright line for a declaration of safe/not safe might itself be based on a statistical calculation of, for example, an upper percentile of exposure, or a tolerance distribution on linear regressions from experimental exposures and dose–response assessments.

Risk assessment for any hazard and/or hazardous event begins with a clear statement of the problem (National Research Council, 2011; USEPA, 2014a; Yoe, 2012). "Planning and scoping is an important first step to ensure that each risk assessment has a clear purpose and well-defined vision. It is critical to producing a sound risk assessment that serves its intended purpose" (USEPA, 2014a). Risk assessments provide information for decision making; thus, the definitions of the problem and scope of the risk assessment are accomplished collaboratively among risk managers and risk assessors. On the other hand, risk assessment research projects, like basic science research, might develop testable hypotheses (as problem statements) after preliminary studies and information gathering. Risk assessments for policy development explore risk scenarios designed to inform trade-off analyses within regulatory boundaries.

Too often the urge to do the work overrides problem definition by interdisciplines because they assume a shared understanding of the problem. Sometimes, recalling the classic triad of scenario, probability, and consequences, and a simple mnemonic device from Kaplan and Garrick (Kaplan and Garrick, 1981) can help interdisciplinary teams to focus on a problem definition that will yield useful risk assessments for decision making. Paraphrasing slightly from Kaplan and Garrick (1981) and Kaplan (1997), it is useful to think of the risk assessment process as posing three simple questions:

- What can happen?
- How is likely (probable) to happen?
- What are the consequences if it does happen?

The first question focuses on describing possible risk scenarios (s) as in hazard identification and exposure assessment (below). The second question estimates or sometimes assigns judgments of probability, p. Finally, the third question links the possibilities to probable consequences, given that the scenario occurs. Given that participants in the risk assessment process likely have differing views of the meaning of "risk" and of phases of risk assessment, every risk assessment can benefit from thinking through the concepts as the risk assessment enters a problem definition and scoping exercise (Figure 14.1).

These questions humanize a more formal view of the scenario (s_i), its probability of realization (p_i), and the consequences (x_i) given that is has happened. The three fundamental questions not only guide the overall risk assessment, but

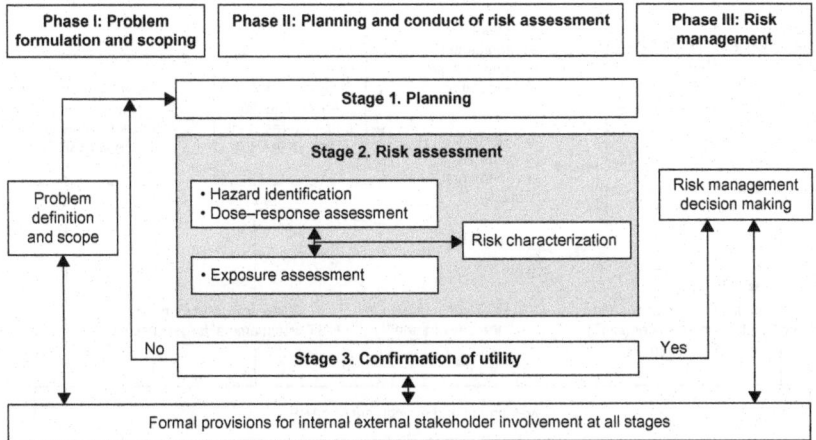

FIGURE 14.1   A framework for human health risk assessment. There are numerous frameworks or paradigms for risk assessment. Most capture similar principles for analysis of the risk(s) of adverse consequences from exposure(s) to hazards. *Source: Adapted from USEPA (2014a) and USEPA and USDA (2012).*

the same logic occurs repeatedly within the phases of the risk assessment. For example, within the exposure assessment phase of a risk assessment, particular exposure scenarios (what can happen) are elucidated and probabilities assigned. The consequences—harms—cannot occur without the combination of hazard and exposure. Dose–response relationships link the doses from exposures to possible adverse consequences.

ANTIMICROBIAL RESISTANCE RISK ASSESSMENT

Microbial (or "microbiological") risk assessment is a reasonable foundation for quantitative antimicrobial resistance risk assessments. When viewed from an epidemiology perspective of infectious disease risk assessment, microbial risk assessment has a very long history for bacteria, parasites, viruses, and fungi. Antimicrobial resistance is a global public health concern for certain microbes from all of these classes (World Health Organization, 2014); however, this chapter emphasizes bacteria in the food chain. Research on food- and waterborne bacteria and viruses have nineteenth century histories in epidemiology (Haas et al., 1999). More recently, Haas developed quantitative risk assessment as a means of setting regulatory safety limits for microbes in drinking water (Haas et al., 1999). A more focused effort on defining foodborne microbial risk assessment processes began in the 1990s (Buchanan and Whiting, 1998; Notermans and Teunis, 1996).

Risk assessments estimate risks indexed by the subpopulations at risk, different pathways of exposure and possibly risk management interventions; thus,

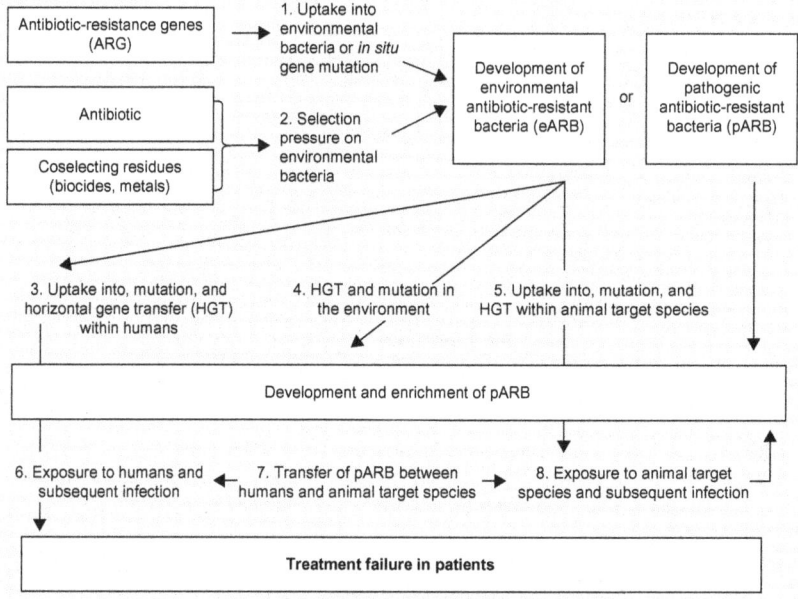

FIGURE 14.2 An example of a conceptual model for transfer of antimicrobial resistance in the animal and human populations. *Source: From Ashbolt et al. (2013). Used by permission of Environmental Health Perspectives.*

it might seem reasonable to think of antimicrobial resistance risk assessment as another index of microbial hazard for "resistant" and "sensitive" fractions of the total foodborne hazard. A general concept model of antimicrobial resistance is useful as a starting point (e.g., Figure 14.2), but underlying the compartments of antimicrobial resistance are the compartments and state transitions of the pathogens or commensal bacteria without resistance. Further complicating a general approach to risk assessment is the fact that the hazardous property in question—antimicrobial resistance—is not necessarily expressed after the original microbial exposure. Moreover, to become "adverse" within the individual who acquired resistant microbial infection a secondary exposure of the antibiotic is necessary. Without selection of the resistant microbes with an antimicrobial drug, it is not likely to have a detectable consequence unless the bacterium is pathogenic.

Hazard Identification

At its most general level of understanding, "[antimicrobial resistance] results in reduced efficacy of antibacterial, antiparasitic, antiviral and antifungal drugs, making the treatment of patients difficult, costly, or even impossible"

TABLE 14.2 Possible Adverse Health Effects Associated with Antimicrobial-Resistant Bacteria

Antibiotic resistance
Microbiological (genotype; phenotype; increased MIC)
Pharmacological (increased MIC at the infection site)
Clinical (established resistant infection)
Epidemiological (community resistance)

Duration of illness

Transfer of resistance determinants to pathogenic strains

Transfer of resistance to commensal strains
Colonization with resistant strains

Mortality from infection

Duration of treatment (with or without antibiotics)
ICU days
Central line catheter
Septicemia

Urinary tract infection (UTI)
Establishment of community resistance to antibiotic

(World Health Organization, 2014). As such the hazard in an antimicrobial resistance risk assessment might be defined broadly as a "property to cause antimicrobial resistance," but the rich complexity of mechanisms of action across microbes and within even narrowly defined hosts and species begs for more precise hazard identification within each risk assessment. For example, antibiotic resistance in bacteria can refer to a reduced *in vitro* susceptibility to an established minimum inhibitory concentration (MIC) of an antibiotic; to pharmacologic resistance if infections are not controlled by antibiotics that exceed the MIC at the site of infection; and to *in vivo* resistance if the resistance results in clinical treatment failures (Verraes et al., 2013).

Types of endpoints for risk assessments include not only the adverse health effects from exposures to pathogenic organisms, but secondary harm arising from the selection for resistant pathogens or commensal bacteria during antimicrobial treatment. Some examples of more explicit hazardous properties of antimicrobial-resistant endpoints are given in Table 14.2. Not all of the endpoints in Table 14.2 might be expressed adversely in the exposed individual, but refer to horizontal transfer of resistance that is viewed as adverse among the population at risk. For example, an individual who acquires resistant commensal bacteria might not experience antimicrobial-resistant disease; however, the individual might horizontally transfer resistance to the community at large. Establishment of community resistance, itself, is generally considered to be an adverse consequence of antimicrobial drug use in both agriculture and human medicine.

Because there are numerous cellular and molecular mechanisms to spread resistance including bacterial conjugation, transduction, and transformation, hazard assessment might logically include descriptions of relevant resistance elements, plasmids, and phages in addition to the biochemical mechanisms of action. Whether or not hazard identification includes detailed study across specific genotypes of resistance or other molecular-genetic classifications depends on the overall purpose of the risk assessment. For instance, a research-level risk assessment might define bacteriophages and plasmids carrying resistance genes as the hazards of interest for endpoints such as, "acquisition of resistance." Although the risk assessment approach can guide scientific processes for mechanistic studies and thereby contribute to the knowledge base, more direct connection with adverse human health endpoints is generally needed for public health risk assessments.

The hazard identification phase of an antimicrobial risk assessment identifies not only the potential health endpoints of concern, but also the factors predicting host susceptibility. A foundational concept in infectious disease biology and epidemiology is the notion that infectious disease depends on a "pathogen–host–environment" triangle of factors (Haas et al., 1999). If the vertices of the pathogen–host–environment are parsed into the organizational framework of risk assessment, then the hazard identification process favors identification of the pathogen–host aspects and the physiological "environment" of the host. The exposure assessment activity of risk assessment is more closely linked with "environment" as part of the infectious disease triad.

The acquisition of resistance in a host, the community, or a bounded subset of a community, for example, a hospital intensive care unit (ICU), might be defined as a potential harm (or "adverse effect") possible from exposures to resistant bacteria in the food chain. Although the hazards of antimicrobial resistance in the food chain are clearly established, the contribution of antimicrobial resistance to morbidity and mortality is more difficult to quantify or specify as part of the hazard identification. For example, in healthy hosts, resistant infections can resolve without intervention similar to sensitive infections. For some commensal bacteria, the prevalence of resistance in the relatively closed system of a poultry house seems to fade soon after removal of the antibiotic from feed can drop with the maintenance of the bedding (McDermott et al., 2005). In sum, endpoints such as "acquisition of resistance" in bacteria or humans after antibiotic need careful definition to link with a definition of a hazard assessment.

Finally, the concept of "medically necessary" recognizes that possibility that antibiotic uses in food animals might select for resistant organisms that can also colonize humans or transfer resistance to human-adapted strains of bacteria. "Medically necessary" is a science policy weighting on the potential harms from antibiotic resistance in the food chain and is used in regulatory risk management of animal antimicrobials (FDA, 2003; World Health Organization, 2001).

Sources of data and information for hazard identification include the research literature, databases on the use of antimicrobials in animals and humans, epidemiologic surveillance and specific studies on the endpoints and particular microbes of interest. Although the risk assessment literature on specific Gram-negative and Gram-positive species and commensal or pathogenic is extensive, it is also clear that risk assessors have yet to compile a single hazard assessment spanning all properties that might be helpful for antimicrobial resistance risk assessments. Given the rich complexity arising from numerous bacteria of interest or concern, the growing variety of drugs that might affect multiple molecular genetics and biochemical pathways in cells, and the host ranges of the bacteria, "antimicrobial resistance hazard" will likely remain a conceptually defined hazard that is more rigorously defined one risk assessment at a time.

Dose–Response Assessment

Whereas hazard identification focuses on the properties of an agent to cause harm—for example, the *possibilities* for risk of harm—dose–response assessment evaluates the quantitative relationship between dose and the probability of response among exposed individuals. Dose–response assessments review animal and human research for mathematical relationships between the dose of the hazardous agent and the incidence of defined adverse outcomes, such as mortality or morbidity. Dose–response relationships are also possible between the dose and measured levels of biomarkers of effect. Biomarkers might be shown to increase or decrease with infections and might eventually serve as surrogate measures of adverse health effects. With respect to antimicrobial resistance risk assessment, it is possible that the host might be colonized with resistant bacteria but with an absence of frank clinical infection.

Reports of dose–response relationships for the probability of illness from foodborne ingestions of known numbers of organisms, viruses, colony-forming units (CFUs), or "infective units" are scarce in the primary literature (Coleman and Marks, 1998; Haas et al., 1999; Verraes et al., 2013). Because the number of cells necessary to cause a focus of infection is theoretically a single cell, oocyst, or virus, large numbers of human volunteers would be necessary to attempt measuring and fitting dose–response relationships after ingesting few cells or viruses. Further limiting resolution in the relevant, low-dose range is that analytical tools cannot accurately determining small numbers of cells in an aliquot of liquid or a portion of food (Haas et al., 1999). Given the unlikely use of direct experimentation to establish a full range of dose–response relationships, most dose–response assessments in risk assessments are derived from basic microbiology theory, empirical studies in animals, or sometimes from exposure–response models fit to human foodborne outbreak data (Varma et al., 2015).

Basic microbiology theory leads to the idea that a single cell can infect an individual, grow into sufficient numbers to cause colonization, illness, or even death. This basic assumption provides support for one-hit formulations of dose–response relationships (Haas et al., 1999; USEPA and USDA, 2012). If r is the probability of infection from one organism and d is the number of infective units in a dose, then the cumulative probability of infection can be described by an exponential, $p = 1 - e^{-rd}$. Although the exponential model is appealingly simple, it does not account for variability in the pathogen–host relationship, or a distribution in infection probabilities for r. Thus, some risk assessors have recommended an approximate beta-Poisson model, where the cumulative probability of infection is given by

$$1 - (1 + d/\beta)^{-\alpha} \qquad (14.1)$$

Here, $\mapsto$ and $\updownarrow$ are parameters of the beta distribution for which ($\updownarrow \gg 1$) and ($\mapsto \ll \updownarrow$) (Coleman and Marks, 1998; Haas et al., 1999; USEPA and USDA, 2012). The parameters, $\mapsto$ and $\updownarrow$, are generally found empirically and not derived from first principles. There are a number of other relevant mathematical models for dose–response relationships which the interested reader can find derivations of and discussion or their applicability to human and animal studies in Coleman and Marks (1998), Cox Jr. (2002), Haas et al. (1999), and USEPA and USDA (2012).

The lack of significant experimental evidence for a nonthreshold response from human and animal experiments argues for modeling a threshold response or, perhaps, estimating a no observed effects level (NOAEL) analogous to chemical dose–response relationships; however, observation that human exposure–response data are sometimes consistent with linear no-threshold dose–response relationships has supported recommendations for the use of nonthreshold models (National Research Council, 2005; USEPA and USDA, 2012). Nonthreshold dose–response relationships are especially relevant for public health agencies that have conservative risk constraints based on risk management policies for controlling pathogenic bacteria in the food chain. Risk assessments performed for research interests might use alternatives to one-hit formulations such as the beta-Poisson model and logistic regression models (Coleman and Marks, 1998; Holcomb et al., 1999; USEPA and USDA, 2012).

Given that antimicrobial resistance risk assessment builds on a foundation of microbial risk assessment, the question arises about where to begin the "antimicrobial resistance" phase of the dose–response assessment. The answer to this question likely depends first on whether or not the hazardous microbes are pathogenic or opportunistically pathogenic. Dose–response assessment for morbidity from resistant bacteria is possibly very similar to that of the antimicrobial-sensitive strains unless there is a substantial energy cost to cells for maintaining resistance. Resistant and sensitive bacteria, in the presence of antimicrobial treatments, express different survival probabilities, r. But given the nearly identical infectious and toxic properties of resistant and sensitive, simply proportioning an ongoing infection by the ratio of resistant:sensitive for a static dose–response model is relatively uninformative. "Resistance" and the proportion of resistant bacteria in an individual, the community, or a confined ICU is substantially more informative for risk assessment when modeled dynamically (Austin and Anderson, 1999; Austin et al., 1997; Bonten et al., 2001).

The addition of drug treatment and subsequent evolution of a resistant infection or colonization from a foodborne infection complicates dose–response assessments by blurring any convenient linkages between administration of an antimicrobial drug to food animals and adverse health events in the human. Changes in probabilities of the bacteria growth, survival, or virulence as a function of the pharmacodynamics and pharmacokinetics of the antibiotic. Clearly, the data suggest that situation-specific risk assessments are much more informative than generalized, static resistant:sensitive proportions even for the "controlled" hospital environments.

In sum, in terms of most health outcomes, an initial dose–response assessment for antimicrobial-resistant pathogens does not differ from a dose–response assessment for the sensitive counterpart. Adverse consequences of antimicrobial resistance, such as increased MIC, generally require a second exposure-like event—antimicrobial drug treatment—before the consequences of antimicrobial resistance can be observed. Defining a classical dose–response relationship for antimicrobial resistance in commensal bacteria is also conceptually challenging: when does carriage of resistance in the community become an adverse health consequence?

Like hazard identification, the principles and methods used in a dose–response assessment can appear more than once in a risk assessment. First, identifying the hazard includes using the available information about dose–response relationships to characterize the hazard in terms of the possible intensity of effects. For example, different types of adverse outcomes are expected at different levels of dose. At the other end of the risk assessment process, dose–response is needed in order to predict the proportion of response to levels anticipated in the exposure assessment.

Exposure Assessment

Exposure assessment often becomes the broadest interdisciplinary undertaking in a risk assessment. For exposure to occur, a minimal exposure assessment consists of characterizing the sources of exposure, transport media, the fate of hazards during transport and, finally, intake or contamination of the receptor. Starting from basic concepts of *sources*, *transport*, *fate*, and *intake*, exposure assessment scenarios gain significant complexity when "the details" for specific pathogens or commensals in addition to features of the hosts are defined within the scope of a risk assessment (USEPA and USDA, 2012).

This chapter focuses on food as the medium of interest for antimicrobial resistance risk assessment in the human population. But even restricting a risk assessment to a broad class of media for exposure assessment raises the possibility of numerous pathways to reach the human receptor—pathways that

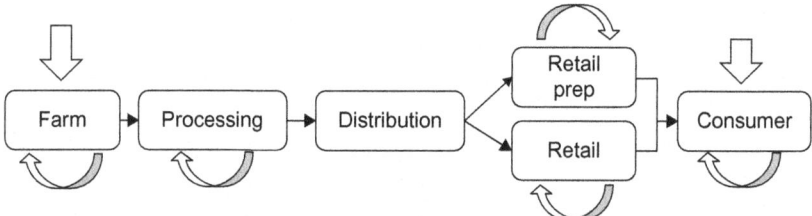

FIGURE 14.3 Antimicrobial treatments occur at both "ends" of a "farm-to-fork" food exposure pathway. The assumption in antimicrobial resistance risk assessment for foodborne exposures is that selection of resistant pathogens occurs from treatment of food animals with antibiotics. Antimicrobial resistance risk assessment usually requires a second treatment of exposures with antimicrobials at the human end of the food chain. The treatment can be *in vivo* in the course of treating infections or *in vitro* for purposes of measuring colonization by resistant bacteria among healthy communities.

can be quite removed from the theoretical direct flow from the food animal to the human (Figure 14.3). The fate and transport of antimicrobial resistance is through possibly numerous indirect pathways including opportunities for cross-contamination at the farm, abattoir, food processors, distribution, preparation in the retail or home environment and human-to-human transmission (Figure 14.4). Moreover, exposures to resistant bacteria are not restricted to an animal→human direction; but also include "backward" flow of human→food animals (Phillips et al., 2004; Verraes, et al., 2013). Each component in the complex exposure scheme could itself be multiplied by various serotypes and host ranges in some bacteria such as *Salmonella* spp. Risk assessments that are truly "farm-to-fork" intrinsically generate complexity for the process of estimating risk, the attribution of specific sources of resistance and the characterization of causal chains of events from source to receptor (Cox Jr., 2002; Hald et al., 2004; Mangen et al., 2010; USEPA and USDA, 2012).

The two sub-environments shown in Figure 14.4 exposure environment suggest that resistance determinants might circulate and essentially establish equilibrium conditions based on the flow of resistance into the environment. Modeling exposures in terms of dynamic flows is inherently more complex than statically measuring the prevalence of resistance at the animal and human "ends" of exposure pathways. There are numerous information and data gaps in the flow of resistance from many species of bacteria and relevant antibiotics that lead to relatively few efforts at quantifying farm-to-fork pathways. Although static measures of prevalence are useful for snapshots of integral exposure at the ends or among intermediate exposure pathways, dynamically modeling exposure assessment is useful for posing testable hypotheses on the transmission, accumulation, and elimination of resistance in the food chain (see below).

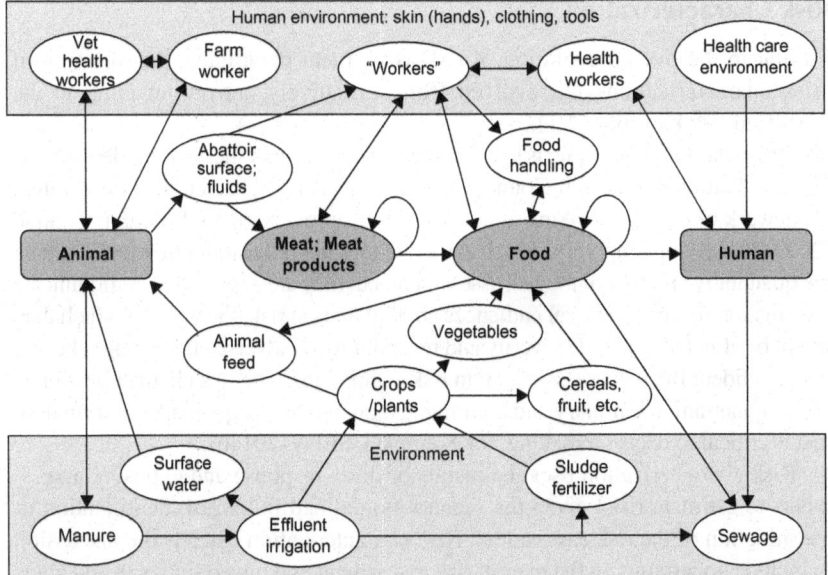

FIGURE 14.4   Possible exposure pathways between animals and humans. For both microbial risk and antimicrobial resistance risks, the exposure pathway between farm and fork might be highly complex due to numerous indirect pathways. Additionally, colonization with a resistant commensal strain complicates exposure scenarios with possible complexity in the time between release at the farm animal source and expression of resistance after antibiotic treatment in the human.

Exposure assessments for antimicrobial resistance risk assessments might be posited as having an exposure assessment for the antimicrobial drug embedded within the exposure assessment for the pathogenic or commensal bacteria. For example, the release assessment for resistant bacteria at the source of exposure (farm animal or plant) is expected to differ depending on the particular farm treatment protocol—growth promotion, treatment of infection, or prevention. The dose–response relationship between the antimicrobial drug and the prevalence of resistant clones or resistant carriage is not always clear (Berge et al., 2005; Gullberg et al., 2010; Oliver et al., 2011). The type of treatment (disease, and prevention or growth promotion) and the period of drug withdrawal further impact the prevalence of antimicrobial resistance and the likelihood of residues remaining at the time of slaughter. The prevalence of resistant colonies among bacteria isolated from food and food animals is often reported as correlated on a population scale but harder to predict on the individual scale. The exposure-relevant measurement of the proportions of contaminated food portions, given specific antibiotic uses on the farm, is lacking in many risk assessments for the foodborne resistance pathway.

Risk Characterization

Similar to the overall evolution of risk assessment paradigms, the meaning of "risk characterization" has evolved from a relatively narrow meaning in the NAS Red Book (Table 14.1).

The evolution of meaning is to a broader interpretation of the analytic-deliberative process (National Research Council, 1996) and sets risk characterization within a framework of decision making about risk, for example National Research Council (2009, 2011). Contemporary risk characterizations not only summarize and evaluate the quantitative finding of the risk assessment, but they are expected to communicate information for more general audiences about who is at risk, the ways in which they might be affected, what the severity and reversibility of adverse effects might be and how confident the risk assessors are in risk estimations. The risk characterization is both a quantitative summary and communication of relevant qualitative information that is critical to decision making, for example USEPA (2014b).

Risk characterization uses the results of dose–response and exposure assessments to estimate risks given the scenario(s) defined in hazard identification or the initiation of the risk assessment. Risk characterization extends the discussion to include uncertainty in the overall risk assessment and uncertainties in and alternatives to the plausible scenarios. This is done within the context of the endpoints that were established by the science policy decisions done in the problem scoping and definition steps. An array of risk estimations are presented for the scenarios, perhaps ranging from death *from* or *associated with* exposure to resistant organisms, additional days spent as an inpatient, to the duration of treatment days.

At least equally important to estimating risks, given the scenarios, is the treatment of uncertainty in the risk assessment. Uncertainty characterizations are typically discussed in the risk characterization; however, the qualitative identification and description of sources of uncertainty is done throughout the risk assessment. "Characterization of uncertainty" spans a wide range of activities from the qualitative to analytical (error propagation) and ultimately to stochastic approaches to estimate "deep uncertainties" (Cox, 2012; Walker et al., 2010). Like dynamic modeling, probabilistic uncertainty assessments and characterization offer quantitatively elegant tools, such as Monte Carlo simulations (e.g., Hurd et al., 2008) for analyzing and communicating uncertainty, especially among risk modelers. However, quantitative risk assessment with visual detailed cumulative risk distributions as output can "overcommunicate" by implying that more is certain about the risk assessment and various scenarios than is true or possible, given the many inputs into the risk assessment. Nevertheless, communicating percentiles of the risk distribution conveys not only a quantitative risk estimate but also the important concept of the uncertainty in the risk estimation. Additionally, risk distributions provide quantitative support for risk management statements about the population weighting of risk using, for example, a 95th percentile of the risk estimates.

Discussion about the uncertainty in causal scenarios is typically within the risk characterization. Causality is notoriously difficult to establish for many types of exposures and human diseases outcomes; thus, is typically narratively argued. The rich complexity of antimicrobial resistance risk assessments creates uncertainty in causal pathways, perhaps best exemplified by the possibilities of alternative sources of resistance in the food chain and human→animal flow of resistance for certain bacteria and antibiotic combinations (Berge et al., 2005; D'Costa et al., 2011; Mayrhofer et al., 2006; Phillips et al., 2004). Logical arguments, such as the Bradford Hill criteria, have been used for decades to support causality; however, these criteria are often necessary but not sufficient in antimicrobial resistance risk assessment (Cox Jr., 2002). Quantitative causality assessment has been proposed (Pearl, 2000); however, formalism is not widely accessible to many stakeholders in antimicrobial risk assessments.

MATHEMATICAL AND OTHER RISK ANALYTICAL APPROACHES

Numerous recent contributions to the research literature on antimicrobial resistance in foodborne and hospital- or community-acquired infections have informed a broad understanding of risk analysis for foodborne antimicrobial resistance risk analysis. For example, basic molecular and cellular biology of resistance in bacteria (discussed elsewhere in this volume) and human responses to resistant infections continue to strengthen hazard assessment and risk characterization. The overall risk characterization and features of exposure assessment are informed by research in dynamic modeling of resistance acquisition in hospitals and the food chain (Austin and Anderson, 1999; Austin et al., 1997; Bonten et al., 2001; Hurd et al., 2008; Leibovici et al., 2001). Finally, food risk attribution methods have informed risk management options for well-defined *Salmonella* hazards supported with adequate surveillance sampling plans (Hald et al., 2004; Mangen et al., 2010).

A major difference between dose–response modeling for microbial risks in general and for antimicrobial/antibiotic risks is the potential dynamics of resistance acquisition in the exposure source and the subsequent infection and colonization of human receptors. To better understand the population dynamics of resistance in the food chain, controlled hospital environments, and the community, Austin and colleagues have developed dynamic compartmental models not only as quantitative descriptions but as useful tools for proposing risk mitigation (Figure 14.5; Austin and Anderson, 1999; Austin et al., 1997). Validations of dynamic models for controlled environments of ICUs and hospital wards attest to the usefulness of dynamic modeling. With commensal organisms in the food chain, including *Enterococcus* and *Escherichia coli* strains, the possibility for asymptomatic carriage of antibiotic resistance after foodborne exposure has fueled much of the concern. The complexity that "silent" carriage brings to infectious disease epidemiology generally is only amplified by the complexities of the food chain when exposures can be removed in both time and space from a more direct exposure pathway (Hammerum and Heuer, 2009; Mayrhofer et al., 2006; Smith et al., 2002).

Definitive attribution of causality continues to challenge risk assessors for adverse health effects from exposures to numerous types of hazards.

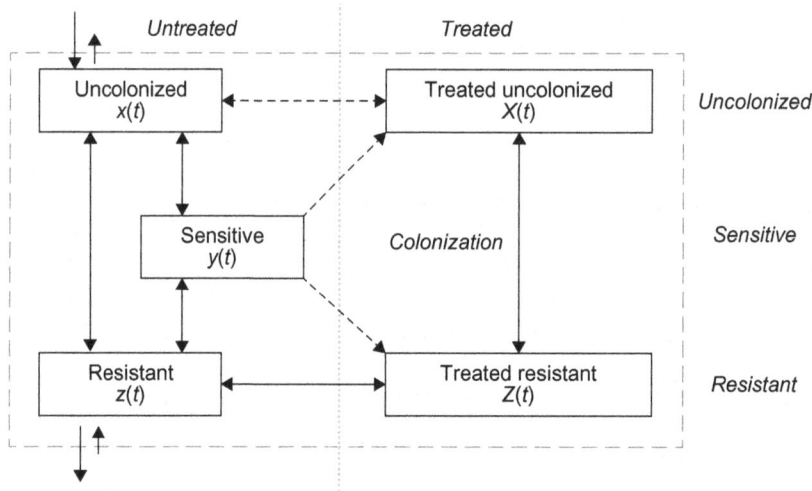

FIGURE 14.5   Basic dynamic model for the acquisition and maintenance of resistance in a defined setting. Dynamic population models have been developed and employed in the identification of risk control methods, particularly for nosocomial infections in ICUs and hospital wards. To apply modeling to food chain exposures, dynamic models similar to the one depicted would be helpful for nearly each component of Figure 14.3 including linking them across the entire pathway.

Antimicrobial resistance risk assessment is no exception. Deterministic attribution of source of risk, helpful for understanding causality, has a long history in disease outbreak studies. Hald et al. (2004) have brought probabilistic risk assessment to source attribution models using Bayesian networks. This method can work for relative risk estimates within a closely defined sampling frame; but it does not provide quantitative information on the risk of new cases. Nevertheless, it will likely to be useful in root cause investigations and in developing risk management options.

Fitting Antimicrobial Resistance Risk Assessment into Public Health Risk Management

A key principle shared by most risk analysis frameworks, is that risk management includes the organization's decision-making activity about controlling, accepting, or mitigating risks. Risk managers weigh the output of risk assessments with programmatic constraints, stakeholder concerns and interests, and strategic risk management goals. Thus, a quantitative risk assessment might be thought of as one input into a multicriteria decision to manage risk. Risk management decisions to accept, mitigate, or eliminate a given risk are made using not only the "objective" input from risk assessment, but inputs of social, political, and economic factors. In fact, recent NAS recommendations to FDA use objective risk information as an *attribute* of a multicriteria decision analysis

TABLE 14.3 Some Factors that Can Generate Complexity and Uncertainty in Antimicrobial Resistance Risk Assessment*

Microbial growth and death
Fitness of resistant versus sensitive strains
Host immunity and susceptibility
Diversity of health endpoints
Duration of infection for resistant versus sensitive infections
Genetic diversity of resistance mechanisms and evolution of microbial strains
Potential for secondary transmission
Relative virulence
Heterogeneous spatial and temporal distribution in the environment
Single exposure health outcome
Modification with antibiotic treatment
Wide range of microbial response to interventions
Detection method sensitivity
Population, community, and ecosystem-level dynamics
Routes of exposure
Probability of resistance
Likelihood that the infected seek treatment
Likelihood that the antimicrobial drug of interest is selected for treatment
Likelihood that a proportion of the affected population will fail treatment (host factors; agent virulence)

Haas et al. (1999).

(MCDA) in a risk characterization framework (National Research Council, 2011). Elsewhere, EPA recommends risk assessments be essentially "fit for purpose" of decision making (USEPA, 2014a,b). Although it is always desirable that risk management remains a distinct process from risk assessment (National Research Council, 1983), it is apparent that the evolution of risk analysis in public health agencies led to a closer integration of risk assessment with risk management's decision making. Fewer in-depth risk assessments for exposure to foodborne hazards are commissioned in favor of quickly accessible meta-analyses and expert elicitation of risks given foodborne hazards rankings. Happily, the parallel evolution in efficient risk assessment methods means that risk managers can continue to rely on science-based risk assessments to quantify human health risks among the risk management decisions (Table 14.3).

ACKNOWLEDGMENT

This chapter represents the thoughts of the author, only, and does not represent the policy of the FDA.

REFERENCES

Ashbolt, N., Amezquita, A., Backhaus, T., Borriello, P., Brandt, K., Collignon, P., et al., 2013. Human Health Risk Assessment (HHRA) for environmental development and transfer of antibiotic resistance. Environ. Health Perspect. 121, 993–1001.

Austin, D.J., Anderson, R.M., 1999. Studies of antibiotic resistance within the patient, hospitals and the community using simple mathematical models. Philos. Trans. R. Soc. Lond. B. Biol. Sci. 354, 721–738.

Austin, D., Kakehashi, M., Anderson, R., 1997. The transmission dynamics of antibiotic-resistant bacteria: the relationship between resistance in commensal organism. Proc. R. Soc. Lond. B 264, 1629–1638.

Berge, A.C.B., Atwill, E.R., Sischo, W.M., 2005. Animal and farm influences on the dynamics of antibiotic resistance in faecal *Escherichia coli* in young dairy calves. Prev. Vet. Med. 69, 25–38.

Bonten, M.J., Austin, D.J., Lipsitch, M., 2001. Understanding the spread of antibiotic resistant pathogens in hospitals: mathematical models as tools for control. Clin. Infect. Dis. 33, 1739–1746.

Buchanan, R.L., Whiting, R.C., 1998. Risk assessment: a means for linking HACCP plans and public health. J. Food Prot. 61, 1531–1534.

Claycamp, H.G., 2007. Perspective on quality risk management of pharmaceutical quality. Drug Inf. J. 41, 351–367.

Claycamp, H., Hooberman, B., 2004. Antimicrobial resistance risk assessment for food safety. J. Food Prot. 67, 2063–2071.

Coleman, M., Marks, H., 1998. Topics in dose–response modeling. J. Food Prot. 61, 1550–1559.

Cox Jr., L., 2002. Risk Analysis: Foundations, Models and Methods. Kluwer Academic Publishers, Boston, MA.

Cox, L., 2012. Confronting deep uncertainties in risk analysis. Risk Anal. 32, 1607–1629.

D'Costa, V., King, C., Kalan, L., Morar, M., Sung, W., Schwartz, C., et al., 2011. Antibiotic resistance is ancient. Nature 477, 457–461.

European Food Safety Authority, and European Centre for Disease Prevention and Control, 2014. The European Union Summary Report on antimicrobial resistance in zoonotic and indicator bacteria from humans, animals and food in 2012. EFSA J. 12, 3590–3926.

FAO, WHO. 2009. Risk Characterization of Microbiological Hazards in Food. FAO and WHO (Eds.), Rome: FAO/WHO Microbiological Risk Assessment Series 17.

FDA, 2003. Guidance for Industry: Evaluating the Safety of Antimicrobial New Animal Drugs with Regard to Their Microbiological Effect on Bacteria of Human Health Concern. US Food and Drug Administration, Center for Veterinary Medicine, Rockville, MD.

Forsythe, S.J., 2002. The Microbiological Risk Assessment of Food. Blackwell Publishing, Oxford, UK.

Gullberg, E., Cao, S., Berg, O., Illbäck, C., Sandegren, L., Hughes, D., et al., 2010. Selection of resistant bacteria at very low antibiotic concentrations. PLoS. Pathog. 7, 1–9.

Haas, C.N., Rose, J., Gerba, C., 1999. Quantitative Microbial Risk Assessment. Wiley & Sons, New York, NY.

Hald, T., Vose, D., Wegener, H., Koupeev, T., 2004. A Bayesian approach to quantify the contribution of animal-food sources to human salmonellosis. Risk Anal. 24, 255–269.

Hammerum, A.M., Heuer, O.E., 2009. Human health hazards from antimicrobial-resistant *Escherichia coli* of animal origin. Clin. Infect. Dis. 48, 916–921.

Holcomb, D., Smith, M., Ware, G., Hung, Y., Brackett, R., Doyle, M., 1999. Comparison of six dose-response models for use with food-borne pathogens. Risk Anal. 19, 1091–1100.

Hurd, H., Enøe, C., Sørensen, L., Wachmann, H., Corns, S., Bryden, K.M., et al., 2008. Risk-based analysis of the Danish pork salmonella program: past and future. Risk Anal. 28, 341–351.

Kaplan, S., 1997. The words of risk analysis. Risk Anal. 17, 407–417.

Kaplan, S., Garrick, B., 1981. On the quantitative definition of risk. Risk Anal. 1, 11–27.

Leibovici, L., Berger, R., Gruenewald, T., Yahav, J., Yehezkelli, Y., Milo, G., et al., 2001. Departmental consumption of antibiotic drugs and subsequent resistance: a quantitative link. J. Antimicrob. Chemother. 48, 535–540.

Mangen, M., Batz, M., Käbohrer, A., Hald, T., Morris, J., Taylor, M., et al., 2010. Integrated approaches for the public health prioritization of foodborne and zoonotic pathogens. Risk Anal. 30, 782–797.

Mayrhofer, S., Paulsen, P., Smulders, F., Hilbert, F., 2006. Antimicrobial resistance in commensal *Escherichia coli* isolated from muscle foods as related to the veterinary use of antimicrobial agents in food-producing animals in Austria. Microb. Drug Resist. 12, 278–283.

McDermott, P., Cullen, P., Huber, S., McDermott, S., Bartholomew, M., Simjee, S., et al., 2005. Changes in antimicrobial susceptibility of native *Enterococcus faecium* in chickens fed virginiamycin. Appl. Environ. Microbiol. 71, 4986–4991.

National Research Council, 1983. Risk Assessment in the Federal Government: Managing the Process. National Academy Press, Washington, DC.

National Research Council, 1996. Understanding Risk: Informing Decisions in a Democratic Society. National Academy Press, Washington, DC.

National Research Council, 2005. Reopening Public Facilities After a Biological Attack: A Decision-Making Framework. National Academies Press, Washington, DC.

National Research Council, 2009. Science and Decisions: Advancing Risk Assessment. National Academies Press, Washington, DC.

National Research Council, 2011. A Risk-Characterization Framework for Decision-Making at the Food and Drug Administration. National Academies Press, Washington, DC.

Notermans, S., Teunis, P., 1996. Quantitative risk analysis and the production of microbiologically safe food: an introduction. Int. J. Food Microbiol. 30, 3–7.

Oliver, S., Murinda, S., Jayarao, B., 2011. Impact of antibiotic use in adult dairy cows on antimicrobial resistance of veterinary and human pathogens: a comprehensive review. Foodborne Pathog. Dis. 8, 337–355.

Pearl, J., 2000. Causality. Models Reasoning, and Inference. Cambridge University Press, Cambridge, UK.

Phillips, I., Casewell, M., Cox, T., De Groot, B., Friis, C., Jones, R., et al., 2004. Does the use of antibiotics in food animals pose a risk to human health? A critical review of published data. J. Antimicrob. Chemother. 53, 28–52.

Salisbury, J.G., Nicholls, T.J., Lammerding, A.M., Turnidge, J., Nunn, M.J., 2002. A risk analysis framework for the long-term management of antibiotic resistance in food-producing animals. Int. J. Antimicrob. Agents 20, 153–164.

Smith, D.L., Harris, A.D., Johnson, J.A., Silbergeld, E.K., Morris, J.G., 2002. Animal antibiotic use has an early but important impact on the emergence of antibiotic resistance in human commensal bacteria. Proc. Natl. Acad. Sci. USA 99, 6434–6439.

USEPA. 2014a. Framework for Human Health Risk Assessment to Inform Decision Making. ed RAF Office of the Science Advisor.

USEPA, 2014b. In: Fowle, J., Dearfield, K. (Eds.), Risk Characterization Handbook Science Policy Council, US Environmental Protection Agency, Washington, DC.

USEPA, USDA. 2012. Microbial Risk Assessment Guideline: Pathogenic Microorganism with Focus on Food and Water (Prepared by the Interagency Microbiological Risk Assessment Guideline Workgroup July 2012).

Varma, J., Greene, K., Ovitt, J., Barrett, T., Medalla, F., Angulo, F., 2015. Hospitalization and anti-microbial resistance in *Salmonella* outbreaks 1984–2002. Emerg. Infect. Dis. 11, 943–946.

Verraes, C., Van Boxstael, S., Van Meervenne, E., Van Coillie, E., Butae, P., Catry, B., et al., 2013. Antimicrobial resistance in the food chain: a review. Int. J. Environ. Res. Public Health 10, 2643–2669.

Vose, D., 1998. The applications of quantitative risk assessment to microbial food safety. J. Food Prot. 61, 640–648.

Walker, W., Marchau, V., Swanson, D., 2010. Addressing deep uncertainty using adaptive policies: introduction to section 2. Technol. Forecast. Soc. Change 77, 917–923.

World Health Organization, 2001. WHO Global Strategy for the Containment of Antimicrobial Resistance. Department of Communicable Disease Surveillance and Response, Geneva, Switzerland.

World Health Organization, 2014. Antimicrobial Resistance: Global Report on Surveillance. World Health Organization, Geneva.

Yoe, C., 2012. Primer on Risk Analysis. Decision Making Under Uncertainty. CRC Press, Taylor & Francis Group, Boca Raton, FL.

Chapter 15

Food Microbial Safety and Animal Antibiotics

Louis Anthony (Tony) Cox

NextHealth Technologies, Cox Associates and University of Colorado, Denver, CO, USA

Chapter Outline

INTRODUCTION

The risk of food poisoning from consumption of food contaminated with disease-causing bacteria and antibiotic-resistant "superbugs" sparks strong political passions, dramatic media headlines, and heated science-policy debates (Chang et al., 2014). A widespread concern is that the use of antibiotics for farm animals creates selection pressures that favor the spread of antibiotic-resistant bacteria such as meticillin-resistant *Staphylococcus aureus* (MRSA), multi-drug resistant *Salmonella*, or *Escherichia coli* (CBS, 2010). Although the most common effects of foodborne illness are diarrhea and possibly fever, vomiting and other symptoms of food poisoning, more serious harm, or death, may occur in vulnerable patients. This is especially likely if foodborne bacterial infections are resistant to usually recommended antibiotic therapies, as might happen if the infections are caused by bacteria from farms where antibiotics are used for purposes of growth promotion or disease prevention. Patients with immune systems compromised by chemotherapy, AIDS, organ transplants, or other sources can have risks hundreds or thousands of times greater than those of consumers with healthy immune systems. Fear that use of animal antibiotics on farms contributes to a rising tide of antibiotic-resistant bacterial infections has spurred

Antimicrobial Resistance and Food Safety. DOI: http://dx.doi.org/10.1016/B978-0-12-801214-7.00015-6

many scientists, physicians, activists, journalists, and members of the public to call for elimination of the use of antibiotics as animal growth promoters.

However, these concerns and calls for action have usually not quantified *how many* excess deaths, treatment failures, or days of illness each year are caused in the United States by antibiotic-resistant bacteria arising specifically from animal antibiotic use, as opposed to other sources. Most human cases of MRSA and other resistant infections are healthcare-associated. They arise, for example, from inadequate hand washing and infection control in hospitals (Kallen et al., 2010). Although media reports sometimes conflate stories on foodborne resistance with statistics reflecting hospital-acquired cases (e.g., CBS, 2010), these are in fact quite distinct etiologies. They can now often be discriminated by identifying specific molecular markers for animal-associated as compared to hospital-associated strains of bacteria, allowing source-tracking based on molecular profiles of the bacteria found in infected patients. How many infections and fatalities per year arise among hospital patients, butchers, slaughterhouse workers, farmers, or the general public from livestock operations, meat handling, and consumption remains a topic of continuing interest, and such source-tracking is providing increasingly powerful molecular biological tools for obtaining answers. Responsible risk management is supported best by understanding how large the human health risks are now, and how much they would be changed by proposed interventions. The size of the risk depends on the care taken to reduce microbial loads by participants throughout the food chain, including use of microbial safety controls during farming, transportation, slaughter, production and packaging, storage, retail, and food preparation and cooking.

This first part of this chapter introduces methods of quantitative risk assessment (QRA) for quantifying the number of adverse human health impacts per year caused by animal antibiotic use. Next, we summarize quantitative estimates and bounds on human health harm obtained by applying these methods to available data for several types of resistant bacteria ("drugs and bugs") of greatest concern for public health in the United States. Finally, we discuss implications of such quantitative estimates for risk management. Throughout the chapter, human health risks are expressed as expected numbers of illnesses, fatalities, illness-days, or quality-adjusted life years (QALYs) lost to illness per year (for population risks) or per capita-year (for individual risks). QRA can help to inform and improve risk management decisions and policies by predicting how changes in the food production process, such as greater or lesser use of antibiotics on the farm, will affect human health risks, including individual and population risks for such subpopulations as well as for the whole population of concern.

METHODS OF QRA

This section reviews methods of quantitative microbial risk assessment (QMRA) and antimicrobial risk assessment. It expands upon and updates the brief summary in Cox (2008).

Farm-to-Fork Risk Simulation Model

When enough data and understanding are available, the effects of alternative risk management actions on risks created by bacteria in food—both antibiotic-resistant and antibiotic-susceptible strains—can be quantified by simulating microbial loads of bacteria along the chain of steps leading from production to consumption for each intervention. If the conditional frequency distribution of microbial loads leaving each step (e.g., slaughter, transportation, processing, storage) can be estimated, given the microbial load entering that step, and given the controls applied (e.g., use of antibiotic sprays, refrigeration), then the effects on microbial loads of alternative risk management policies can be quantified and compared. Microbial loads are typically expressed in units such as colony-forming units (CFUs) of bacteria per unit (e.g., per pound, per carcass) of food. If, in addition, dose-response relations are available to convert microbial loads in ingested foods to corresponding probabilities of illnesses, together with measures of illness severity (e.g., illness-days, QALYs lost, fatalities), then the effects of alternative risk management policies on human health can also be estimated and compared. As an example, Figure 15.1 shows how the frequency distribution of illnesses per year caused by *Vibrio parahaemolyticus* in oysters are predicted to change if refrigeration time requirements that accomplish different degrees of reduction in microbial loads are implemented. The underlying QMRA model simulates the changes in microbial loads at successive stages from harvesting to consumption; its main logical structure and data inputs are shown in Figure 15.2.

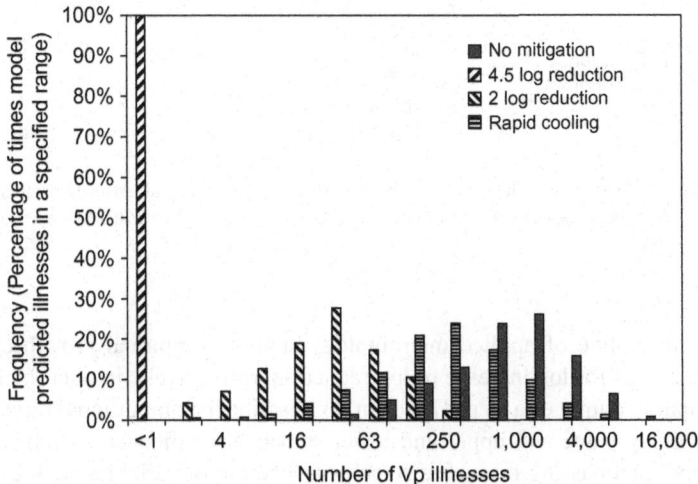

FIGURE 15.1   Frequency distributions of number of *V. parahaemolyticus* (Vp) illnesses per year from oysters with and without mitigation from cooling requirements. *Source: www.fda.gov/Food/FoodScienceResearch/RiskSafetyAssessment/ucm184074.htm.*

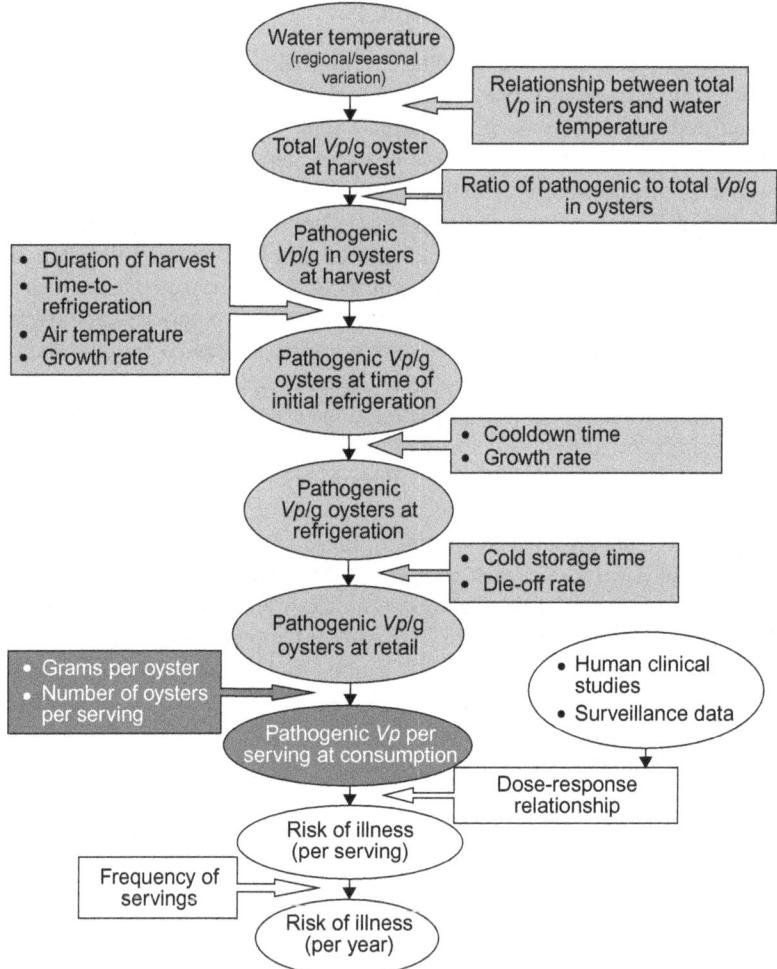

FIGURE 15.2 Structure of the QMRA model that allows quantitative risk estimates such as those in Figure 15.1 to be made. *Source: www.fda.gov/Food/FoodScienceResearch/RiskSafetyAssessment/ ucm185190.htm.*

The discipline of applied microbiology supplies empirical growth curves and kill curves for log increase or log reduction, respectively, in microbial load from input to output of a step. These curves describe the output:input ratio (e.g., a most likely value and upper and lower statistical confidence limits) for the microbial load passing through a stage as a function of variables such as temperature, pH, and time.

A "farm-to-fork" simulation model can be constructed by modeling successive steps representing stages in the food production process. Each step receives

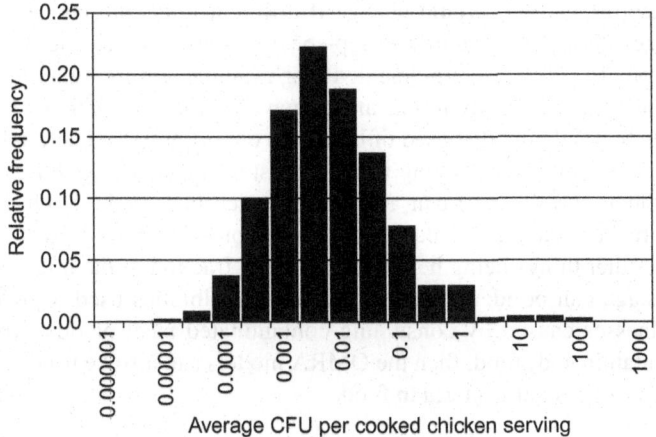

FIGURE 15.3 Average dose (CFU *Salmonella*) per serving in meals prepared from contaminated broilers. *Source: www.fao.org/docrep/005/y4392e/y4392e0r.htm.*

a microbial load from its predecessor. It produces as output a microbial load value sampled from the conditional frequency distribution of the output microbial load, given the input microbial load, as specified by the microbial growth model describing that stage. Measured frequency distributions of microbial loads on animals (or other units of food) leaving the farm provide the initial input to the whole model. The key output from the model is a frequency distribution of the microbial load, x, of pathogenic bacteria in servings of food ingested by consumers.

Risk-reducing factors such as antimicrobial sprays and chilling during processing, freezing or refrigeration during storage, and cooking before serving are often modeled by corresponding reduction factors for microbial loads. (These may be represented as random variables, for example, with log-normal distributions and geometric means and variances estimated from data.) The complete model is then represented by a product of factors that increase or decrease microbial loads, applied to the empirical frequency distribution of initial microbial loads on which the factors act. Running the complete farm-to-fork model multiple times produces a final distribution of microbial loads on servings eaten by consumers. Some farm-to-fork exposure models also consider effects of cross-contamination in the kitchen, if pathogenic bacteria are expected to be spread to other foods by poor kitchen hygiene practices (e.g., failure to wash a cutting board after use). As an example of the output from such a model, Figure 15.3 shows an example of the distribution of microbial loads of *Salmonella* in servings of chicken.

In summary, farm-to-fork simulation models can estimate the frequency distributions of microbial loads ingested by consumers in servings of food. As already illustrated in Figure 15.1, QMRA can also estimate how these frequency distributions would change if different interventions (represented by changes in one or more of the step-specific factors increasing or decreasing microbial load) were implemented. For example, enforcing a limit on the maximum time that ready-to-eat meats may be stored at delis or points of retail sale before being disposed of limits the opportunity for bacterial growth prior to consumption. Changing processing steps (such as scalding, chilling, antimicrobial sprays, etc.) can also reduce microbial loads. Such interventions shift the cumulative frequency distribution of microbial loads in food leftward, other things being held equal. If some fraction of the microbial load at each stage can be identified as resistant to antibiotics used to treat foodborne illnesses caused by consuming contaminated meat or other (possibly cross-contaminated) food, then the QMRA models can also be used to predict exposures to resistant bacteria in food.

Dose-Response Models for Foodborne Pathogens

Once a serving of food (e.g., chicken, oysters, hamburger, deli meats) reaches a consumer, the probability that an ingested dose will cause infection and illness is described by dose-response models. Several parametric statistical models have been developed to describe the relation between quantity of bacteria ingested in food and resulting probability of illness. One of the simplest is the following exponential dose-response relation:

$$r(x) = \text{Pr(illness} \mid \text{ingest microbial load} = x\text{CFUs)} = 1 - e^{-\lambda x}.$$

This model gives the probability that an ingested dose of xCFUs of a pathogenic bacterium will cause illness. $r(x)$ denotes this probability. The function $r(x)$ is a *dose-response curve*. λ is a parameter reflecting the potency of the exposure in causing illness. Sensitive subpopulations have higher values of λ than the general population.

More complex dose-response models (especially, the widely used Beta-Poisson model) have two or more parameters, for example, representing the population distribution of individual susceptibility parameter values and the conditional probability of illness given a susceptibility parameter. The standard statistical tasks of estimating model parameters, quantifying confidence intervals or joint confidence regions, and validating fitted models can be accomplished using standard statistical methods such as maximum likelihood estimation and resampling methods. The excellent monograph by Haas et al. (1999) provides details and examples. It notes that "It has been possible to evaluate and compile a comprehensive database on microbial dose-response models." Chapter 9 of this monograph provides a compendium of dose-response data and dose-response curves, along with critical evaluations and results of validation studies, for the following: *Campylobacter jejuni* (based on human feeding study data),

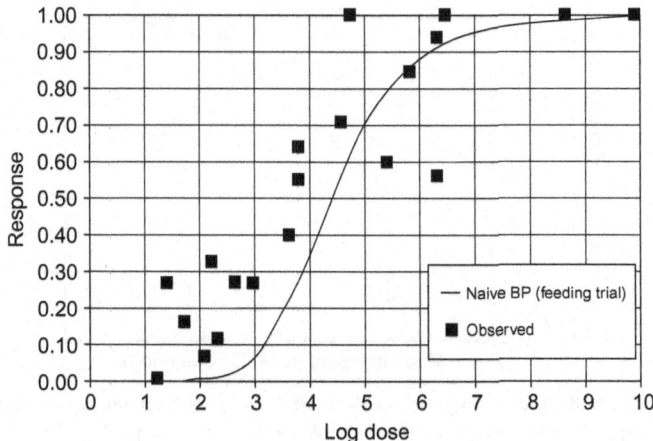

FIGURE 15.4   Even the best-fitting dose-response model in a specified parametric family, such as the Beta-Poisson (BP) family, may provide a biased description of data. Here, a best-fitting dose-response model for *Salmonella* data systematically underestimates risk at low doses. *Source: www. fao.org/DOCREP/005/Y4392E/y4392e10.gif.*

Cryptosporidium parvum, pathogenic *E. coli*, *E. coli* O157:H7 (using *Shigella* species as a surrogate), *Giardia lamblia*, non-typhoid *Salmonella* (based on human feeding study data), *Salmonella typhi*, *Shigella dystenteriae*, *S. flexneri*, *Vibrio cholerae*, Adenovirus 4, Coxsackie viruses, Echovirus 12, Hepatitis A virus, Poliovirus I (minor), and rotavirus. Thus, for many foodborne and water-borne pathogens of interest, dose-response models and assessments of fit are readily available.

Despite this base of relatively well-developed and validated dose-response models, however, two important challenges remain in developing dose-response models for specific strains of pathogenic bacteria, including antibiotic-resistant strains. Figure 15.4 illustrates the first, and Figure 15.5 the second. In Figure 15.4, the best-fitting model in a specific class of parametric models (the "naïve" Beta-Poisson model, which provides a widely used approximate mathematical model for response probabilities for different doses (CFUs) ingested) provides a clearly biased description of the observed feeding trial data, underestimating all observed response probabilities for log dose <5. Figure 15.5 illustrates the problem of low-dose extrapolation, in which the dose-response relation at doses far below the range of observed data depends greatly on which specific model is assumed.

Because of these challenges, dose-response models may be highly uncertain for specific strains of pathogens, and hence risk projections based on them may also be very uncertain. Characterizing this uncertainty is a key step in QMRA that uses dose-response models.

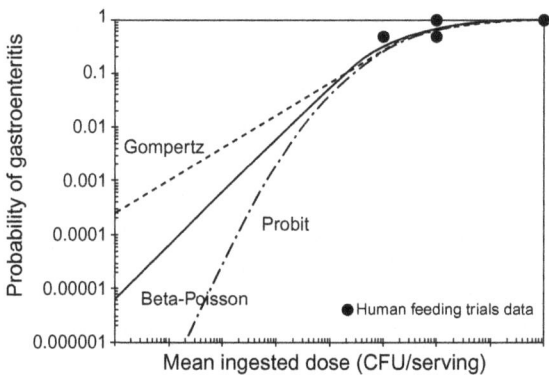

FIGURE 15.5 Multiple dose-response models that fit the available experimental data equally well may make very different predictions for risks outside the range of observed data. *Source: www.fda. gov/Food/FoodScienceResearch/RiskSafetyAssessment/ucm185177.htm.*

Quantitative Risk Characterization for QMRA and Risk Management

Sampling values of exposures x from the frequency distribution predicted by a farm-to-fork model (expressed in units of bacteria-per-serving) and then applying the dose-response relation $r(x)$ to each sampled value of x produces a frequency distribution of the risk-per-serving in an exposed population. This information can be displayed in various ways to inform risk management decision-making.

For example, Figure 15.6 shows how the (base 10 logarithm of) risk-per-serving of chicken for salmonellosis is reduced by a mitigation strategy that encourages consumers to cook chicken properly before eating it, based on the exposure sub-model in Figure 15.3, the Beta-Poisson dose-response model in Figure 15.5, and assumptions about how mitigation measures will affect the distribution of cooking practices.

Other displays showing the expected number of illnesses per year in a population, expected illnesses per capita-year in the overall population and for members of sensitive subpopulations, and frequency distributions or upper and lower confidence limits around these expected values are typical outputs of risk characterization. Suppose that particular risk management decisions are being considered, such as setting new standards for the maximum times and/or temperatures at which ready-to-eat meats can be stored before being disposed of. Then plotting expected illnesses per year against the decision variables (i.e., maximum times or temperatures, in this example) provides the quantitative links between alternate decisions and their probable health consequences needed to guide effective risk management decision-making. Figure 15.7 illustrates the

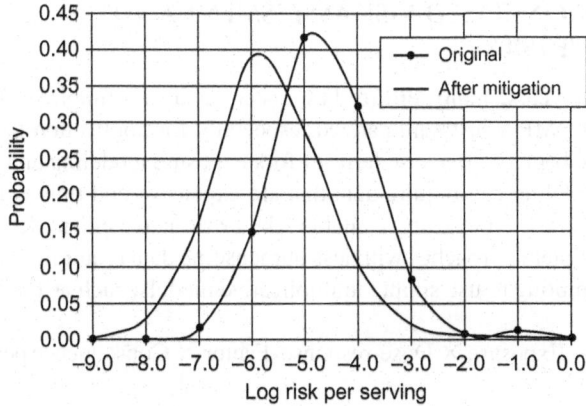

FIGURE 15.6 A display showing how salmonellosis risk per serving of chicken is reduced by better cooking practices. *Source: www.fao.org/docrep/005/y4392e/y4392e0r.htm#bm27.*

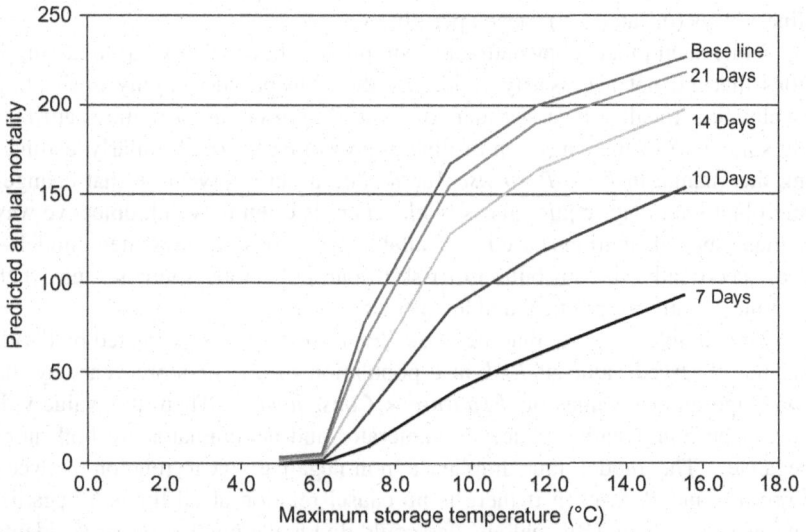

FIGURE 15.7 Expected elderly mortalities per year from *Listeria monocytogenes* if different maximum storage times and temperatures are allowed. *Source: www.fda.gov/Food/FoodScienceResearch/RiskSafetyAssessment/ucm197644.htm.*

key concept of informing risk management decisions by showing how a measure of risk (here, expected deaths per year) varies with decisions about maximum allowed storage times and temperatures. Such displays, linking actions to their probable consequences, provide the essential information needed to inform risk management decisions.

ATTRIBUTION-BASED RISK ASSESSMENT AND CONTROVERSIES

In recent years, many efforts have been made to simplify the standard approach to QMRA just summarized, especially for application to antimicrobial-resistant bacteria. Because farm-to-fork exposure modeling and valid dose-response modeling can require data that are expensive and time consuming to collect, or that are simply not available when risk management decisions must be made, simpler approaches with less burdensome data requirements are desirable. It is tempting to use simple multiplicative models, such as the following:

$$\text{Risk} = \text{Exposure} \times \text{Dose-response Factor} \times \text{Consequence per case}$$

where *Risk* is the expected number of excess illness-days per year, *Exposure* is measured in potentially infectious meals ingested per year in a population, *Dose-Response Factor* is the expected number of illnesses caused per potentially infectious meal ingested, and *Consequence per case* is measured in illness-days (or fatalities) caused per illness.

While such models have attractive simplicity, they embody strong assumptions that are not necessarily valid, and thus can produce highly misleading results. Specifically, the assessment of *Dose-Response Factor* requires attributing some part of the causation of illness-days to *Exposure*. Similarly, estimating the change in *Dose-Response Factor* due to an intervention that changes microbial load may require guess-work. There is often no valid, objective way to make such attributions based on available data. The risk assessment model—and, specifically, the attribution of risk to particular food sources—may then become a matter of political and legal controversy.

For example, suppose that the *Dose-Response Factor* is estimated by dividing the observed value of *Risk* in a population in one or more years by the contemporaneous values of (*Exposure* × *Consequence*). Then, this value will *always* be non-negative (since its numerator and denominator are both non-negative). The model thus implies a non-negative linear relation between *Exposure* and *Risk*, even if there is no causal relation at all (or is a negative one) between them. (By analogy, one could divide the number of car accidents in Florida in a year by the number of oranges consumed in Florida that year, but the resulting "car accidents per orange consumed" ratio, although certainly positive, would not in any way imply a causal relation, or that reducing consumption of oranges would reduce car accidents per capita-year. Replacing car accidents with foodborne illnesses such as antibiotic-resistant salmonellosis and oranges with chicken servings improves the intuitive plausibility but not the logic or credibility of such calculations.) In addition, it is a frequent observation that some level of exposure to bacteria in food protects against risk of foodborne illnesses, for example, by stimulating acquired immunity. Thus, the use of a simple multiplicative model implying a necessarily non-negative linear relation between *Exposure* and *Risk* may be incorrect, producing meaningless results (or, more optimistically, extreme upper bounds on estimated risks) if the true relation is negative or non-linear.

Unfortunately, past estimates of risk of antibiotic-resistant illnesses caused by consumption of foods contaminated with resistant bacteria attributed to farm use of antibiotics have often simply assumed that some fraction of total resistant infections is caused by farm use of antibiotics (e.g., Barza and Travers, 2002) or that the ratio of estimated excess resistant cases per year to servings of food per year could be interpreted causally (on a logical par with the car-accidents-per-orange-consumed example above). For example, Chang et al. (2014) describe a case in which the US Food and Drug Administration (FDA) used an assumption that excess cases of fluoroquinolone (FQ)-resistant campylobacteriosis were proportional to consumption of chicken exposed to FQ on the farm to estimate that between 4,960 and 14,370 patients per year could have compromised treatment with ciprofloxacin (Bartholomew et al., 2005, cited in Chang et al., 2014). This model was used to support a risk management decision to withdraw FQ use in poultry in the United States. But the subsequent decade of experience showed that the withdrawal had no detectable causal impact in reducing levels of ciprofloxacin resistance in the United States (Chang et al., 2014). As in the car crashes per orange analogy, FDA had interpreted a positive ratio as causal, and discovered only after the fact that reducing the denominator had no real-world effect on reducing the numerator.

Thus, great caution should be taken when using such simplified risk assessment models. In general, they may be useful in making rapid calculations of plausible *upper bounds* in certain situations (e.g., if the true but unknown dose-response relation between exposure and risk is convex, or upward-curving), but should not be expected to produce accurate risk estimates unless they have been carefully validated (Cox, 2006).

EMPIRICAL UPPER-BOUNDING

An alternative method is available for quantifying upper bounds on the adverse human health consequences per year that could be prevented by reducing or eliminating antibiotic uses in agriculture. This method, which has advantages of simplicity, logical soundness, and reliance only on readily available data, does not attempt to simulate in detail the microbial loads traversing different pathways (e.g., food, water, environment, co-selection in other bacteria) or to quantify the relevant dose-response relations. Instead, it begins with the total number of adverse events per year that might be caused by antibiotic use (e.g., total treatment failures caused by antibiotic resistance), and then uses molecular biological data to estimate upper bounds on the fraction of all such cases that might be caused by (and preventable by eliminating) animal antibiotic uses. We call this the empirical upper-bounding approach. We will illustrate it in some detail using ampicillin-resistant *Enterococcus faecium* (AREF) bacteria, and then summarize the results from applying the approach to other drug–bug pairs.

Case Study: AREF Bacteria

This section, adapted from Cox et al. (2009), illustrates the empirical upper-bounding approach for potential risks to human health from the use of penicillin drugs in agriculture. It illustrates how to do QRA when neither all pathways from farm to consumer (or patient) nor relevant dose-response relations are known with enough confidence to permit useful simulation of microbial loads and illnesses.

Penicillin-based drugs are approved for use in food animals in the United States to treat, control, and prevent diseases and, to a lesser extent, to improve growth rates (AHI, 2006; FDA-CVM, 2007; Sechen, 2006). Concerns that penicillin use might increase the risk of antibiotic resistance in human enterococcal infections from non-human sources, thus leading to increased morbidity and mortality (WHO, 2005), have made approved feed usages of penicillins in food animals a controversial topic for several decades in the United States (FDA, 2000, 2003; IOM, 1989). The following sections develop a plausible upper bound on the potential for continued use of penicillin drugs in food animals to harm human health by increasing the number of antibiotic-resistant enterococcal infections in human patients. After summarizing relevant background for the hazard of greatest concern—infection of intensive care unit (ICU) patients with AREF bacteria—the following sections focus on quantifying the fraction of such resistant infections that might be prevented by discontinuing the use of penicillin drugs in food animals.

Risk to human health arises because some strains of enterococci may become opportunistic pathogens, potentially resistant to multiple drugs, that infect patients who are already seriously ill (typically in ICUs) with immune systems weakened by organ transplants, chemotherapy, AIDS, or other causes. Indeed, enterococcal infection is the second most common hospital-acquired infection in the United States (Varman et al., 2006). These infections can prolong illness and increase patient mortality. Vancomycin-resistant enterococci (VRE) are of particular concern because of their virulence and resistance to even some recently developed antibiotics. Vancomycin-resistant *E. faecium* (VREF) can cause serious and often fatal disease in vulnerable populations, such as liver transplant patients and patients with hematologic malignancies (Rice, 2001).

Although many enterococcal infections, including VRE, resolve without antimicrobial treatment (Rice, 2001; Varman et al., 2006), in severe cases for which antimicrobial treatment is provided, penicillin and ampicillin are often the leading choices. Most *E. faecium* infections in ICU patients in the United States are now resistant to vancomycin (Edmond et al., 1999; Jones et al., 2004). Patients with VREF have worse outcomes than those with vancomycin-susceptible strains—longer hospital stays and higher mortality (Webb et al., 2001). As noted by Rice (2001), *virtually all VREF are also ampicillin resistant:* "More than 95% of VRE recovered in the United States are *E. faecium*; virtually all are resistant to high levels of ampicillin." Hence, our risk assessment treats VREF as being (at least approximately) a subset of AREF. Since most VREF are AREF (although many AREF are not VREF), and assuming that changes in animal *penicillin* use would not significantly affect *vancomycin* resistance (consistent with historical data), we focus on human (ICU patient) infections with *vancomycin-susceptible* strains of AREF. Presumably, this is the subpopulation that might experience decreased ampicillin resistance if discontinuing animal penicillin drugs were to replace some AREF cases with ampicillin-susceptible cases. For patients with VREF, we assume that AREF would persist (due to the observed co-occurrence of AREF in VREF strains), so that no benefit from reduced AREF would be achieved for these patients.

Recognizing that a farm-to-fork model is not practical for AREF, due to data and knowledge gaps in release, exposure, and dose-response relations, we instead start with more readily available human data on ICU caseloads and resistance rates, similar to the approach in Cox and Popken (2004). We then work backward to estimate a plausible upper bound on the annual number of human patient mortalities that might be prevented by discontinuing penicillin use in food animals.

For purposes of conservative (i.e., upper-bound) risk assessment, we define a *potentially preventable mortality* to occur whenever the following conditions hold: (1) an ICU patient dies, following (2) an *E. faecium* infection that (3) is resistant to ampicillin (AREF) (and hence might have benefited had ampicillin resistance been prevented). The infection was: (4) vancomycin-susceptible (and hence might have also been ampicillin-susceptible, had it not been for penicillin use in food animals); (5) not known to have been contracted from the hospital environment (and hence might have been prevented by actions external to the hospital, such as elimination of AREFs from food animals); (6) could have come from food animals (i.e., has a genotype or resistance determinants of the types found in food animals). (7) The patient tolerated penicillin (i.e., was not allergic, and hence might have benefited from ampicillin, had it not been for resistance). We propose that the conjunction of these seven conditions should be interpreted as *necessary* for a mortality to have been caused (with non-negligible probability) by resistance due to use of penicillin in food animals, even though it is not *sufficient* (e.g., the infecting strain might have had some other origin than food animals, or the patient might have died anyway, even if the infection had been ampicillin-susceptible). Accordingly, the following sections estimate a plausible upper bound on annual preventable mortalities from AREF infections based on the following product of factors:

Preventable AREF mortalities per year ≤ (Total number of ICU infections per year) × (fraction caused by *E. faecium*) × (fraction of ICU *E. faecium* infections that are AREF and exogenous, i.e., not known to be of nosocomial origin) × (fraction of these exogenous AREF cases that are vancomycin-susceptible) × (fraction of vancomycin-susceptible exogenous AREF cases that might have come from food animals) × (fraction of these cases that are penicillin-tolerant) × (excess mortality rate for AREF cases compared to ASEF cases).

That is, we first quantify the expected annual number of AREF cases in the United States that might benefit from ampicillin treatment if food animal uses of penicillin were halted (i.e., cases that are penicillin-tolerant and vancomycin-susceptible and that might have been caused by resistance determinants from food animals). Then, we multiply this number by the excess mortality rate for resistant as opposed to susceptible cases. Each of the foregoing factors can be estimated from available data, as discussed in detail in Cox et al. (2009) and summarized in Table 15.1.

Table 15.1 shows key parameter estimates, calculations, assumptions, and resulting risk estimates. When presenting point estimates, it is customary to also present interval estimates to inform decision-makers about the plausible range of estimated values. In this analysis, however, the key uncertainties have little to do with statistical sampling error, and they are not adequately characterized by confidence limits. Rather, they arise from uncertainty about the validity and conservatism of the assumptions in Table 15.1. Qualitatively, the main uncertainty is about whether a non-zero risk to human health exists from animal use of penicillin drugs. We have assumed that there is, but there is no clear empirical proof that the risk is non-zero. To bridge this knowledge gap, Table 15.1 incorporates several conservative qualitative assumptions that jointly imply that the risk is non-zero. Other quantitative parameter values presented, and their implied risk estimate of less than or equal to 0.135 excess mortalities per year, are intended to be realistic, data-driven values (rather than extreme upper-bounds or 95% upper confidence limits) *contingent* on these conservative qualitative assumptions. The most important conservative elements in Table 15.1 are the following qualitative assumptions:

- *Transfer of ampicillin resistance from food animal bacteria to bacteria infecting human patient occurs.* The assumption that ampicillin-resistant strains and/or determinants are transferred from strains in food animals to human ICU patients is fundamental to the assessment in Table 15.1. Such transfer has never been shown to occur, but may be possible.
- *Withdrawing animal drug use would immediately and completely prevent the problem.* Table 15.1 assumes that halting penicillin use in food animals would immediately eliminate all ampicillin resistance from the cases in Table 15.1. This is a deliberately extreme assumption. In reality, halting use might have little or no impact on the already very low levels of ampicillin resistance.
- *Resistance increases patient mortality.* The assumption that ampicillin resistance causes an increase in the mortality rates of the patients in Table 15.1 is made even though, in reality, no statistically significant difference in mortality rates has been found between resistant and non-resistant cases (Fortun et al., 2002).

With these assumptions, the calculations in Table 15.1 predict that excess mortalities per year in the entire United States population could be as high

TABLE 15.1 Summary of AREF Risk Calculation Using Empirical Upper Bounding

Factor	More Conservative Value	Less Conservative Value	Source
N=ICU infections per year	N=315,000	N=104,372.5	FDA-CVM (2004)
P_{ent}=fraction of ICU infections caused by *Enterococcus*	0.10	0.09 (Wisplinghoff et al., 2004)	FDA-CVM (2004)
$P_{EF, ent}$=fraction of enterococcal infections caused by *E. faecium*	0.25		FDA-CVM (2004)
Fraction of enterococcal infections caused by *E. faecium* that are exogenous (non-nosocomial)	≤0.17		Cox and Popken (2004) (may be smaller now due to spread of CC-17)
Fraction of exogenous cases that are ampicillin-resistant	0.187		Willems et al. (2005)
Fraction of exogenous ampicillin-resistant cases that are vancomycin-susceptible	0.155		Jones et al. (2004)
Fraction of exogenous ampicillin-resistant vancomycin-susceptible cases possibly from food animals	0–0.069 (0.069 assumed)		Data of Leavis et al. (2006)
Fraction of exogenous ampicillin-resistant cases with penicillin-tolerant host	0.844		Lee et al. (2000)
Fraction of these cases that would become ampicillin-susceptible if penicillin use in food animals were terminated	0.00–1.00 (1 is assumed)		Conservative assumption
Increase in mortality risk per case, due to ampicillin resistance	0.00–0.06 (0.06 is assumed)		Fortun et al. (2002), conservative assumption
RISK= ≤ 0.135 potential excess mortalities per year	315000*0.10*0.25*0.17*0.187*0.14*0.069*0.844*0.06 = 0.135	104372.5*0.09*0.25*0.17*0.187*0.155*0.069*0.844*0.06 ≈0.04 mortalities per year	Product of preceding factors

Source: From Cox et al. (2009).

as 0.135, or about one excess mortality expected once every 7–8 years on average, if current conditions persist. This risk is concentrated among ICU patients already at high risk of such infections. With less conservative assumptions, the estimated risk falls to about 0.04 excess mortalities per year, that is, about one excess mortality every 25 years in the United States under current conditions. The multiplicative calculation in Table 15.1 makes sensitivity analysis of these results to changes in the values of specific factors especially straightforward: the final risk estimate is directly proportional to each factor listed.

The more conservative risk estimate of 0.135 excess mortalities per year equates to an average individual risk rate in the most at-risk group (ICU patients) of approximately $0.135/315,000 = 4.3 \times 10^{-7}$ excess mortalities per ICU patient. For the US population as a whole, this corresponds to an average individual risk of approximately $0.135/300E6 = 4.5 \times 10^{-10}$ excess fatalities per person-year, or a lifetime risk of about $80 \times (6 \times 10^{-10}) = 3.6 \times 10^{-8}$ excess risk of mortality per lifetime (for an assumed 80-year lifetime). If the less conservative risk estimate of 0.04 excess mortalities per year is used, these individual and population risks are reduced by a factor of 0.04/0.135, or more than three-fold. If one or more of the key qualitative assumptions listed above are violated, then the true risk could be as low as zero.

The main conclusion from these calculations is that not more than 0.04 excess mortalities per year (under conservative assumptions) to 0.14 excess mortalities per year (under very conservative assumptions) might be prevented in the whole US population if current use of penicillin drugs in food animals were discontinued, and if this successfully reduced the prevalence of antibiotic-resistant *E. faecium* infections among ICU patients. The true risk could well be zero, if one or more of the conservative assumptions above is false.

Summary of Results from Applying Empirical Upper-Bounding Risk Assessment to Other Antibiotic-Resistant Bacteria

Antimicrobial risk analyses have now been completed for several antimicrobial-resistant bacteria of public health concern using empirical upper-bounding approaches. Among the results now available are the following:

- For streptogramins, banning virginiamycin has been estimated to prevent from 0 to less than 0.06 statistical mortalities per year in the entire US population (Cox and Popken, 2004; see also FDA-CVM, 2004). More data tend to reduce such upper bounds, which in part reflect uncertainties in the data available at the time of the study.
- For macrolide-resistant campylobacter, Hurd and Malladi (2008) concluded that "the predicted risk of suboptimal human treatment of infection with *C. coli* from swine is only 1 in 82 million; with a 95% chance it could be as high as 1 in 49 million. Risks from *C. jejuni* in poultry or beef are even less." (This analysis followed the FDA approach of interpreting simple ratios as if they applied that reducing exposures in the denominator would

proportionally reduce cases in the numerator. Thus, the results may have no predictive validity if this assumption turns out to be incorrect, similar to the case of FQs discussed by Chang et al., 2014.)

- For tetracyclines, Cox and Popken (2010) concluded that "As a case study, examining specific tetracycline uses and resistance patterns suggests that there is no significant human health hazard from continued use of tetracycline in food animals. Simple hypothetical calculations suggest an unobservably small risk (between 0 and 1.75E-11 excess lifetime risk of a tetracycline-resistant infection), based on the long history of tetracycline use in the United States without resistance-related treatment failures."

- For MRSA, Cox and Popken (2014) "construct a conservative (plausible upper bound) probability estimate for the actual human health harm (MRSA infections and fatalities) arising from [livestock-associated] ST398-MRSA from pigs. The model provides plausible upper bounds of approximately one excess human infection per year among all U.S. pig farm workers, and one human infection per 31 years among the remaining total population of the United States. These results assume the possibility of transmission events not yet observed, so additional data collection may reduce these estimates further."

Such quantitative risk estimates suggest that banning agricultural uses of these antibiotics might create small human health benefits (perhaps reducing compromised treatments due to resistance by a few cases per century), but are unlikely to make any measurable difference in improving public health. This finding disagrees with the passionate convictions of many experts who advocate prompt bans as urgently needed to slow the spread of resistance (Chang et al., 2014).

Since the empirical upper-bounding approach was originally developed in the early 2000s with support from the animal antibiotic industry, results such as those just cited are sometimes viewed with suspicion (*ibid.*) A virtue of QRA in helping to inform (and perhaps occasionally resolve) politically charged debates over what to do is that the logic, data sources, and calculations used are completely transparent and easy to summarize, as in Table 15.1, so that anyone interested can check the logic and conclusions and experiment with varying the assumptions. However, even if QRA proves to be too controversial to support trusted conclusions, it is often still possible to manage risks pragmatically using principles discussed next.

MANAGING UNCERTAIN FOOD RISKS VIA QUALITY PRINCIPLES: HACCP

Even without QRA, it is often possible to apply process quality improvement ideas to control the microbial quality of food production processes—including both susceptible and resistant bacteria. This approach has been developed and deployed successfully (usually on a voluntary basis) using the *Hazard Analysis and*

TABLE 15.2 Summary of Seven HACCP Principles

- *Analyze hazards.* Potential hazards associated with a food and measures to control those hazards are identified. The hazard could be biological, such as a microbe; chemical, such as a toxin; or physical, such as ground glass or metal fragments
- *Identify critical control points.* These are points in a food's production—from its raw state through processing and shipping to consumption by the consumer—at which the potential hazard can be controlled or eliminated. Examples are cooking, cooling, packaging, and metal detection
- *Establish preventive measures with critical limits for each control point.* For a cooked food, ... this might include... minimum cooking temperature and time required to ensure the elimination of any harmful microbes
- *Establish procedures to monitor the critical control points.* Such procedures might include determining how and by whom cooking time and temperature should be monitored
- *Establish corrective actions to be taken when monitoring shows that a critical limit has not been met*—for example, reprocessing or disposing of food if the minimum cooking temperature is not met
- *Establish procedures to verify that the system is working properly*—for example, testing time-and-temperature recording devices to verify that a cooking unit is working properly
- *Establish effective recordkeeping to document the HACCP system.* This would include records of hazards and their control methods, the monitoring of safety requirements and action taken to correct potential problems. Each of these principles must be backed by sound scientific knowledge: for example, published microbiological studies on time and temperature factors for controlling foodborne pathogens

Source: From USDA/FDA (2004); http://www.cfsan.fda.gov/~lrd/bghaccp.html.

Critical Control Points (HACCP) approach summarized in Table 15.2. The main idea of HACCP is to first identify steps or stages in the food production process where bacteria can be controlled, and then to apply effective controls at those points, regardless of what the ultimate quantitative effects on human health risks may be. Reducing microbial load at points where it can be done effectively has proved very successful in reducing final microbial loads and improving food safety.

DISCUSSION AND CONCLUSIONS

This chapter has introduced and illustrated key ideas used to quantify and manage human health risks from food contaminated by bacteria, both antibiotic-resistant and antibiotic-susceptible. Somewhat similar ideas apply to other foodborne hazards, from pesticide residues to mad cow disease, that is, risk assessment can be carried out by modeling the flow of contaminants through the food production process (together with any increases or decreases at different steps), resulting in levels of exposures in ingested foods or drinks. These are then converted into quantitative risks using dose-response functions.

The practical successes of the HACCP approach provide a valuable reminder that QRA is not always a prerequisite for effective risk management. It may not be necessary to quantify a risk in order to reduce it. Reducing exposures at critical control points throughout the food production process can reduce exposure-related risk even if the size of the risk is unknown.

Where QRA can make a crucial contribution is in situations where there is doubt about whether an intervention is worthwhile. For example, QRA can reveal whether expensive risk-reducing measures are likely to produce correspondingly large health benefits. It may be a poor use of resources to implement expensive risk-reducing measures if the quantitative size of risk reduction procured thereby is very small. QRA methods such as farm-to-fork exposure modeling and dose-response modeling (Haas et al., 1999), or empirical upper-bounding approaches based on multiplicative models (Cox, 2006), can then be valuable in guiding effective risk management resource allocations, by revealing the approximate sizes of the changes in human health risks caused by alternate interventions.

REFERENCES

Animal Health Institute (AHI), 2006. Animal Health Companies Meet Increase Market Need for Antibiotics. News Release from the Animal Health Institute, Washington, DC. Available at: <http://www.ahi.org/mediacenter/documents/Antibioticuse2005REVISED.pdf> (accessed 12.10.06.).

Bartholomew, M.J., Vose, D.J., Tollefson, L.R., Travis, C.C., 2005. A linear model for managing the risk of antimicrobial resistance originating in food animals. Risk Anal. 25 (1), 99–108.

Barza, M., Travers, K., 2002. Excess infections due to antimicrobial resistance: the "Attributable Fraction". Clin. Infect. Dis. 34 (Suppl. 3), S126–S130.

CBS, 2010. Animal Antibiotic Overuse Hurting Humans? Katie Couric Investigates Feeding Healthy Farm Animals Antibiotics. Is it Creating New Drug-Resistant Bacteria? CBS Special News Report: Katie Couric Investigates. Available at: <http://www.cbsnews.com/stories/2010/02/09/eveningnews/main6191530.shtml> (accessed 16.06.10).

Chang, Q., Wang, W., Regev-Yochay, G., Lipsitch, M., Hanage, W.P., 2014. Antibiotics in agriculture and the risk to human health: how worried should we be? Evol. Appl. Available from: <http://onlinelibrary.wiley.com/doi/10.1111/eva.12185/abstract>.

Cox Jr, L.A., 2006. Quantitative Health Risk Analysis Methods: Modeling the Human Health Impacts of Antibiotics Used in Food Animals. Springer, New York, NY.

Cox, Jr L.A., 2008. Managing Food-Borne Risks. *Wiley Encyclopedia of Quantitative Risk Analysis and Assessment.*

Cox, L.A., Popken, D.A., 2004. Quantifying human health risks from virginiamycin used in chickens. Risk Anal. 24 (1), 271–288.

Cox Jr., L.A., Popken, D.A., 2010. Assessing potential human health hazards and benefits from subtherapeutic antibiotics in the United States: tetracyclines as a case study. Risk Anal. 30 (3), 432–457.

Cox Jr, L.A., Popken, D.A., Mathers, J., 2009. Human health risk assessment of penicillin/aminopenicillin resistance in enterococci due to penicillin use in food animals. Risk Anal. 29 (6), 796–805.

Edmond, M.B., Wallace, S.E., McClish, D.K., Pfaller, M.A., Jones, R.N., Wenzel, R.P., 1999. Nosocomial bloodstream infections in United States hospitals: a three-year analysis. Clin. Infect. Dis. 29 (2), 239–244.

Fortun, J., Coque, T.M., Martin-Davila, P., Moreno, L., Canton, R., Loza, E., et al., 2002. Risk factors associated with ampicillin resistance in patients with bacteraemia caused by *Enterococcus faecium*. J. Antimicrob. Chemother. 50 (6), 1003–1009.

Haas, C.N., Rose, J.B., Gerba, C.P., 1999. Quantitative Microbial Risk Assessment. John Wiley & Sons, New York, NY.

Hurd, H.S., Malladi, S., 2008. A stochastic assessment of the public health risks of the use of macrolide antibiotics in food animals. Risk Anal. 28 (3), 695–710.

Institute of Medicine (IOM), 1989. Human Health Risks with the Subtherapeutic Use of Penicillin or Tetracyclines in Animal Feed. Report by the Committee of Human Health Risk Assessment of Using Subtherapeutic Antibiotics in Animal Feeds, Institute of Medicine, IOM-88-89. National Academy Press, Washington, DC.

Jones, M.E., Draghi, D.C., Thornsberry, C., Karlowsky, J.A., Sahm, D.F., Wenzel, R.P., 2004. Emerging resistance among bacterial pathogens in the intensive care unit—a European and North American surveillance study (2000–2002). Ann. Clin. Microbiol. Antimicrob. 3 (14) Online journal article available at: <http://www.pubmedcentral.nih.gov/articlerender.fcgi?artid=509280> .

Kallen, A.J., Mu, Y., Bulens, S., Reginold, A., Petit, S., Gershman, K., et al., 2010. Health care-associated invasive MRSA, 2005–2008. JAMA 304 (6), 641–648.

Leavis, H.L., Bonten, M.J., Willems, R.J., 2006. Identification of high-risk enterococcal clonal complexes; global dispersion and antibiotic resistance. Curr. Opin. Microbiol. 9 (5), 454–460.

Lee, C.E., Zembower, T.R., Fotis, M.A., Postelnick, M.J., Greenberger, P.A., Peterson, L.R., et al., 2000. The incidence of antimicrobial allergies in hospitalized patients: implications regarding prescribing patterns and emerging bacterial resistance. Arch. Intern. Med. 160 (18), 2819–2822.

Rice, L.B., 2001. Emergence of vancomycin resistant enterococci. Emerg. Infect. Dis. 7 (2), 183–187.

Sechen, S., 2006. The review of animal production drugs by FDA. FDA Veterinarian 21 (1), 8–11. Available at: <http://www.fda.gov/cvm/Documents/FDAVetVolXXINo1.pdf>.

US Food and Drug Administration—Center for Veterinary Medicine (FDA-CVM), 2000. Review of Agricultural Antibiotics Policies. Available at: <http://www.fda.gov/cvm/HRESP106_157.htm#nrdc>.

US Food and Drug Administration—Center for Veterinary Medicine (FDA-CVM), 2003. Guidance for Industry # 152: Evaluating the Safety of Antimicrobial New Animal Drugs with Regard to their Microbiological Effects on Bacteria of Human Health Concern. US Department of Health and Human Services, Food and Drug Administration, Center for Veterinary Medicine., Available at: <http://www.fda.gov/cvm/Guidance/fguide152.pdf>.

US Food and Drug Administration—Center for Veterinary Medicine (FDA-CVM), 2004. Risk Assessment of Streptogramin Resistance in *Enterococcus faecium* Attributable to the Use of Streptogramins in Animals. Draft for Comment. Available at: <http://www.fda.gov/cvm/Documents/SREF_RA_FinalDraft.pdf> (accessed 23.11.04).

US Food and Drug Administration—Center for Veterinary Medicine (FDA-CVM), 2007. FDA Database of Approved Animal Drug Products. FDA Center for Veterinary Medicine, VMRCVM Drug Information Lab. Available at: <http://dil.vetmed.vt.edu/>.

Varman, M., Chatterjee, A., Abuhammour, W., Johnson, W.C., 2006. Enterococcal infection. Emedicine.com. Online article available at: <http://www.emedicine.com/ped/topic2703.htm> (last edited 26.07.06).

Webb, M., Riley, L.W., Roberts, R.B., 2001. Cost of hospitalization for and risk factors associated with vancomycin-resistant *Enterococcus faecium* infection and colonization. Clin. Infect. Dis. 33 (4), 445–452.

Willems, R.J., Top, J., van Santen, M., Robinson, D.A., Coque, T.M., Baquero, F., et al., 2005. Global spread of vancomycin-resistant *Enterococcus faecium* from distinct nosocomial genetic complex. Emerg. Infect. Dis. 11 (6), 821–828.

Wisplinghoff, H., Bischoff, T., Tallent, S.M., Seifert, H., Wenzel, R.P., Edmond, M.B., 2004. Nosocomial bloodstream infections in US hospitals: analysis of 24,179 cases from a prospective nationwide surveillance study. Clin. Infect. Dis. 39 (3), 309–317.

World Health Organization (WHO), 2005. Critically Important Antibacterial Agents for Human Medicine for Risk Management Strategies of Non-Human Use. Report of a WHO Working Group Consultation, 15–18 February, 2005. Canberra, Australia. Available at: <http://www.who.int/entity/food-borne_disease/resistance/FBD_CanberraAntibacterial_FEB2005.pdf>.

Chapter 16

Antibiotic-Resistant Bacteria and Resistance Genes in the Water–Food Nexus of the Agricultural Environment

Pei-Ying Hong

Water Desalination and Reuse Center, Division of Biological and Environmental Sciences and Engineering, King Abdullah University of Science and Technology (KAUST), Thuwal, Saudi Arabia

Chapter Outline

INTRODUCTION

The production of food is intricately linked to water consumption. While each person drinks on average 2–4 L of water per day, they can consume up to 2,000–5,000 L of virtual water from the three meals that they have on a daily basis. The Virtual Water project estimates that a single 300 g of steak on a dining

Antimicrobial Resistance and Food Safety. DOI: http://dx.doi.org/10.1016/B978-0-12-801214-7.00016-8

table would mean that up to 4,650 L of water have been consumed. Even an apple would incur a 70 L water footprint (http://virtualwater.eu/). Summing up all these estimates, it would not be far-fetched to conclude that agricultural production remains the thirstiest sector relative to the needs for domestic and industrial use. A total of 70–80% of freshwater supplies are withdrawn to feed the never-ending thirst for water of the agriculture sector. To further compound the problem, most of these waters are lost into the environment through evapotranspiration, and cannot be recovered back for reuse and recycling (Shiklomanov, 1998).

Despite this, the production of food to feed 7 billion people remains crucial. Coupled with the exponential increase in population growth that is projected to hit 9 billion in the year 2050, balancing food and water security at the same time remains a key concern in the minds of governments and stakeholders. A feasible alternative is to utilize treated wastewater as a resource for agricultural irrigation. Simply put, treated effluent from municipal wastewater treatment plants can be used as an alternative water supply for agricultural irrigation. This would enable a large portion of the pristine groundwater supplies to be freed up for use in other sectors. However, before one can consider the use of treated wastewater as an alternative water source for agricultural irrigation, it is important to first assess the risk involved in using this treated wastewater.

These risks can arise from the occurrence of antibiotic-resistant bacteria and antibiotic resistance genes. In 2013, the US Centers for Disease Control and Prevention (CDC) issued an Antimicrobial Resistance Threat Report that outlines the health concerns involving antimicrobial resistance (CDC, 2013). Antimicrobial resistance was estimated to cause up to 2 million illnesses and 23,000 deaths in the United States. Unlike developed countries such as United States, extensive monitoring is not implemented in developing countries, and therefore such estimates of antimicrobial resistance threats are not easily available. However, it does not mean that antimicrobial resistance threats are not prevalent in developing countries.

A recent survey of river sediment samples collected upstream and downstream from an Indian pharmaceutical wastewater treatment plant revealed significantly higher concentrations of antibiotics and antibiotic resistance genes in the samples that were collected downstream from the discharge site (Kristiansson et al., 2011). As such, the conventional notion that antimicrobial resistance threats are of more detrimental concern in hospital settings than in community settings is no longer valid. There is increasing evidence that community-acquired resistance can occur, particularly when the infected individuals work in occupations that require the use of antibiotics or live in an environment that is contaminated with antibiotic resistance genes. To illustrate, a novel metallo beta-lactamase (*NDM-1*) gene was first discovered in a Swedish patient of Indian origin who had traveled to New Delhi and was admitted into a local hospital (Yong et al., 2009). Since the first discovery, it was observed that not all patients who were infected with bacteria carrying the *NDM-1* gene have a history of hospital admission in India. An environmental survey of seepage, tap water, and sewage effluent samples found that a small subset of these samples (i.e., 12 of 171 seepage samples and two of 50 water samples) were tested positive for bacteria that carries the *NDM-1* gene on mobile plasmids (Walsh et al., 2011). In another study, the *NDM-1* genes were detected throughout the treatment schematic of a municipal wastewater treatment plant in northern China (Luo et al., 2014).

It is therefore likely that acquisition of the *NDM-1*-positive bacterium in the patients without any hospital admission history were through exposure to contaminated water supplies or in discharged wastewater that was insufficiently treated. Summing up these scientific evidences, there is understandably a cause for concern when using wastewater for agricultural irrigation.

METHODS TO DETERMINE ANTIBIOTIC-RESISTANT BACTERIA AND RESISTANCE GENES

Currently, wastewater quality is regulated by guidelines devised mainly based on organic matter contents, total, and fecal coliforms. No guidelines on the permissible concentration of antibiotic-resistant bacteria and resistance genes in discharge effluent exist. As antimicrobial resistance threats become more prevalent in the near future, monitoring, and evaluation of their abundance and occurrence would be required to understand some fundamental questions related to (i) how much of these antibiotic-resistant bacteria and genes are present in the food samples and how they might impact public health, (ii) the effectiveness of current sanitation approaches in removing these contaminants, and (iii) the fate and persistence of these contaminants in the environment.

Standardized Method to Assess Antimicrobial Resistance or Susceptibility

Assessing whether a bacterial isolate is resistant to a specific type of antibiotic can be done by a standardized protocol as detailed by the Clinical and Laboratory Standards Institute (CLSI).

The CLSI method is a standardized test to evaluate the antibiotic susceptibility of clinically relevant *Enterobacteriaceae* (including *Klebsiella*, *Escherichia*), *Pseudomonas aeruginosa*, *Staphylococcus* spp., *Enterococcus* spp., *Acinetobacter* spp., *Burkholderia cepacia*, *Stentrophomonas maltophila*, and other non-*Enterobacteriaceae* that are defined to include *Pseudomonas* spp. and other nonfastidious, glucose-nonfermenting, Gram-negative bacilli (CLSI, 2014).

To conduct the test, Mueller–Hinton (MH) agar plates are prepared by pouring the sterile medium into petri dishes to a depth of 4 mm. The bacterial suspension can then be diluted in 2 mL of sterile saline until its optical density at 600 nm wavelength (OD_{600}) is approximately 0.08 to 0.13. A sterile swab is then dipped into the inoculum tube and spread evenly across the MH agar plate.

The antibiotic-infused disks are then placed on the surface of the agar in a manner that ensures the disks are more than 24 mm apart. The prepared plates are then placed into a 37°C incubator for the bacterium to grow for at least 18 h. After overnight incubation, the zone size, including the diameter of the disk, is then measured and compared to the CLSI table to determine whether the bacterium exhibits resistance to the antibiotics (Hudzicki, 2009). Quality control measures to ensure that the MH agar plates are well prepared can be carried out using specific strains (e.g., *Escherichia coli* ATCC 25922, *E. coli* ATCC 35218, *Staphylococcus aureus* ATCC 25923) that are known for their resistance/susceptibility towards defined concentrations of antibiotics.

The CLSI method however has its limitations in assessing antibiotic susceptibility of environmental bacterial isolates. Some bacteria isolated from the environment require more than 24–48 h to reach the exponential phase of their growth. Furthermore, the resistant profiles of most environmental isolates are not specified in the CLSI standards. A possible modification of the protocol would involve first harvesting the bacterial strain to be tested at the exponential phase (i.e., when it is actively growing), diluting the bacterial culture with saline buffer to an OD_{600} of approximately 0.1, and then inoculating this diluted bacteria suspension into individual wells on a 96-well plate that contains either no antibiotics (i.e., control) or antibiotics. The inoculated suspension is then incubated for growth for 24–48 h, depending on the rate of growth for the bacterium in the substrate medium. Measurements of the optical density at 600 nm for the bacterium in substrate medium with antibiotics is then normalized against that obtained from the inoculated bacterium in substrate medium without antibiotics. The minimum inhibitory concentration of the antibiotics can be further evaluated by inoculating the bacterium in growth medium that is spiked with dilution series of antibiotics, generally at concentrations ranging from 0 to 512 μg/mL. One can then categorize the bacterium to be either resistant or susceptible to the tested antibiotics based on a cutoff percentage. For example, if the OD_{600} of the bacterium in substrate medium with antibiotics is more than 90% of the OD_{600} of the bacterium in antibiotic-free substrate, one can conclude that the bacterium's growth is only minutely affected by the presence of antibiotics and that this bacterium is resistant to the tested antibiotic (Singer et al., 2004; Tunney et al., 2004). These methods would allow a systematic way to assess whether bacterial isolates are resistant to certain antibiotics and at what inhibitory concentration of antibiotics.

Quantitative Polymerase Chain Reaction

Besides isolating and characterizing antibiotic-resistant bacteria, one can determine the total abundance of antibiotic resistance genes using molecular approaches like quantitative polymerase chain reaction (i.e., qPCR). In qPCR, primers targeting known antibiotic resistance genes are used along with a fluorophore reporter (e.g., SYBR green or Taqman assays) for PCR amplification.

As the amplification progresses, fluorescence increases with the abundance of the amplified gene products. qPCR relies on a threshold cycle (C_t) number to calculate the copy numbers of the target gene. Threshold cycle number is defined as the cycle number required for detection of a signal that is above the baseline signal. It is usually arbitrarily defined at the midpoint of the initial exponential phase in an amplification cycle. For every unknown sample that is to be determined for its copy number of genes, a standard has to be generated using a serial dilution of known copies of that particular gene target.

To illustrate, if one is interested to know the abundance of tetracycline resistance gene *tetZ* in a wastewater sample, dilutions of known concentrations of *tetZ* standards (e.g., 10^{10}, 10^8, 10^6, 10^4, and 10^2 copies/mL) would have to be prepared and run alongside in the qPCR thermal cycler. For each concentration of *tetZ*, a corresponding C_t value would be generated (e.g., C_t value of 12 for 10^{10} copies/mL, C_t value of 15 for 10^8 copies/mL). This would allow plotting of the C_t values against the log copy numbers to generate a linear line. The theoretical best-fit slope would be -3.32, which would denote an amplification factor of 2 or an amplification efficiency of 100%. Using the equation generated from the standards, the C_t value obtained from the sample containing unknown copies of the genes can then be calculated accordingly for the log copy number of genes. The copy number of genes is then normalized against either (i) the amount of genomic DNA, or (ii) the 16S rRNA gene copy numbers, or (iii) the volume of sample that was extracted for the genomic DNA.

Although primer sets to quantify a large suite of antibiotic resistance genes are already available, antibiotic resistance genes exhibit diverse nucleotide sequences and there remains no universal primer sets available to target a family of resistance genes. Therefore, if one is interested in a large suite of antibiotic resistance genes, as is often the case in dealing with environmental samples, qPCR can become rather costly and time-consuming.

PAST MONITORING SURVEYS—DETAILING THE KNOWLEDGE GAPS

Despite the availability of methods to identify, characterize, and quantify the antibiotic-resistant bacteria and antibiotic resistance genes, limited databases on the baseline microbial quality of irrigation water have been compiled to date (Gerba and Choi, 2006). In one survey done in Canada during 1991–1992, it was found that 21% of the wells on about 1,200 farms were contaminated with fecal coliforms and that 14% of the wells had nitrate contents that exceeded 10 mg/L (Goss and Barry, 1995). Similarly, a survey of 136 stream water and 143 groundwater samples conducted by the US Geological Survey (USGS) showed that total coliforms, *E. coli* and *Clostridium perfringens* were detected in 99%, 97%, and 73%, respectively of the stream water samples. In comparison, the groundwater quality was significantly better than that of the stream water, and that less than 1% of the groundwater samples were positive for *E. coli* (Francy et al., 2000). However, it seems that localized contamination of the groundwater supplies can be an issue in rural places, particularly those that were in close proximity to apparent sources of contamination (e.g., livestock production farms or sewage outfall). Long-term monitoring survey of groundwater wells near swine confinement facilities showed that up to 13.3% of water samples collected in one of the wells were positive for fecal coliforms, and as many as 66.7% of these samples were positive for fecal *Streptococcus* (Krapac et al., 2002). Furthermore, qPCR-based approaches to monitor the antibiotic resistance genes in these groundwater wells over a decade have consistently shown that some of the water samples were sporadically contaminated by tetracycline and erythromycin resistance genes (Chee-Sanford et al., 2001; Hong et al., 2013b; Koike et al., 2007, 2010). Groundwater contamination is equally prevalent in less developed countries. A 5-year survey performed from 1984 till 1989 showed that 6% of the groundwater wells were contaminated with ammonium and fecal coliforms (Alaa el-Din et al., 1994), with a range of 380–1,133 MPN/mL of fecal coliforms detected in the groundwater samples (Eldin et al., 1993).

Surprisingly, antibiotic-resistant Gram-negative bacteria can be recovered from rural groundwater supplies that had no apparent sources of contamination (Mckeon et al., 1995). Gram-negative bacteria can also be found in organic and conventional fruit and vegetables even though no antibiotics were used in these farm sites (Ruimy et al., 2010). As such, it is likely that antibiotic-resistant genes and antibiotic-resistant bacteria may be ubiquitous in the environment, even in seemingly pristine environment.

A lot of these monitoring surveys were conducted more than a decade ago and the water systems in today's context may be different from that time. For example, over a prolonged period of nonrenewable use of groundwater supplies, groundwater intrusion, or infiltration of contaminated surface runoffs can perturb the quality of groundwater supplies. It is also evident from these studies that monitoring effort remains limited to the quantification of fecal coliforms, and does not address antibiotic resistance genes and antibiotic-resistant bacteria. For example, microbiological analyses were only added to monitoring framework of the USGS National Water Quality Assessment (NAWQA) program in 1997, and were restricted to the analysis of total coliforms, *E. coli*, *C. perfringens*, and somatic and F-specific coliphages. Other pathogens like *Giardia* and *Cryptosporidium* were recognized for their health risks but cannot be routinely monitored on a large scale because of the lack of appropriate analytical methods (Francy et al., 2000). Similarly, the lack of detection of antibiotic-resistant bacteria and their genes in large-scale sampling surveys is mainly because these contaminants remain unrecognized for their potential health risks. This is despite the fact that up to 23,000 deaths were reported to be due to infection by antibiotic-resistant bacteria and fungi (CDC, 2013).

From these surveys, one can conclude that the current sources of water (i.e., both surface and ground waters) used for agricultural irrigation can be contaminated with microorganisms which would in turn impose a certain extent of public health risks. Yet, no standardized regulations are devised to mandate the microbial quality of agricultural irrigation water. The Food Agriculture Organization (FAO) devised guidelines for irrigation water quality but only listed salinity, sodium, chloride, boron, and other trace elements, as well as nitrogen and bicarbonate (Ayers and Westcot, 1994) as parameters to measure. On the contrary, because of concerns related to the use of treated wastewater for agricultural irrigation, USEPA has a clear defined guidance on the acceptable quality before treated wastewater can be used. To illustrate, secondary treatment and disinfection processes are required to achieve water quality ≤ 10 mg/L biochemical oxygen demand (BOD), ≤ 2 NTU (i.e., an indication to the turbidity of water) and with no detectable fecal coliforms per 100 mL before the water can be applied for irrigating food crops (Asano et al., 2007b).

Clearly, all available guidelines fail to consider imposing a permissible level of antibiotic-resistant bacteria and genes as there remains limited availability of information on how antibiotic-resistant bacteria and genes would result in detrimental health impact. Nevertheless, numerous reports related to the abundance of antibiotic resistance genes in the treated effluent, as summarized in an earlier review paper by Hong et al. (2013a) are already available. An estimated abundance of up to $10^{5.61}$ copies of beta-lactamase genes bla_{TEM} can be found in 1 mL of wastewater that is treated by the conventional schematics (Lachmayr et al., 2009). Likewise, approximately $10^{3.63}$ copies of tetW are present in the same volume of treated wastewater (Munir et al., 2011). Therefore, these abundances of antibiotic resistance genes are anticipated to be present if this water were to be used for agricultural irrigation. However, epidemiological studies to illustrate how much antibiotic-resistant bacteria and genes are needed to illicit a host response are clearly lacking. Scientific observations, as illustrated in the subsequent sections, do suggest that antibiotic-associated contaminants can impose a detrimental impact on agricultural soils, crops, and public health.

IMPACT ON SOILS, CROPS, AND HEALTH

Impact on Agricultural Soils

Soil microbiota generally harbor a more diverse microbial richness than that found in water samples (Hong et al., 2013b). Several ecological theories suggest that an ecosystem with higher biodiversity would be more resilient to invasive species that are inadvertently introduced during anthropogenic contamination events (Stachowicz et al., 1999; van Elsas et al., 2012). As such, it was generally thought that the soil microbiota may be less perturbed despite being exposed to treated wastewater. Hong et al. tracked the microbial communities in groundwater and soil samples that were exposed to animal fecal contamination, and found that unlike the groundwater samples, soil samples did not show any apparent changes in the soil microbial communities at the phylum level after manure application. However, the abundance of tetracycline resistance genes and integrase genes in the soil samples increased after manure application, suggesting an introduction of resistance genes that originated from the manure. Given that the manure was applied over a prolonged period of time, the antibiotic-resistant genes and bacteria can accumulate in the soil, particularly because these high levels of resistance and mobile genetic elements can continue to persist for up to 12 months (Hong et al., 2013b). A similar observation was made by Fahrenfeld et al. (2013) as they reported that the abundance of *sul-1* and *sul-2* genes increased in soil microcosms that were repeatedly irrigated with secondary wastewater effluent or applied with manure (Fahrenfeld et al., 2014). These observations were in contrast to those made by Negreanu et al. (2012). These authors assessed the abundance of antibiotic-resistant bacteria and resistance genes in soil samples irrigated with either freshwater or treated wastewater, and concluded that there were no observable differences in the abundance of the resistant bacteria and their genes between soils exposed to treated wastewater and soils irrigated with fresh water.

The contradictory observations may stem from the varying environmental factors that are unique to each of the three studies. Antibiotic-resistant bacteria and resistance genes that are disseminated into the environment from the treated wastewater can be inactivated by ultraviolet (UV) irradiation (Biswal et al., 2014; Guo et al., 2013b; Huang et al., 2013; McKinney and Pruden, 2012) or adsorbed onto natural organic matters (Lu et al., 2012). Given that Negreanu et al. conducted their study in Israel, which is in the subtropical region with a consistent exposure of sunlight irradiation throughout the year, it is likely that the contaminants in the treated wastewater were rapidly inactivated by environmental factors and no longer detectable by molecular-based approaches.

The lack of detection by molecular-based methods can be due to the low detection sensitivity of these methods. It does not definitively indicate that no antibiotic-resistant bacteria and resistance genes are present. Remnant antibiotic-resistant genes or antibiotic-resistant bacteria may persist in the environment, and if given sufficient time, horizontal gene transfer can occur among microorganisms. Horizontal gene transfer refers to the exchange of mobile genetic elements like integrases, transposases, or plasmids. In a recent study, Forsberg et al. (2012) made use of high-throughput sequencing approaches to characterize the metagenomes of soil microbiota and human pathogens. The sequences were then assembled for contigs (i.e., long assembled sequences of more than 1,000 bp in length), and bioinformatically annotated for antibiotic resistance genes or resistance-related genes in the contigs. These genes were further aligned against amino acid entries in GenBank and were observed to share more than 90% identity with resistance genes associated with human pathogens. Furthermore, several of these resistance genes were associated with mobile genetic elements like class 1 integrases or transposases, suggesting that there was shared antibiotic resistome between soil bacteria and human pathogens. Although it remains uncertain what would be the timeframe required to promote the horizontal gene transfer, their findings suggest that given sufficient time, there exists a likelihood for horizontal gene transfer interactions to happen.

Impact on Agricultural Crops

A more direct impact on agricultural crops would be the uptake of antibiotics by crops that are irrigated by tertiary treated wastewater. Antimicrobial reagents like triclosan and triclocarban were found to concentrate in root tissues and translocated into beans (Wu et al., 2010). In a separate study, Boxall et al. spiked 1 mg/kg of antibiotics into loamy soil to represent a worst-case scenario of what might happen if crops were irrigated with treated wastewater that was not sufficiently removed of the antibiotic contaminants (Boxall et al., 2006). It was determined that different types of crops can bio-attenuate different types of antibiotics. For example, only florfenicol, levamisole, and trimethoprim were detected in lettuce leaves whereas diazinon, enrofloxacin, florfenicol, and trimethoprim were detected in carrot root (Boxall et al., 2006). The findings suggest that more types of crops would have to be tested for their abilities to bio-attenuate antibiotics, and the subsequent effect it may have on hosts as they consume these crops. Marti et al. detected the positive presence of gene targets associated with antibiotic resistance from vegetables grown in nonmanured soil. This suggests that vegetables are able to uptake gene elements that are naturally present in the soils. When the vegetables were grown in manure-fertilized soil, there was a positive detection of several additional antibiotic resistance genes (Marti et al., 2013).

The above-mentioned studies illustrate how crops can be contaminated when exposed to manure-applied soils. However, contact with manure-applied soil is not the only route for crops to become contaminated. To illustrate, crops can also be exposed to pathogens during irrigation events. There are different irrigation modes, namely by overhead spraying, subsurface drip irrigation, furrow irrigation, that are widely practiced. The furrow irrigation requires application of water to the soil surface, and therefore results in direct contact between the aboveground portion of the plant with the irrigation water. The subsurface irrigation introduces water directly to the root system of the plant, in turn conserving water and ensuring minimal contact between the water and the aboveground portions of the crops (Stine et al., 2005). In the instance when irrigation water is applied overhead, contaminants that may be present in the water can adhere onto the leafy surfaces.

Upon surface contamination, pathogens can be internalized into the plants (Schikora et al., 2012), albeit this phenomenon usually happens at a low frequency and that the required concentration of pathogens would need to be exceedingly high (i.e., at concentrations $>10^6$ CFU/g or 10^6 CFU/mL). For example, certain serovars of *Salmonella enterica*, namely Typhimurium, Enteritidis, and Senftenberg, have been found to colonize seeds, leaves, and fruits (Gu et al., 2011) via the natural apertures (i.e., leaf stomata, lateral junctions of roots) or through damaged tissues (Erickson, 2012). Besides *Salmonella* spp., several other pathogens such as *E. coli* O157:H7 (Solomon et al., 2002) and *Cryptosporidium parvum* (Macarisin et al., 2010) were also able to internalize into plant hosts.

Currently there are no studies that evaluate the antibiotic resistance and susceptibility of these internalized pathogens within the plants. It remains unknown if the same serovar of pathogen would be able to successfully internalize the plants at the same rate if it carries a mobile genetic content within its genome (i.e., larger genome content). Pathogens that carry with them a larger genome are hypothesized to be metabolically burdened and might not have the competitive edge in their growth over those without the mobile genome (Andersson and Levin, 1999; Moran, 2002). Furthermore, upon internalization into the plant hosts where it could gain ready access to nutrients, pathogens with antibiotic resistance traits encoded on mobile genetic elements may lose these genetic traits due to the absence of selective pressure. Further studies would be required to look into whether the concern arising from the internalization of antibiotic-resistant bacteria would be valid. Nevertheless, a worrying observation is that upon infection by *Salmonella* Typhimurium via the stomata, the fruits exhibit no apparent disease symptoms except for a slight decrease in plant growth (Gu et al., 2011). This would mean that consumers are not able to discern the quality of the produce they are purchasing from just the mere looks. The onus would therefore be on the relevant regulatory bodies to make sure that necessary safety checks are in place, and that these checks are conducted regularly so that fresh produce sold in the open markets is safe for consumption.

Impact on Occupational and Public Health

In the United States, monitoring of food safety is performed either by the United States Department of Agriculture (USDA) or the US Food and Drug Administration (FDA). The USDA Food Safety and Inspection Service (FSIS) performs extensive monitoring efforts arising from pathogens in meat, poultry, and egg products while all other food would fall under the monitoring jurisdiction of the FDA (USDA, 2014). The sheer amount of chemicals and contaminants to be tested often means that detection would only concentrate on those contaminants with known risks. Simply put, regulatory agencies would only be motivated to pursue monitoring of certain contaminants when conclusive evidence of the imposed threats is present. It is generally presumed that only a portion of pathogenic microorganisms that are ingested along with food would survive their passage through the gastrointestinal tract (Kelly et al., 1997), and that colonization of the gastrointestinal tract may be only transient. In the unfortunate event that a human host is infected by a foodborne pathogen, doctors prescribe antibiotics in a bid to kill off the pathogen and then monitor if the host would react favorably to the prescribed antibiotics. This wait-and-see approach might not work if the causative agent is highly resistant to the usual treatment of antibiotics. In fact, antimicrobial resistance threats account for up to 23,000 deaths annually in the United States alone (CDC, 2013). Clearly, the threat to public health would increase as more and more bacterial isolates gain resistance to antibiotics.

For now, adequate cooking seems to be the best solution as it is widely accepted that the heat from cooking would destroy foodborne pathogens, including antibiotic-resistant bacteria. However, it is also important to consider dissemination of antibiotic-resistant bacteria from livestock animals to human hosts through other routes, for example, daily contact with these animals, or living in close proximity to livestock production farms. To illustrate, numerous studies showed that a small subset of farm workers who come in close contact with the animals were positive for antibiotic-resistant *E. coli* isolates that share the same pulsed-field gel electrophoresis (PFGE) fingerprinting patterns as the isolates retrieved from the livestock, although the rate of transfer and infection is infrequent (Parsonnet and Kass, 1987; van den Bogaard et al., 2001; Zhao et al., 2010).

PFGE is a standard method which utilizes the periodic changing of voltage direction to achieve DNA fragment separation that is based on the electric charge and size. As such, DNA from *E. coli* isolates exhibiting the same PFGE patterns was presumed to be clonal cells or of the same strain, and likely to originate from the same source. However, similar to many other methods, PFGE cannot establish clonal homogeneity of the bacterial isolates with 100% accuracy. Resolution of the technique is dependent on the number of isolates that are studied, and based on how well the DNA is separated to result in the predominant bands shown in gels (Cantor et al., 1988). A poor separation of large DNA molecules on PFGE can result in similar fingerprinting patterns that are indistinguishable, in turn generating false-positive results on clonal homogeneity. This false-positive result overestimates the true extent of bacterial dissemination among the different hosts.

QUANTITATIVE MICROBIAL RISK ASSESSMENT

Quantitative microbial risk assessment (QMRA) is defined as a framework and approach that brings information and data together with mathematical models to address the spread of microbial agents through environmental exposures, and to characterize the nature of the adverse outcomes (CAMRA, 2013). Simply put, QMRA can be used to better evaluate the extent of risks involved in using treated wastewater for agricultural irrigation. To achieve this aim, information pertaining to the types of microbial hazard, dose–response and exposure assessment would have to be first made available prior to risk characterization. The inclusion of a small subsection of QMRA within this chapter serves to provide readers with a brief introduction. Readers are encouraged to read the QMRA wikipage (CAMRA, 2013) or the corresponding textbook to learn more on QMRA (Haas et al., 1999).

Hazard Identification

To illustrate, microbial hazards include but are not limited to the six pathogens that collectively account for over 90% of estimated food-related deaths in the United States: *Salmonella* (31%), *Listeria* (28%), *Toxoplasma* (21%), Norwalk-like viruses (7%), *Campylobacter* (5%), and *E. coli* O157:H7 (3%) (Mead et al., 1999). In particular, at the species or serovar level, these would include *S. enterica* serovar Typhimurium, *Listeria monocytogenes*, *Campylobacter coli*, and *Campylobacter jejuni*. After identifying the microbial agents of interest, appropriate methods like qPCR can then be performed to determine the abundance of these microbial contaminants in the food and water samples.

By first extracting for the genomic DNA from food or water samples, the genomic DNA can then be used as a DNA template for subsequent qPCR reactions that would assist in determining the total copy number of genes per gram of food sample or per ml of water sample. To further convert the total copy number of genes into cell numbers, prior knowledge on the number of gene copies per cell would be required. For example, a database that provides the copy number of 16S rRNA or 23S rRNA genes per cell is curated by the Schmidt Laboratory (Klappenbach et al., 2001; Lee et al., 2009). From this database, one would be able to determine the copy number of rRNA genes per cell of a particular bacterium. Therefore, given that qPCR determines the total copy number of 16S rRNA genes, one can then divide with the 16S rRNA copies per cell to determine the total number of cells present in a particular sample.

Alternatively, one can also utilize a culture-based method to perform a plate count of a particular pathogenic species, albeit the accuracy of the counts would be dependent on whether the bacterium is able to grow in the media. Commercially available media like *Listeria* mono differential agar (HiMedia Laboratories, India) can be used to selectively enrich and differentiate *L. monocytogenes*. However, most of the media like *Campylobacter* selective agar (Oxoid, UK) or *Salmonella* chromogen agar (Sigma-Aldrich, MO, USA) are only useful to select for the bacteria at the genus level but are not sufficiently selective enough to target a particular pathogenic species. The bacterial isolates growing on the plates would therefore require further testing (e.g., serotype testing) to determine the identity of the pathogenic species. Furthermore, unlike molecular-based methods that only involve handling of genomic DNA (i.e., noninfectious agents); culture-based methods involve enriching the number of pathogenic species, sometimes to levels that might reach an infectious dose. Therefore, the experiment should only be done in a biosafety-secure laboratory with appropriate handling and disposal protocols.

Exposure Assessment

The actual dose at which the host is exposed to would be dependent on the exposure routes and the probability of transmission of the causative agent from the contamination source to the host. Examples of exposure routes would include ingestion of contaminated food and exposure through an open dermal wound. The USEPA exposure factors handbook details statistical data on various factors that can in turn be used to calculate human exposure to contaminants (USEPA, 2011). To illustrate, one can assume that 33% of the total 15,310 cm^2 body surface area will be exposed during irrigation, with 0.14635 mL/cm^2 aqueous particulates from water to adhere to exposed body surface. These numerical constants can be assumed based on the statistical data provided from the exposure factors handbook. Thereafter, the exposure concentration of pathogenic species can be obtained by multiplying the exposed surface area with (i) concentration of adhered water particulates, (ii) concentration of pathogenic species in water sample, and (iii) the probability of transmission, which is often an assumed probability or derived from a larger set of epidemiological study that determines the probability of transmission in a subgroup population.

Dose–Response

Dose–response models are response curves obtained by plotting the probability of a response outcome (e.g., infection, illness, or death) against the known dose of causative pathogen through a particular transmission route. To obtain dose–response models, dosing experiments are generally performed on animal models, where one spikes a known concentration of pathogens to animals and then observes for the response outcome. The obtained data are then fitted into exponential or beta-Poisson models to obtain numerical constants that would allow calculation of probability of death or infection. The concentration of pathogens required to initiate a response in 50% of the tested population would be termed as either lethal dose-50 (LD$_{50}$) or infectious dose-50 (ID$_{50}$).

For starters, the QMRA wiki is a good resource to access the dose–response models pertaining to some of the common pathogens (CAMRA, 2013). However, dose–response models for antibiotic-resistant pathogens remain unavailable, and it remains uncertain whether the dose–response might vary between the antibiotic-resistant and nonresistant strains. As of now, determining the dose–response models for the large suite of foodborne pathogens remains a work in progress.

Given that most dose–response models are obtained from animal experiments, one of the major limitations is that the models might not be directly applicable to ascertain the dose–response of human hosts. Even if the dose–response can be applied to human hosts, one has to note that every individual might have differences in their response towards a causative agent. These differences may be due to genetics, gender, weight, health status, and so on. Therefore, even though the dose–response models provide a quick and easy way for the assessor to determine the threshold dose of pathogens, this dose might be applicable only for the general population but not for at-risk groups (e.g., young children, pregnant women, immune-compromised patients). As such, there still remains a certain unknown level of risk for these individuals.

Risk Characterization

One can then estimate the risk arising from an exposed dose of causative pathogen via a particular transmission route assuming a certain dose–response model. The determined risk can then be compared to an acceptable level of risk. Generally, the annual risk acceptable for drinking water quality is 10^{-4} (Smeets et al., 2009). This acceptable risk is significantly lower than the water quality imposed for recreational activities. The USEPA BEACH Act rule stated that recreational water quality should correspond to an illness rate of 0.8% for swimmers in freshwater (i.e., 8×10^{-3}) and 1.9% for swimmers in marine waters (i.e., 1.9×10^{-2}) (USEPA, 2006a). In contrast, the risks incurred if wastewater were to be used for irrigation are not yet fully understood (Fatta-Kassinos et al., 2011), and therefore an annual risk of 10^{-4} can be used as a benchmark since the fresh produce that is exposed to treated wastewater is often ingested raw like drinking water.

QMRA is not just restricted in its use for risk determination. Instead, QMRA can be used as a tool to guide engineers or practitioners on what would be the level of pathogens permissible in treated wastewater. To illustrate, Stine et al. (2005) performed a case study to determine the permissible level of enteric pathogens in irrigation water. These enteric pathogens can impose infection when contaminated fresh produce was ingested. It was first determined that the average percentage of transferring *E. coli* from the irrigation water to the leaf surfaces of fresh produce via irrigation would range from 3.1×10^{-2} to 6×10^{-7}, depending on the type of fresh produce and the irrigation method. These transfer rates were also independently assessed in a separate study, which reported an average transfer rate of bacteria from the water to the surface of fruit ranging from 0.00021% to 9.4% (i.e., 9.4×10^{-2} to 2.1×10^{-6}) (Gerba and Choi, 2006).

Stine et al. then considered the amount of the fresh produce that was consumed each year per capita, and performed QMRA to determine that the permissible abundance of *E. coli* in the treated wastewater should range from 7.4 to 6.2×10^6 CFU/100 mL. Furthermore, the fresh produce should only be harvested at least 1 day after using treated wastewater for irrigation (Stine et al., 2005). This QMRA-based finding seems to agree with the WHO guidelines which recommend that no more than 10^5 fecal coliforms/100 mL should be present when adult farm workers are exposed to spray irrigation (Blumenthal et al., 2000).

As such, the QMRA approach not only serves to provide an estimation of the risk from a causative pathogen, it can also facilitate better decision-making processes for the relevant stakeholders. Stakeholders can determine what would be the lowest permissible level of pathogens that are allowed in a particular food or water sample, and then provide technological solutions to reach that targeted concentration. This prevents unnecessary implementation of expensive technologies that over-treats the wastewater, and yet does not bring substantial reduction to microbial risk. It is anticipated that QMRA will become an essential tool in the pursuit of having treated wastewater as a safe and sustainable alternative water source.

TECHNOLOGICAL SOLUTIONS

The general public perception is that using reclaimed wastewater for agricultural irrigation would compromise microbiological food safety. Therefore, as more water-scarce countries start to use wastewater for irrigation, public confidence in food safety will increasingly be challenged. Regulatory agencies have the responsibility to ensure that the use of treated wastewater does not impair food safety. A major challenge is that most agricultural sites are located off the infrastructure grid, and oftentimes, the treated wastewater has to be pumped from the centralized wastewater treatment plant through long distances of purple pipelines to reach the agricultural sites. Microbial regrowth can therefore occur within the purple pipelines. Depending on the treatment schematic, treated wastewater generally contains about 30 mg/L of BOD. BOD is the amount of dissolved oxygen needed by aerobic microorganisms to break down organic matter present in the wastewater, and correlates to the amount of organic matter that still remains in the treated wastewater. It is important to note that no existing wastewater treatment process can achieve total removal of microorganisms. Simply put, microorganisms still remain after the treatment process, and when distributed into the pipeline networks, can utilize the remaining organic matter to grow and multiply within the purple pipelines. Over time, the planktonic-state microorganisms can establish biofilms on the pipelines, and these biofilm matrices provide safe harbor for antibiotic-resistant bacteria by protecting the bacterial cells from disinfectant residuals (Jefferson, 2004).

Therefore, bringing down the risks associated with these pathogens would require a multifaceted effort that includes the implementation of appropriate wastewater treatment systems at the point of agricultural irrigation.

Chlorine and UV Disinfection

A proposed solution would be to utilize chlorine and UV disinfection to inactivate microbial agents and therefore suppress the extent of microbial regrowth. To illustrate, chlorine can be introduced into the purple pipeline to achieve inactivation of antibiotic-resistant bacteria. The mode of action for chlorine to disinfect microbial agents would be through the formation of chlorinated derivatives with nucleotide bases of DNA and amino groups in proteins (Davies et al., 1993; Hawkins and Davies, 1998; McDonnell and Russell, 1999; Russell, 2003). The oxidative damage to DNA and proteins by chlorine is incompatible with life as they exert detrimental effects on DNA synthesis and enzyme activities.

Despite the obvious advantage of using chlorination to inactivate pathogens, incessant use of chlorination should be discouraged. This is because chlorine not only reacts with DNA and proteins of microbial agents; it can also react with other organic and inorganic matter that may be present in the water. For example, chlorine is able to react with ammonium to form monochloramines, which is a milder disinfectant than chlorine. Monochloramine along with other organic nitrogen-containing compounds can be precursors for the formation of carcinogenic N-nitrosodimethylamine (NDMA) (Mitch and Sedlak, 2002, 2004). In the presence of pharmaceutical compounds including ranitidine, minocycline, doxepin, and amitriptyline, additional carcinogenic/toxic nitrogenous disinfection byproducts can also occur (Le Roux et al., 2011).

Besides the formation of carcinogenic and toxic disinfection byproducts, studies suggest that a sublethal concentration of chlorination is an oxidative stress that can result in an unintended increase in several types of stress response genes (Berry et al., 2010; Bodet et al., 2012; Wang et al., 2009). For example, it was observed that the relative abundance of genes encoding for resistance against ampicillin, gentamicin, beta-lactams, tetracycline, and erythromycin (i.e., *ampC*, *aphA2*, *bla$_{TEM-1}$*, *tetA*, *tetG*, *ermA*, and *ermB* genes) increased with respect to the total 16S rRNA genes (Shi et al., 2013). In particular, both *tetA* and *tetG* encode for tetracycline resistance via efflux pumps (Aminov et al., 2002). This suggests that bacterial strains that possess these genes activate the efflux pumps to maintain an optimal intracellular concentration of chlorine. Other bacteria like *P. aeruginosa* upregulate the production of capsular extracellular polysaccharide to protect themselves within the biofilm matrix (Xue et al., 2013), therefore decreasing their susceptibility to chlorine.

Given that chlorination can result in the formation of toxic byproducts, an alternative disinfection approach for the antibiotic-resistant bacteria would be UV irradiation (Guo et al., 2013a,b; McKinney and Pruden, 2012). UV exposure was thought to result in the dimerization of adjacent pyrimidines (i.e., thymine and cytosine residues within the DNA or uracil and cytosine residues within the RNA) and affect cellular growth (Asano et al., 2007a; Oguma et al., 2001). It was further observed that bacteria with larger genomes were more effectively inactivated by UV exposure than those with smaller genomes (McKinney and Pruden, 2012). This is presumably because larger genomes would have more sites that are available to be targeted by UV for dimerization. While UV exposure was effective in achieving up to 5-log inactivation of antibiotic-resistant bacteria at doses between 10 and 20 mJ/cm^2, the method was ineffective in achieving the same log inactivation of the resistance genes (McKinney and Pruden, 2012). Instead, a high dose of 200–400 mJ/cm^2 would be required to achieve 3- to 4-log damage to the antibiotic resistance genes (McKinney and Pruden, 2012). This amount is higher than the recommended dose of 186 mJ/cm^2 required to achieve a 4-log inactivation of viruses (USEPA, 2006b). These findings suggest that it would seem impractical from the operational viewpoint to solely rely on UV disinfection to tackle the antibiotic resistance genes.

Anaerobic Membrane Bioreactor as a Decentralized Treatment Technology

Since both chlorine and UV disinfection have their limitations in effectively tackling antibiotic-resistant bacteria and their genes, a proposed solution would be to adopt the membrane separation process. One example of the membrane separation process would be the anaerobic membrane bioreactor, which is also suitable to be utilized as a decentralized treatment technology. Decentralized treatment technologies are standalone unit processes that are not reliant on the infrastructure grids and able to serve a community at the localized site. For the purpose of agricultural irrigation, these technologies should ideally be able to convert wastewater to clean reclaimed water while imposing minimal energy costs and retaining the nitrogen and phosphorus content of the wastewaters.

Among the various proposed technologies, anaerobic processes demonstrate potential in being an energy-neutral or energy-positive treatment technology (Shoener et al., 2014; Wei et al., 2014). Anaerobic processes rely on the establishment of anaerobic fermentative bacteria and methane-producing archea to break down the organic constituents that are present in the wastewater. Methane generated is a greenhouse gas but can be burnt off to provide energy for heating and cooking purposes. Furthermore, anaerobic processes generally do not produce as much sludge as a typical aerobic process and would significantly decrease the volume of solid waste that needs to be disposed of. Considering that most of the antibiotics and antibiotic-resistant bacteria are associated with the biomass fraction of sludge, the reduced solid waste volume from an anaerobic process would mean that there is reduced disposal frequency and a lower likelihood of disseminating these contaminants into the environment.

Anaerobic bioprocesses can be further retrofitted with a membrane filtration unit to form the anaerobic membrane bioreactor. By coupling the membrane separations to the anaerobic process, better permeate quality can be achieved. There are various types of membranes that are used to treat wastewater, namely microfiltration, ultrafiltration, and nanofiltration membranes. These membranes are named according to the pore sizes on the membranes. Microfiltration membranes have pore sizes ranging from 0.1 to 10 μm; pore sizes on ultrafiltration membranes range from 0.001 to 0.1 μm while those on a nanofiltration membrane would range from 0.1 to 1 nm. Depending on the pore sizes of the used membranes, substances that are smaller than the stated pore sizes would permeate through the membranes. Ultrafiltration membranes were found to achieve up to 5.9-log removal of antibiotic-resistant genes when colloidal materials were present (Breazeal et al., 2013), while it was possible to achieve more than 4-log removal of viruses by means of coagulation–microfiltration (Madaeni et al., 1995; Zhu et al., 2005). In both cases, the permeate can be used for nonpotable reuse purposes with reduced risks arising from antibiotic-resistant bacteria and horizontal transfer of antibiotic resistance genes via transduction.

Summing up, membrane-based separation processes demonstrate high potential in mitigating the risks arising from antibiotic-resistant bacteria and antibiotic resistance genes by providing efficient removal of these contaminants.

CONCLUSIONS

Moving forward, water will remain a critical resource and a major bottleneck in meeting the food production targets. Reclaimed water can be used as a renewable source of water to irrigate agricultural crops. While there is a need to be concerned about emerging contaminants, a multifaceted effort that involves (i) regular monitoring of potentially impacted areas and (ii) implementing appropriate treatment technologies would serve to contain the risks, particularly antibiotic-associated risks.

REFERENCES

Alaa el-Din, M.N., Madany, I.M., Al-Tayaran, A., Al-Jubair, A.H., Gomaa, A., 1994. Trends in water quality of some wells in Saudi Arabia, 1984–1989. Sci. Total Environ. 143, 173–181.

Aminov, R.I., Chee-Sanford, J.C., Garrigues, N., Teferedegne, B., Krapac, I.J., White, B.A., et al., 2002. Development, validation, and application of PCR primers for detection of tetracycline efflux genes of gram-negative bacteria. Appl. Environ. Microbiol. 68, 1786–1793.

Andersson, D.I., Levin, B.R., 1999. The biological cost of antibiotic resistance. Curr. Opin. Microbiol. 2, 489–493.

Asano, T., Burton, F.L., Leverenz, H.L., Tsuchihashi, R., Tchobanoglous, G., 2007a. Water Reuse: Issues, Technologies, And Applications, first ed. McGraw-Hill, New York, NY, pp. 604–605.

Asano, T., Burton, F.L., Leverenz, H.L., Tsuchihashi, R., Tchobanoglous, G., 2007b. Water Reuse: Issues, Technologies, And Applications, first ed. McGraw-Hill, New York, NY, pp. 954–955.

Ayers, R.S., Westcot, D.W., 1994. Water quality for agriculture. <http://www.fao.org/DOCReP/003/T0234e/T0234E01.htm#ch1.4> (accessed 28.06.14).

Berry, D., Holder, D., Xi, C., Raskin, L., 2010. Comparative transcriptomics of the response of *Escherichia coli* to the disinfectant monochloramine and to growth conditions inducing monochloramine resistance. Water Res. 44, 4924–4931.

Biswal, B.K., Khairallah, R., Bibi, K., Mazza, A., Gehr, R., Masson, L., et al., 2014. Impact of UV and PAA disinfection on the prevalence of virulence and antimicrobial resistance genes in uropathogenic *Escherichia coli* in wastewater effluents. Appl. Environ. Microbiol. 80, 3656–3666.

Blumenthal, U.J., Mara, D.D., Peasey, A., Ruiz-Palacios, G., Stott, R., 2000. Guidelines for the microbiological quality of treated wastewater used in agriculture: recommendations for revising WHO guidelines. Bull. World Health Organ. 78, 1104–1116.

Bodet, C., Sahr, T., Dupuy, M., Buchrieser, C., Hechard, Y., 2012. *Legionella pneumophila* transcriptional response to chlorine treatment. Water Res. 46, 808–816.

Boxall, A.B.A., Johnson, P., Smith, E.J., Sinclair, C.J., Stutt, E., Levy, L.S., 2006. Uptake of veterinary medicines from soils into plants. J. Agric. Food Chem. 54, 2288–2297.

Breazeal, M.V., Novak, J.T., Vikesland, P.J., Pruden, A., 2013. Effect of wastewater colloids on membrane removal of antibiotic resistance genes. Water Res. 47, 130–140.

CAMRA, 2013. Quantitative Microbial Risk Assessment (QMRA) wiki. <http://qmrawiki.msu.edu/index.php?title=Quantitative_Microbial_Risk_Assessment_(QMRA)_Wiki> (accessed 02.07.14).

Cantor, C.R., Smith, C.L., Mathew, M.K., 1988. Pulsed-field gel electrophoresis of very large DNA molecules. Annu. Rev. Biophys. Biophys. Chem. 17, 287–304.

CDC, 2013. Antimicrobial resistance threat report 2013. <http://www.cdc.gov/drugresistance/threat-report-2013/> (accessed 02.07.14).

Chee-Sanford, J.C., Aminov, R.I., Krapac, I.J., Garrigues-Jeanjean, N., Mackie, R.I., 2001. Occurrence and diversity of tetracycline resistance genes in lagoons and groundwater underlying two swine production facilities. Appl. Environ. Microbiol. 67, 1494–1502.

CLSI, 2014. Performance Standards for Antimicrobial Susceptibility Testing; Twenty-Fourth Informational Supplement. Clinical and Laboratory Standards Institute, Wayne, PA.

Davies, J.M., Horwitz, D.A., Davies, K.J., 1993. Potential roles of hypochlorous acid and N-chloroamines in collagen breakdown by phagocytic cells in synovitis. Free Radic. Biol. Med. 15, 637–643.

Eldin, M.N.A., Madany, I.M., Altayaran, A., Aljubair, A.H., Gomaa, A., 1993. Quality of water from some wells in Saudi-Arabia. Water Air Soil Pollut. 66, 135–143.

Erickson, M.C., 2012. Internalization of fresh produce by foodborne pathogens. Annu. Rev. Food Sci. Technol. 3, 283–310.

Fahrenfeld, N., Ma, Y., O'Brien, M., Pruden, A., 2013. Reclaimed water as a reservoir of antibiotic resistance genes: distribution system and irrigation implications. Front. Microbiol. 4, 130.

Fahrenfeld, N., Knowlton, K., Krometis, L.A., Hession, W.C., Xia, K., Lipscomb, E., et al., 2014. Effect of manure application on abundance of antibiotic resistance genes and their attenuation rates in soil: field-scale mass balance approach. Environ. Sci. Technol. 48, 2643–2650.

Fatta-Kassinos, D., Kalavrouziotis, I.K., Koukoulakis, P.H., Vasquez, M.I., 2011. The risks associated with wastewater reuse and xenobiotics in the agroecological environment. Sci. Total Environ. 409, 3555–3563.

Forsberg, K.J., Reyes, A., Wang, B., Selleck, E.M., Sommer, M.O., Dantas, G., 2012. The shared antibiotic resistome of soil bacteria and human pathogens. Science 337, 1107–1111.

Francy, D.S., Helsel, D.R., Nally, R.A., 2000. Occurrence and distribution of microbiological indicators in groundwater and stream water. Water Environ. Res. 72, 152–161.

Gerba, C.P., Choi, C.Y., 2006. Role of irrigation water in crop contamination by viruses. In: Goyal, S.M. (Ed.), Viruses in Foods Springer, pp. 345.

Goss, M.J., Barry, D.A.J., 1995. Groundwater quality: responsible agriculture and public perceptions. J. Agric. Environ. Ethics 8, 52–64.

Gu, G., Hu, J., Cevallos-Cevallos, J.M., Richardson, S.M., Bartz, J.A., van Bruggen, A.H., 2011. Internal colonization of *Salmonella enterica* serovar Typhimurium in tomato plants. PLoS ONE 6, e27340.

Guo, M.T., Yuan, Q.B., Yang, J., 2013a. Microbial selectivity of UV treatment on antibiotic-resistant heterotrophic bacteria in secondary effluents of a municipal wastewater treatment plant. Water Res. 47, 6388–6394.

Guo, M.T., Yuan, Q.B., Yang, J., 2013b. Ultraviolet reduction of erythromycin and tetracycline resistant heterotrophic bacteria and their resistance genes in municipal wastewater. Chemosphere 93, 2864–2868.

Haas, C.N., Rose, J.B., Gerba, C.P., 1999. Quantitative Microbial Risk Assessment. Wiley, New York, NY.

Hawkins, C.L., Davies, M.J., 1998. Hypochlorite-induced damage to proteins: formation of nitrogen-centred radicals from lysine residues and their role in protein fragmentation. Biochem. J. 332 (Pt 3), 617–625.

Hong, P.Y., Al-Jassim, N., Ansari, M.I., Mackie, R.I., 2013a. Environmental and public health implications of water reuse: antibiotics, antibiotic resistant bacteria, and antibiotic resistance genes. Antibiotics 2, 367–399.

Hong, P.Y., Yannarell, A.C., Dai, Q., Ekizoglu, M., Mackie, R.I., 2013b. Monitoring the perturbation of soil and groundwater microbial communities due to pig production activities. Appl. Environ. Microbiol. 79, 2620–2629.

Huang, J.J., Hu, H.Y., Wu, Y.H., Wei, B., Lu, Y., 2013. Effect of chlorination and ultraviolet disinfection on tetA-mediated tetracycline resistance of *Escherichia coli*. Chemosphere 90, 2247–2253.

Hudzicki, J., 2009. Kirby-Bauer disk diffusion susceptibility test protocol. <http://www.microbelibrary.org/component/resource/laboratory-test/3189-kirby-bauer-disk-diffusion-susceptibility-test-protocol> (accessed 23.06.14).

Jefferson, K.K., 2004. What drives bacteria to produce a biofilm? FEMS Microbiol. Lett. 236, 163–173.

Kelly, C.P., Chetham, S., Keates, S., Bostwick, E.F., Roush, A.M., Castagliuolo, I., et al., 1997. Survival of anti-*Clostridium difficile* bovine immunoglobulin concentrate in the human gastrointestinal tract. Antimicrob. Agents Chemother. 41, 236–241.

Klappenbach, J.A., Saxman, P.R., Cole, J.R., Schmidt, T.M., 2001. rrndb: the ribosomal RNA operon copy number database. Nucleic Acids Res. 29, 181–184.

Koike, S., Krapac, I.G., Oliver, H.D., Yannarell, A.C., Chee-Sanford, J.C., Aminov, R.I., et al., 2007. Monitoring and source tracking of tetracycline resistance genes in lagoons and groundwater adjacent to swine production facilities over a 3-year period. Appl. Environ. Microbiol. 73, 4813–4823.

Koike, S., Aminov, R.I., Yannarell, A.C., Gans, H.D., Krapac, I.G., Chee-Sanford, J.C., et al., 2010. Molecular ecology of macrolide-lincosamide-streptogramin B methylases in waste lagoons and subsurface waters associated with swine production. Microb. Ecol. 59, 487–498.

Krapac, I.G., Dey, W.S., Roy, W.R., Smyth, C.A., Storment, E., Sargent, S.L., et al., 2002. Impacts of swine manure pits on groundwater quality. Environ. Pollut. 120, 475–492.

Kristiansson, E., Fick, J., Janzon, A., Grabic, R., Rutgersson, C., Weijdegard, B., et al., 2011. Pyrosequencing of antibiotic-contaminated river sediments reveals high levels of resistance and gene transfer elements. PLoS ONE 6, e17038.

Lachmayr, K.L., Kerkhof, L.J., Dirienzo, A.G., Cavanaugh, C.M., Ford, T.E., 2009. Quantifying nonspecific TEM beta-lactamase (blaTEM) genes in a wastewater stream. Appl. Environ. Microbiol. 75, 203–211.

Le Roux, J., Gallard, H., Croue, J.P., 2011. Chloramination of nitrogenous contaminants (pharmaceuticals and pesticides): NDMA and halogenated DBPs formation. Water Res. 45, 3164–3174.

Lee, Z.M., Bussema III, C., Schmidt, T.M., 2009. rrnDB: documenting the number of rRNA and tRNA genes in bacteria and archaea. Nucleic Acids Res. 37, D489–D493.

Lu, N., Mylon, S.E., Kong, R., Bhargava, R., Zilles, J.L., Nguyen, T.H., 2012. Interactions between dissolved natural organic matter and adsorbed DNA and their effect on natural transformation of *Azotobacter vinelandii*. Sci. Total Environ. 426, 430–435.

Luo, Y., Yang, F.X., Mathieu, J., Mao, D., Wang, Q., Alvarez, P.J., 2014. Proliferation of multidrug-resistant New Delhi metallo-beta-lactamase genes in municipal wastewater treatment plants in northern China. Environ. Sci. Technol. Lett. 1, 26–30.

Macarisin, D., Bauchan, G., Fayer, R., 2010. *Spinacia oleracea* L. leaf stomata harboring *Cryptosporidium parvum* oocysts: a potential threat to food safety. Appl. Environ. Microbiol. 76, 555–559.

Madaeni, S.S., Fane, A.G., Grohmann, G.S., 1995. Virus removal from water and waste-water using membranes. J. Memb. Sci. 102, 65–75.

Marti, R., Scott, A., Tien, Y.C., Murray, R., Sabourin, L., Zhang, Y., et al., 2013. Impact of manure fertilization on the abundance of antibiotic-resistant bacteria and frequency of detection of antibiotic resistance genes in soil and on vegetables at harvest. Appl. Environ. Microbiol. 79, 5701–5709.

McDonnell, G., Russell, A.D., 1999. Antiseptics and disinfectants: activity, action, and resistance. Clin. Microbiol. Rev. 12, 147–179.

Mckeon, D.M., Calabrese, J.P., Bissonnette, G.K., 1995. Antibiotic-resistant Gram-negative bacteria in rural groundwater supplies. Water Res. 29, 1902–1908.

McKinney, C.W., Pruden, A., 2012. Ultraviolet disinfection of antibiotic resistant bacteria and their antibiotic resistance genes in water and wastewater. Environ. Sci. Technol. 46, 13393–13400.

Mead, P.S., Slutsker, L., Dietz, V., McCaig, L.F., Bresee, J.S., Shapiro, C., et al., 1999. Food-related illness and death in the United States. Emerg. Infect. Dis. 5, 607–625.

Mitch, W.A., Sedlak, D.L., 2002. Formation of N-nitrosodimethylamine (NDMA) from dimethyl-amine during chlorination. Environ. Sci. Technol. 36, 588–595.

Mitch, W.A., Sedlak, D.L., 2004. Characterization and fate of N-nitrosodimethylamine precursors in municipal wastewater treatment plants. Environ. Sci. Technol. 38, 1445–1454.

Moran, N.A., 2002. Microbial minimalism: genome reduction in bacterial pathogens. Cell 108, 583–586.

Munir, M., Wong, K., Xagoraraki, I., 2011. Release of antibiotic resistant bacteria and genes in the effluent and biosolids of five wastewater utilities in Michigan. Water Res. 45, 681–693.

Negreanu, Y., Pasternak, Z., Jurkevitch, E., Cytryn, E., 2012. Impact of treated wastewater irrigation on antibiotic resistance in agricultural soils. Environ. Sci. Technol. 46, 4800–4808.

Oguma, K., Katayama, H., Mitani, H., Morita, S., Hirata, T., Ohgaki, S., 2001. Determination of pyrimidine dimers in *Escherichia coli* and *Cryptosporidium parvum* during UV light inactivation, photoreactivation, and dark repair. Appl. Environ. Microbiol. 67, 4630–4637.

Parsonnet, K.C., Kass, E.H., 1987. Does prolonged exposure to antibiotic-resistant bacteria increase the rate of antibiotic-resistant infection. Antimicrob. Agents Chemother. 31, 911–914.

Ruimy, R., Brisabois, A., Bernede, C., Skurnik, D., Barnat, S., Arlet, G., et al., 2010. Organic and conventional fruits and vegetables contain equivalent counts of Gram-negative bacteria expressing resistance to antibacterial agents. Environ. Microbiol. 12, 608–615.

Russell, A.D., 2003. Similarities and differences in the responses of microorganisms to biocides. J. Antimicrob. Chemother. 52, 750–763.

Schikora, A., Garcia, A.V., Hirt, H., 2012. Plants as alternative hosts for *Salmonella*. Trends Plant. Sci. 17, 245–249.

Shi, P., Jia, S., Zhang, X.X., Zhang, T., Cheng, S., Li, A., 2013. Metagenomic insights into chlorination effects on microbial antibiotic resistance in drinking water. Water Res. 47, 111–120.

Shiklomanov, I.A., 1998. World Water Resources—A New Appraisal and Assessment for the 21st Century. In: United Nations Educational, Scientific and Cultural Organization (Ed.), Paris, France.

Shoener, B.D., Bradley, I.M., Cusick, R.D., Guest, J.S., 2014. Energy positive domestic wastewater treatment: the roles of anaerobic and phototrophic technologies. Environ. Sci. Process Impacts 16, 1204–1222.

Singer, R.S., Patterson, S.K., Meier, A.E., Gibson, J.K., Lee, H.L., Maddox, C.W., 2004. Relationship between phenotypic and genotypic florfenicol resistance in *Escherichia coli*. Antimicrob. Agents Chemother. 48, 4047–4049.

Smeets, P.W.M.H., Medema, G.J., Van Dijk, J.C., 2009. The Dutch secret: how to provide safe drinking water without chlorine in the Netherlands. Drink. Water Eng. Sci. 2, 1–14.

Solomon, E.B., Yaron, S., Matthews, K.R., 2002. Transmission of *Escherichia coli* O157:H7 from contaminated manure and irrigation water to lettuce plant tissue and its subsequent internalization. Appl. Environ. Microbiol. 68, 397–400.

Stachowicz, J.J., Whitlatch, R.B., Osman, R.W., 1999. Species diversity and invasion resistance in a marine ecosystem. Science 286, 1577–1579.

Stine, S.W., Song, I.H., Choi, C.Y., Gerba, C.P., 2005. Application of microbial risk assessment to the development of standards for enteric pathogens in water used to irrigate fresh produce. J. Food Prot. 68, 913–918.

Tunney, M.M., Ramage, G., Field, T.R., Moriarty, T.F., Storey, D.G., 2004. Rapid colorimetric assay for antimicrobial susceptibility testing of *Pseudomonas aeruginosa*. Antimicrob. Agents Chemother. 48, 1879–1881.

USDA, 2014. Food Safety and Inspection Service (FSIS)—Associated Agencies and Partnerships. <http://www.fsis.usda.gov/wps/portal/informational/aboutfsis> (accessed 21.08.14).

USEPA, 2006a. Acceptable risk levels in Great Lake waters. <http://water.epa.gov/lawsregs/lawsguidance/beachrules/bacteria-risk-level-factsheet.cfm> (accessed 02.07.14).

USEPA, 2006b. Ultraviolet disinfection guidance manual for the final long term 2 enhanced surface water treatment rule. In: USEPA (Ed.), Washington, DC.

USEPA, 2011. Exposure Factors Handbook 2011. In: USEPA (Ed.), Washington, DC.

van den Bogaard, A.E., London, N., Driessen, C., Stobberingh, E.E., 2001. Antibiotic resistance of faecal *Escherichia coli* in poultry, poultry farmers and poultry slaughterers. J. Antimicrob. Chemother. 47, 763–771.

van Elsas, J.D., Chiurazzi, M., Mallon, C.A., Elhottova, D., Kristufek, V., Salles, J.F., 2012. Microbial diversity determines the invasion of soil by a bacterial pathogen. Proc. Natl. Acad. Sci. U.S.A. 109, 1159–1164.

Walsh, T.R., Weeks, J., Livermore, D.M., Toleman, M.A., 2011. Dissemination of NDM-1 positive bacteria in the New Delhi environment and its implications for human health: an environmental point prevalence study. Lancet Infect. Dis. 11, 355–362.

Wang, S., Deng, K., Zaremba, S., Deng, X., Lin, C., Wang, Q., et al., 2009. Transcriptomic response of *Escherichia coli* O157:H7 to oxidative stress. Appl. Environ. Microbiol. 75, 6110–6123.

Wei, C.H., Harb, M., Amy, G., Hong, P.Y., Leiknes, T., 2014. Sustainable organic loading rate and energy recovery potential of mesophilic anaerobic membrane bioreactor for municipal wastewater treatment. Bioresour. Technol. 166, 326–334.

Wu, C., Spongberg, A.L., Witter, J.D., Fang, M., Czajkowski, K.P., 2010. Uptake of pharmaceutical and personal care products by soybean plants from soils applied with biosolids and irrigated with contaminated water. Environ. Sci. Technol. 44, 6157–6161.

Xue, Z., Hessler, C.M., Panmanee, W., Hassett, D.J., Seo, Y., 2013. *Pseudomonas aeruginosa* inactivation mechanism is affected by capsular extracellular polymeric substances reactivity with chlorine and monochloramine. FEMS Microbiol. Ecol. 83, 101–111.

Yong, D., Toleman, M.A., Giske, C.G., Cho, H.S., Sundman, K., Lee, K., et al., 2009. Characterization of a new metallo-beta-lactamase gene, bla(NDM-1), and a novel erythromycin esterase gene carried on a unique genetic structure in *Klebsiella pneumoniae* sequence type 14 from India. Antimicrob. Agents Chemother. 53, 5046–5054.

Zhao, J.J., Chen, Z.L., Chen, S., Deng, Y.T., Liu, Y.H., Tian, W., et al., 2010. Prevalence and dissemination of oqxAB in *Escherichia coli* isolates from animals, farmworkers, and the environment. Antimicrob. Agents Chemother. 54, 4219–4224.

Zhu, B., Clifford, D.A., Chellam, S., 2005. Virus removal by iron coagulation-microfiltration. Water Res. 39, 5153–5161.

Chapter 17

Development and Application of Novel Antimicrobials in Food and Food Processing

Yangjin Jung and Karl R. Matthews

Department of Food Science, School of Environmental and Biological Sciences, Rutgers, The State University of New Jersey, New Brunswick, NJ, USA

Chapter Outline

INTRODUCTION—WHY NOVEL ANTIMICROBIALS ARE NEEDED

Raw and processed foods are subject to microbial spoilage from preparation to storage and in consumer homes, reducing shelf-life and increasing costs to the consumer. Many foods are at risk for contamination with pathogens that can have devastating human health implications. Globally consumers desire foods that are ready-to-eat and minimally processed. The food industry, in an effort to meet consumer demands, may use an array of synthetic preservatives or food antimicrobials independently or in conjunction with other treatments including thermal processing, high-pressure processing, or temperature control. The underlying need for use of antimicrobial agents in food is to overcome challenges with food safety and quality. Microbial communities associated with foods can be complex and include yeasts, bacteria, and molds that are capable of growth or survival under a wide range of intrinsic and extrinsic conditions.

The safety of the food supply greatly influences consumers globally. In developed countries, consumers desire, even demand, products year-round regardless of the growing season of those commodities. A plethora of factors

Antimicrobial Resistance and Food Safety. DOI: http://dx.doi.org/10.1016/B978-0-12-801214-7.00017-X

come into play when attempting to ensure the safety of the food supply. Food safety typically relates to ensuring that food is free of pathogenic microorganisms such as *Escherichia coli* O157:H7, *Clostridium botulinum*, *Listeria monocytogenes*, *Salmonella*, and *Staphylococcus aureus* that can negatively impact human health. Development and application of natural antimicrobials are particularly appealing for use on raw agricultural commodities. Raw agricultural commodities such as fresh and fresh-cut fruits and vegetables are perishable by nature and support the growth of human pathogens. The number of reported foodborne disease outbreaks associated with fresh fruits and vegetables has made consumers wary and regulators frustrated. Consumption of contaminated fruits and vegetables during 2009–2010 was associated with 59 outbreaks and 2,074 cases of illnesses according to Centers for Disease Control and Prevention (CDC) data (CDC Foodborne Outbreak Online Database, 2014). The prevalence of outbreaks was greater for vegetables than for fruits; the principal problem commodities were lettuce, other leafy greens, sprouts, tomatoes, and melons (Painter et al., 2013). Such products are generally not treated with antimicrobials that will control microbial outgrowth prior to reaching the consumer. The number of outbreaks and cases of foodborne illness linked to the consumption of contaminated produce continues to escalate globally (Hoelzer et al., 2012a,b; Painter et al., 2013; Teplitski et al., 2011; Critzer and Doyle, 2010).

Globalization of food trade is apparent as the importation of food continues to increase in the United States and other developed countries. Novel antimicrobials deemed safe for use in food in one country may not be approved for use in food in other countries. In 2009, imports accounted for 17% of the food consumed in the United States. In the United States ~80% of the fish and shellfish consumed is imported, while nearly 34% of fruits and vegetables consumed are imported (USDA ERS, 2012). The continued increase in imports is associated with growing ethnic diversity and consumer preference for a wider selection of food products such as premium coffee, cheeses, processed meats, and tropical fruits (USDA ERS, 2012). A similar import pattern has emerged in the European Union (EU) (Jaud et al., 2013).

The need to evolve to keep pace with consumer demands is reflected in the vast number of items carried in a supermarket. In 2013, the average number of items carried by a supermarket was 43,844 (Food Marketing Institute, 2014). The change in availability of highly perishable products has also increased. In 1998, a typical grocery store carried 345 produce items compared with just 173 in 1987 (Calvin et al., 2001). Studies suggest that per capita expenditure on fresh fruits and vegetables will increase more than for any other product group through 2020 (Blisard et al., 2003). Changes in technology from the field to retail outlets ensure fresh quality products can be made available to consumers. Advances in controlled atmosphere technologies have significantly improved the shelf-life of perishable products, permitting shipment globally (Huang, 2004). Controlled atmosphere can slow ripening, retard discoloration, and maintain freshness. Even with these technological advances, each year billions of dollars in losses occur due to spoilage of fruits and vegetables. Losses of fruits and vegetables to waste and spoilage at the retail level for 2005 and 2006 averaged 11.4% for fresh fruits and 9.7% for fresh vegetables (Buzby et al., 2009). Depending on how fragile the commodity, losses can be >20% (Buzby et al., 2011). The diverse microorganisms associated with fresh fruits and vegetables are capable of growth under a wide range of conditions (pH, oxygen level, moisture, nutrient) such that controlled atmosphere, temperature, and improved handling practices may only limit and not prevent spoilage.

In recent years consumers have requested and even demanded that certain food antimicrobials or preservatives be removed from foods (Tiwari et al., 2009). Food manufacturers have used an array of antimicrobial agents to reduce spoilage and control the growth of human pathogens in foods. Indeed, having "preservatives" in the ingredient list can be detrimental to the sale of a product. Research suggests potential links between use of sodium nitrite, propylparaben, and sodium propionates and hypersensitivity, asthma, and increased death from Parkinson's disease and type 2 diabetes (reviewed in Anand and Sati, 2013). Some preservatives, for example acetic acid and benzoic acid, have been used globally for years by many major food manufacturers. Without the use of food antimicrobials many products would have an unacceptable shelf-life and may even present significant human health risks. Many consumers are concerned with nitrates added to cured meat products, but fail to recognize they are added to inhibit the growth of *C. botulinum*. The traditional food antimicrobials are very effective, cost-appropriate, and applicable to a wide range of foods.

The desire to seek novel food antimicrobials is not out of fear that microbes in food have developed resistance to traditional food antimicrobials. Many studies have demonstrated that under defined laboratory conditions yeasts, molds, and bacteria may exhibit tolerance to traditional and natural antimicrobials (reviewed in Davidson and Harrison, 2002). However, when evaluated under real food processing/handling conditions antimicrobial resistance does not appear to be a major phenomenon. Traditional food antimicrobials are proven and remain extremely effective in achieving shelf-life and food safety goals. Using natural food antimicrobials as alternatives to traditional food antimicrobials means food manufacturers may be faced with reformulation of products which may change intrinsic properties of the product.

Researchers have been studying antimicrobials derived from nature for years, but few have been exploited as food antimicrobials/preservatives on a commercial scale. There exist many challenges in achieving regulatory approval for use of natural food antimicrobials and the costs of manufacture is often greater than traditional food antimicrobials. Typically, natural antimicrobials are categorized by the source from which they are derived—microorganisms, animals, herbs, and other plants. Several review papers on natural antimicrobials have recently been published; however, challenges in application and potential for development of tolerance were not emphasized (Juneja et al., 2012; Crozier-Dodson et al., 2004; Lucera et al., 2012). This chapter is not inclusive of all novel antimicrobials, but rather devoted to review of recent novel antimicrobials and their application.

ESSENTIAL OILS

Plant extracts such as flavonoids, alkaloids, terpenes, coumarins, phenolics, and polyphenols have shown antimicrobial effects on undesirable microorganisms in food. In particular, activity *in vitro* and in food matrices of various essential oils mostly derived from terpenes, terpenoids, and phenylpropenes have been evaluated and shown to exhibit broad-spectrum antimicrobial action against fungi, bacteria, and virus (Belletti et al., 2008; Hyldgaard et al., 2012; Solórzano-Santos and Miranda-Novales, 2012). Silva et al. (2013) investigated the antimicrobial effects of essential oils from Mediterranean aromatic plants (thyme, oregano, rosemary, verbena, basil, peppermint, pennyroyal, and mint), demonstrating their bacteriostatic effects against foodborne pathogens and spoilage bacteria. Essential oils from oregano, clove, and zataria exhibited antimicrobial activity against *E. coli* O157:H7 and endogenous bacteria associated with commercial baby-leaf salads (Azizkhani et al., 2013). Hernández-Ochoa et al. (2014) recently investigated the use of essential oil from cumin, clove, and elecampane on meat; reporting a 3.6- to 3.78-log reduction of foodborne pathogens. The antimicrobial activity of oregano-essential oil against ciprofloxacin-resistant *Campylobacter* spp. *in vitro* and multidrug-resistant *Salmonella* Typhimurium DT104 in ground pork has been demonstrated (Aslim and Yucel, 2008; Chen et al., 2013). Ravishankar et al. (2010) reported the inhibition of antibiotic-resistant *Salmonella enterica* on celery and oysters when exposed to 1% of cinnamaldehyde and carvacrol. *S. enterica* was reduced to below detection limits on day 0 following carvacrol treatment, while treatment of celery with 1% cinnamaldehyde resulted in a 1- to 2.3-log reduction on day 0 and day 3 posttreatment, respectively. In addition, the same antimicrobials decreased the population of *S. enterica* by ∼5-log on oysters by day 3 posttreatment. Orhan et al. (2011) reported the inhibitory activity of several essential oils (mostly terpene and aromatic types) against ten strains of extended-spectrum beta-lactamase-producing *Klebsiella pneumoniae*. Spray application of cinnamon oil (0.3% and 0.5% v/v) on organic romaine lettuce, iceberg lettuce, organic baby and mature spinach, inoculated with a strain of multidrug-resistant *Salmonella* Newport reduced the population of the pathogen on all commodities, and exhibited residual inhibitory effects during storage (Todd et al., 2013).

According to the FDA code of federal regulations, crude oils from, for example, clove, oregano, thyme, basil, mustard, and cinnamon are classified as generally recognized as safe (GRAS), and are considered to present no health risks to consumers. In spite of the broad antimicrobial activity of essential oils, there are some limitations for the application in food. Based on the existing research-essential oil would need to be used at a high concentration to achieve equivalent results to traditional preservatives. Use of a high concentration of essential oil may have adverse effects on the organoleptic properties of food and be potentially toxic to humans. Many of the studies demonstrating efficacy of essential oils at low concentration were conducted *in vitro*; activity may be substantially reduced by components of food, pH, temperature, and the commensal microbial population (Cava-Roda et al., 2010; Rattanachaikunsopon and Phumkhachorn, 2010; Somolinos et al., 2010; Hyldgaard et al., 2012). Loss of efficacy and the need for a high concentration of essential oil in a food system may be overcome through use of micro- and nano-encapsulation technology or the application through packaging films.

CHITOSAN

Among animal origin antimicrobials, the application of chitosan in foods has been widely investigated since it is biodegradable, nontoxic, has unique cationic character, and exhibits broad-spectrum antimicrobial activity against bacteria, yeasts, and molds (Kong et al., 2010; Juneja et al., 2012). Chitosan is a biopolymer obtained by the deacetylation of chitin. Many researchers have demonstrated the antimicrobial activity of chitosan on postharvest fruits and vegetables, such as citrus fruit (Chien et al., 2007), strawberry (Campaniello et al., 2008), fresh-cut papaya (González-Aguilar et al., 2009), and fresh-cut broccoli (Moreira et al., 2011). Moreover, chitosan has been used in meat and meat products, fish, and dairy products to maintain quality and inhibit the growth of microorganisms (Zivanovic et al., 2007; Alishahi and Aïder, 2012; Kanatt et al., 2013). Dipping of ready-to-cook meat samples for 2 min in a chitosan solution (2% w/v) resulted in an immediate 2- to 3-log reduction of Gram-positive bacteria (*S. aureus* and *Bacillus cereus*), and a 1- to 2-log reduction in Gram-negative bacteria (*Pseudomonas fluorescens* and *E. coli*). Chhabra et al. (2006) reported the antimicrobial properties of chitosan (0.5%, 1.0%, and 2.0% v/v) against *S. aureus*, *Salmonella* Typhimurium, and *Vibrio vulnificus* associated with raw oysters. Following chitosan treatment, *S. aureus* and *V. vulnificus* populations significantly declined over a 12-day storage period at 4°C. The population of *Salmonella* Typhimurium was not significantly affected by the treatment. Chitosan effectively prevented formation of microbial biofilm on the surface of Plexiglas (polymethylmethacrylate) coupon by *S. epidermidis*, *S. aureus*, *K. pneumoniae*, *P. aeruginosa*, and *Candida albicans* (Carlson et al., 2008). Kong et al. (2010) proposed that factors including intrinsic microbial properties, intrinsic properties of chitosan molecules, chitosan physical state, and environmental or extrinsic factors each influence the efficacy of chitosan in a food matrix. Prior to considering the use of chitosan in a food, all factors must be carefully considered.

ANTIMICROBIAL PEPTIDES

Antimicrobial peptides are produced by animals, plants, and bacteria as natural defense responses. The reasons why antimicrobial peptides can be a promising alternative to conventional chemical preservatives in food are many and include but are not limited to: (1) less likely to induce resistance and cross-resistance to conventional antibiotics, (2) less toxic to human cells, and (3) many exhibit broad-spectrum antimicrobial activity (Gordon and Romanowski, 2005).

Plant-derived antimicrobial peptides, which are cysteine-rich compounds, are classified based on structure and amino acid sequence characteristics, and include thionin, plant defensins, lipid transfer proteins (LTPs), havein- and knottin-like peptides, snakins, and cyclodies (Koo et al., 2002). It is known that the antibacterial effects of most plant antimicrobial peptides are driven by the interaction with cell membranes (Jenssen et al., 2006). According to PhytAMP database, only 39.5% of reported plant peptides were evaluated for their biological activity (Hammami et al., 2009). Among the plant peptides evaluated, 51% exhibited antifungal activity and 33% exhibited antibacterial activity. Wu et al. (2013a,b) investigated the antimicrobial activity of the plant-derived peptide Ib-AMP1 and its mode-of-action against enteric foodborne pathogens. In serial studies, Ib-AMP1 exhibited bactericidal activity on *S. aureus*, *E. coli* O157:H7, *Salmonella* Newport, *P. aeruginosa*, and *B. cereus in vitro*. The bactericidal effects of Ib-AMP1 against *E. coli* O157:H7 was associated with disruption of outer- and inner-membrane integrity and interference with intracellular biosynthesis of DNA, RNA, and protein.

Bacteriocins are ribosomally produced proteinaceous substances produced by both Gram-positive and Gram-negative bacteria. The use of bacteriocin-producing lactic acid bacteria (LAB) in food has gained considerable interest for food preservation applications (Arthur et al., 2014). Gálvez et al. (2007) delineated properties of bacteriocins as food preservatives: (1) GRAS substance, (2) nontoxic to eukaryotic cells, (3) likely inactivated by digestive proteases, (4) no induction of cross-resistance to antibiotics due to their mode-of-action, and (5) suitable to genetic manipulation (Settanni and Corsetti, 2008). Among bacteriocins, nisin produced by *Lactococcus lactis* is approved by FDA for food applications, and has been approved as a food additive in over 45 countries (Settanni and Corsetti, 2008). Although other bacteriocins such as pediocin, have proven potent antimicrobial activity, they are not currently approved as antimicrobial food additives (Naghmouchi et al., 2007). Pediocin produced by *Pediococcus acidilactici* exhibits antimicrobial activity against Gram-positive bacteria over a wide pH range, and has been used as a starter culture. ε-Poly-L-lysine isolated from *Streptomyces albulus* is also considered as a promising bacteriocin for use in foods due to its broad-spectrum activity against Gram-negative and Gram-positive bacteria, yeasts, and molds. ε-Poly-L-lysine can be used at concentrations of up to 50 mg/kg in food (GRAS No. 000135). Although the efficacy and properties of bacteriocins have been widely reported, the practical and widespread use of bacteriocins has been limited, in part due to narrow effective range of activity, the possibility of the development of resistance, and legal restrictions (Settanni and Corsetti, 2008; Hiron et al., 2011).

BACTERIOPHAGE

Bacteriophages are viruses that only infect and lyse bacterial cells. Soon after their discovery by Ernest Hankin and Frederick Twort, bacteriophages were quickly adopted for therapy against bacterial infections in humans. Their use was rapidly eclipsed by the discovery of antibiotics and the subsequent development of technology to chemically synthesize antimicrobials (Sulakvelidze et al., 2001). After the concern associated with antibiotic resistance surfaced, bacteriophage treatment of infectious bacterial diseases entered back into the spotlight of novel antimicrobials. The potential application of bacteriophages to control pathogens of human concern has reached all corners of the food industry. Bacteriophages can be used in therapy to reduce foodborne pathogen colonization in food-producing animals; in biosanitation programs to disinfect equipment and food contact surfaces; in biocontrol on beef carcasses, raw agricultural products, and RTE products; and, in biopreservation of finished products (Greer, 2005; García et al., 2008; Sillankorva et al., 2012). Research has increased substantially on the use of bacteriophages to control foodborne pathogens, including *E. coli* O157:H7, *Salmonella*, *Campylobacter*, *L. monocytogenes*, *Cronobacter sakazaki*, and *S. aureus*. Researchers have focused on the oral administration of bacteriophage preparations during production to cattle, swine, and poultry for the control *E. coli* O157:H7, *Salmonella*, and *Campylobacter* (Callaway et al., 2008; Niu et al., 2009). Postharvest bacteriophage treatments have been used on fresh produce, meat products, dairy products, RTE foods, and food contact surfaces. Leverentz et al. (2001, 2003) applied *Salmonella*-specific and *Listeria*-specific bacteriophages on fresh-cut honeydew melon and achieved a 3.5-log and 2.0- to 4.6-log reduction, respectively, of each pathogen. The investigators indicated that effectiveness of specific bacteriophage treatment was likely influenced by the pH of the honeydew melon. *E. coli* O157:H7-specific phage effectively inactivated the pathogen on hard surfaces, tomatoes, spinach, broccoli, ground beef, lettuce, and cantaloupes (Abuladze et al., 2008; Sharma et al., 2009). Patel et al. (2011) spray applied *E. coli* O157:H7-specific bacteriophages on spinach harvesting equipment. The bacteriophage treatment resulted in a 4.5-log reduction of the pathogen on the blade used for cutting the spinach. Siringan et al. (2011) observed a 1- to 3-log CFU/cm^2 reduction of *Campylobacter* biofilms after application of a bacteriophage treatment. Collectively, bacteriophages have potential as an antimicrobial treatment of crops, animal products, and minimally processed foods. Cautious optimism for broad-spectrum application must be exercised since most of the research demonstrating the potential of bacteriophages was conducted using model systems, not under production conditions (Mahony et al., 2011). A limited number of commercially available bacteriophage-based products are available for use by the food industry. Bacteriophage preparations have received regulatory clearance from both the Food and Drug Administration and United States Department of Agriculture,

Food Safety Inspection Service (USDA FSIS). In 2006, FDA approved a bacteriophage product as an additive for RTE foods (21 CFR §172.785) to control *L. monocytogenes*. FDA provided Food Contact Notification, in 2011, for a bacteriophage product to control *E. coli* O157:H7 in ground meat. GRAS recognition from the FDA of a bacteriophage-based product to control *Salmonella* in red meat and poultry products was achieved in 2013.

NATURAL ANTIMICROBIAL COMBINATIONS

A major limitation in the use of many natural antimicrobials is that high concentrations are required to achieve efficacy comparable to conventional preservatives. This has led to a wave of research to determine synergistic or antagonistic effects of combinations of natural antimicrobials. Combining two or more natural antimicrobials may actually lower costs and eliminate potential off-odor and flavor effects linked with the use of high concentrations of some natural antimicrobials (e.g., essential oils).

Combined essential oils, such as carvacrol, eugenol, thymol, or cinnamon resulted in synergistic activity against Gram-positive and Gram-negative bacteria (Turgis et al., 2012; Abdollahzadeh et al., 2014). Gutierrez et al. (2009) evaluated the combined effects of plant-essential oils and found synergistic effects with the combination of thyme and oregano oils. Techathuvanan et al. (2014) compared *in vitro* the efficacy of the commercially available natural antimicrobials: white mustard-essential oil, citrus flavonoid and acid blend, olive extract, and lauric arginate, against foodborne pathogens and spoilage microorganisms. A range of "additive," "synergistic," or "antagonistic" effects were observed depending on the different combinations of each antimicrobial.

There is ample evidence demonstrating the combined effects of chitosan and plant extracts on control of microflora on products such as application of chitosan and oregano oil on chicken (Petrou et al., 2012; Khanjari et al., 2013), and chitosan and rosemary oil on turkey meat (Vasilatos and Savvaidis, 2013). Alvareza et al. (2013) investigated the efficacy of chitosan combined with essential oils to enhance the microbial safety of fresh-cut broccoli. The chitosan treatment alone significantly reduced populations of mesophilic and psychrotrophic bacteria on the fresh-cut broccoli compared to untreated controls at day 2 and day 7 posttreatment. The greatest reduction in bacterial populations was achieved on broccoli treated with a combination of chitosan and essential oil (tea tree, pollen, and propolis).

Nisin has been evaluated in combination with a vast array of other antimicrobials in an effort to achieve a broader spectrum of activity and greater range of applications. Traditionally, combinations of the bacteriocin with chemical chelators, such as EDTA, disodium pyrophosphate, or trisodium phosphate, have proven effective against Gram-negative bacteria *in vitro* and in foods; however, major interest rests in using nisin in combination with other natural antimicrobials. A mixture of nisin (6,400 IU/mL) and grape seed extracts

(1%) reduced populations of *L. monocytogenes* to undetectable levels in both a nutrient broth at 37°C and on turkey frankfurters at 4°C and 10°C (Sivarooban et al., 2007). Research demonstrated that a combination of lactoferrin and nisin acted synergistically to inhibit the growth of *E. coli* O157:H7 and *L. monocytogenes* (Murdock et al., 2007). Govaris et al. (2010) demonstrated the bactericidal effect of a combination of oregano-essential oil (0.6% or 0.9%) and nisin (500 or 1,000 IU/g) against *Salmonella* Enteritidis in minced sheep meat stored at 10°C. Nisin treatment (1000 IU/g) alone of minced sheep meat was insufficient to influence *Salmonella* Enteritidis growth or survival. The combination of nisin (1,000 IU/g) and thyme essential oil (0.6%) was effective against *E. coli* O157:H7 in minced beef stored at 10°C (Solomakos et al., 2008).

Magnone et al. (2013) recently investigated a produce wash which combined bacteriophage and levulinic acid to control *E. coli* O157:H7, *Salmonella*, and *Shigella* on contaminated fresh broccoli, cantaloupe, and strawberries. In water with high organic matter content the produce wash containing the combined antimicrobials demonstrated a significantly greater reduction of pathogen populations compared to treatment with 200-ppm free available chlorine. A 5-log reduction of *E. coli* O157:H7 occurred within 10 min on leaf tissue treated with a combination of bacteriophage and essential oil (*trans*-cinnamaldehyde), whereas independently the essential oil or bacteriophage resulted in a 3- and 1-log CFU reduction, respectively (Viazis et al., 2011).

SYNERGISTIC ACTIVITY OF NOVEL ANTIMICROBIALS AND NOVEL TECHNOLOGY

The limited antimicrobial activity of many antimicrobial dips or sprays for treatment of food contact surfaces, and even raw and processed food, has led to the development of various antimicrobial materials. Edible films and coatings containing novel antimicrobials, such as chitosan, essential oils, natural plant extracts, or bacteriocins have shown promise in preventing and delaying the growth of undesirable microorganisms that may be associated with food. In addition, the emergence of nanotechnology and other innovative technologies has contributed to enhancing activity of novel antimicrobials.

Antimicrobial substances can be delivered through polymers, polysaccharides, and nano- or microparticles. Comprehensive reviews on chitosan-based antimicrobial films and bacteriocin delivery systems have been recently published (Dutta et al., 2009; Arthur et al., 2014). Chitosan–polysaccharide (tapioca starch, hydroxypropyl methylcellulose, pectin laminated) and chitosan–protein (round scad protein, whey protein, fish gelatin) based films have been investigated for their potential food application. Utilizing antimicrobials in micro- or nanoparticle format in biopolymers can improve the performance of the antimicrobial. Silver and zinc oxide nanoparticles exhibit strong antimicrobial activity. A chitosan/methyl cellulose film with incorporated vanillin was applied to fresh-cut cantaloupe and pineapple that was then stored at 10°C.

The treatment effectively inhibited growth of *E. coli* associated with each commodity (Sangsuwan et al., 2008). Sun et al. (2014) developed chitosan coatings containing 0.5% of one of the following essential oils: carvacrol, cinnamaldehyde, and *trans*-cinnamaldehyde. Microbial populations were effectively controlled on fresh blueberries treated with the chitosan essential oil coatings. Control of microbial populations on strawberries was achieved using a combined treatment of dipping into a chitosan (1% solution, w/v) solution and modified atmosphere packaging (Campaniello et al., 2008). Pullulan films containing rosemary and oregano oils, and silver and zinc oxide nanoparticles exhibited antimicrobial activity against *S. aureus*, *Salmonella* Typhimurium, *L. monocytogenes*, and *E. coli* O157:H7 (Morsy et al., 2014).

Nanotechnology has proven to be an effective tool for the development of delivery systems for antimicrobial peptides. Bi et al. (2011) described the use of a stabilizing carbohydrate emulsion to extend the efficacy of nisin against foodborne pathogens. Silver nanoparticles capped with bacteriocin effectively inactivated foodborne pathogens (Sharma et al., 2012).

There is a rising interest in the use of photosensitizers for food surface decontamination and in the treatment of raw agricultural commodities (Luksiene, 2005). Photosensitization requires a combination of three nontoxic agents: photosensitizer, light, and oxygen. Buchovec et al. (2009) reported on the use of 5-aminolevulinic acid (ALA) as a photosensitizer to control multidrug-resistant *Salmonella*. ALA-based photoinactivation was used to control *Listeria* biofilm, achieving a 3.1-log reduction in the pathogen. Chlorophyllin-based photosensitization of strawberries reduced ~2-log of natural flora composed of yeast, mold, and mesophilic microbes (Luksiene and Paskeviciute, 2011a). The researchers also demonstrated that chlorophyllin could be used to effectively decontaminate surfaces (Luksiene and Paskeviciute, 2011b). Other natural compounds have been identified as potential photosensitizers, such as hypercin, curcumin, and alpha-terthienyl (Luksiene and Brovko, 2013).

CONSIDERATIONS WHEN SELECTING NOVEL ANTIMICROBIALS

Bacteria can develop resistance or decreased susceptibility to antimicrobials as the result of exposure to sublethal concentrations of an antimicrobial or through cross protection when the organism is exposed to other stresses (To et al., 2002; Riazi and Matthews, 2010; Bergholz et al., 2013). Some stressors, including exposure to acid and salt, are known to contribute to the resistance of *L. monocytogenes* to nisin (Bonnet and Montville, 2005; Bergholz et al., 2013). Sublethal concentrations of the essential oil of *Origanum vulgare* L. and its major compound carvacrol caused injury to the cytoplasmic membrane and outer membrane of *Salmonella* Typhimurium from which the cell could recover (Luz Ida et al., 2014). When used appropriately, it is unlikely that bacteria associated with food would develop resistance to novel antimicrobials. Numerous research studies on adaptive responses of bacteria to novel antimicrobials have been published. Becerril et al. (2012) evaluated bacterial resistance to oregano and cinnamon-essential oils. Each essential oil exhibited high antimicrobial activity against each of 60 Gram-negative bacteria isolates, but only the oregano oil induced cross-resistance to antibiotics. In contrast, continuous exposure to eugenol and citral did not induce cross-resistance to antibiotics or to the substance itself in *S. aureus*, methicillin-resistant *S. aureus*, and *L. monocytogenes* (Apolónio et al., 2014). Regardless, consideration should be given to the impact that stresses to the bacterial cell may have on development of decreased susceptibility or cross-resistance to antimicrobials.

Novel antimicrobials that interfere with bacterial efflux pumps and quorum sensing activity are of interest (Savoia, 2012; Sultanbawa, 2011). Researchers have identified new natural antimicrobials that act as efflux pump inhibitors of Gram-positive and Gram-negative bacteria (Garvey et al., 2011; Fiamegos et al., 2011). Klančnik et al. (2012) evaluated the role of *Alpina katsumadai* ethanolic extract as a putative efflux pump inhibitor of *Campylobacter*. The extract was active against antibiotic-sensitive and multi-antibiotic-resistant *Campylobacter* isolates and exhibited resistance-modifying activity as an efflux pump inhibitor. Kuete et al. (2011) also suggested that xanthone, naphthoquinone, flavonoid, and coumarin can be attractive candidates for the development of antimicrobials against multi-antibiotic-resistant bacteria in combination with efflux pump inhibitors. Grapefruit juice, furocoumarins, and citrus flavonoids inhibited biofilm formation of *E. coli* O157:H7, *Salmonella* Typhimurium and *P. aeruginosa*, *V. harveyi* BB120, and *E. coli* O157:H7 by inhibition of autoinducers AI or AI-2 (Girennavar et al., 2008; Vikram et al., 2010). Kerekes et al. (2013) reported the anti-biofilm-forming and anti-quorum-sensing activity of clary sage, juniper, lemon, and marjoram-essential oils. Adukwu et al. (2012) observed the biofilm inhibitory activity of lemongrass and grapefruit-essential oils against five *S. aureus* strains.

CONCLUSION AND FUTURE TRENDS

The use of natural or nontraditional antimicrobials in foods to minimize spoilage and enhance microbial safety continues to grow. A wealth of research has been conducted on natural antimicrobials, especially bacteriophages, chitosan, bacteriocins, organic acids, and essential oils. Although many of these agents exhibit antimicrobial activity at low concentrations *in vitro*, efficacy in food systems may be limited. Studies in food systems often demonstrated that the concentration required for microbial control resulted in off-odors and flavors, or because of the high concentrations needed it would be commercially cost-prohibitive.

The mode of action of many natural antimicrobials is either not known or only partially elucidated. Understanding the mode of action of these compounds would facilitate targeted use and increase the potential for combining one or more natural compounds that would effectively act against a wide range of spoilage and pathogenic microorganisms. In conjunction, research must delineate how intrinsic properties (e.g., pH, cations) of a food influence efficacy. The thermal stability of natural antimicrobials should also be determined; at elevated temperatures some novel antimicrobials may be partially or completely inactivated.

The investigation of novel plant antimicrobials and bacteriophages for use in foods should be further encouraged. Plant-derived antimicrobials exhibit broad-spectrum activity against Gram-negative and Gram-positive bacteria, yeasts, and molds. The compounds are active across a wide pH spectrum, temperature range, and at low concentrations. They can also be commercially synthesized. Aside from use as food preservatives, they have potential application as antimicrobial growth promoters in animal feeds. Bacteriophages have potential application in a broad spectrum of raw and processed foods. The specificity of bacteriophages may be considered a drawback, but cocktails of bacteriophages that target a wide range of pathogenic bacteria can be developed. As with any of the novel antimicrobials discussed, regulatory approval must be achieved, labeling guidelines established, and consumer acceptance met.

REFERENCES

Abdollahzadeh, E., Rezaei, M., Hosseini, H., 2014. Antibacterial activity of plant essential oils and extracts: the role of thyme essential oil, nisin, and their combination to control *Listeria monocytogenes* inoculated in minced fish meat. Food Control 35, 177–183.

Abuladze, T., Li, M., Menetrez, M.Y., Dean, T., Senecal, A., Sulakvelidze, A., 2008. Bacteriophages reduce experimental contamination of hard surfaces, tomato, spinach, broccoli, and ground beef by *Escherichia coli* O157:H7. Appl. Environ. Microbiol. 74, 6230–6238.

Adukwu, E.C., Allen, S.C., Phillips, C.A., 2012. The anti-biofilm activity of lemongrass (*Cymbopogon flexuosus*) and grapefruit (*Citrus paradisi*) essential oils against five strains of *Staphylococcus aureus*. J. Appl. Microbiol. 113, 1217–1227.

Alishahi, A., Aïder, M., 2012. Applications of chitosan in the seafood industry and aquaculture: a review. Food Bioprocess Technol. 5, 817–830.

Alvareza, M., Ponce, A.G., Moreira, M.R., 2013. Antimicrobial efficiency of chitosan coating enriched with bioactive compounds to improve the safety of fresh cut broccoli. LWT Food Sci. Technol. 50, 78–87.

Anand, S.P., Sati, N., 2013. Artificial preservatives and their harmful effects: looking toward nature for safer alternatives. Int. J. Pharm. Sci. Res. 4, 2496–2501.

Apolónio, J., Faleiro, M.L., Miguel, M.G., Neto, L., 2014. No induction of antimicrobial resistance in *Staphylococcus aureus* and *Listeria monocytogenes* during continuous exposure to eugenol and citral. FEMS Microbiol. Lett. 354, 92–101.

Arthur, T.D., Cavera, V.L., Chikindas, M.L., 2014. On bacteriocin delivery systems and potential applications. Future Microbiol. 9, 235–248.

Aslim, B., Yucel, N., 2008. *In vitro* antimicrobial activity of essential oil from endemic *Origanum minutiflorum* on ciprofloxacin-resistant *Campylobacter* spp. Food Chem. 107, 602–606.

Azizkhani, M., Elizaquível, P., Sánchez, G., Selma, M.V., Aznar, R., 2013. Comparative efficacy of *Zataria multiflora* Boiss., *Origanum compactum* and *Eugenia caryophyllus* essential oils against *E. coli* O157:H7, feline calicivirus and endogenous microbiota in commercial baby-leaf salads. Int. J. Food Microbiol. 166, 249–255.

Becerril, R., Nerín, C., Gómez-Lus, R., 2012. Evaluation of bacterial resistance to essential oils and antibiotics after exposure to oregano and cinnamon essential oils. Foodborne Pathog. Dis. 8, 699–705.

Belletti, N., Lanciotti, R., Patrignani, F., Gardini, F., 2008. Antimicrobial efficacy of citron essential oil on spoilage and pathogenic microorganisms in fruit-based salads. J. Food Sci. 73, M331–M338.

Bergholz, T.M., Tang, S., Wiedmann, M., Boor, K.J., 2013. Nisin resistance of *Listeria monocytogenes* is increased by exposure to salt stress and is mediated via LiaR. Appl. Environ. Microbiol. 79, 5682.

Bi, L., Yang, L., Bhunia, A.K., Yao, Y., 2011. Carbohydrate nanoparticle-mediated colloidal assembly for prolonged efficacy of bacteriocin against food pathogen. Biotechnol. Bioeng. 108, 1529–1536.

Blisard, N., Variyam, J.N., Cromartie, J., 2003. Food Expenditures by U.S. Households: looking ahead to 2020. ERS USDA Report No. 821.

Bonnet, M., Montville, T.J., 2005. Acid-tolerant *Listeria monocytogenes* persist in a model food system fermented with nisin-producing bacteria. Lett. Appl. Microbiol. 40, 237–242.

Buchovec, I., Vaitonis, Z., Luksiene, Z., 2009. Novel approach to control *Salmonella enterica* by modern biophotonic technology: photosensitization. J. Appl. Microbiol. 106, 748–754.

Buzby, J.C., Wells, H.F., Axtman, B., Mickey, J., 2009. Supermarket loss estimates for fresh fruit, vegetables and their use in the ERS loss adjusted food availability data. ERS USDA eib44.

Buzby, J.C., Hyman, J., Stewart, H., Wells, H.F., 2011. The value of retail and consumer level fruit and vegetable losses in the United States. J. Consum. Aff. Fall, 492–515.

Callaway, T.R., Edrington, T.S., Brabban, A.D., Anderson, R.C., Rossman, M.L., Engler, M.J., et al., 2008. Bacteriophage isolated from feedlot cattles can reduce *Escherichia coli* O157:H7 populations in ruminant gastrointestinal tracts. Foodborne Pathog. Dis. 5, 183–191.

Calvin, L., Cook, R., Denbaly, M., Dimitri, C., Glaser, L., Handy, C., et al. 2001. U.S. fresh fruit and vegetable marketing: Emerging trade practices, trends and issues. USDA, ERS. Agriculture Information Bulletin No. 795.

Campaniello, D., Bevilacqua, A., Sinigaglia, M., Corbo, M.R., 2008. Chitosan: antimicrobial activity and potential applications for preserving minimally processed strawberries. Food Microbiol. 25, 992–1000.

Carlson, R.P., Taffs, R., Davison, W.M., Stewart, P.S., 2008. Anti-biofilm properties of chitosan-coated surfaces. J. Biomater. Sci. Polym. Ed. 19, 1035–1046.

Cava-Roda, R.M., Taboada-Rodríguez, A., Valverde-Franco, M.T., Marin-Iniesta, F., 2010. Antimicrobial activity of vanillin and mixtures with cinnamon and clove essential oils in controlling *Listeria monocytogenes* and *Escherichia coli* O157:H7 in milk. Food Bioprocess Technol. Available from: http://dx.doi.org/10.1007/s11947-010-0484-4.

CDC Foodborne Outbreak Online Database. 2014. <http://www.cdc.gov/foodsafety/outbreaks/> (accessed June 2014).

Chen, C.H., Ravishankar, S., Marchello, J., Friedman, M., 2013. Antimicrobial activity of plant compounds against *Salmonella* Typhimurium DT104 in ground pork and the influence of heat and storage on the antimicrobial activity. J. Food Prot. 76, 1264–1269.

Chhabra, P., Huang, Y.W., Frank, J.F., Chmielewski, R., Gates, K., 2006. Fate of *Staphylococcus aureus, Salmonella enterica* serovar typhimurium, and *Vibrio vulnificus* in raw oysters treated with chitosan. J. Food Prot. 69, 1600–1604.

Chien, P.J., Sheu, F., Lin, H.R., 2007. Coating citrus (Murcott tangor) fruit with low molecular weight chitosan increases postharvest quality and shelf life. Food Chem. 100, 1160–1164.

Critzer, F.J., Doyle, M.P., 2010. Microbial ecology of foodborne pathogens associated with produce. Curr. Opin. Biotechnol. 21, 125–130.

Crozier-Dodson, B.A., Carter, M., Zheng, Z., 2004/2005. Formulating food safety: an overview of antimicrobial ingredients. Food Saf. Mag. <http://www.foodsafetymagazine.com/magazine-archive1/december-2004january-2005/formulating-food-safety-an-overview-of-antimicrobial-ingredients/>.

Davidson, P.M., Harrison, M.A., 2002. Resistance and adaptation to food antimicrobials, sanitizers, and other process controls. Food Technol. 56, 69–78.

Dutta, P.K., Tripathia, S., Mehrotraa, G.K., Duttab, J., 2009. Perspectives for chitosan based antimicrobial films in food applications. Food Chem. 114, 1173–1182.

Fiamegos, Y.C., Kastritis, P.L., Exarchou, V., Han, H., Bonvin, A.M.J.J., Vervoort, J, et al., 2011. Antimicrobial and efflux pump inhibitory activity of caffeoylquinic acids from *Artemisia absinthium* against Gram-positive pathogenic bacteria. PLoS One 6, 1–12.

Food Marketing Institute. 2014. Supermarket Facts—industry overview. <http://www.fmi.org/research-resources/supermarket-facts> (accessed June 2014).

Gálvez, A., Abriouel, H., López, R.L., Ben Omar, N., 2007. Bacteriocin-based strategies for food biopreservation. J. Food Microbiol. 120, 51–70.

García, P., Martínez, B., Obeso, J.M., Rodríguez, A., 2008. Bacteriophages and their application in food safety. Lett. Appl. Microbiol. 47, 479–485.

Garvey, I.M., Rahman, M.M., Gibbons, S., Piddock, J.V.L., 2011. Medicinal plant extracts with efflux inhibitory activity against Gram-negative bacteria. Int. J. Antimicrob. Agents 37, 145–151.

Girennavar, B., Cepeda, M.L., Soni, K.A., Vikram, A., Jesudhasan, P., Jayaprakasha, G.K., et al., 2008. Grapefruit juice and its furocoumarins inhibits autoinducer signaling and biofilm formation in bacteria. Int. J. Food Microbiol. 125, 204–208.

González-Aguilar, G.A., Valenzuela-Soto, E., Lizardi-Mendoza, J., Goycoolea, F., Martínez-Téllez, M.A., Villegas-Ochoa, M.A., 2009. Effect of chitosan coating in preventing deterioration and preserving the quality of fresh-cut papaya "Maradol." J. Sci. Food Agric. 89, 15–23.

Gordon, Y.J., Romanowski, E.G., 2005. A review of antimicrobial peptides and their therapeutic potential as anti-infective drugs. Curr. Eye Res. 30, 505–515.

Govaris, A., Solomakos, N., Pexara, A., 2010. The antimicrobial effect of oregano essential oil, nisin and their combination against *Salmonella* Enteritidis in minced sheep meat during refrigerated storage. Int. J. Food Microbiol. 137, 175–180.

Greer, G.G., 2005. Bacteriophage control of foodborne bacteria. J. Food Prot. 68, 1102–1111.

Gutierrez, J., Barry-Ryan, C., Bourke, P., 2009. Antimicrobial activity of plant essential oils using food model media: efficacy, synergistic potential and interaction with food components. Food Microbiol. 26, 142–150.

Hammami, R., Ben Hamida, J., Vergoten, G., Fliss, I., 2009. PhytAMP: a database dedicated to antimicrobial plant peptides. Nucl. Acid Res. 37, D963–D968.

Hernández-Ochoa, L., Aguirre-Prieto, Y.B., Nevárez-Moorillón, G.V., Gutierrez-Mendez, N., Salas-Muñoz, E., 2014. Use of essential oils and extracts from spices in meat protection. J. Food Sci. Technol. 51, 957–963.

Hiron, A., Falord, M., Valle, J., Débarbouillé, M., Msadek, T., 2011. Bacitracin and nisin resistance in *Staphylococcus aureus*: a novel pathway involving the BraS/BraR two-component system (SA2417/SA2418) and both the BraD/BraE and VraD/VraE ABC transporters. Mol. Microbiol. 81, 602–622.

Hoelzer, K., Pouillot, R., Egan, K., Dennis, S., 2012a. Produce consumption in the United States—an analysis of consumption frequencies, serving sizes, processing forms, and high-consuming population subgroups for microbial risk assessments. J. Food Prot. 75, 328–340.

Hoelzer, K., Pouillot, K., Dennis, S., 2012b. *Listeria monocytogenes* growth dynamics on produce: a review of the available data for predictive modeling. Foodborne Pathog. Dis. 9, 661–673.

Huang, W. 2004. Global trade patterns in fruits and vegetables. ERS USDA Agriculture and trade report#WRS-04-06.

Hyldgaard, M., Mygind, T., Meyer, R.L., 2012. Essential oils in food preservation: mode of action, synergies, and interactions with food matrix components. Front. Microbiol. 3, 12–25.

Jaud, M.J., Cadot, O., Suwa-Eisenmann, A., 2013. Do food scares explain supplier concentration? An analysis of EU agri-food imports. Eur. Rev. Agric. Econ. 38, 1–18.

Jenssen, H., Hamill, P., Hancock, R.E., 2006. Peptide antimicrobial agents. Clin. Microbiol. Rev. 19, 491–511.

Juneja, V.K., Dwivedi, H.P., Yan, X., 2012. Novel natural food antimicrobials. Annu. Rev. Food Sci. Technol. 3, 381–403.

Kanatt, S.R., Rao, M.S., Chawla, S.P., Sharma, A., 2013. Effects of chitosan coating on shelf-life of ready-to-cook meat products during chilled storage. LWT Food Sci. Technol. 53, 321–326.

Kerekes, E.B., Deák, É., Takó, M., Tserennadmid, R., Petkovits, T., Vágvölgyi, C., et al., 2013. Anti-biofilm forming and anti-quorum sensing activity of selected essential oils and their main components on food-related micro-organisms. J. Appl. Microbiol. 115, 933–942.

Khanjari, A., Karabagias, I.K., Kontominas, M.G., 2013. Combined effect of *N, O*-carboxymethyl chitosan and oregano essential oil to extend shelf life and control *Listeria monocytogenes* in raw chicken meat fillets. LWT Food Sci. Technol. 53, 94–99.

Klančnik, A., Gröblacher, B., Kovač, J., Bucar, F., Možina, S.S., 2012. Anti-*Campylobacter* and resistance-modifying activity of *Alpinia katsumadai* seed extracts. J. Appl. Microbiol. 113, 1249–1262.

Kong, M., Chen, X.G., Xing, K., Park, H.J., 2010. Antimicrobial properties of chitosan and mode of action: a state of the art review. Int. J. Food Microbiol. 144, 51–63.

Koo, J.C., Chun, H.J., Park, H.C., Kim, M.C., Koo, Y.D., Koo, S.C., et al., 2002. Over-expression of a seed specific hevein-like antimicrobial peptide from *Pharbitis nil* enhances resistance to a fungal pathogen in transgenic tobacco plants. Plant Mol. Biol. 50, 441–452.

Kuete, V., Alibert-Franco, S., Eyong, K.O., Ngameni, B., Folefoc, G.N., Nguemeving, J.R., et al., 2011. Antibacterial activity of some natural products against bacteria expressing a multidrug-resistant phenotype. Int. J. Antimicrob. Agents 37, 156–161.

Leverentz, B., Conway, W.S., Alavidze, Z., Janisiewicz, W.J., Fuchs, Y., Camp, M.J., et al., 2001. Examination of bacteriophage as a biocontrol method for *Salmonella* on fresh-cut fruit: a model study. J. Food Prot. 64, 1116–1121.

Leverentz, B., Conway, W.S., Camp, M.J., Janisiewicz, W.J., Abuladze, T., Yang, M., et al., 2003. Biocontrol of *Listeria monocytogenes* on fresh-cut produce by treatment with lytic bacteriophages and a bacteriocin. Appl. Environ. Microbiol. 69, 4519–4526.

Lucera, A., Costa, C., Conte, A., Del Nobile, M.A., 2012. Food applications of natural antimicrobial compounds. Front. Microbiol. 3, 1–13.

Luksiene, Z., 2005. New approach to inactivate harmful and pathogenic microorganisms: photosensitization. Food Technol. Biotech. 43, 1–8.

Luksiene, Z., Brovko, L., 2013. Antibacterial photosensitization-based treatment for food safety. Food Eng. Rev. 5, 185–199.

Luksiene, Z., Paskeviciute, E., 2011a. Novel approaches to the microbial decontamination of strawberries: chlorophyllin-based photosensitization. J. Appl. Microbiol. 110, 1274–1283.

Luksiene, Z., Paskeviciute, E., 2011b. Microbial control of food-related surfaces: Na-chlorophyllin-based photosensitization. J. Photochem. Photobiol. B. 105, 69–74.

Luz Ida, S., de Melo, A.N., Bezerra, T.K., Madruga, M.S., Magnani, M., de Souza, E.L., 2014. Sublethal amounts of *Origanum vulgare L.* essential oil and carvacrol cause injury and changes in membrane fatty acid of *Salmonella* Typhimurium cultivated in a meat Broth. Foodborne Pathog. Dis. 11, 357–361.

Magnone, J.P., Marek, P.J., Sulakvelidze, A., Senecal, A.G., 2013. Additive approach for inactivation of *Escherichia coli* O157:H7, *Salmonella*, and *Shigella* spp. on contaminated fresh fruits and vegetables using bacteriophage cocktail and produce wash. J. Food Prot. 76, 1308–1479.

Mahony, J., McAuliffe, O., Ross, R.P., van Sinderen, D., 2011. Bacteriophages as biocontrol agents of food pathogens. Curr. Opin. Biotechnol. 22, 157–163.

Moreira, M.R., Ponce, A., Ansorena, R., Roura, S.I., 2011. Effectiveness of edible coatings combined with mild heat shocks on microbial spoilage and sensory quality of fresh cut broccoli (*Brassica oleracea* L.). J. Food Sci. 76, M367–M374.

Morsy, M.K., Khalaf, H.H., Sharoba, A.M., El-Tanahi, H.H., Cutter, C.N., 2014. Incorporation of essential oils and nanoparticles in pullulan films to control foodborne pathogens on meat and poultry products. J. Food Sci. 79, M675–M684.

Murdock, C.A., Cleveland, J., Matthews, K.R., Chikindas, M.L., 2007. The synergistic effect of nisin and lactoferrin on the inhibition of *Listeria monocytogenes* and *Escherichia coli* O157:H7. Lett. Appl. Microbiol. 44, 255–261.

Naghmouchi, K., Kheadr, E., Lacroix, C., Fliss, I., 2007. Class I/Class IIa bacteriocin cross-resistance phenomenon in *Listeria monocytogenes*. Food Microbiol. 24, 718–727.

Niu, Y.D., McAllister, T.A., Xu, Y., Johnson, R.P., Stephens, T.P., Stanford, K., 2009. Prevalence and impact of bacteriophages on the presence of *Escherichia coli* O157:H7 in feedlot cattle and their environment. Appl. Environ. Microbiol. 75, 1271–1278.

Orhan, I.E., Ozcelik, B., Kan, Y., Kartal, M., 2011. Inhibitory effects of various essential oils and individual components against extended-spectrum beta-lactamase (ESBL) produced by *Klebsiella pneumoniae* and their chemical compositions. J. Food Sci. 76, M538–M546.

Painter, J.A., Hoekstra, R.M., Ayers, T., Tauxe, R.V., Braden, C.R., Angulo, F.J., et al., 2013. Attribution of foodborne illnesses, hospitalizations, and deaths to food commodities by using outbreak data, United States, 1998–2008. Emerg. Infect. Dis. 19, 407–415.

Patel, J., Sharma, M., Millner, P., Calaway, T., Singh, M., 2011. Inactivation of *Escherichia coli* O157:H7 attached to spinach harvester blade using bacteriophage. Foodborne Pathog. Dis. 8, 541–546.

Petrou, S., Tsiraki, M., Giatrakou, V., Savvaidis, I.N., 2012. Chitosan dipping or oregano oil treatments, singly or combined on modified atmosphere packaged chicken breast meat. Int. J. Food Microbiol. 156, 264–271.

Rattanachaikunsopon, P., Phumkhachorn, P., 2010. Antimicrobial activity of basil (*Ocimum basilicum*) oil against *Salmonella* Enteritidis *in vitro* and in food. Biosci. Biotechnol. Biochem. 74, 1200–1204.

Ravishankar, S., Zhu, L., Reyna-Granados, J., Law, B., Joens, L., Friedman, M., 2010. Carvacrol and cinnamaldehyde inactivate antibiotic-resistant *Salmonella enterica* in buffer and on celery and oysters. J. Food Prot. 73, 212–404.

Riazi, S., Matthews, K.R., 2010. Failure of foodborne pathogens to develop resistance to sanitizers following repeated exposure to common sanitizers. Int. Biodeterior. Biodegradation 65, 374–378.

Sangsuwan, J., Rattanapanoneb, N., Rachtanapun, P., 2008. Effect of chitosan/methyl cellulose films on microbial and quality characteristics of fresh-cut cantaloupe and pineapple. Postharvest Biol. Technol. 49, 403–410.

Savoia, D., 2012. Plant-derived antimicrobial compounds: alternatives to antibiotics. Future Microbiol. 7, 979–990.

Settanni, L., Corsetti, A., 2008. Application of bacteriocins in vegetable food biopreservation. Int. J. Food Microbiol. 121, 123–138.

Sharma, M., Patel, J.R., Conway, W.S., Ferguson, S., Sulakvelidze, A., 2009. Effectiveness of bacteriophages in reducing *Escherichia coli* O157:H7 on fresh-cut cantaloupes and lettuce. J. Food Prot. 72, 1481–1485.

Sharma, T.K., Sapra, M., Chopra, A., Sharma, R., Patila, S.D., Malikc, R.K., et al., 2012. Interaction of bacteriocin-capped silver nanoparticles with food pathogens and their antibacterial effect. Int. J. Green Nanotechnol. 4, 93–110.

Sillankorva, S.M., Oliveira, H., Azeredo, J., 2012. Bacteriophages and their role in food safety. Int. J. Microbiol. 2012, 1–13.

Silva, N., Alves, S., Gonçalves, A., Amaral, J.S., Poeta, P., 2013. Antimicrobial activity of essential oils from Mediterranean aromatic plants against several foodborne and spoilage bacteria. Food Sci. Technol. Int. 19, 503–510.

Siringan, P., Connerton, P.L., Payne, R.J., Connerton, I.F., 2011. Bacteriophage-mediated dispersal of *Campylobacter jejuni* biofilms. Appl. Environ. Microbiol. 77, 3320–3326.

Sivarooban, T., Hettiarachchy, N.S., Johnson, M.G., 2007. Inhibition of *Listeria monocytogenes* using nisin with grape seed extract on turkey frankfurters stored at 4 and 10°C. J. Food Prot. 70, 1017–1020.

Solomakos, N., Govaris, A., Koidis, P., Botsoglou, N., 2008. The antimicrobial effect of thyme essential oil, nisin, and their combination against *Listeria monocytogenes* in minced beef during refrigerated storage. Food Microbiol. 25, 120–127.

Solórzano-Santos, F., Miranda-Novales, M.G., 2012. Essential oils from aromatic herbs as antimicrobial agents. Curr. Opin. Biotechnol. 23, 136–141.

Somolinos, M., García, D., Condón, S., Mackey, B., Pagán, R., 2010. Inactivation of *Escherichia coli* by citral. J. Appl. Microbiol. 108, 1928–1939.

Sulakvelidze, A., Alavidze, Z., Morris Jr, J.G., 2001. Bacteriophage therapy. Antimicrob. Agents Chemother. 45, 649–659.

Sultanbawa, Y., 2011. Plant antimicrobials in food applications: mini-review. In: Méndez-Vilas, A. (Ed.), Science Against Microbial Pathogens: Communicating Current Research and Technological Advances, pp. 1084–1093.

Sun, X., Narciso, J., Wang, Z., Ference, C., Bai, J., Zhou, K., 2014. Effects of chitosan-essential oil coatings on safety and quality of fresh blueberries. J. Food Sci. 79, M955–M960.

Techathuvanan, C., Reyes, F., David, J.R., Davidson, P.M., 2014. Efficacy of commercial natural antimicrobials alone and in combinations against pathogenic and spoilage microorganisms. J. Food Prot. 77, 269–275.

Teplitski, M., Warriner, K., Bartz, J., Schneider, K.R., 2011. Untangling metabolic and communication networks: interactions of enterics with phytobacteria and their implications in produce safety. Trends Microbiol. 19, 121–127.

Tiwari, B.K., Valdramidis, V.P., O'Donnell, C.P., Muthukumarappan, K., Bourke, P., Cullen, P.J., 2009. Application of natural antimicrobials for food preservation. J. Agric. Food Chem. 22, 5987–6000.

To, M.S., Favrin, S., Romanova, N., Griffiths, M.W., 2002. Post adaptational resistance to benzalkonium chloride and subsequent physicochemical modifications of *Listeria monocytogenes*. Appl. Environ. Microbiol. 68, 5258–5264.

Todd, J., Friedman, M., Patel, J., Jaroni, D., Ravishankar, S., 2013. The antimicrobial effects of cinnamon leaf oil against multi-drug resistant *Salmonella* Newport on organic leafy greens. Int. J. Food Microbiol. 166, 193–199.

Turgis, M., Vua, K.D., Dupont, C., Lacroix, M., 2012. Combined antimicrobial effect of essential oils and bacteriocins against foodborne pathogens and food spoilage bacteria. Food Res. Int. 48, 696–702.

USDA, ERS. 2012. Import share of consumption. Available from: <http://www.ers.usda.gov/topics/international-markets-trade/us-agricultural-trade/import-share-of-consumption.aspx> (accessed June 2014).

Vasilatos, G.C., Savvaidis, I.N., 2013. Chitosan or rosemary oil treatments, singly or combined to increase turkey meat shelf-life. Int. J. Food Microbiol. 166, 54–58.

Viazis, S., Akhtar, M., Feirtag, J., Diez-Gonzalez, F., 2011. Reduction of *Escherichia coli* O157:H7 viability on leafy green vegetables by treatment with a bacteriophage mixture and trans-cinnamaldehyde. Food Microbiol. 28, 149–157.

Vikram, A., Jayaprakasha, G.K., Jesudhasan, P.R., Pillai, S.D., Patil, B.S., 2010. Suppression of bacterial cell-cell signalling, biofilm formation and type III secretion system by citrus flavonoids. J. Appl. Microbiol. 109, 515–527.

Wu, W.H., Di, R., Matthews, K.R., 2013a. Activity of the plant-derived peptide Ib-AMP1 and the control of enteric foodborne pathogens. Food Control. 33, 142–147.

Wu, W.H., Di, R., Matthews, K.R., 2013b. Antibacterial mode of action of Ib-AMP1 against *Escherichia coli* O157: H7. Probiotics Antimicrob. Prot. 5, 131–141.

Zivanovic, S., Li, J., Davidson, P.M., Kit, K., 2007. Physical, mechanical, and antibacterial properties of chitosan/PEO blend films. Biomacromolecules 8, 1505–1510.

Chapter 18

Database Resources Dedicated to Antimicrobial Peptides

Guangshun Wang

Department of Pathology and Microbiology, University of Nebraska Medical Center, Omaha, NE, USA

Chapter Outline

INTRODUCTION

The discovery of penicillin by Alexander Fleming has improved human life quality substantially. Meanwhile, pesticides were used to increase crop yield in agriculture. After many years of successful practice, the world has realized that artificial chemicals could also do harm to the environment as well as human health. The wide use of artificial chemicals had another shortcoming—microbes

Antimicrobial Resistance and Food Safety. DOI: http://dx.doi.org/10.1016/B978-0-12-801214-7.00018-1

have developed resistance, rendering many antibiotics no longer effective. Thus, there is a strong desire to return to green chemistry and traditional practice so that we can maintain the healthy state of our ecosystem. A classic and useful approach is to identify potent compounds from nature.

Antimicrobial peptides (AMPs) are natural compounds that protect the host from microbial invasion (Boman, 2003; Hancock and Sahl, 2006; Lehrer, 2007; Zasloff, 2002). Most of the known AMPs are cationic, making them perfect to recognize and disrupt anionic bacterial membranes (Epand and Vogel, 1999; Wang, 2008). Some AMPs can also penetrate membranes and act on intracellular molecular targets (Brogden, 2005). It is likely that AMPs work by multiple mechanisms, a feature perhaps responsible for long-lasting potency over millions of years.

AMPs have been identified from bacteria, protists, fungi, plants, and animals (Epand and Vogel, 1999; Tossi and Sandri, 2002; Wang et al., 2009; Zasloff, 2002). In the bacterial kingdom, AMPs are referred to as bacteriocins to distinguish them from antibiotics used to treat people. Nisin, the first- and best-studied bacteriocin, has been utilized as a food preservative for over 40 years. This natural compound differs from artificial additives and is regarded as safe. Remarkably, significant resistance to nisin has not been observed (Delves-Broughton et al., 1996). The expression of a bacteriocin involves a cluster of genes that encode the peptide, the ABC transporter, modifying enzymes, and immunity proteins to avoid toxicity to self. However, some bacterial peptides such as gramicidin are synthesized by a multiple enzyme system. Bacteriocins are diverse in terms of amino acid sequence and 3D structure. They can be linear, sidechain-modified lantibiotics, sidechain-backbone connected lassos, or even circular (i.e., the N- and C-termini are linked by forming a peptide bond). In addition, some bacteriocins are composed of two polypeptide chains that work synergistically for optimal activity (Bierbaum and Sahl, 2009). Nisin is known to target lipid II followed by pore formation (Breukink et al., 1999). The combination of high affinity of nisin for lipid II (cell wall precursor) with pore formation may be responsible for its nanomolar activity.

Amoebapore A, a pore-forming AMP from protozoa, consists of five helices with three disulfide bridges (Leippe M et al., 2005). Another four protozoan AMPs (amoebapores B, C, Naegleriapores A, and B) also belong to the saposin-like protein family. Porcine NK-lysin showed sequence homology to ancient pore-forming protozoan AMPs (Leippe, 1995). Evidently, AMPs with this type of fold are universal in nature. The six protozoan AMPs collected in the APD (Wang et al., 2009) all belong to the helical family. Amino acid composition analysis of these protozoan AMPs (Wang and Wang, 2004) reveals the abundance of residues Ile, Val, Leu, Cys, Ala, and Lys, accounting for ~10% each. On average, they are 75.3 amino acids in length with a net charge of -2.3. The negative net charge could be important for these AMPs to fold into a pore structure.

In contrast to the α-helical peptides from protozoa, AMPs from the plant kingdom normally form β-sheet structures stabilized by multiple disulfide bonds (Egorov et al., 2005). Typical examples are defensins (including γ-thionins) and circular cyclotides. Disulfide-bridged defensins have also been found in fungi and animals, including humans (Wilmes et al., 2011). They are classified into α-, β-, and θ-defensins based on the disulfide bond pattern (Lehrer, 2007; Seebah et al., 2007). There are also successful examples to reduce pathogenic infection via the expression of AMPs in plants (Goyal and Mattoo, 2014). As of June 2014, two dozen AMPs that were expressed in transgenic plants to reduce infection are annotated in the APD (Wang et al., 2009).

In the animal kingdom, AMPs from amphibians frequently possess a helical conformation upon interaction with membranes or their mimics (Conlon and Mechkarska, 2014; Epand and Vogel, 1999). Only in a rare case is distinctin helical in aqueous solution since two copies of the disulfide-linked two-polypeptide chain can form a four-helix bundle structure (Batista et al., 2001). In contrast, AMPs from insects, birds, fish, reptiles, or humans can adopt various 3D structures (Bulet and Stocklin, 2005; Cerovsky and Bem, 2014; Masso-Silva and Diamond, 2014; van Hoek, 2014; Wang, 2014; Zhang and Sunkara, 2014). The structural difference may determine the mechanism of action of these peptides. Pro-rich AMPs from insects usually do not have defined α-helical or β-sheet structures. They work by binding to heat-shock proteins (Rozgonyi et al., 2009). Some defensins are known to bind to lipid II, thereby blocking cell wall synthesis (Wilmes et al, 2011). It is clear that AMPs are diverse in terms of structure and mechanisms of action. They are also universal because they have been identified in all life domains.

The number of AMPs discovered per year (e.g., during 1970–2012) has been increasing rapidly (Wang, 2013), since the discovery of magainins, cecropins, and defensins in the 1980s (Selsted et al., 1985; Steiner et al., 1981; Zasloff, 1987). The pace of AMP discovery is accelerated due to the development and use of genomic and proteomic methods (Li et al., 2007; Yang et al., 2012). To effectively manage AMPs, 23 databases have been established, mostly since 2004 (Table 18.1). These resources vary in peptide scope, mode, and type of information annotation, capability for peptide search and prediction approach. This chapter discusses known AMP databases as of June 2014. They can be classified into general and specialized AMP databases based on the database scope. A general database has a broader range of entries than a specialized one. While the Antimicrobial Peptide Database (APD) is a general resource dedicated to the AMP field (Wang and Wang, 2004; Wang et al., 2009), complementary features from specialized databases, as well as other general databases, are described (Table 18.1). Subsequently, select web servers developed to predict AMPs are also discussed.

GENERAL DATABASES FOR AMPs

Antimicrobial Sequence Database

Antimicrobial Sequence Database (AMSDb) is probably the first database for AMPs built by the Alex Tossi laboratory (online 1998). The SWISS-PROT data format was maintained with modifications. The last accessible version

TABLE 18.1 Databases for AMPs as of June 2014

Year	Database	URL (http://)	Scope	Pep #	Country
1998	AMSDb (inactive)	http://www.bbcm.univ.trieste.it/~tossi/amsdb.html	Plant/animal AMPs	895	Italy
2002	SAPD (inactive)	http://oma.terkko.helsinki.fi:8080/~SAPD/	Synthetic AMPs	~200	Finland
2004	Peptaibols	www.cryst.bbk.ac.uk/peptaibol/home.shtml	Fungal peptaibols	317	England
2004	APD	aps.unmc.edu/AP/	AMPs	2408	USA
2004	DAMPD	apps.sanbi.ac.za/dampd/	AMPs	1232	S. Africa
2006	PenBase	penbase.immunaqua.com	Shrimp AMPs	36	France
2006	Cybase	research1t.imb.uq.edu.au/cybase/	Circular proteins	813	Australia
2007	Defensins	defensins.bii.a-star.edu.sg/	Defensins	566	Singapore
2007	BACTIBASE	bactibase.pfba-lab-tun.org/main.php	Bacteriocins	220	Canada
2008	RAPD	faculty.ist.unomaha.edu/chen/rapd/index.php	Recombinant AMPs	179	USA
2009	PhytAMP	phytamp.pfba-lab-tun.org/main.php	Plant AMPs	273	Tunisie
2010	CAMP	www.bicnirrh.res.in/antimicrobial	AMPs (predicted)	5040	India
2012	DADP	split4.pmfst.hr/dadp/	Frog peptides	2571	Croatia
2012	THIOBASE	db-mml.sjtu.edu.cn/THIOBASE	Bacteria thiopeptides	39	China
2012	EnzyBase	biotechlab.fudan.edu.cn/database/EnzyBase/home.php	Lytic enzymes	1144	China
2012	YADAMP	yadamp.unisa.it/	Helical AMPs	2525	Italy
2013	LAMP	biotechlab.fudan.edu.cn/database/lamp/guide.php	AMP links	5547	China

(Continued)

TABLE 18.1 (Continued)

Year	Database	URL (http://)	Scope	Pep #	Country
2013	HIPdb	crdd.osdd.net/servers/hipdb	Anti-HIV peptides	981	India
2014	Hemolytik	crdd.osdd.net/raghava/hemolytik/	Hemolytic peptides	1750	India
2014	MilkAMP	/milkampdb.org	Milk AMPs	371	Canada
2014	AVPdb	http://crdd.osdd.net/servers/avpdb	Antiviral peptides	2683	India
2014	DBAASP	www.biomedicine.org.ge/dbaasp/	AMPs (Synthetic)	4395	Georgia
2014	ParaPep	crdd.osdd.net/raghava/parapep/	Anti-parasitic peptides	519	India

compiled 895 AMPs, proteins, and their precursors from plants and animals (Tossi and Sandri, 2002). This database is currently not available online (accessed in June 2014).

Antimicrobial Peptide Database

The Antimicrobial Peptide Database (APD), built in 2003, is dedicated to all types of AMPs (Wang and Wang, 2004). The 2004 version contained 525 peptide entries. There were 1,228 entries in the second version (Wang et al., 2009). Since then, this database has been constantly validated, improved, updated, and expanded (Wang, 2015). As of June 2014, 2,408 peptides were collected with a focus on natural AMPs that have been demonstrated to possess antimicrobial activity *in vitro*. Peptide information can be searched alone or in combination by using APD ID, peptide name, amino acid sequence, peptide motif, chemical modification, length, charge, hydrophobic content, PDB ID, 3D structure, methods for structural determination, peptide source organism, peptide family name, source life domains, antimicrobial activity, synergistic effects, target microbe, molecular target, mechanism of action, contributing author, and year of publication. A unique pipeline design of the APD search engine enables the users to filter the information to the details at their will. Furthermore, the AMP output can be sorted based on net charge, length, and database ID. In addition to antibacterial, antiviral, antifungal, and antiparasitic activities, these peptides may have other biological functions (e.g., hemolytic, chemotactic, anticancer, insecticidal, spermicidal, enzyme inhibition, and antioxidant). Of the 17 types of activities annotated in the APD, antibacterial, antifungal, antiparasitic, and insecticidal activities are more related to agriculture. In addition, 24 types of chemical modifications are annotated (Wang, 2012a, 2015), making the APD most comprehensive.

Peptide sources were classified in the APD based on the five kingdoms (Wang, 2010). Currently, there are 236 AMPs from bacteria, six from protists, 12 from fungi, 308 from plants, and 1803 from animals. A sixth class "synthetic" is reserved for synthetic peptides, which can be derivatives of natural parent peptides or *de novo* designed peptides. One can also obtain the peptides from the three life domains: 236 peptides from bacteria, two from archaea, and 2,131 from eukaryota. Peptide basic properties such as net charge, length, hydrophobic percentage, and Boman index are provided for each peptide. About 90% of the AMPs are cationic (net charge >0) with less than 50 amino acids. The average net charge, hydrophobic content, and length of all the peptides are 3.18, 42.02%, and 32.36, respectively. Additional peptide parameters can be calculated using the updated peptide calculator in the prediction interface of the APD. These include molecular weight, molecular formula, extinction coefficient, GRAVY, Wimley-White whole-residue hydrophobicity, Boman index, and amino acid composition.

Three-dimensional (3D) structures of AMPs are also annotated in the APD (Wang and Wang, 2004). 3D structures are separated into four classes: α, β, αβ, and non-αβ (Wang, 2010). Structures in each class can further be classified into families based on the number of α-helices or β-strands/sheets (Wang, 2015). Such a systematic structural classification in the APD greatly facilitated users to obtain a set of AMPs with a defined structure class (e.g., see Fjell et al., 2011). The structural information annotated in the APD includes structural determination methods (NMR and X-ray diffraction), structural classes, secondary structural regions, critical residues, and the type of membrane-mimetic environments used for structural determination by NMR spectroscopy. Recently, circular dichroism (CD) is also included as it provides a clear indication for helical structure. When the structure is deposited in the Protein Data Bank (PDB), users can rotate and view the 3D structure using Jmol in the PDB directly via the APD links in the additional information field. Many other features of the AMP structure can also be explored in the PDB. These include structural summary, secondary structures along a peptide sequence, domain annotation, sequence similarity, 3D similarity, literature, biology and chemistry report, additional information for structural determination method, structural geometry, and links to other structure-related programs (Rose et al., 2013). When there are multiple structures determined either by different methods or under different conditions, the major APD link is usually given to the structure with the highest resolution. Other structures for the same peptide determined by X-ray or NMR methods can be viewed via additional links in the APD and PDB. Note that AMP structures that are not deposited into the PDB but reported in the literature are also annotated in the APD.

The APD also provides detailed statistics for AMPs from a defined domain or those from a defined structural class. The amino acid composition of AMPs, first programmed in the APD, appears to possess structural, functional, and evolutional implications (Mishra and Wang, 2012a; Wang and Wang, 2004). For example, cysteines are abundant in plant AMPs, consistent with the dominance in β-sheet structures. In contrast, glycines, leucines, lysines, and alanines are most abundant in amphibian AMPs, consistent with their dominance in AMPs with known helical structures. Such abundant amino acids identified from the amino acid profiles are sufficient for *de novo* design of novel helical AMPs (Mishra and Wang, 2012b; Wang et al., 2009). For additional information not found in the APD, users may visit other databases via the APD links (Wang, 2007, 2010). In the following, we highlight the complementary features from other databases.

Dragon Antimicrobial Peptide Database

Dragon antimicrobial peptide database (DAMPD) is a transformed version of ANTIMIC initially published in 2004 (Brahmachary et al., 2004). While ANTIMIC reported 1,700 entries, DAMPD contains 1,232 peptides (Seshadri Sundararajan et al., 2012). It is not anticipated that the number of AMPs in the new version is reduced compared to the original version. The DAMPD can be searched based on taxonomy, species, AMP family, citation, keywords, and a combination of search terms and fields (advanced search). A number of tools such as Blast, ClustalW, HMMER, SignalP, AMP predictor were also programmed. Users can view the total GOOGLE citations for each AMP database via this database links.

Collections of AMPs

Collections of AMPs (CAMP), published in 2010, were updated to a second version in 2014. It collected 5,040 peptides (2,438 no activity data) in the sequence database (Waghu et al., 2014). In addition, 1,716 sequences from patents are listed in a separate database. This database appears to have the largest number of sequences. However, whether the 2,438 predicted sequences are AMPs remains to be established experimentally. In addition, inclusion of such uncharacterized sequences into statistics led to a list of abundant amino acids different from the consistent list already published (Mishra and Wang, 2012a; Wang, 2010; Wang and Wang, 2004; Wang et al., 2009).

A separate database for structure was also built in the 2014 version of the CAMP by duplicating the "3D view" and "sequence" web features of the PDB. In contrast, the APD (Wang and Wang, 2004) avoided such duplicated efforts by establishing links to the PDB that enable users to directly view the structure (see the APD section for details). The CAMP has 682 entries (270 unique) by including multiple deposited structures for each AMP solved under different conditions. On the contrary, the APD contains 319 unique structures covering both deposited and non-deposited AMPs from the literature.

Links AMPs

A database links AMPs (Zhao et al., 2013). This database collected 5,547 sequences, including a significant number of peptides without any biological activity. The aim of the LAMP is to create links for each peptide entry to the same AMP in other databases. Such links may allow users to get a more complete view on a particular peptide when there is complementary information in existing AMP databases (Table 18.1).

Database of Antimicrobial Activity and Structure of Peptides

The database name was abbreviated from Database of Antimicrobial Activity and Structure of Peptides (DBAASP). This latest database has 4,395 peptides (1,536 ribosomally synthesized, three non-ribosomally synthesized, and 2,856 synthetic peptides) (Gogoladze et al., 2014). Complementary to the APD, users can search 2,856 synthetic peptides in the DBAASP.

SPECIALIZED DATABASES FOR AMPs

Peptides from Defined Life Domains

Peptaibols

This unique database collects peptaibols from fungi (Whitmore and Wallace, 2004). Peptaibols are peptide antibiotics with a high proportion of non-standard amino acids, especially aminoisobutyric acid (Aib).

PenBase

This database has 28 penaeidins (AMPs from shrimps). It provides information on peptide properties, function, diversity, and nomenclature (Gueguen et al., 2006).

BACTIBASE

This specialized database was built in 2007 to provide more detailed information for bacterial AMPs (i.e., bacteriocins). The first version of BACTIBASE had 113 entries (Hammami et al., 2007). The second version contains 220 entries, similar to 236 bacteriocins in the APD (Wang et al., 2009). Useful bacteriocin parameters in this database include hydropathy index, instability index, aliphatic index, peptide half-life, extinction coefficient, and UV absorbance at 280 nm. Peptide properties (general data, physiochemical data, structural data, literature, and taxonomy) can be searched in four interfaces in the BACTIBASE. Furthermore, the peptides were registered in a different manner. In the APD (Wang et al., 2009), the two polypeptide chains of bacteriocins (e.g., lichenicidin and Staphylococcin C55) are registered as one entry, because both chains are usually required for activity. However, each two-chain peptide occupies two entries (e.g., Bac194-Bac197 for lichenicidin and Staphylococcin C55) in the BACTIBASE. Although the database has been updated (Hammami et al., 2010), the Boman index (taken from the APD) was incorrectly programmed in the BACTIBASE (e.g., −9.39 for halocin C8; the correct value is −0.2 kcal/mol in the APD).

PhytAMP

The same group also duplicated the computer codes of the BACTIBASE and established a specialized database for plant AMPs. PhytAMP provides molecular formula, mass, pI, and amino acid composition for each entry. Currently, this database contains 273 entries (Hammami et al., 2009), while the APD has 308 AMPs from plants (Wang et al., 2009).

Database for Anuran Defense Peptides

DADP is a Database for Anuran Defense Peptides. It contains 2,571 entries with a total of 1,923 unique bioactive sequences (Novkovic et al., 2012). Similar to 936 amphibian AMPs in the APD (Wang et al., 2009), only 921 frog peptides were demonstrated to be antimicrobial. In addition, signal sequences are also provided and classified into six types. This database may be used to study the evolutionary relationship of bioactive peptides from frogs.

THIOBASE

This unique database contains 39 thiopeptides (Li et al., 2012), which are essentially not collected in other databases. Interest in these highly modified compounds results from the observed effects of thiopeptides on antibiotic-resistant superbugs. These peptides are now recognized as a new family of bacteriocins (Kelly et al., 2009).

EnzyBase

While small AMPs are collected in several databases, there are also proteins that can lyse bacteria. EnzyBase is a unique database for lytic enzymes and antimicrobial proteins. Published in 2012, this database contains 1,144 entries (Wu et al., 2012).

Peptides from Defined Tissues or Geographic Regions

The APD also annotated the tissue origins from which AMPs are expressed or isolated. For instance, at least 11 plant AMPs were isolated from flowers, five from leaves and stem, and two from roots. The majority of amphibian AMPs were isolated from skin. In addition, the APD annotated the geographic location of the amphibian species from which the AMPs were identified. Of 936 amphibian peptides, 71, 292, 414, 93, and 51 AMPs were found from frogs from Africa, America, Asia, Australia, and Europe, respectively (Wang et al., 2009). There is a specialized database for AMPs from milk.

MilkAMP

As reflected in the name, this database was created for milk peptides. It covers both active and inactive peptides (Theolier et al., 2013). Of 371 entries, 299 peptides are antimicrobial. They originate mainly from cattle (244) and humans (105). These peptides were produced in a variety of ways (e.g., synthetic, recombinant, enzymatic hydrolysis, chemical processing, and physical processing). Readers interested in such peptides may also refer to a relevant review on dairy-derived AMPs (Akalin, 2014).

Peptides with Defined Activity

Recently, new databases have also been established for antiviral or antiparasitic peptides (natural + synthetic). Note that many of these peptides in these databases are not AMPs for host defense.

HIPdb

This database has 981 natural and synthetic (10% from phage display) sequences that displayed HIV-inhibitory activities. These peptides were tested in at least one of the 35 different cell lines. Targets of these peptides include fusion inhibitors (29%), integrase (29%), virus entry (11%), protease inhibitors (8%), and reverse transcriptase (4%) (Qureshi et al., 2013). AMPs with anti-HIV activity are annotated in the APD (Wang et al., 2009) and have been systematically discussed in a recent review (Wang, 2012b).

Hemolytik

Reducing peptide hemolytic ability is an important goal in engineering molecules for therapeutic use. Hemolytic assays are useful to measure peptide cytotoxicity. Hemolytik was built to document detailed hemolytic data. For example, 17 types of red blood cell were utilized in those assays. The same peptides evaluated using different erythrocytes are listed as separate entries (2,970); only 1,750 sequences are unique (Gautam et al., 2014). Based on statistical analysis in the APD, hemolytic AMPs tend to have a higher hydrophobic content than other AMPs (Wang, 2010).

AVPdb

This database contains information for 2,683 experimentally tested antiviral peptides (2,059 normal sequences and 624 modified). Many of these peptides were derived from viral proteins. The target microbes of these antiviral peptides include over 60 clinically important viruses such as influenza, HSV, and RSV. Antiviral assays were tested in at least one of the 85 cell lines (Qureshi et al., 2014).

ParaPep

Likewise, the same group generated a database for 519 antiparasitic peptides. It includes information on peptide sequence, chemical modifications, stereo-chemistry, antiparasitic activity, origin, nature of peptide, assay types, type of parasite, mode of action, and hemolytic activity (Mehta et al., 2014).

Peptides with Defined Chemical Bonding Pattern

In addition to a systematic 3D structure-based classification of AMPs mentioned above (Wang, 2010), we have recently classified AMPs into four classes based on the chemical bonding pattern of polypeptides (Wang, 2015). Class I is composed of linear peptides which do not contain any chemical bonds between different residues; Class II peptides contain chemical bonds between two side chains ($i \neq j$); Class III peptides possess a chemical bond between backbone and side chain ($i \neq j$); and Class IV peptides are circular with a covalent bond between the N- and C-termini of the peptide backbone (i.e., backbone to backbone bonding). The following sections describe other databases for AMPs with a defined structural pattern.

Defensins Knowledgebase

This database contains 566 defensins (mainly disulfide-linked class II peptides above) from different sources. Common defensins such as θ-defensins, α-defensins, and β-defensins contain three pairs of disulfide bonds that stabilize the peptide structure. This database also compiled information on "Defensins in the News", "literature", "clinical studies", "patents", "commercial entities", "defensins labs", and "defensins interaction" (Seebah et al., 2007).

CyBase

CyBase was created to list circular proteins (class IV peptides above) from bacteria, plants, and animals (Wang et al., 2008). There are 813 proteins in CyBase (532 cyclotides). Because of the lack of activity data, it is not clear how many are AMPs. The APD has 104 circular peptides with known antimicrobial activity (Wang, 2015).

Yet Another Database for AMPs

YADAMP (yet another database for AMPs) has 2,525 peptides (294 synthetic) with a special focus on short helical peptides (Piotto et al., 2012). It provides a classic helical wheel view for peptides. This database tabulated MIC values for AMPs against defined bacterial species. For example, there are 1,015 peptides active against *E. coli* and 957 against *Staphylococcus aureus*. It provides detailed information on peptide properties such as net charges at pH 5, 7, and 9. This should be useful when the net charge of a peptide (e.g., histidine-containing AMPs) varies with pH. Unlike the APD, the effect of chemical modifications on net charge of a peptide has not been considered in the YADAMP. For example, due to C-terminal amidation, the net charge of aurein 1.2 at pH 7 should be +1 (as in the APD) rather than zero (as in the YADAMP). Note that peptide sequences for some defensins are incomplete in the YADAMP probably due to a focus on helical regions.

Peptides Produced by Different Methods

In nature, peptides are usually synthesized ribosomally. Bacteria also make peptides by a multienzyme system. While most of the databases in Table 18.1 contain ribosomally synthesized AMPs, Peptaibol database consists of nonribosomally synthesized peptides (natural peptides). Peptides can also be made in laboratories (synthetic peptides). In addition, long peptides or proteins (>100 amino acids) are preferably produced by expressing them in bacteria, yeast, plants, or cell-free systems (recombinant peptides). There are two databases in this category.

SAPD

This is an early database for ~200 synthetic peptides (Wade and Englund, 2002). This resource is outdated and not accessible at the moment.

RAPD

This database compiled 179 entries for recombinant AMPs (Li and Chen, 2008).

WEBSITES DEDICATED TO AMP PREDICTION

Traditionally, AMPs were isolated and characterized from natural sources (Selsted et al., 1985; Steiner et al., 1981; Zasloff, 1987). This is an important approach still in use today. In certain cases, the isolated peptides yielded definitive support to the mature sequence of a new AMP (see, e.g., Wang, 2014). Because of low expression levels, isolation of AMPs from natural sources is not always feasible. With the sequencing of more and more genomes, it becomes attractive to identify potential AMP regions at the gene level. Many potential candidates have been identified in this manner. Whichever approach is used, the identified candidates need to be tested to verify their activity.

Based on the information content used in prediction, the prediction methods of AMPs were classified into five types (Wang, 2010). The first type uses only mature peptide sequences, which we will discuss below. The second prediction method involves the sequence information from the pro-peptide region. For example, cathelicidin precursors all share a highly conserved "cathelin" domain in the N-terminal region (i.e., pro-peptide) and highly variable C-terminal antimicrobial regions. Thus, additional AMPs in the cathelicidin family were identified based on sequence similarity in the pro-peptide region (Zanetti, 2005).

The third prediction type considers the protein precursor (i.e., pro-peptide and mature AMP). Torrent et al. (2012) developed a program to locate potential AMP region in a precursor protein. The fourth method employs the sequence similarity of the conserved modifying enzymes to identify new bacteriocins (McClerren et al., 2006). Finally, the fifth method makes prediction based on genomic information. The BAGEL is such an example for predicting bacteriocins from bacterial genome (de Jong et al., 2010). Some of the online prediction programs are given in Table 18.2 and can be accessed via the APD links (http:// aps.unmc.edu/AP/links.php).

Each prediction category above can be achieved in different ways. For example, there are different approaches to predictions of AMPs based on mature peptide sequences. The first AMP prediction program was established in the APD (Wang and Wang, 2004). In the updated version, it makes predictions based on the peptide parameters defined by all the AMPs in the database (Wang, 2015). Thus, an input sequence will be rejected if its calculated parameters fall outside the database-defined parameter ranges for AMPs. In addition, the APD program also provides five peptides most similar to the user's input based on multiple sequence alignment. Using the AMP sequences in the original APD (Wang and Wang, 2004), Lata et al. (2007) first tested some known prediction algorithms. A second version was also published (Lata et al., 2010) after the APD update in 2009 (Wang et al., 2009). Likewise, the CAMP programed similar algorithms for peptide prediction (Waghu et al., 2014). Most of these methods are machine-learning, which requires both positive and negative data sets. While it is easy to download the positive data set from the existing AMP databases, it is difficult to obtain a true negative data set. Yet, these programs claimed to reach predictive

TABLE 18.2 Select Websites for AMP Prediction

Name	Website (http://)	Prediction Basis (Ref)
APD	aps.unmc.edu/AP	Peptide properties and sequence alignment (Wang and Wang, 2004)
BAGLE	bioinformatics.biol.rug.nl/websoftware/bagel/bagel_start.php	Genome-prediction of bacteriocins (de Jong et al., 2010)
AntiBP	www.imtech.res.in/raghava/antibp	ANN, QM, SVM (Lata et al., 2007, 2010)
AMPer	marray.cmdr.ubc.ca/cgi-bin/amp.pl	Peptide clustering (Fjell et al., 2007)
CAMP	www.camp.bicnirrh.res.in/	SVM, RF, ANN, DA (Waghu et al., 2014)
AMPA	tcoffee.crg.cat/apps/ampa	locate antimicrobial regions (Torrent et al., 2012)
iAMP-2L	www.jci-bioinfo.cn/iAMP-2L	Two Tier Prediction (Xiao et al., 2013)

accuracy up to 99%. We tested these programs using ten new AMPs. These new peptides were registered into the APD during May 2014 and have not been included into the training data set for the above programs. Using support vector machine (SVM) programed in AntiBP2 (Lata et al., 2010), 70% of the tested AMPs were correctly predicted (Table 18.3). The SVM, also programed in the CAMP (Waghu et al., 2014), gave a similar prediction. We also tested RF (random forest), ANN (artificial neural network), and DA (discriminant analysis). While RF succeeded in predicting eight out of ten (i.e., 80% accuracy), DA and ANN showed 70% and 60% success rates, respectively. Overall, these algorithms were able to correctly predict 60–80% of the tested sequences (Table 18.3).

AMPer was set up to classify AMPs into different clusters based on sequence similarity (Fjell et al., 2007). Since it was developed based on the AMSDb with limited sequences of mature peptides and precursors, the program was unable to classify 50% of the above ten newly identified AMPs into the known clusters.

A recent two-tier prediction method iAMP-2L (Xiao et al., 2013) was developed based on multiple functions of AMPs annotated in the APD (Wang et al., 2009). The program first predicts whether the input is potentially an antibacterial peptide. It also predicts other potential functions such as antifungal and antiviral activities. The ten new AMPs were again tested and 80% were predicted as possible antibacterial peptides. However, functional predictions (tier 2) by iAMP-2 were not yet reliable with about 50% matching the activity spectrum annotated in the APD. Further studies are needed to improve such predictions.

DATABASE-BASED PEPTIDE DESIGN

The APD also enabled the identification of novel AMPs. For example, both database screening and *de novo* design were utilized or developed to identify active peptides against difficult-to-kill human immune-deficiency virus type 1 (HIV-1) or meticillin-resistant *Staphylococcus aureus* (MRSA) (Menousek et al., 2012; Mishra and Wang, 2012b; Wang et al., 2009, 2010). This topic has recently been reviewed (Wang, 2013).

CONCLUDING REMARKS

This chapter covered 23 databases constructed for (or related to) AMPs (Table 18.1). AMSDb is an earlier database but currently not available. The APD and ANTIMIC are the two general databases published in 2004. The ANTIMIC (Brahmachary et al., 2004) has been replaced by DAMPD (Seshadri Sundararajan et al., 2012). The APD is the "most popular" and "most comprehensive" database in the AMP field (Fjell et al., 2011; Fox, 2013; Hammami et al., 2010). It influenced the construction of more recent databases to varying degrees. In addition, the peptide entries in the APD were used not only to build other databases such as the ParaPep, CAMP, LAMP, and YADAMP, but also to develop prediction tools for AMPs.

TABLE 18.3 Prediction Results Using Newly Identified AMPs from the APD

APD ID	Peptide Name	SVM (Lata)	RF (Waghu)	SVM (Waghu)	ANN (Waghu)	DA (Waghu)	iAMP-2L
2408	Microbisporicin A2	yes	yes	yes	yes	yes	yes
2407	Hispidalin	no	no	no	no	no	yes
2406	Mytichitin-A	yes	yes	yes	yes	yes	yes
2405	Plantaricin CS	yes	yes	yes	yes	yes	yes
2404	Taromycin A	no	no	no	no	no	no
2403	Hp1404	no	yes	yes	yes	yes	yes
2402	Garvicin	yes	yes	yes	no	no	yes
2401	BTM-P1	yes	yes	yes	yes	yes	yes
2400	CFBD-1	yes	yes	yes	yes	yes	no
2399	Sil	yes	yes	yes	no	yes	yes
	Success rate%	*70%*	*80%*	*80%*	*60%*	*70%*	*80%*

The APD was developed with both research and education in mind. For educational purposes, web pages for glossary, AMP timeline, classification, nomenclature, APD facts, and links have been created. For example, the timeline lists lysozyme as the first AMP based on the suggestion of Robert Lehrer and the publication of Richard Gallo (Gallo, 2013; Ganz and Lehrer, 1997). The APD is the first AMP database built using LAMP, a bundle of free and open-source software, including the Linux operating system, the Apache web server, MySQL relational database system, and server-side scripting language (PHP). This database first classified AMP sources based on the five kingdoms or three life domains (Whittaker, 1969; Woese and Fox, 1977). The APD also first adopted a new all-inclusive classification scheme for 3D structures of AMPs (Wang, 2010). It provides most comprehensive information for peptide chemical modifications (23 types) and AMP activities or functions (17 types) (Wang, 2015). Because peptides were collected based on a set of criteria, the APD provides a clean set of mature sequences with demonstrated activity. Our practice allows us to estimate how many natural AMPs have been discovered and experimentally verified. According to the APD, at least 2,200 natural peptides have been characterized as AMPs (i.e., with known antimicrobial activities). This estimation of unique natural AMPs is important considering the various peptide numbers in other databases (Table 18.1).

Additional information for AMPs can be found in other databases. In particular, specialized databases, including PenBase, Defensins Knowledgebase, BACTIBASE, PhytAMP, and DADP, may contain more detailed information. Peptaibols, THIOBASE, and EnzyBase provide unique peptide/protein information. While users can view 1,716 patented and 2,438 predicted peptides in the CAMP, they can search 2,856 synthetic peptides in the DBAASP. In addition, some recent databases expanded the anti-HIV, antiviral, and antiparasitic peptides beyond AMPs by including peptides designed based on viral protein targets. Although the APD is regularly updated, users may also find most recent literature on structure and function of AMPs in public search engines such as PubMed, PDB, and SWISS-Prot (Rose et al., 2013; Wheeler et al., 2002; Wu et al., 2006).

Based on the established databases, multiple prediction programs (mostly machine learning) are available for AMP prediction (Table 18.2). It is encouraging that a simple test run of these programs using newly identified AMPs revealed 60–80% accuracy, further validating their usefulness (Table 18.3). Of note, all the tested methods, either machine learning or two-tier prediction, failed to predict taromycin A, a marine lipopeptide similar to daptomycin. This may indicate that these programs have not been trained using lipopeptides recently collected into the APD database. Due to their unique properties, a separate program is needed to predict antimicrobial activity of lipopeptides.

In conclusion, the APD is the most comprehensive database for timeline, glossary, nomenclature, classification, search, prediction, and design of AMPs (Wang et al., 2009). Together with other AMP databases (Table 18.1), they constitute a set of useful tools for developing new antimicrobial compounds to combat resistant microbes that could compromise food quality during production, processing, transportation, storage, and consumption.

ACKNOWLEDGMENTS

This study was supported by the grants from the NIH (1R56AI105147-01; 1R01AI105147-01A1) and the State of Nebraska to GW.

REFERENCES

Akalin, A.S., 2014. Dairy-derived antimicrobial peptides: action mechanisms, pharmaceutical uses and production proposals. Trends Food Sci. Technol. 36, 79–95.

Batista, C.V., Scaloni, A., Rigden, D.J., Silva, L.R., Rodrigues Romero, A., Dukor, R., et al., 2001. A novel heterodimeric antimicrobial peptide from the tree-frog *Phyllomedusa distincta*. FEBS Lett. 494, 85–89.

Bierbaum, G., Sahl, H.G., 2009. Lantibiotics: mode of action, biosynthesis and bioengineering. Curr. Pharm. Biotechnol. 10, 2–18.

Boman, H.G., 2003. Antibacterial peptides: basic facts and emerging concepts. J. Intern. Med. 254, 197–215.

Brahmachary, M., Krishnan, S.P., Koh, J.L., Khan, A.M., Seah, S.H., Tan, T.W., et al., 2004. ANTIMIC: a database of antimicrobial sequences. Nucleic Acids Res. 32, D586–D589.

Breukink, E., Wiedemann, I., van Kraaij, C., Kuipers, O.P., Sahl, H.G., de Kruijff, B., 1999. Use of the cell wall precursor lipid II by a pore-forming peptide antibiotic. Science 286, 2361–2364.

Brogden, K.A., 2005. Antimicrobial peptides: pore formers or metabolic inhibitors in bacteria? Nat. Rev. Microbiol. 3, 238–250.

Bulet, P., Stocklin, R., 2005. Insect antimicrobial peptides: structures, properties and gene regulation. Protein Pept. Lett. 12, 3–11.

Cerovsky, V., Bem, R., 2014. Lucifensins, the insect defensins of biomedical importance: the story behind maggot therapy. Pharmaceuticals (Basel) 7, 251–264.

Conlon, J.M., Mechkarska, M., 2014. Host-defense peptides with therapeutic potential from skin secretions of frogs from the family pipidae. Pharmaceuticals (Basel) 7, 58–77.

de Jong, A., van Heel, A.J., Kok, J., Kuipers, O.P., 2010. BAGEL2: mining for bacteriocins in genomic data. Nucleic Acids Res. 38, W647–W651.

Delves-Broughton, J., Blackburn, P., Evans, R.J., Hugenholtz, J., 1996. Application of bacteriocin nisin. Antonie Van Leeuwenhoek 69, 193–202.

Egorov, T.A., Odintsova, T.I., Pukhalsky, V.A., Grishin, E.V., 2005. Diversity of wheat anti-microbial peptides. Peptides 26, 2064–2073.

Epand, R.M., Vogel, H.J., 1999. Diversity of antimicrobial peptides and their mechanisms of action. Biochim. Biophys. Acta. 1462, 11–28.

Fjell, C.D., Hancock, R.E., Cherkasov, A., 2007. AMPer: a database and an automated discovery tool for antimicrobial peptides. Bioinformatics 23, 1148–1155.

Fjell, C.D., Hiss, J.A., Hancock, R.E., Schneider, G., 2011. Designing antimicrobial peptides: form follows function. Nat. Rev. Drug Discov. 11, 37–51.

Fox, J.L., 2013. Antimicrobial peptides stage a comeback. Nat. Biotechnol. 31, 379–382.

Gallo, R.L., 2013. The birth of innate immunity. Exp. Dematol. 22, 517.

Ganz, T., Lehrer, R.I., 1997. Antimicrobial peptides of leukocytes. Curr. Opin. Hematol. 4, 53–58.

Gautam, A., Chaudhary, K., Singh, S., Joshi, A., Anand, P., Tuknait, A., et al., 2014. Hemolytik: a database of experimentally determined hemolytic and non-hemolytic peptides. Nucleic Acids Res. 42, D444–D449.

Gogoladze, G., Grigolava, M., Vishnepolsky, B., Chubinidze, M., Duroux, P., Lefranc, M.P., et al., 2014. DBAASP: database of antimicrobial activity and structure of peptides. FEMS Microbiol. Lett. Available from: http://dx.doi.org/10.1111/1574-6968.12489.

Goyal, R.K., Mattoo, A.K., 2014. Multitasking antimicrobial peptides in plant development and host defense against biotic/abiotic stress. Plant Sci. Available from: http://dx.doi.org/10.1016/j.plantsci.2014.05.012.

Gueguen, Y., Garnier, J., Robert, L., Lefranc, M.P., Mougenot, I., de Lorgeril, J., et al., 2006. PenBase, the shrimp antimicrobial peptide penaeidin database: sequence-based classification and recommended nomenclature. Dev. Comp. Immunol. 30, 283–288.

Hammami, R., Zouhir, A., Ben Hamida, J., Fliss, I., 2007. BACTIBASE: a new web-accessible database for bacteriocin characterization. BMC Microbiol. 7, 89.

Hammami, R., Ben Hamida, J., Vergoten, G., Fliss, I., 2009. PhytAMP: a database dedicated to antimicrobial plant peptides. Nucleic Acids Res. 37, D963–D968.

Hammami, R., Zouhir, A., Le Lay, C., Ben Hamida, J., Fliss, I., 2010. BACTIBASE second release: a database and tool platform for bacteriocin characterization. BMC Microbiol. 10, 22.

Hancock, R.E., Sahl, H.G., 2006. Antimicrobial and host-defense peptides as new anti-infective therapeutic strategies. Nat. Biotechnol. 24, 1551–1557.

Kelly, W.L., Pan, L., Li, C., 2009. Thiostrepton biosynthesis: prototype for a new family of bacteriocins. J. Am. Chem. Soc. 131, 4327–4334.

Lata, S., Sharma, B.K., Raghava, G.P., 2007. Analysis and prediction of antibacterial peptides. BMC Bioinform. 8, 263.

Lata, S., Mishra, N.K., Raghava, G.P., 2010. AntiBP2: improved version of antibacterial peptide prediction. BMC Bioinform. 11 (Suppl. 1), S19.

Lehrer, R.I., 2007. Multispecific myeloid defensins. Curr. Opin. Hematol. 14, 16–21.

Leippe, M., 1995. Ancient weapons: NK-lysin, is a mammalian homolog to pore-forming peptides of a protozoan parasite. Cell 83, 17–18.

Leippe, M., Bruhn, H., Hecht, O., Grötzinger, J., 2005. Ancient weapons: the three-dimensional structure of amoebapore A. Trends Parasitol. 21, 5–7.

Li, J., Xu, X., Xu, C., Zhou, W., Zhang, K., Yu, H., et al., 2007. Anti-infection peptidomics of amphibian skin. Mol. Cell Proteomics 6, 882–894.

Li, J., Qu, X., He, X., Duan, L., Wu, G., Bi, D., et al., 2012. ThioFinder: a web-based tool for the identification of thiopeptide gene clusters in DNA sequences. PLoS ONE 7, e45878.

Li, Y., Chen, Z., 2008. RAPD: a database of recombinantly-produced antimicrobial peptides. FEMS Microbiol. Lett. 289, 126–129.

Masso-Silva, J.A., Diamond, G., 2014. Antimicrobial peptides from fish. Pharmaceuticals (Basel) 7, 265–310.

McClerren, A.L., Cooper, L.E., Quan, C., Thomas, P.M., Kelleher, N.L., van der Donk, W.A., 2006. Discovery and in vitro biosynthesis of haloduracin, a two-component lantibiotic. Proc. Natl. Acad. Sci. USA 103, 17243–17248.

Mehta, D., Anand, P., Kumar, V., Joshi, A., Mathur, D., Singh, S., et al., 2014. ParaPpep: a web resource for experimentally validated antiparasitic peptide sequences and their structures. Nucleic Acids Res. http://dx.doi.org/10.1093/database/bau051

Menousek, J., Mishra, B., Hanke, M.L., Heim, C.E., Kielian, T., Wang, G., 2012. Database screening and in vivo efficacy of antimicrobial peptides against methicillin-resistant *Staphylococcus aureus* USA300. Int. J. Antimicrob. Agents 39, 402–406.

Mishra, B., Wang, G., 2012a. The importance of amino acid composition in natural AMPs: an evolutional, structural, and functional perspective. Front. Immunol. 3, 221.

Mishra, B., Wang, G., 2012b. *Ab initio* design of potent anti-MRSA peptides based on database filtering technology. J. Am. Chem. Soc. 134, 12426–12429.

Novković, M., Simunić, J., Bojović, V., Tossi, A., Juretić, D., 2012. DADP: the database of anuran defense peptides. Bioinformatics 28, 1406–1407.

Piotto, S.P., Sessa, L., Concilio, S., Iannelli, P., 2012. YADAMP: yet another database of antimicrobial peptides. Int. J. Antimicrob. Agents 39, 346–351.

Qureshi, A., Thakur, N., Kumar, M., 2013. HIPdb: a database of experimentally validated HIV inhibiting peptides. PLoS ONE 8, e54908.

Qureshi, A., Thakur, N., Tandon, H., Kumar, M., 2014. AVPdb: a database of experimentally validated antiviral peptides targeting medically important viruses. Nucleic Acids Res. 42, D1147–D1153.

Rose, P.W., Bi, C., Bluhm, W.F., Christie, C.H., Dimitropoulos, D., Dutta, S., et al., 2013. The RCSB Protein Data Bank: new resources for research and education. Nucleic Acids Res. 41, D475–D482.

Rozgonyi, F., Szabo, D., Kocsis, B., Ostorházi, E., Abbadessa, G., Cassone, M., et al., 2009. The antibacterial effect of a proline-rich antibacterial peptide A3-APO. Curr. Med. Chem. 16, 3996–4002.

Seebah, S., Suresh, A., Zhou, S., Choong, Y.H., Chua, H., Chuon, D., et al., 2007. Defensins knowledgebase: a manually curated database and information source focused on the defensins family of antimicrobial peptides. Nucleic Acids Res. 35, D265–D268.

Selsted, M.E., Harwig, S.S., Ganz, T., Schilling, J.W., Lehrer, R.I., 1985. Primary structures of three human neutrophil defensins. J. Clin. Invest. 76, 1436–1439.

Seshadri Sundararajan, V., Gabere, M.N., Pretorius, A., Adam, S., Christoffels, A., Lehväslaiho, M., et al., 2012. DAMPD: a manually curated antimicrobial peptide database. Nucleic Acids Res. 40, D1108–D1112.

Steiner, H., Hultmark, D., Engström, Å., Bennich, H., Boman, H.G., 1981. Sequence and specificity of two antibacterial proteins involved in insect immunity. Nature 292, 246–248.

Theolier, J., Fliss, I., Jean, J., Hammami, R., 2013. MilkAMP: a comprehensive database of antimicrobial peptides of dairy origin. Dairy Sci. Technol. Available from: http://dx.doi.org/10.1007/s13594-013-0153-2.

Torrent, M., Di Tommaso, P., Pulido, D., Nogués, M.V., Notredame, C., Boix, E., et al., 2012. AMPA: an automated web server for prediction of protein antimicrobial regions. Bioinformatics 28, 130–131.

Tossi, A., Sandri, L., 2002. Molecular diversity in gene-coded, cationic antimicrobial polypeptides. Curr. Pharm. Des. 8, 743–761.

van Hoek, M.L., 2014. Antimicrobial peptides in reptiles. Pharmaceuticals (Basel) 7, 723–753.

Wade, D., Englund, J., 2002. Synthetic antibiotic peptides database. Protein Pept. Lett. 9, 53–57.

Waghu, F.H., Gopi, L., Barai, R.S., Ramteke, P., Nizami, B., Idicula-Thomas, S., 2014. CAMP: collection of sequences and structures of antimicrobial peptides. Nucleic Acids Res. 42, D1154–D1158.

Wang, C.K., Kaas, Q., Chiche, L., Craik, D.J., 2008. Cybase: a database of cyclic protein sequences and structures, with applications in protein discovery and engineering. Nucleic Acids Res. 36, D206–D210.

Wang, G., 2007. Tool developments for structure-activity studies of host defense peptides. Protein Pept. Lett. 14, 57–69.

Wang, G., 2008. Structures of human host defense cathelicidin LL-37 and its smallest antimicrobial peptide KR-12 in lipid micelles. J. Biol. Chem. 283, 32637–32643.

Wang, G., 2010. Antimicrobial Peptides: Discovery, Design and Novel Therapeutic Strategies. CABI, England.

Wang, G., 2012a. Chemical modifications of natural antimicrobial peptides and strategies for peptide engineering. Curr. Biotechnol. 1, 72–79.

Wang, G., 2012b. Natural antimicrobial peptides as promising anti-HIV candidates. Curr. Top. Pept. Protein Res. 13, 93–110.

Wang, G., 2013. Database-guided discovery of potent peptides to combat HIV-1 or superbugs. Pharmaceuticals 6, 728–758.

Wang, G., 2014. Human antimicrobial peptides and proteins. Pharmaceuticals (Basel) 7, 545–594.

Wang, G., 2015. Peptide database: improved methods for classification, prediction and design of antimicrobial peptides. Methods Mol. Biol. 1268, 43–66.

Wang, G., Li, X., Wang, Z., 2009. The updated antimicrobial peptide database and its application in peptide design. Nucleic Acids Res. 37, D933–D937.

Wang, G., Watson, K.M., Peterkofsky, A., Buckheit Jr., R.W., 2010. Identification of novel human immunodeficiency virus type 1 inhibitory peptides based on the antimicrobial peptide database. Antimicrob. Agents Chemother. 54, 1343–1346.

Wang, Z., Wang, G., 2004. APD: the antimicrobial peptide database. Nucleic Acids Res. 32, D590–D592.

Wheeler, D.L., Church, D.M., Lash, A.E., Leipe, D.D., Madden, T.L., Pontius, J.U., et al., 2002. Database resources of the National Center for Biotechnology Information: 2002 update. Nucleic Acids Res. 30, 13–16.

Whitmore, L., Wallace, B.A., 2004. The peptaibol database: a database for sequences and structures of naturally occurring peptaibols. Nucleic Acids Res. 32, D593–D594.

Whittaker, R.H., 1969. New concepts of kingdoms of organisms. Science 163, 150–160.

Wilmes, M., Cammue, B.P., Sahl, H.G., Thevissen, K., 2011. Antibiotic activities of host defense peptides: more to it than lipid bilayer perturbation. Nat. Prod. Rep. 28, 1350–1358.

Woese, C.R., Fox, G.E., 1977. Phylogenetic structure of the prokaryotic domain: the primary kingdoms. Proc. Natl. Acad. Sci. USA 74, 5088–5090.

Wu, C.H., Apweiler, R., Bairoch, A., Natale, D.A., Barker, W.C., Boeckmann, B., et al., 2006. The Universal Protein Resource (UniProt): an expanding universe of protein information. Nucleic Acids Res. 34, D187–D191.

Wu, H., Lu, H., Huang, J., Li, G., Huang, Q., 2012. EnzyBase: a novel database for enzybiotic studies. BMC Microbiol. 12, 54.

Xiao, X., Wang, P., Lin, W.Z., Jia, J.H., Chou, K.C., 2013. iAMP-2L: a two-level multi-label classifier for identifying antimicrobial peptides and their functional types. Anal. Biochem. 436, 168–177.

Yang, X., Lee, W.H., Zhang, Y., 2012. Extremely abundant antimicrobial peptides existed in the skins of nine kinds of Chinese odorous frogs. J. Proteome Res. 11, 306–319.

Zanetti, M., 2005. The role of cathelicidins in the innate host defenses of mammals. Curr. Issues Mol. Biol. 7, 179–196.

Zasloff, M., 1987. Magainins, a class of antimicrobial peptides from *Xenopus* skin: isolation, characterization of two active forms, and partial cDNA sequence of a precursor. Proc. Natl. Acad. Sci. USA 84, 5449–5453.

Zasloff, M., 2002. Antimicrobial peptides of multicellular organisms. Nature 415, 359–365.

Zhang, G., Sunkara, L.T., 2014. Avian antimicrobial host defense peptides: from biology to therapeutic applications. Pharmaceuticals (Basel) 7, 220–247.

Zhao, X., Wu, H., Lu, H., Li, G., Huang, Q., 2013. LAMP: a database linking antimicrobial peptides. PLoS One 8, e66557.

Chapter 19

Metabolic Network Analysis-Based Identification of Antimicrobial Drug Target in Pathogenic Bacteria

Vinayak Kapatral

Igenbio, Inc., Chicago, IL, USA

Chapter Outline

INTRODUCTION

Conventional drug discovery program is a five-step process beginning with target selection, hit molecule discovery, lead molecule identification and optimization followed by pre-clinical and clinical studies. This process of new drug molecule to file new drug entity takes 12–15 years with huge costs accompanied by a high risk of failure. This is an underlying problem with drug companies spending nearly 40% of their resources in the first three steps of a drug discovery program. Hence, new target identification becomes the most critical step in keeping a healthy pipeline of targets. Traditional methods depend on knowledge gathered from experimental research that identifies targets, which are believed to be critical for cell survival, thus a "druggable candidate" for a new molecule discovery. On the other hand, genome-based computational methods allow a panel of target identification based on metabolic pathways. This approach utilizes flux computations of well-annotated genomes, metabolic pathways and reaction networks of single or multiple sequenced organisms. These computationally based predicted targets provide rapid, cost-effective and strain-specific metabolic targets. In addition, these methods also facilitate the identification of single essential and synthetic lethals (pairs of enzymes in the same or different pathways) metabolic enzymes as targets. These methods triumph experimentally

Antimicrobial Resistance and Food Safety. DOI: http://dx.doi.org/10.1016/B978-0-12-801214-7.00019-3

based gene essentiality, as they are often growth conditional dependent, costly, and time consuming.

Genome sequence variations manifest in their metabolic or physiological characteristics of a given organism. Identification of these variations defines their unique biochemical capabilities that allow growth and survival in a specific ecological niche. However, recognizing a common core metabolic network allows identification of metabolic enzymes as the target for anti-infective discovery. In this chapter, I illustrate examples of novel multiple metabolic targets discovery in human and animal bacterial pathogens based on annotations, metabolic constructions, and flux balance methods, which can be used for plant pathogens as well.

BACKGROUND

Bacterial pathogens have a unique ability to adapt, evolve in a rapidly changing host environment and excessive use of antibiotics. This rapid evolution alters genotype, metabolic changes including drug resistance and such a high emergence of resistant strains poses a huge challenge to controlling infection and spread. Low costs in sequencing technology coupled with advanced computational methods along with genome-wide expression or essentiality studies have provided a vast genomic imprint of a single strain or group of strains. This has renewed interest not only in diagnostics but also in identifying new targets and molecules effective against multiple drug-resistant bacteria. I describe a few examples of these "*in silico*" approaches that have yielded several new metabolic enzymes as potential targets in the section below.

METABOLIC TARGET IDENTIFICATION OF *STAPHYLOCOCCUS AUREUS*

Staphylococcus aureus is a major hospital/community-acquired opportunistic human pathogen. The incidence of multidrug resistance is on the rise not only among the general population but among the immune-compromised and AIDS patients in the United States (Pray, 2008). It is known to cause bacteremia, pneumonia, endocarditis, meningitis, and toxic-shock syndrome in adult humans and skin lesions, impetigo, and abscesses in children. Currently, *S. aureus* infections are treated with β-lactam antibiotics, sulfa drugs, tetracycline, or clindamycin and as a last resort, methicillin and vancomycin are used to treat drug-resistant strains. Unfortunately, to date, very few new drug candidates, such as platensimycin (yet to be clinically approved), daptomycin, ceftaroline have been found to be effective against a few multidrug-resistant strains (Antibiotic Resistant Threats in the United States, CDC: http://www.cdc.gov/drugresistance/threat-report-2013). This grim outlook presents an urgent need and a challenge for a sustained pipeline of new anti-infective molecules. The rise in multidrug resistance is attributed to several environmental factors, global travel in addition to the over-use of antibiotics in developed countries and/or natural emergence of newer strains in underdeveloped countries (Pray, 2008). Genome sequence variability often provides contradicting results in identifying target or virulence factor discovery. Comparative genomic studies have not only confirmed clonality of the genomic backbone but have also identified ~20% sequence variation attributable to the occurrence of prophages and pathogenicity islands. Several virulence and drug resistance markers have been identified; they provide the molecular basis for diagnostics only but do not aid in the discovery of new anti-infectives.

Since the seminal *in silico* work by Edwards and Palsson (2000) on metabolic mapping and flux balance analysis (FBA) computations of *E. coli* MG1655, several other models including experimental reaction inclusions have been considered (Almaas et al., 2004, 2005). These methods progressively improved as assumptions changed, computational algorithms improved and more genome sequence data became available. Studies by Becker and Palsson (2005) using the annotated *S. aureus* strain, predicted 518 metabolic reactions containing 571 metabolites. In this study, four (L-alanine, L-arginine, L-proline, and L-glycine) amino acids were considered to be essential and were found to concur with the experimental studies. However, other metabolites such as cytidine and uridine were not considered to be required for growth but were later predicted to be essential in a genome-scale essentiality model (Joyce and Palsson, 2008). A second study using experimental data, Heinemann et al. (2005) improved the metabolic reconstruction of *S. aureus* and found 774 metabolic reactions from 394 unique enzymes. Both these studies were carried out on a laboratory strain and were limited in scope and application to target discovery to naturally evolved multidrug-resistant strains. These studies also did not consider the transport reactions, efflux proteins, or plasmid-encoded enzymes. Second, several annotations were inaccurate and enzyme functions were not accurately mapped with the Enzyme Commission leading to incorrect reactions and metabolite network. A comprehensive approach factoring the above limitations and the use of several diverse methicillin-resistant *S. aureus* (MRSA) were considered by Lee et al. (2009). Initially, Lee et al. (2009) also used a laboratory strain, N315, as a model organism for metabolic reconstruction and FBA studies to test their assumption. Their algorithms also included transport or exchange reactions and reversibility of enzyme actions. Further, this approach was tested on a number of diverse strains such as *S. aureus* Mu50, *S. aureus* MW2, *S. aureus* subsp. *aureus* COL, *S. aureus* EMRSA-16 Strain 25, *S. aureus* MSSA Strain 476, *S. aureus* subsp. *aureus* JH1, *S. aureus* subsp. *aureus* JH9, *S. aureus* RF122 , *S. aureus* subsp. *aureus* COL, *S. aureus* subsp. *aureus* USA300, *S. aureus* subsp. aureus USA300_TCH1516, and *S. aureus* subsp. *aureus* str. Newman (Lee et al., 2009).

Using the ERGO genomic database, Lee et al. (2009) identified 2,593 ORFs in the *S. aureus* N135 chromosome and 31 ORFs on its plasmid. Nearly, 2,505 ORFs were assigned with a functional role and a total of 906 ORFs were identified as having an enzymatic role as described (Kapatral et al., 2002; Overbeek et al., 2003). About 668 ORFs with complete EC numbers and 61 ORFs with partial EC numbers and 1,493 identified metabolic reactions were identified. Of these nearly 22% of reactions were active, that is, carry nonzero fluxes. The flux-balance analysis computations identified only 410 reactions that are active in the rich growth medium, in which all the included transport reactions could occur without rate limitation. The procedure of improved FBA is well described in Lee et al. (2009). In an effort to identify candidate targets, Lee et al. (2009) initially performed a *in silico* single enzyme deletion study for growth for all the *S. aureus* strains in the simulated rich medium. Seventy enzymes were determined to be lethal across all 13 strains and 44 of them were strain-independent. In *S. aureus* N315 strain, 63 enzymes were identified as essential compared to 135 single essential enzymes in the Becker and Palsson (2005) model. Thirty metabolic enzymes were common in both the studies, of which 27 were strain-independent essential enzymes (Lee et al., 2009). A larger number of single essential enzymes identified in the Becker and Palsson (2005) study can be attributed partially to the smaller number of metabolic reactions and the assumption of irreversibility of enzyme reactions in their FBA computations. Examples of such common unconditionally (i.e., essential for growth in conditions) essential enzymes were found in the amino-sugar synthesis pathway. Two enzymes, UDP-*N*-acetylglucosamine 1-carboxyvinyltransferase (EC 2.5.1.7) and UDP-*N*-acetyl-enolpyruvoylglucosamine reductase (EC 1.1.1.158), which are involved in the conversion of UDP-*N*-acetyl-D-glucosamine to UDP-*N*-acetylmuramate through UDP-*N*-acetyl-3-*O*-(1-carboxyvinyl)-D-glucosamine, were predicted to be unconditionally essential. Inactivation of either of these two enzymes leads to failure to generate the key cell wall components UDP-*N*-acetylmuramate and UDP-*N*-acetyl-D-glucosamine. Lee et al. (2009) compared computationally derived enzyme with the list of ORFs that were identified to be experimentally essential for *S. aureus* subspecies RN450 and RN4220. Of the 44 common single essential enzymes that were identified as essential, six matched the experimentally identified enzymes, namely, transketolase (EC 2.2.1.1), hydroxylmethylbilane synthase (EC 2.5.1.61), methionine adenosyltransferase (EC 2.5.1.6), UDP-*N*-acetyl glucosamine 1-carboxyvinyltransferase (EC 2.5.1.7), and protein-*N* (pi)-phospho-histidine-sugar phosphotransferase (EC 2.7.1.69), and acetyl-CoA carboxylase (EC 6.4.1.2). About 17 enzymes matched other experimentally determined genes from other pathogenic bacteria (Lee et al., 2009).

Many enzymes were predicted to be dispensable for growth, that is, biomass could be generated in their absence, as alternate pathway routes are present. Lee et al. (2009) tested the hypothesis, that if a pair of enzymes independently play similar functional roles in the metabolism, then deleting both enzymes would lead to growth inhibition, that is, inactivation of any two enzymes through FBA would affect survival. Fifty-four such synthetic lethal pairs of enzymes were identified in all the 13 subspecies whose deletions could prevent biomass production. Of these, 10 synthetic lethal pairs were found to be common to all the 13 subspecies. Six lethal pairs belong to amino sugars/peptidoglycan metabolic pathways that are involved in the cell wall metabolic subsystem; two pairs belong to the amino-acid subsystem; only one pair of each enzyme for nucleotide and carbohydrate subsystems. In *S. aureus*, there are two pathways to generate UDP-*N*-acetylglucosamine, the precursor of the UDP-*N*-acetylmuramate, one is from D-glucosamine-1P through *N*-acetyl-D-glucosamine-1P, and the other is from *N*-acetyl-D-mannosamine (Lee et al., 2009). Synthetic lethal pairs exhibit even stronger strain dependence; less than 20% were found in all the strains and 40% of them were found in only one of each specific strain. These pairs allow discovery of a common molecule that can inhibit enzymatic activity.

In the above section, I have described methods and results of computationally predicted target discovery for a single bacterium and diverse strains of the same genus. However, this methodology can now be applied to identify common targets across different genera. Category A pathogenic agents include *Bacillus anthracis* (the causative agent of anthrax), *Yersinia pestis* (the causative agent of bubonic plague), *Francisella tularensis* (the causative agent of tularemia), all belonging to different genera (Pohanka and Skladal, 2009) and there is recognition of threat and therapeutic challenges (Gilligan, 2002). The FBA computational *in silico* approaches are not only safe but also cost-effective, particularly for these warfare agents. Examples of a common target discovery for a new class of anti-infective molecules are described in the following section.

METABOLIC TARGETS IN *BACILLUS ANTHRACIS*

B. anthracis, a Gram-positive agent causative of anthrax, is naturally found in animals and in soil worldwide (Pilo and Frey, 2011). Due to its survival ability in both aerobic and anaerobic conditions, and its inherent ability to form heat-resistant spores, thus it is an ideal agent for biological warfare. Like other bacteria, its genomic sequences vary depending on geographical isolates as determined by DNA microarray experiments and whole-genome sequencing. Currently, antibiotics such as ciprofloxacin or doxycycline are used both prophylactically and for treatment, however there are no other targets or small molecules in the treatment pipeline.

Using similar FBA methods described for *S. aureus*, Ahn et al. (2014) identified 35 metabolic enzymes as essential for growth and biomass production in eight strains of *B. anthracis*. These geographically diverse isolates included *B. anthracis* A1055, *B. anthracis* Ames ancestor, *B. anthracis* CNEVA-9066, *B. anthracis* Ames, *B. anthracis* Str Strene, *B. anthracis* Str Kruger B, *B. anthracis* A2012, and *B. anthracis* Str Vollum. Ahn et al. (2014) used a *B. subtilis* model to compare biomass and essentiality genes (Kobayashi et al., 2003), as there are no such studies in *B. anthracis*. The occurrence of essential enzymes in the eight strains and their relationship are given in Figure 19.1. In general, the majority of the essential enzymes identified in *B. anthracis* belonged to the amino acids, vitamins, nucleotides, or cofactor biosynthesis pathways. For example, enzymes involved in histidine biosynthesis (HisD, HisG, HisB, and HisA), L-methionine biosynthesis

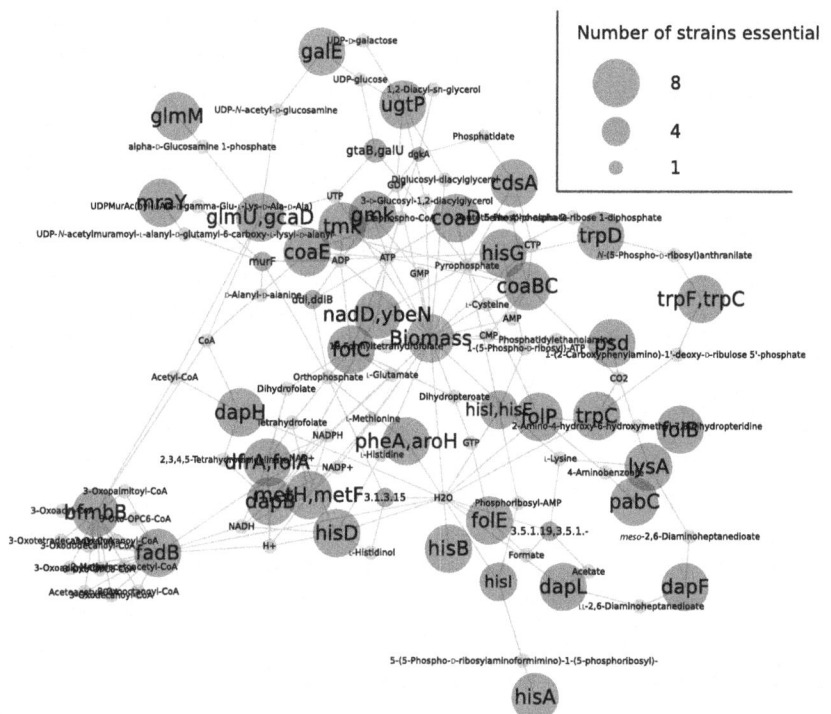

FIGURE 19.1 A network illustration of essential enzymes identified as drug targets common to diverse strains of *B. anthracis*. These metabolic enzymes are essential for the biomass generation. The size of the circle represents the number of strains.

(MetH, MetF), L-tryptophan biosynthesis (TrpD, TrpC, and TrpF) were identified as essential for growth. Pathways for the synthesis of L-lysine and LL-diaminopimelate from L-aspartate catalyzed by enzymes encoded by *dapH, dapB, dapL, lysA*, and *dapF* were determined to be essential. Diaminopimelate is used for both the synthesis of lysine and peptidoglycan. Other enzymes involved in L-phenylalanine and L-tyrosine biosynthesis (AroA, AroH) were also identified as essential.

Among vitamins, enzymes involved in the folate biosynthesis pathway (PabC DfrA, FolB, FolC, FolE, and FolP), coenzyme A biosynthesis (CoaE, CoaD), pantothenate (Dfp), and *de novo* biosynthesis/salvage of NAD and NADPH (NadD) were identified as essential for growth. Only two enzymes involved in the nucleotide biosynthesis, guanylate kinase (Gmk) and thymidine nucleotide biosynthesis dTMP kinase (Tmk), were identified as essential for biomass production.

Enzymes involved in the CDP–diglyceride biosynthesis, a major component for the phosphatidyl group of phospholipids such as diacylglycerol beta glucosyltransferase (UgtP) and phosphatidate cytidylyl-transferase (CdsA) were also identified as essential. Another key enzyme, phosphatidylserine decarboxylase (Psd), involved in phospholipid (phosphatidylethanolamine) biosynthesis, was found to be essential. Phosphoethanolamine head groups of phosphatidylethan-olamine are transferred and attached to the LPS core sugars and to periplasmic membrane-derived oligosaccharides. Among the enzymes involved in cell wall biosynthesis, phospho-*N*-acetylmuramoyl-pentapeptide-transferase (MraY) involved in peptidoglycan biosynthesis is essential, as in *E. coli* (Boyle and Donachie, 1998). A second enzyme, glucosamine-1-phosphate acetyltransferase (GlmU), is involved in UDP-*N*-acetyl-D-glucosamine biosynthesis, an essential precursor of peptidoglycan. A third UDP-glucose 4-epimerase (GalE), which is involved in galactose, amino sugar, and nucleotide sugar metabolism, was also identified as an essential enzyme in the cell wall biosynthesis. UDP-D-galactose is a building block for colonic acid and mycolyl-arabinogalactan-peptidoglycan complex biosynthesis. Phosphoglucosamine mutase (GlmM), involved in cell-wall peptidoglycan and LPS biosynthesis, was determined to be essential. It is interesting to note that none of the fatty acid biosynthetic enzymes (FabA, FabB, FabI, FabG, etc.), unlike in *S. aureus*, were found to be essential (Lee et al., 2009; Shen et al., 2010).

METABOLIC TARGETS IN *FRANCISELLA TULARENSIS*

F. tularensis, a Gram-negative, facultative, intracellular mammalian pathogen, is a causative agent of tularemia. Despite wide geographical occurrence, the genome-wide comparisons indicate only a limited genetic diversity of less than 4% among these species but with allelic variation (Johansson et al., 2001; Oyston et al., 2004). However, there is an extensive allelic variation due to the presence of short sequences and tandem repeats. *F. tularensis* can be grouped into four distinct subspecies; *F. tularensis* spp. *tularensis*, *F. tularensis* spp. *holartica*, *F. tularensis mediasatica*, and *F. tularensis* spp. *novicida*. *F. tularensis* (Biovar type A) is highly virulent and occurs predominantly in North America. *F. tularensis* spp. *holarctica* (Biovar type B) is the primary cause of tularemia in Europe and is relatively nonpathogenic to humans (Titball and Petrosino, 2007). Comparative virulence and pathogenic features due to large-scale sequence rearrangements among virulence species have been carefully identified (Oyston et al., 2004; Champion et al., 2009). Except for *F. t. novicida*, there is no sys-tematic identification of essential genes as targets for drug discovery among this group of bacteria (Gallagher et al., 2007). Antibiotics against Gram-negative bacteria such as streptomycin or gentamycin are used as primary therapeutic choice; doxycycline or ciprofloxacin are usually recommended for prophy-lactic treatment. However, the emergence of drug resistance or the intentional release of multidrug-resistant engineered strains is a potential threat to human life. Recent identification of erythromycin-resistant *F. tularensis* spp. *holarctica* emphasizes the need for newer targets and drug identification. There is also an increased effort for human vaccine development, as the current LVS strain is ineffective in certain populations (Pechous et al., 2009).

Using genome annotations and reconstructions, Ahn et al. (2014), performed FBA analysis for *F. philomiragia* (ATCC 25017), *F. t. tularensis* SCHU S4, *F. t. mediasiactica*, *F. t. holarctica* OSU18, *F. t. novicida* U112, *F. t. tularensis* FSC198, and *F. t. holarctica* (LVS). Although there are genome-wide sequence variations among these strains, there is a significant variation in the number of unique enzymes and metabolic reactions. Ahn et al. (2014) identified a total of 46 single essential enzymes across the seven species and the majority of them were in the vitamins, cofactors, and cell-wall biosynthesis pathways. An enzyme involved in lysine biosynthesis (MurF), which is required for the synthesis of peptidoglycan biosynthesis, is also essential. Several enzymes in the vitamin sub-system involved in the biosynthesis of porphyrin, heme and tetrapyrrole (HemF, HemC, HemE, HemB, HemD, HemH, and HemL1) were identified as essential. In *F. t. tularensis*, two metabolic synthetic routes of NAD biosynthesis have been identified, but none of the genes involved in the NAD biosynthesis pathway were identified as essential except for NAD kinase, a key step in the phosphorylation of NAD to form NADP (Sorci et al., 2009). Neither NMN synthetase nor NAD synthetases were identified to be essential in our study (Ahn et al., 2014). Several enzymes in the cofactor and vitamin biosynthesis pathways were identified as essential for survival: coenzyme A biosynthesis pathway (CoaE, CoaD, and CoaBC), riboflavin (FMN) biosynthesis (RibC, RibF and RibD), folate biosynthesis pathway (FolE, FolB, and FolC), nicotinate/nicotinamide biosynthesis (PpnK), and ubiquinone biosynthesis pathway (UbiC). Isoprenoids necessary for ubiquinone production are synthesized using the nonmevalonate pathway in *F. tularensis*, several enzymes encoded by genes such as *isp*H, *isp*D, *isp*F, *dxs*, *upp*S, *isp*E, and *dxr* were determined as essential. The intermediate metabolite undecaprenyl diphosphate is also a precursor of glycosyl carrier lipid, which is involved in the biosynthesis of bacterial cell wall polysaccharide components such as peptidoglycan and lipopolysaccharide. Only three genes in the phospholipid biosynthesis pathway (CdsA, PssA, and Psd) were identified as essential for survival and growth.

Several enzymes involved in the cell wall synthesis were identified to be essential among all the *F. tularensis* strains. These include lipid A and peptidoglycan biosynthesis pathways and enzymes encoded by ORFs such as *lpx*A, *lpx*B, *mur*G, *mur*B, *kds*A, *mur*A, *lpx*K, *kds*B, *mra*Y, *mur*F, *mur*D, *mur*I, and *glm*M. In purine and pyrimidine metabolism (synthesis, degradation, and salvage) thymidine kinase (TmK), dTMP kinase converts dTMP to dTDP using ATP, which is a well-known target for host cells that are infected with herpes virus, was also identified. Other enzymes involved in nucleotide and deoxynucleotide metabolism such as guanylate kinase (GmK) and purine nucleoside phosphorylase (DeoD) were identified as essential. Guanylate kinase converts GMP to GDP using ATP for the synthesis of nucleotide diphosphates such as ADP and GDP.

METABOLIC TARGETS IN *YERSINIA PESTIS*

Y. pestis, the causative agent of the plague, is a Gram-negative enteric bacterium that caused one of the deadliest epidemics in human history in Europe in the 14th century and is a biological warfare agent. *Y. pestis* is significantly diverse and is divided into three major branches, Branch 0 (Microtus and Pestiodes isolates), Branch 1 (Orientalis, African Antiqua), and Branch 2 (Medievalis and Asian isolates) (Achtman et al., 2004). The first line of antibiotics against an infection is streptomycin or gentamycin, others such as doxycycline, ciprofloxacin, or chloramphenicol are also administered as prophylactic drugs of choice (Wong et al., 2000). Despite these choices, there is always a threat of an epidemic and a drug-resistant strain outbreak.

Recently, Charusanti et al. (2011) have built a metabolic model and experimentally identified several potential targets of a clinical *Y. pestis* CO92 isolate. However, the sequence diversity among various *Y. pestis* geographical isolates is high and is reflected in their varied metabolic capabilities, hence demands the need for strain-specific target identification. Among the four *Y. pestis* genomes studied, *Y. pestis* CO92, *Y. pestis* Angola, *Y. pestis* biovar microtus str. 91001, and *Y. pestis* KIM, about 37 single essential metabolic enzymes were common to all the four geographically diverse strains (Ahn et al., 2014). The majority of these enzymes belonged to vitamins, cofactors, cell wall biosynthesis pathways, and 24 enzymes that were considered to be essential matched those experimentally identified as essential in the *Y. pestis* CO92 strain (Charusanti et al., 2011). In contrast, nine enzymes encoded by *metH*, *rhaB*, *lyx*, *kdsC*, *mtnN*, *lysA*, *dapF*, *hidD*, and *kdsD* genes were identified as essential, but were found to be dispensable (not essential) in the experimental studies in the *Y. pestis* CO92 strain.

The enzymes identified as essential by computational approaches were in the cofactor pathways including folate (FolP, FolE, FolB, FolC, and PabC), coenzyme A (CoaB, CoaC, CoaD, and CoaE), riboflavin (RibA, RibC, and RibD), FMN and pantothenate biosynthesis (*dfp*), and nicotinate/nicotinamide biosynthesis (PpnK). Seven enzymes involved in the lipid A and peptidoglycan biosynthesis were identified (L*px*A, L*px*B, L*px*K, K*ds*A, K*ds*C, K*ds*B, K*ds*D, and H*ldD*). Two enzymes, GmK (nucleotide biosynthesis) and TmK (thymidine nucleotide biosynthesis) were also identified as essential in all the *Y. pestis* genomes. Surprisingly, only two enzymes, phosphatidylserine decarboxylase (Psd) and CDP-diacylglycerol-serine *O*-phosphatidyl-transferase (PssA) were determined as essential for biomass production.

Three enzymes in the carbohydrate metabolism, starch synthase (GlgA) involved in bacterial glycogen, ramnulokinase (RhaB) involved in pentose degradation, and L-xylulokinase (Lyx) involved in the breakdown of pentose sugars such as L-lyxose and L-xylulose were identified as essential for biomass and growth. These enzymes may have a specific role in metabolism in insect or human hosts under specific conditions. Finally, arabinose 5-phosphate isomerase (KdsD), which is involved in the synthesis of ribulose-5 phosphate that is necessary for nucleotides, was also determined as essential.

Taken together, Ahn et al. (2014) have identified nine metabolic enzymes as being essential in all 19 strains spanning three genera of the three Category A bioterrorism agents (Table 19.1). In the absence of whole-genome essentiality experiments of these bacteria, the predicted targets were compared to experimentally validated essentiality enzymes from other organisms in the DEG 5.0 database (http://tubic.tju.edu.cn/deg/) (Zhang and Lin, 2009). These common essential enzymes belong to the cofactor synthesis pathway, including the Coenzyme A biosynthesis [phosphopantothenoyl cysteine decarboxylase (CoaB), phosphopantothenate cysteine ligase (CoaC), pantetheine-phosphate adenylyltransferase (CoaD), dephospho-CoA kinase (CoaE)] and folate biosynthesis pathways [dihydroneopterin aldolase (FolB), dihydrofolate synthase/tetrahydrofolate synthase (FolC), GTP cyclohydrolase I (FolE)]. CoaE is essential for growth in *E. coli* and nine other bacterial species (Table 19.1). Phosphatidyl serine decarboxylase (Psd), involved in phosphatidyl-ethanolamine synthesis, is essential in other Gram-negative bacteria such as *F. tularensis, E. coli,* and *S. enterica.* Deletions of two key enzymes in the nucleic acid pathways, guanylate kinase (Gmk) and thymidylate kinase (Tmk), is lethal and is therefore essential for survival in several bacteria (Chaperon, 2006), including *F. t. novicida* (Kraemer et al., 2009) and *Y. pestis* (Charusanti et al., 2011) (Table 19.1). Interestingly, only two enzymes, CoaD and Tmk, were experimentally identified as essential in a broad range of pathogens including *Vibrio* spp., *Helicobacter* spp., *Salmonella* spp., *Staphylococcus* spp., *Bacillus* spp., and so on (Table 19.1).

To develop antibiotics only against Category A bioterrorism agents may not be economically feasible. An alternative solution would be the development of combinatorial treatment protocols using existing antibiotics that are used to treat common bacterial infections. A potential approach toward this goal is to combine the use of common antibiotics already used to treat *B. anthracis, Y. pestis,* or *F. tularensis* infections with the use of drugs that target enzymes unconditionally essential in all strains of Category A bioterrorism agents. Of these, only one, folylpolyglutamate synthase (EC 6.3.2.17), is known to be inhibited by two antimicrobials, trimethoprim (Mathieu et al., 2005) and Rab1 (Bourne et al., 2010). Trimethoprim can block dihydrofolate reductase enzymatic activity directly, and its folypolyglutamate synthetase activity indirectly through the accumulation of a potent inhibitor, dihydrofolate (Barrow et al., 2007). However, *in vivo* it is only effective against *Y. pestis,* while *B. anthracis* and *F. tularensis* are fully and partially resistant against it, respectively. In contrast, Rab1 blocks the enzyme's activity and is active against all three Category A bioterrorism agents as well as against both methicillin-sensitive and -resistant *S. aureus* strains. The nine common essential enzymes identified in the three Category A organisms are shared among several metabolic pathways, but three of them are clustered in the dihydrofolate and tetrahydrofolate biosynthesis pathway. The other enzymes catalyze steps in the coenzyme A, cell wall, and phospholipid metabolism pathways, and nucleic acid subsystem. Of these, thymidylate

TABLE 19.1 Comparisons of Predicted Essential Enzyme Encoding Genes Shared by All Three Category A Agents to Experimentally Identified Essential Genes Across Various Pathogenic Bacteria

Sl	Enzyme	Gene	Vc	Bs	Ec	Hi	Sp	Fn	Ab	Pa	Se	Sa	Mg	Mt	Mp	Hp
1	4.1.1.36/ 6.3.2.5	dfp (coaB, coaC)	-	-	+	-	+	+	+	-	+	-	-	+	-	-
2	2.7.7.3	coaD	+	-	+	+	+	+	+	+	+	+	+	-		-
3	2.7.1.24	coaE	+	-	+	+	+	-	-	+	+	+	-	+		-
4	4.1.2.25	folB	-	+	-	-	-	+	+	+	+	+	-	+	-	+
5	6.3.2.12/ 6.3.2.17	folC	-	-	+	-	-	-	+	-	+	+	-	+		-
6	3.5.4.16	folE	+	+	+	-	-	-	+	-	+	+	-	+		+
7	2.7.4.8	gmk	-	+	+	+	-	+	+	-	-	+	+	-	+	-
8	2.7.4.9	tmk	+	+	+	+	-	-	-	+	+	+	+	-	+	-
9	4.1.1.65	psd	-	-	+	-	-	+	+	+	+	-	-	-		-

Vc: Vibrio cholera N116961, Bs: Bacillus subtilis 168, Ec: E. coli MG1655, Hi: Haemophilus influenzae Rd KW20, Sp: Streptococcus pneumonia, Fn: Francisella novicida U112, Ab: Acinetobacter baylyi ADP1, Pa: Pseudomonas aeruginosa UCBPP-PA14, Se: Salmonella enterica serv Typhi, Sa: Staphylococcus aureus N315, Mg: Mycoplasma genitalium G37, Mt: Mycobacterium tuberculosis H37Rv, Mp: Mycoplasma pulmonis UAB CTIP, Hp: Helicobacter pylori 26695. (+: Essential, −: Nonessential).

kinase, which is known to catalyze the conversion of dTMP to ddTMP, is essential for the viability of *E. coli* and has been considered as a drug target in other bacteria. Interestingly, none of the fatty acid metabolic enzymes were identified as essential in any of the three genera studied, unlike in *E. coli* (Almaas et al., 2004) or *S. aureus* (Lee et al., 2009). This clearly demonstrates such methods work best in pathogen-specific or multiple infections. Recently, there has been a comprehensive study quantitatively assessing individual metabolic reactions and enzymes that are necessary to synthesize all biomass components for a composite group of bacteria in a given environment (Barve et al., 2012). This method has identified 124 reactions required in all metabolic networks of several bacteria and is defined as "super essential". Barve et al. (2012) conclude that these super essential enzymes can serve as a potential drug target for a broad-spectrum anti-infective. When this list of super essential enzymes was compared to the nine uniquely identified essentials by Ahn et al. (2014), only four enzymes, CoaD (EC2.7.7.3), FolB (EC 4.1.2.25), FolC (EC 6.3.2.12/EC 6.3.2.17), and Gmk (EC 2.7.4.8), were found to be common. This underscores the importance of understanding and integrating results from different computational approaches for identifying a common target(s) using metabolic network.

TARGET TO HIT MOLECULE GENERATION

As described above, target identification is a first step in the five-step process of drug development. A computationally identified target provides a cost-effective way to generate a pipeline of targets, especially from different genera and geographically distinct species. Irrespective of the methods used, a combination of computational and experimental methods to identify new targets is necessary for a portfolio of molecules against both emerging bacterial infections and drug-resistant pathogens. In this section, I briefly illustrate examples of hit molecule discovery from the FBA results from *E. coli* MG1655 and *S. aureus* (Shen et al., 2010). Taking into consideration genome-scale experimental gene deletion experiments which predicted 38 indispensable reactions and FBA-based computational target prediction, Shen et al. (2010) identified seven indispensable reactions in a single pathway shared among *E. coli* MG1655 and *S. aureus* strains. One such pathway is fatty acid synthesis (FAS II), whose first step is catalyzed by malonyl-CoA-acyl carrier protein transacylase (FabD). Using the ZINC lead library that contains over a million small molecules, 15 potential inhibitors were identified by virtual screening of the FabD protein crystal structure. Interestingly, these inhibitors were active against bacterial FabD enzyme but not human mitochondrial homologs. Subsequently inhibitors for all the steps in the FAS II pathway were screened for all the enzymes in the pathway (FabH, FabB/F, FabG, FabA/Z, and FabI) either by using ligand-bound crystal structures or homology models (Shen et al., 2010). These hit molecules had varying degrees of inhibition, which now could be used to identify lead molecule generation and optimization.

CONCLUSION

Despite sequencing of diverse geographical isolates, there has been only very limited new target discovery and virtually no specific drug development. Identifying gene essentiality by experimental approaches either using transposon mutagenesis or RNA silencing is time-consuming and expensive, and the results tend to be strain-specific. In contrast, computational-based methods provide an alternate approach for the identification of single essential and synthetic lethal metabolic enzymes that can be simultaneously tested for multiple strains. These approaches can also be tested simultaneously under several growth conditions to identify organism/strain-specific single gene essential metabolic enzymes that can be tested as targets for drug discovery.

Availability of a complete genome sequence not only provides identification of gene variability, pathogenic features, and antibiotic resistance determinants but also provides valuable metabolic information for *in silico* metabolic networks. Although several virulence genes individually have been characterized to a great extent, generally they tend to be not feasible drug target "non-druggable" candidates as often these are not universal and tend to be strain-specific or prophage-borne. On the other hand, metabolic enzymes tend to be common to all the strains and are broader target candidates irrespective of bacteria or their host-specificity. Using a metabolic reaction network, one can perform FBA and identify unconditionally essential metabolic enzymes for any pathogen.

I conclude that computational identification of single essential enzymes as targets from a metabolic network of geographically distinct isolates across several genera is a rapid and cost-effective approach for antimicrobial research. Common essential enzymes should be considered as targets with the criteria that it should exhibit sufficient structural similarity in their catalytic sites and their human orthologs should be structurally dissimilar such that the identified chemical inhibitor is not toxic to the host.

REFERENCES

Achtman, M., Morelli, G., Zhu, P., Wirth, T., Diehl, I., Kusecek, B., et al., 2004. Microevolution and history of the plague bacillus, *Yersinia pestis*. Proc. Natl. Acad. Sci. USA 51, 17837–17842.

Ahn, Y.Y., Lee, D.S., Burd, H., Blank, W., Kapatral, V., 2014. Metabolic network analysis-based identification of antimicrobial drug targets in category A bioterrorism agents. PLoS One 9 (1), e85195.

Almaas, E., Kovacs, B., Vicsek, T., Oltvai, Z.N., Barabasi, A.L., 2004. Global organization of metabolic fluxes in the bacterium *Escherichia coli*. Nature 427, 839–843.

Almaas, E., Oltvai, Z.N., Barabasi, A.L., 2005. The activity reaction core and plasticity of metabolic networks. PLoS Comput. Biol. 1, e68.

Barrow, E.W., Dreier, J., Reinelt, S., Bourne, P.C., Barrow, W.W., 2007. In vitro efficacy of new antifolates against trimethoprim-resistant *Bacillus anthracis*. Antimicrob. Agents Chemother. 51, 4447–4452.

Barve, A., Rodrigues, J.F., Wagner, A., 2012. Superessential reactions in metabolic networks. Proc. Natl. Acad. Sci. USA 109 (18), E1121–E1130.

Becker, S.A., Palsson, B.O., 2005. Genome-scale reconstruction of the metabolic network in *Staphylococcus aureus* N315: an initial draft to the two dimensional annotation. BMC Microbiol. 5, 8–20.

Bourne, C.R., Barrow, E.W., Bunce, R.A., Bourne, P.C., Berlin, K.D., Barrow, W.W., 2010. Inhibition of antibiotic-resistant *Staphylococcus aureus* by the broad-spectrum dihydrofolate reductase inhibitor RAB1. Antimicrob. Agents Chemother. 54, 3825–3833.

Boyle, D.S., Donachie, W.D., 1998. mraY is an essential gene for cell growth in *Escherichia coli*. J. Bacteriol. 180, 6429–6432.

Champion, M.D., Zeng, Q., Nix, E.B., Nano, F.E., Keim, P., Kodira, C.D., et al., 2009. Comparative genomic characterization of *Francisella tularensis* strains belonging to low and high virulence subspecies. PLoS Pathog. 5 (5), e1000459.

Chaperon, D.N., 2006. Construction and complementation of in frame deletions of the essential *Escherichia coli* thymidylate kinase gene. Appl. Environ. Microbiol. 72, 1288–1294.

Charusanti, P., Chauhan, S., McAteer, K., Lerman, J.A., Hyduke, D.R., Motin, V.L., et al., 2011. An experimentally-supported genome-scale metabolic network reconstruction for *Yersinia pestis* CO92. BMC Syst. Biol. 5, 163.

Edwards, J.S., Palsson, B.O., 2000. The *Escherichia coli* MG1655 in silico metabolic genotype: its definition, characteristics, and capabilities. Proc. Natl. Acad. Sci. USA 97, 5528–5533.

Gallagher, L.A., Ramage, E., Jacobs, M.A., Kaul, R., Brittnacher, M., Manoil, C., 2007. A comprehensive transposon mutant library of *Francisella novicida*, a bioweapon surrogate. Proc. Natl. Acad. Sci. USA 104, 1009–1014.

Gilligan, P.H., 2002. Therapeutic challenges posed by bacterial bioterrorism threats. Curr. Opin. Microbiol. 5, 489–495.

Heinemann, M., Kummel, A., Ruinatscha, R., Panke, S., 2005. "In silico" genome-scale reconstruction and validation of the *Staphylococcus aureus* metabolic network. Biotechnol. Bioeng. 92, 850–864.

Johansson, A., Goransson, I., Larsson, P., Sjostedt, A., 2001. Extensive allelic variation among *Francisella tularensis* strains in a short-sequence tandem repeat region. J. Clin. Microbiol. 39, 3140–3146.

Joyce, A., Palsson, B.O., 2008. Predicting gene essentiality using genome-scale *in silico* models. Methods Mol. Biol. 416, 433–457.

Kapatral, V., Anderson, I., Ivanova, N., Reznik, G., Los, T., Lykidis, A., et al., 2002. Genome sequence and analysis of the oral bacterium *Fusobacterium nucleatum* strain ATCC 25586. J. Bacteriol. 184, 2005–2018.

Kobayashi, K., Ehrlich, S.D., Albertini, A., Amati, G., Andersen, K.K., Arnaud, M., et al., 2003. Essential *Bacillus subtilis* genes. Proc. Natl. Acad. Sci. USA 100, 4678–4683.

Kraemer, P.S., Mitchell, A., Pelletier, M.R., Gallagher, L.A., Wasnick, M., Rohmer, L., et al., 2009. Genome-wide screen in *Francisella novicida* for genes required for pulmonary and systemic infection in mice. Infect. Immun. 77, 232–244.

Lee, D.S., Burd, H., Liu, J., Almaas, E., Wiest, O., Barabási, A.L., et al., 2009. Comparative genome-scale metabolic reconstruction and flux balance analysis of multiple *Staphylococcus aureus* genomes identify novel antimicrobial drug targets. J. Bacteriol. 191, 4015–4024.

Mathieu, M., Debousker, G., Vincent, S., Viviani, F., Bamas-Jacques, N., Mikol, V., 2005. *Escherichia coli* FolC structure reveals an unexpected dihydrofolate binding site providing an attractive target for anti-microbial therapy. J. Biol. Chem. 280, 18916–18922.

Overbeek, R., Larsen, N., Walunas, T., D'Souza, M., Pusch, G., Selkov Jr., E., et al., 2003. The ERGO genome analysis and discovery system. Nucleic Acids Res. 31 (1), 164–171.

Oyston, P.C., Sjostedt, A., Titball, R.W., 2004. Tularemia: bioterrorism defence renews interest in *Francisella tularensis*. Nat. Rev. Microbiol. 2, 967–978.

Pechous, R.D., McCarthy, T.R., Zahrt, T.C., 2009. Working toward the future: insights into *Francisella tularensis* pathogenesis and vaccine development. Microbiol. Mol. Biol. Rev. 73, 684–711.

Pilo, P., Frey, J., 2011. *Bacillus anthracis*: molecular taxonomy, population genetics, phylogeny and patho-evolution. Infect. Genet. Evol. 11, 1218–1224.

Pohanka, M., Skladal, P., 2009. *Bacillus anthracis, Francisella tularensis* and *Yersinia pestis*. The most important bacterial warfare agents. Folia Microbiol. 54, 263–272.

Pray, L., 2008. Antibiotic resistance, mutation rates and MRSA. Nat. Edu. 1 (1), 30.

Shen, Y., Liu, J., Estiu, G., Isin, B., Ahn, Y.Y., Lee, D.S., et al., 2010. Blueprint for antimicrobial hit discovery targeting metabolic networks. Proc. Natl. Acad. Sci. USA 107 (3), 1082–1087.

Sorci, L., Martynowski, D., Rodionov, D.A., Eyobo, Y., Zogaj, X., Klose, K.E., et al., 2009. Nicotinamide mononucleotide synthetase is the key enzyme for an alternative route of NAD biosynthesis in *Francisella tularensis*. Proc. Natl. Acad. Sci. USA 106, 3083–3088.

Titball, R.W., Petrosino, J.F., 2007. *Francisella tularensis* genomics and proteomics. Ann. NY Acad. Sci. 1105, 98–121.

Wong, J.D., Barash, J.R., Sandfort, R.F., Janda, J.M., 2000. Susceptibilities of *Yersinia pestis* strains to 12 antimicrobial agents. Antimicrob. Agents Chemother. 44, 1995–1996.

Zhang, R., Lin, Y., 2009. DEG 5.0 a database of essential genes in both prokaryotes and eukaryotes. Nucleic Acids Res. 37, 455–458.

Chapter 20

Application of Metagenomic Technologies for Antimicrobial Resistance and Food Safety Research and Beyond*

Chin-Yi Chen[1], Xianghe Yan[1], Siyun Wang[2] and Charlene R. Jackson[3]

[1]*US Department of Agriculture, Agricultural Research Service, Eastern Regional Research Center, Wyndmoor, PA, USA,* [2]*Food, Nutrition and Health, Faculty of Land and Food Systems, The University of British Columbia, Vancouver, BC, Canada,* [3]*US Department of Agriculture, Agricultural Research Service, Russell Research Center, Athens, GA, USA*

Chapter Outline

* CYC and XY contributed equally to this review. The opinions expressed in this review are entirely those of the authors and do not represent those of the USDA.

Antimicrobial Resistance and Food Safety. DOI: http://dx.doi.org/10.1016/B978-0-12-801214-7.00020-X

INTRODUCTION

Metagenomics is a relatively new and ever-expanding field that addresses the collective genetic material and functional composition of a microbial community without bias and the necessity of culturing its inhabitants (Galbraith et al., 2004; Marco, 2008). The term "metagenome" first appeared in a publication by Handelsman et al. (1998) on the identification of natural products from soil bacteria without culturing. Almost all metagenomics studies begin at the same step: nucleotide sequencing of the whole metagenome of the microbial organisms from a specific environment, such as soil microbial communities or gastrointestinal systems of human or animals. As such, metagenomics is not simply large-scale DNA sequencing; it consists of compositional and/or function-based analyses of all the microbial genomes achieved by assembling and annotating millions of random DNA sequence reads. Metagenomics requires interdisciplinary training in molecular biology, cell biology, genetics, informatics, computer science, and ecology. Metagenomics includes, in a broad sense, the areas of meta-transcriptomics (Raes and Bork, 2008), metabolomics (McHardy et al., 2013), and meta-proteomics (Askenazi et al., 2010).

Metagenomics has the potential to further advance modern food safety research by focusing on microbial communities, including foodborne pathogens, associated with food or food animals. One major advantage of this technique is that isolation of bacteria, including native microbiota and pathogens, from food and environmental samples can be omitted, thus facilitating studies on unculturable, fastidious, or slow-growing bacteria (e.g., *Yersinia enterocolitica*), microaerophilic or anaerobic bacteria, or obligate symbionts/parasites (Schloss and Handelsman, 2005; Messelhausser et al., 2011; Hongoh, 2011). Traditional culture-dependent methods have several limitations including lack of appropriate growth media for some bacteria, time-consuming isolation procedures, and the potential of underestimating the microbial community due to injured or unculturable bacteria.

To date, many successful research studies using metagenomics have been achieved by (i) amplifying ribosomal RNA gene sequences (16S or others) from complex clinical, food, and environmental communities (Wang and Qian, 2009); (ii) applying "targeted metagenomics" approaches with specific molecular signature sequences (Gloux et al., 2011); and (iii) profiling with shotgun sequences by aligning with reference genomes (Oh et al., 2011; Zhu et al., 2010). Metagenomics has been applied to many fields within the life sciences; some examples include acid mine drainage biofilm exploration (Yelton et al., 2013), marine microbiota identification (Martin et al., 2014), animal feed and nutrition utilization (Singh et al., 2008), microbial detection and identification (Petrosino et al., 2009), taxonomy research (Siqueira and Rocas, 2010; Clemente et al., 2010), pathogen monitoring and control (Culligan et al., 2009, 2014), emerging and re-emerging pathogen detection (Miller et al., 2013), identification of novel antimicrobial resistance genes on spinach (Berman and Riley, 2013), molecular identification of symbiotic bacteria in the termite hindgut (Warnecke et al., 2007), elucidation of the relationship between microbial community composition and obesity in mice and humans (Ley et al., 2006; Turnbaugh and Gordon, 2009), human nutrition and diabetes studies (Cani, 2013; Price et al., 2009; Ohlrich et al., 2010), and microbiota in malnourished children (Ghosh et al., 2014). Metagenomic approaches can also be used for the discovery of genes involved in survival and persistence of pathogens, of beneficial bacteria such as probiotics in gastrointestinal tracts (Joeres-Nguyen-Xuan et al., 2010; Marcinakova et al., 2010), and on the assessment of bacterial diversity in cattle (Durso et al., 2010), sheep, goats, swine, rodents, rabbits, humans, poultry, and food products.

META-OMIC WORKFLOW

Metagenomics research involves molecular techniques, methodologies, and computational/bioinformatic analysis algorithms and software. Figure 20.1 shows the workflow of typical meta-omics projects, from sample collection, sequencing, data processing and assembly, to computational analysis (phylogenic, functional, or other classifications). More specifically for metagenomic research, after the proper experimental design and collection of samples, the following three main aspects are involved.

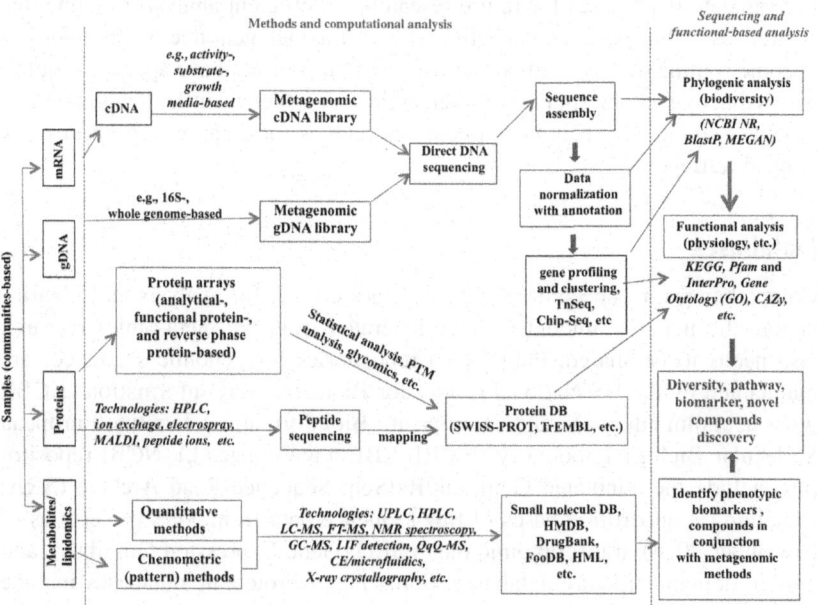

FIGURE 20.1   Flow diagram of a typical metagenomics/metaproteomics/metabolomics project.

DNA Sequencing Technologies

Major "next-generation sequencing (NGS)" technologies currently in use include Illumina/Solexa, Ion Torrent (Life Technologies), Pacific BioSciences single-molecule real-time sequencing, 454 pyrosequencing, and SOLiD (Applied Biosystems); with Illumina and Ion Torrent being the current major players in metagenomic applications. The traditional Sanger sequencing also plays an integral role, although more for the validation/verification steps. Other newer, emerging sequencing technologies often involve nanopore, single-molecule sequencing and may not be suitable for metagenomic applications. Selection of sequencing technologies will depend on the needs and goals of the projects.

Bioinformatics Analysis Software and Pipeline

Computational analysis is a bottleneck for any NGS or metagenomic application. Raw sequence data are processed and assembled using assembly algorithms/software such as K-mer, Bowtie, and Velvet, or various commercial packages. Assembled sequences are then compared to other known sequences in the database for identifying the target of interest, such as pathogens, specific genes/pathways, etc. BLAST (Altschul et al., 1990) is the most popular alignment tool for gene annotation; in addition, other specialized software/algorithms may be used. The results can then be further analyzed to derive conclusions or used to formulate hypotheses for future research. An efficient analysis pipeline that is fast, less laborious, and objective is vital for metagenomic technologies to become routine testing methods. Most research projects have specific requirements that may not be applied to other studies, thus there is little, if any, standard pipelines suitable for routine applications such as for diagnostic purposes in a clinical setting.

Databases

Comprehensive or specialized databases are crucial for analysis and identification; the massive amount of data generated from the meta-omics research also needs to be shared. Public data repositories for genomic sequences are maintained by the US National Center for Biotechnology Information (NCBI; www.ncbi.nlm.nih.gov) and European Bioinformatics Institute/European Molecular Biology Laboratory (EMBL-EBI; www.ebi.ac.uk). NCBI repositories include the annotated GenBank/RefSeq, Sequence Read Archive (SRA; containing 3 quadrillion bases of raw sequence data from NGS projects as of November 2014), transcriptomic data (GEO: Gene Expression Omnibus), and protein domain/structure databases. Some major proteomic databases include PRoteomics IDEntification database (PRIDE) at the EMBL-EBI (www.ebi.ac.uk/pride/), PeptideAtlas (www.peptideatlas.org; Institute for Systems

Biology Seattle Proteome Center), and the Global Proteome Machine Database (GPMDB; gpmdb.thegpm.org). A metabolomic database, MetaboLights, is also maintained by EMBL. Other research institutes, such as the Welcome Trust Sanger Institute (www.sanger.au.uk) and various sequence centers/consortia, house data associated with specific projects.

EXAMPLES OF METAGENOMIC RESEARCH

In June 2012, the US National Institutes of Health announced the completion of phase I (2007–2012) of the Human Microbiome Project (HMP). The study sampled from up to 18 body sites ("habitats") of 242 healthy adults and utilized 16S rRNA gene analysis as well as paired-end shotgun metagenomic reads to catalog the microbial communities, generated >5,000 unique microbial taxonomic profiles, and over 3.5 Tbp of metagenomic sequences. Analysis results showed that there were wide variations among different individuals and body sites on the diversity and abundance of signature microbes, with the oral and stool communities exhibiting the highest diversity. Conversely, the metabolic pathways remained stable within the (healthy) population (HMPC, 2012a, 2012b). Additionally, over a dozen papers were published in PLoS journals in association with the project in 2012. HMP Data Analysis and Coordination Center (DACC) released more sequences in April 2014, bringing the total data to over 14 terabytes; protocols, bioinformatics analysis software, tutorials, and links to genomes of the reference strains are available on the DACC website (http://hmpdacc.org). Phase II (2013–2015) of the HMP was aimed at creating the integrated datasets of biological properties from both the microbiome and host from studies of microbiome-associated diseases using multi-"omics" technologies (http://commonfund.nih.gov/hmp/index; accessed 11/10/2014).

This coordinated interdisciplinary research effort not only provided insights into the human microbiome, but also contributed greatly to the development of the standard operating procedures, protocols, bioinformatic analysis software for metagenomic research, and served as the driving force on sequencing technologies and computer/informatics infrastructures. A good overview on how to conduct a microbiome study has been recently published (Goodrich et al., 2014a).

In the past few years, the Centers for Disease Control and Prevention (CDC) and the Food and Drug Administration (FDA) have applied whole-genome sequencing (WGS) for outbreak source tracking and surveillance of clinical, food, and environmental isolates, and for retrospective studies of historical samples. GenomeTrakr network (http://www.fda.gov/Food/FoodScienceResearch/WholeGenomeSequencingProgramWGS/ucm363134.htm) was set up for collecting and sharing WGS of foodborne pathogens, with all labs using standardized procedures and sequencing methods. As of December 2014, the GenomeTrakr project had recorded >13,700 BioSamples, with >9,000 *Salmonella* sequences deposited in the NCBI SRA, contributed by FDA, state public health labs, and other sequencing centers internationally. FDA also partnered with CDC on the effort of sequencing all *Listeria monocytogenes* isolates from clinical, food, and environmental samples; currently ~1,200 sequences are available on the NCBI website. NCBI is also developing a pipeline for pathogen WGS assembly using different assemblers including *de novo* assembly and assembly against reference; K-mer-based tree analysis is used to determine clusters and distance while Genome Workbench is used for genome viewing, but not analysis.

This chapter will present some recent examples of metagenomic/meta-omic studies related to evolution of antimicrobial resistant clones, identification of new antibiotics, pathogen detection and surveillance, human and animal microbiomes, and probiotics research that we thought relevant and may be of general interest to a broader audience. This is by no means intended to be a thorough review of the current field.

Antibiotic Resistance Development/Adaptation and Identification of Novel Antibiotics

Some metagenomic or population/evolution projects related to antibiotic research in the NCBI SRA database are listed in Table 20.1; when possible, the publications associated with the studies are included. Many studies focused on how antibiotic treatments affected the development of resistance in bacteria or affected the gut microbiota. Some of the relevant studies are briefly discussed below.

Antibiotic Treatment Resulted in Dominance of Resistant Pathogens

One of the earlier studies using 16S rRNA pyrosequencing by Ubeda and colleagues (2010) showed that the intestinal microbiota was altered after antibiotic treatment and that vancomycin-resistant *Enterococcus* was able to dominate in the intestines of humans and mice. Some bacterial populations failed to recover after withdrawal of the antibiotics. Mice were also shown to be more susceptible to *Salmonella* colonization after antibiotic treatment. The study showed the impact that antibiotic treatments could bring to patients already undergoing treatment for infections.

Adaptation to Antibiotics

A metagenomic approach was used to assess the evolution and adaptation of 63 *Escherichia coli* populations to antibiotics at various dosage-treatment combinations of doxycycline and erythromycin (Laehnemann et al., 2014). Sequence changes including single nucleotide variations, insertions/deletions (indels), structural variations, and large-scale amplification were analyzed. The results suggested convergent evolution from independently evolved populations. Certain mutation sites were preferred, and different types of mutations may accumulate. A large sequence amplification (316-kb) encoding a multidrug efflux pump, AcrA-AcrB-TolC, was significantly enriched by specific treatments; such amplification

TABLE 20.1 Select Metagenomic Studies Related to Antibiotic or Antibiotic Resistance in the NCBI Short Read Archive

SRA Identifier	Title	Year	Reference or Sequence Submitter
SRP033483	*Escherichia coli* Evolution under Antibiotic Regimes	2014	Fridman et al. (2014)
SRP045373	Experimentally Evolved Multiple Antibiotic Resistance in *Staphylococcus aureus*	2014	Harvard Medical School, USA
SRP045811	Studies of Intestinal Microbiome and Infection Resistance	2014	Memorial Sloan Kettering Cancer Center, USA
ERP002192	Antibiotics Impact on the Human Intestinal Microbiota	2014	Centre for Public Health Research, Spain (http://www.csisp.gva.es/web/csisp)
ERP003998	Genomics of Rapid Adaptation to Antibiotics: Convergent Evolution and Scalable Sequence Amplification	2013	Laehnemann et al. (2014)
SRP013429	Community Analysis of Chronic Wound Bacteria Using 16S rRNA Gene-Based Pyrosequencing: Impact of Diabetes and Antibiotics on Chronic Wound Microbiota	2012	Translational Genomics Research Institute, USA (Price et al., 2009)
SRP013856	The Effects of Subtherapeutic Antibiotics on the Murine Colonic Microbiome and Early Life Adiposity	2012	J. Craig Venter Institute, USA
SRP015747	GSE40864: Investigating the sRNA and mRNA Transcriptional Response to Antibiotics in Methicillin-Resistant *Staphylococcus aureus* Using Illumina RNAseq	2012	Howden et al. (2013)
ERP000450	Stepwise Mutation to Multiple Drug Resistance in *Mycobacterium tuberculosis* in an Otherwise Stable Genetic Background Does Not Support Models of High Mutation in the Host and Emphasizes the Need for Maintained Effective Concentrations of Multiple Antibiotics	2011	Saunders et al. (2011)
SRP003766	Vancomycin-Resistant *Enterococcus* Domination of Intestinal Microbiota Is Enabled by Antibiotic Treatment of Mice and Precedes Bloodstream Invasion in Humans	2010	Ubeda et al. (2010)

was unstable and quickly diminished after culturing in the absence of antibiotics. The authors' conclusion, "...distinct selective challenges are countered by different genomic response mechanisms.... Thus, increased antibiotic stress does not necessarily lead to bacterial elimination but rather causes a change in the set of genomic adaptations," underscored the importance of judicious use of antibiotics.

Development of Antibiotic-Resistant Clones in Biofilms

Meticillin-resistant *Staphylococcus aureus* (MRSA) infections result in an ~20% mortality rate and are commonly treated with vancomycin. There are reports of the emergence of vancomycin-intermediate *S. aureus* (VISA) which are resistant to moderate levels of vancomycin, thus presenting a threat to the last-resort MRSA treatment. A study was conducted to evaluate the diversification of *S. aureus* communities *in vivo* and in biofilm after exposure to Bsa bacteriocin (Koch et al., 2014). Phenotypes (staphyloxanthin pigment production, biofilm-formation, hemolysis, and antibiotic resistance) were correlated with the gene expression levels (by qRT-PCR) of the staphyloxanthin gene (*crt*), biofilm-related genes (*ica* and *spa*), and the hemolytic toxin gene (*hla*) in different variants. RNA-seq was also used to evaluate genome-wide gene expression. VISA-like mutations were found in three operons (*graRS*, *vraRS*, and *walKR*) that synergistically regulate cell wall synthesis, and their emergence was apparently unrelated to the vancomycin treatment. Although not a true metagenomic study, this report suggested that competition within biofilms may play an important role in population diversification, thus encouraging the development of antibiotic-resistant variants.

Identification of New Antibiotics

Donia and colleagues (2014) identified over 3,000 small-molecule biosynthesis gene clusters from human microbiome using a systematic approach. The group analyzed 752 HMP metagenomic samples, searching for gene clusters involved in small molecule biosynthesis. A class of thiopeptide antibiotics was found to be widely distributed in the metagenomes of human microbiota. A particular thiopeptide, lactocillin, was purified and the structure solved. Lactocillin was shown to have potent antibacterial activities against some Gram-positive vaginal pathogens. Metatranscriptomic (RNA-seq) data analysis revealed that the gene cluster was actively expressed in humans. Their findings not only suggested that thiopeptides may play an integral role in microbe–microbe and microbe–host interactions, but showed great promise for identification and isolation of new antibiotics/therapeutics from the human microbiome.

Pathogen Identification and Surveillance

Identification, Typing, and Surveillance of Pathogens

Identifying serogroups, serotypes, and genotypes of foodborne pathogens is important to epidemiological classification, disease surveillance, and outbreak source-tracking. Approaches for bacterial pathogen molecular typing and sub-typing using sequencing technologies include 16S rRNA classification for genus/species-level identification, single nucleotide polymorphism (SNP) analysis and whole-genome multilocus sequence typing (wgMLST) for closely related isolates; additionally, polymerase chain reactions (PCR) targeting specific virulence or other genes of interest are also frequently used. Particular contribution of NGS-based technologies has enabled Cummings and his co-workers (2010) to conduct whole-genome typing of microbial pathogens in parallel to detect strain-specific polymorphisms in *Bacillus anthracis* and *Yersinia pestis*. The conclusions of this report suggested that the possibility offered by NGS technology during a forensic or epidemiological investigation could facilitate accurate, rapid and high-throughput detection of strains. Morelli and colleagues (2010) applied both conventional Sanger sequencing methods and NGS technologies to identify meaningful patterns of global phylogenetic diversity through the comparison of 17 whole genomes of *Yersinia pestis* isolates from global sources. Chen et al. (2010) compared NGS data from three pathogenic and eight non-pathogenic members of the *Yersinia* genus and identified 100 regions within the genome of *Yersinia enterocolitica* that represented potential candidates for the design of nucleotide sequence-based assays for unique detection of the pathogen.

It is apparent that metagenomics and the NGS technologies will provide systematic information on rapid evaluation of specific regions or biomarkers of the genome of a microorganism to determine to which genus, species, and/or strain it belongs and its potential pathogenicity. For example, accurate determination of the serotype and genotype based on O-antigen and virulence genes is important for the identification of Shiga toxin-producing *E. coli* (STEC). *E. coli* serotype O157:H7 and the "Top Six" non-O157 STEC serogroups, O26, O45, O103, O111, O121, and O145, are important foodborne pathogens that cause similar illnesses such as hemorrhagic colitis or hemolytic uremic syndrome. The classification of STEC is traditionally based on phenotypic analyses and/or PCR-based molecular typing targeting specific biomarkers such as O-antigen and Shiga toxin (*stx*) genes. These procedures are often time-consuming and inaccurate. The power of NGS technologies and bioinformatics can be used to facilitate *E. coli* genotyping and detection based on computational O-antigen gene cluster and virulence gene analysis. We have sequenced, annotated, and/or deposited over 95 different serotypes of *E. coli* O-antigen clusters in NCBI GenBank, and manually curated other *E. coli* serotypes from public databases. All O-antigen cluster sequences of the known *E. coli* serotypes are now available in the public and/or in-house database (Iguchi et al., 2014; Yan et al., personal communication).

Pipeline for Rapid Pathogen Identification

To be practical for real-world diagnostic applications (timely and less labor-intensive analysis), a rapid and preferably automated analysis pipeline to identify pathogens or virulence genes is crucial. Many issues need to be considered: host genome subtraction, reference genome database, analysis software, and comparison criteria/algorithms. Many analysis pipelines published to date utilized simulated data sets, or were limited in scope, thus should be considered "proof of concept" exercises. Naccache and colleagues (2014) developed and tested a cloud-compatible bioinformatics pipeline for sequence-based ultrarapid pathogen identification (SURPI). SURPI can be operated in two modes, fast (nucleotide-based alignment) and comprehensive (amino acid-based alignment), utilizing SNAP and RAPSearch alignment tools for routine and novel pathogen identification, respectively. SURPI was tested using 157 clinical samples multiplexed in five independent datasets and was able to identify bacterial and parasitic species, as well as viruses. Results can be obtained within minutes to hours, so clinical actions can be taken to treat patients in a timely manner.

Gut Microbiomes of Humans and Animals

Foodborne pathogens generally result in gastrointestinal symptoms; the infection and the antibiotic treatments often affect the balance of gut microbiomes. The animal gastrointestinal tract is a complex ecosystem, and its microbiota reflects the co-evolution of the resident microorganisms with their host environment (Ley et al., 2006). To understand the composition of microbiota in different animal species and their roles in metabolism, nutrient utilization, and immunological functions, three fundamental questions need to be addressed: (i) the morphological and physicochemical heterogeneity of the gut systems in humans and animals; (ii) the taxonomic composition of the microbiota; and (iii) the functional capacities of these microorganisms.

Before the debut of modern molecular approaches, the identification and characterization of the rumen microbiome were mainly based on traditional methods involving isolation and enumeration with different media. More than 200 culturable bacterial species were identified from rumen by Russell and colleagues (1981). Later on, culture-independent small-subunit (SSU)-rRNA gene-based techniques began to be widely used to identify microbes through phylogenetic sequence analysis. The 16S rRNA gene clone libraries constructed using universal PCR primers were adopted to explore diversity of microbes in the rumen ecosystem (Reilly and Attwood, 1998; Kittelmann et al., 2013). Various molecular fingerprinting methods were used, including PCR-denaturing gradient gel electrophoresis (PCR-DGGE), temperature gradient gel electrophoresis (TGGE), and restriction fragment length polymorphisms (RFLP). It was found that the culturable bacteria may not reflect the actual proportion of the total bacterial genera/species in the rumen microbial community. For example, *Treponema* spp. were found to account for a very limited proportion of the rumen microbial community (0.02%) based on earlier cultivation-based studies (Paster et al., 1991). However, the 16S rRNA gene-based analysis demonstrated that the presence of *Treponema* spp. was highly underestimated, due to the fact that the most abundant species in this group are "unculturable". *Treponema* spp. were therefore identified as a core member of the rumen microbial community, and they play an important role in the degradation of soluble fibers (Bekele et al., 2011). Through the 16S rRNA gene PCR and sequencing of clone libraries, Kong and colleagues (2010) identified 616 operational taxonomic units belonging to 32 genera, most in the phyla Firmicutes and Bacteroidetes, in rumens of cows fed with different diets.

Human Gut Microbiota

The intestinal microbiota of the human is important and essential for human health, as it possesses many metabolic and protective responses and activities including nutrient processing, antimicrobial defenses (Lievin-Le Moal and Servin, 2006), immune system development (Backhed et al., 2004, 2005), and essential vitamin biosynthesis (Backhed et al., 2005; Guarner and Malagelada, 2003). The bacterial population and phylotype compositions in the human gut populations are remarkably consistent between individuals (Zoetendal et al., 1998), implying that mechanisms exist to avoid blooms of subpopulations and to maintain a similar microbiome/metabolome network. Moreover, 552,700 unique, novel genes have been found by comparing culture-independent sequence data, which are estimated to contain over nine million genes from 202 human gut bacteria with publically available complete genomes (Yang et al., 2009). In order to have a better understanding of the functional relationship among the genetic composition, the physiological role of microbial consortia, and the physical features within human guts, investigators in food and nutrition sciences need to highlight the uniqueness of the human gut among mammals.

Gut Microbiome Influenced by Host Genetics

Host genetics was shown to affect the gut microbiome in twin studies in the United Kingdom (Goodrich et al., 2014b). Over 1,000 fecal samples collected from 416 twin pairs were analyzed, and the abundance of microbial taxa was shown to be influenced by host genetics. The Christensenellacease family, the most heritable taxon frequently co-occurring with other heritable bacteria and methanogenic Archaea, was found to be enriched in people with low body mass index. When amended with *Christensenella minuta* in obese mice, the recipient mice showed reduced weight gain and had altered microbiome composition. The results suggested that the gut microbiome may impact host metabolism.

Bovine Microbiota

Beef producers require accurate feed analyses, as well as a basic understanding of the ruminal digestive system and uniqueness of its microbiota composition. Bovine, along with other ruminants such as sheep, goats, and deer, have a digestive system which allows them to utilize roughage (e.g., hay, grass) as a major source of nutrients; non-ruminants (e.g., pigs, dogs, human) are not able to efficiently digest cellulose. One main characteristic of ruminant animals is that they have four stomachs: the rumen, reticulum, omasum, and abomasums. The rumen is a large fermentation chamber containing a very high population of microorganisms (bacteria and protozoa), which is responsible for much of the initial digestion of feed. Fat-soluble vitamins A, D, and E are absorbed with the help of dietary fats (lipids) in the intestinal tract and are then stored in the liver until needed. In small ruminants, nine water-soluble vitamins (eight B vitamins and vitamin C) and vitamin K are synthesized in the rumen or in body tissues. Microbes in the reticulorumen include eubacteria, protozoa, fungi, archaea, and viruses. In the omasum and abomasums, feedstuffs are continually exposed to microbial fermentation. Therefore, the influence of a complex microbiota composed of a large number of predominantly anaerobic bacteria, protozoa, and fungi located in the gastrointestinal tract has to be studied. Moreover, the transmission of foodborne pathogens has been linked to many common bovine diseases (Mathijs et al., 2012; Hussein and Omaye, 2003). Massive depth metagenomic sequencing techniques and "SEED" annotation (based on function roles; Overbeek et al., 2005) were used to gain detailed characterization of the bovine rumen microbiome (Brulc et al., 2009). Many members in the rumen microbial community can degrade cellulosic plant material, but most of them are unculturable. To characterize biomass-degrading genes and genomes, scientists also sequenced and analyzed more than 2 GB of metagenomic DNA from microbes adherent to plant fiber incubated in bovine rumen. More than 20,000 carbohydrate-active genes and 90 candidate proteins were identified to be potentially responsible for the deconstruction of cellulosic biomass (Hess et al., 2011).

Poultry Microbiota

Applying metagenomics to poultry microbiota research and animal production has received much interest in recent years, particularly due to restrictions and/or bans on farm use of antibiotic growth promoters (AGPs) in the European Union. The misuse and overuse of antibiotics not only affect animal growth, health, and well-being, but also may affect the safety and quality of food products. A recent publication on the distribution of bacterial phyla between chicken and human microbiomes demonstrated that *Bacteroides* species are more abundant in chicken and human distal gut microbiomes (Lamendella et al., 2011), implicating that *Bacteroides* may share similar cytotoxic and immune-stimulatory activities in both species. It is expected that a fundamental understanding of molecular/cellular mechanisms on the effects of dietary antibiotics on development of the chicken immune system will be greatly enhanced by metagenomic research.

Microbiota of Other Animals

Compared to the human microbiota, the majority of the mouse gut population is unique; however, the distal gut microbiota of mouse and human are very similar on the division level (Ley et al., 2005), which provides the best model for study-ing diabetes and obesity at the molecular and functional level. By analyzing the 16S rRNA genes, the gut microbiota of genetically obese mice showed reduced abundance (50%) in Bacteroidetes with a corresponding increase in the proportion of Firmicutes (Ley et al., 2005). It was recently reported that the gut microbiota of obese dogs are similar to that of humans, with the phylum Proteobacteria being the predominant group (76%), in contrast to the Firmicutes (85%) in the lean group (Park et al., 2014a,b). Currently, diabetes and obesity are major health risk factors for humans in developed countries. Better under-standing of the gut microbiota dynamics and their roles in nutrient utilization and influence on disease outcome in different animals and humans could greatly enhance future prospects of human and animal health in general.

Probiotics and Microbiota on Foods

Probiotics are defined by the Food and Agriculture Organization of the United Nations and the World Health Organization as "live microorganisms, which, when administered in adequate amounts, confer a health benefit on the host" (FAO/WHO, 2001). The probiotic benefits to the host may include: anti-infectious properties, immune modulation, metabolic effects, and alteration of intestinal mobility or function (Walker and Buckley, 2006). Over 40,000 bacteria are estimated to exist in the gastrointestinal microbiome of humans (Forsythe et al., 2010). Metagenomic applications in probiotic research will gain new insights into understanding the maintenance of an appropriate healthy gut microbiota composition and for the identification of novel probiotic compounds, which could be beneficial as a diet or food supplement (Parvaneh et al., 2014; Lodemann et al., 2008; Davis et al., 2008; Culligan et al., 2009; Saulnier et al., 2009). A recent study showed that gut microbes, particularly the Clostridia, can regulate immune responses by inducing interleukin-22 in mice and protect the host against challenge of peanut allergens and cholera toxin (Stefka et al., 2014).

Microbial Distribution on Dairy Products

By applying a metagenomic approach, Sun and colleagues took a phylogenetic sur-vey of the bacterial and fungal populations from tarag, the naturally fermented dairy products thought to have medicinal values in addition to being a food (Sun et al., 2014). They found a total of 47 bacterial genera (four phyla) with *Lactobacillus* being the predominant genus (43.7–67.2% of sequence reads); 14 genera were common in all samples. Two phyla of fungi were identified: Ascomycota (73.8–98.8%) and Basidiomycota (1–26%). The microbial diversity varied by the geographic regions where samples were collected. The rapidly expanding micro-bial genomic information obtained from other metagenomic studies can be applied to food microbiota to identify underrepresented genera/species carrying specific genes/products or even for discovery of novel probiotics strain(s).

Probiotic Dietary and Feed Supplements

Probiotic strains can be used as dietary and feed supplements to replace AGPs in animal diets, or as bio-therapeutic agents in cases of antibiotic-associated diarrhea with traveler's diarrhea, childhood diarrhea, and other bacterial gastrointestinal illnesses (Salim et al., 2013). The use of AGPs has been implicated with the emergence of antibiotic-resistant bacteria (Emborg et al., 2003). Using a culture-independent metagenomic-based approach, the studies on the impact of AGPs on the intestinal microbiota of different animals, including beef cattle, poultry, and swine have been initiated, which should greatly improve our understanding on the alterations of the distal intestinal microflora in response to AGPs (Lin, 2014; Kim et al., 2012). In the United States, most probiotic products are available as dietary supplements or foods, but are not used as a main source in food supplements (McFarland, 2014). The data obtained from metagenomic experiments related to probiotic applications suggest an impressive range of potential benefits (Gueimonde and Collado, 2012). However, for many of the potential benefits, only preliminary results are available due to limited research, and the effects can only be attributed to the particular strain(s) tested. Testing results of a specific supplement may not be extrapolated to any other strain of the same species, and do not imply that comparable benefits will be imparted from other probiotics. A comprehensive, unbiased metagenomic approach could be used to identify known or novel probiotic strains and study the effects of probiotic supplements on animal disease control and nutrition utilization efficiency.

Integration of metagenomics with other -omics technologies, such as metabolomics, transcriptomics, and proteomics, will be critical in furthering the field of food science as well as in other life sciences, achieving the goal of a better understanding of nutrient utilization and food safety. Metabolomics is the study of small molecule metabolites in an organism (Skogerson et al., 2010; Nicholson et al., 2008; Mashego et al., 2007). These small molecules include metabolic intermediates such as sugars, organic acids, essential amino acids, peptides, extra- and intracellular signaling molecules, and secondary metabolites, e.g., polyphenols, food additives, and phytochemicals in foods. Metabolomics can be used for quantitatively characterizing food/feedstock nutrition composition, food adulteration, quality and safety assessment, diet monitoring, measuring phenotype/genotype relationships arising from diet and microflora changes, and identifying novel biomarkers for bacterial typing and surveillance. Technically, metabolomics has revolutionized the application of traditional biochemical and biomedical research. Unlike the situation with genomics and metagenomics, metabolomics is not nearly as developed for food safety research. Nevertheless, mass spectrometry-based metabolomics could provide a unique opportunity to explore a systematic genotype–phenotype relationship and develop an indicative phenotyping and envirotype modeling program due to the specificity, sensitivity, and predictive value of these small molecule metabolites (Ideker et al., 2001; Han et al., 2014; Cascante and Marin, 2008). However, a key limitation of metabolomics for food safety research is that not all foodborne pathogen metabolomes are well characterized or studied due to the difficulty in the detection and identification of all of the metabolites and the genotypic diversity of foodborne pathogens (Dunn et al., 2005).

CHALLENGES AND FUTURE OUTLOOK

Since the establishment of the National Antimicrobial Resistance Monitoring System (NARMS; a coordinated effort between the state and local health departments, CDC, FDA, and USDA in the United States) in 1996, public awareness on antimicrobial resistance and resistant pathogens has increased. Just within the last two years (2013–2014) WHO, government agencies, professional societies, and interest groups published special reports on issues related to antimicrobial resistance to advocate responsible antibiotic stewardship (some examples include DeWaal and Grooters, 2013; CDC, 2013; WHO, 2014). A new journal "Antibiotics" focused on all aspects of antibiotics, including resistance, was established in 2012. Several specialized symposia and workshops on alternative antibiotics and alternative treatment applications were organized to explore new options, in addition to the regularly planned society meetings on antibiotic-related subjects. FDA announced the first progress report on its strategy to promote judicious use of antimicrobials in food-producing animals on June 30, 2014 (Guidance for Industry #213; http://www.fda.gov/AnimalVeterinary/NewsEvents/CVMUpdates/ucm403285.htm). In addition, the US White House issued an Executive Order on "Combating Antibiotic-Resistant Bacteria" in September 2014 (www.whitehouse.gov/the-press-office/2014/09/18/exccutive-order-combating-antibiotic-resistant-bacteria). The "National Strategy for Combating Antibiotic Resistant Bacteria" presented the road map for future antibiotic resistance research; the five goals are (i) slow the development of resistant bacteria and prevent the spread of resistant infections; (ii) strengthen national One-Health surveillance efforts to combat resistance; (iii) advance development and use of rapid and innovative diagnostic tests for identification and characterization of resistant bacteria; (iv) accelerate basic and applied research and development for new antibiotics, other therapeutics, and vaccines; (v) improve international collaboration and capacities for antibiotic resistance prevention, surveillance, control, and antibiotic research and development (www.whitehouse.gov/sites/default/files/docs/carb_national_strategy.pdf).

Challenges

The major challenges we are currently facing in antimicrobial resistance research (or food safety in general) can be divided into three categories:

1. **Spread and transmission of resistance genes and resistant bacteria.** Industrialization and modernization of societies is resulting in centralized

commercial food-processing practices. Globalization of the economy promotes import and export of food and other commodities worldwide, as well as frequent (global) traveling for business and leisure. All of these factors could contribute to the rapid spread of pathogens, as well as resistance genes/resistant bacteria.

2. **Research and development for new antimicrobials or alternative treatment options**. Few new antibiotics or antimicrobials are in development, and even fewer are approved for use. Research funding and opportunities are limited especially in developing countries where resistance appears to be rampant, thus requiring international collaboration and cooperation. The gap between laboratory research and real-world applications is often insurmountable.

3. **Information management and communication.** Surveillance/monitoring, risk assessment and modeling, as well as collection and analyses of all the "omics" and NGS data all require fast, and reliable network infrastructure for information exchange and sharing. Standardization of guidelines and procedures for data collection, analysis and validation, establishment of comprehensive databases, computational and bioinformatic analysis software, as well as fast, real-time communication of research results are also desirable.

Directions for Future Research

The future of antimicrobial resistance research should take a more holistic approach to understand the interaction between resistant bacteria and their environments in a community setting, ideally suited for applying the "meta-omics" technologies. The communication of surveillance and research results are also crucial. Some specific subject areas that would benefit include:

- Utilization of alternative treatments for infections or interventions for food processing—to reduce the dependence on traditional antibiotics. Nonconventional procedures or treatments such as phage therapy, non-traditional antimicrobials, and quorum sensing inhibitors may be useful.
- Application of probiotics/prebiotics in food animals and in humans—aimed to promote general health and well-being of animals or humans, thus reducing the necessity of antimicrobials in disease treatment or growth promotion.
- Development of new/alternative antimicrobials: natural antimicrobials, synthetic or "designer" antimicrobial compounds and peptides.
- Application of new sequencing and metagenomic technologies in detection, typing, and surveillance—to increase the knowledge base and build better, more comprehensive databases, as well as develop pipelines for rapid identification and analysis.
- Establishment of better network in data-sharing and communicating outbreaks and (multi)drug-resistant pathogen tracking, preferably near real-time.
- Effective use of "omics" research data: building comprehensive databases, developing computational analysis algorithms.

- Understanding the interaction between resistance genes/resistant bacteria and the environment—a thorough holistic evaluation could be possible with the availability and accessibility of meta-omic technologies.

Looking Forward: Preparing for the Future

Never before has molecular and genomic research been so promising for the life sciences. Technical advances in next-generation sequencing, informatics, and computer science present unprecedented opportunities for developing novel applications to improve human and animal health. This chapter summarizes studies utilizing metagenomics in development of antibiotic resistance, discovery of novel antimicrobials, and characterization of probiotics and gut microbiome research that demonstrated the potential of metagenomic research for the coming decades.

The application of meta-omic technologies will not only facilitate the identification of resistance genes/products, but will enhance the surveillance of pathogens and help outbreak investigations. It can be useful for projecting/predicting the evolution or transfer of antibiotic resistance and further enabling studies on the previously unidentified antibiotic resistance genetic elements, and potential routes/mechanisms of transmission, as well. This relatively new technology has the potential to provide an unbiased tool for understanding the interactions and relationships between microbes, their hosts, and the environment. For instance, progress in metagenomic research has demonstrated that the microbial diversity in nature is far greater than that reflected in laboratories. Moreover, traditional methods for searching/testing for new antimicrobial drugs are time-consuming and expensive. Using the vast amount of metagenomics data, one may apply appropriate algorithm and searching criteria *in silico* to identify potential enzymes/pathways that may lead to discovery of new antimicrobial compounds, or identify new drug targets. Finally, the application of metagenomic approaches in different ecological niches, such as foods, is a promising way to assess the potential of food microbiota for specific functions and to orient strategies for assessing the functionality of probiotic potentials. This may lead, in turn, to minimizing antibiotic usage in treating infections and/or in growth promotion of food animals.

ACKNOWLEDGMENT

We thank Drs. Joshua Gurtler, Robert Li, and Pina Fratamico for their comments and edits.

REFERENCES

Altschul, S.F., Gish, W., Miller, W., Myers, E.W., Lipman, D.J., 1990. Basic local alignment search tool. J. Mol. Biol. 215, 403–410.

Askenazi, M., Marto, J.A., Linial, M., 2010. The complete peptide dictionary—a meta-proteomics resource. Proteomics 10, 4306–4310.

Backhed, F., Ding, H., Wang, T., Hooper, L.V., Koh, G.Y., Nagy, A., et al., 2004. The gut microbiota as an environmental factor that regulates fat storage. Proc. Natl. Acad. Sci. USA. 101, 15718–15723.

Backhed, F., Ley, R.E., Sonnenburg, J.L., Peterson, D.A., Gordon, J.I., 2005. Host–bacterial mutualism in the human intestine. Science 307, 1915–1920.

Bekele, A.Z., Koike, S., Kobayashi, Y., 2011. Phylogenetic diversity and dietary association of rumen *Treponema* revealed using group-specific 16S rRNA gene-based analysis. FEMS Microbiol. Lett. 316, 51–60.

Berman, H.F., Riley, L.W., 2013. Identification of novel antimicrobial resistance genes from microbiota on retail spinach. BMC Microbiol. 13, 272.

Brulc, J.M., Antonopoulos, D.A., Miller, M.E., Wilson, M.K., Yannarell, A.C., Dinsdale, E.A., et al., 2009. Gene-centric metagenomics of the fiber-adherent bovine rumen microbiome reveals forage specific glycoside hydrolases. Proc. Natl. Acad. Sci. USA. 106, 1948–1953.

Cani, P.D., 2013. Gut microbiota and obesity: lessons from the microbiome. Brief Funct. Genomic. 12, 381–387.

Cascante, M., Marin, S., 2008. Metabolomics and fluxomics approaches. Essays Biochem. 45, 67–81.

Centers for Disease Control and Prevention (CDC), 2013. Antibiotic resistance threats in the United States, 2013 report. <www.cdc.gov/drugresistance/threat-report-2013/pdf/ar-threats-2013-508.pdf>.

Chen, P.E., Cook, C., Stewart, A.C., Nagarajan, N., Sommer, D.D., Pop, M., et al., 2010. Genomic characterization of the *Yersinia* genus. Genome Biol. 11, R1.

Clemente, J.C., Jansson, J., Valiente, G., 2010. Accurate taxonomic assignment of short pyrosequencing reads. Pac. Symp. Biocomput. 3–9.

Culligan, E.P., Hill, C., Sleator, R.D., 2009. Probiotics and gastrointestinal disease: successes, problems and future prospects. Gut Pathog 1, 19.

Culligan, E.P., Sleator, R.D., Marchesi, J.R., Hill, C., 2014. Metagenomics and novel gene discovery: promise and potential for novel therapeutics. Virulence 5, 399–412.

Cummings, C.A., Bormann Chung, C.A., Fang, R., Barker, M., Brzoska, P., Williamson, P.C., et al., 2010. Accurate, rapid and high-throughput detection of strain-specific polymorphisms in *Bacillus anthracis* and *Yersinia pestis* by next-generation sequencing. Investig. Genet. 1, 5.

Davis, M.E., Parrott, T., Brown, D.C., de Rodas, B.Z., Johnson, Z.B., Maxwell, C.V., et al., 2008. Effect of a *Bacillus*-based direct-fed microbial feed supplement on growth performance and pen cleaning characteristics of growing-finishing pigs. J. Anim. Sci. 86, 1459–1467.

DeWaal, C.S., Grooters, S.V., 2013. Antibiotic resistance in foodborne pathogens—a Center for Science in the Public Interest White Paper. <cspinet.org/new/pdf/outbreaks_antibiotic_resistance_in_foodborne_pathogens_2013.pdf>.

Donia, M.S., Cimermancic, P., Schulze, C.J., Wieland Brown, L.C., Martin, J., Mitreva, M., et al., 2014. A systematic analysis of biosynthetic gene clusters in the human microbiome reveals a common family of antibiotics. Cell 158, 1402–1414.

Dunn, W.B., Bailey, N.J., Johnson, H.E., 2005. Measuring the metabolome: current analytical technologies. Analyst 130, 606–625.

Durso, L.M., Harhay, G.P., Smith, T.P., Bono, J.L., Desantis, T.Z., Harhay, D.M., et al., 2010. Animal-to-animal variation in fecal microbial diversity among beef cattle. Appl. Environ. Microbiol. 76, 4858–4862.

Emborg, H.D., Andersen, J.S., Seyfarth, A.M., Andersen, S.R., Boel, J., Wegener, H.C., 2003. Relations between the occurrence of resistance to antimicrobial growth promoters among *Enterococcus faecium* isolated from broilers and broiler meat. Int. J. Food. Microbiol. 84, 273–284.

FAO/WHO, 2001. Evaluation of Health and Nutritional Properties of Probiotics in Food, Including Powder Milk with Live Lactic Acid Bacteria. Food and Agriculture Organization of the United Nations and World Health Organization Expert Consultation Report.

Forsythe, P., Sudo, N., Dinan, T., Taylor, V.H., Bienenstock, J., 2010. Mood and gut feelings. Brain Behav. Immun. 24, 9–16.

Fridman, O., Goldberg, A., Ronin, I., Shoresh, N., Balaban, N.Q., 2014. Optimization of lag time underlies antibiotic tolerance in evolved bacterial populations. Nature 513, 418–421. Available from: http://dx.doi.org/10.1038/nature13469.

Galbraith, E.A., Antonopoulos, D.A., White, B.A., 2004. Suppressive subtractive hybridization as a tool for identifying genetic diversity in an environmental metagenome: the rumen as a model. Environ. Microbiol. 6, 928–937.

Ghosh, T.S., Gupta, S.S., Bhattacharya, T., Yadav, D., Barik, A., Chowdhury, A., et al., 2014. Gut microbiomes of Indian children of varying nutritional status. PLoS One 9, e95547.

Gloux, K., Berteau, O., El Oumami, H., Beguet, F., Leclerc, M., Dore, J., 2011. A metagenomic beta-glucuronidase uncovers a core adaptive function of the human intestinal microbiome. Proc. Natl. Acad. Sci. USA. 108 (Suppl. 1), 4539–4546.

Goodrich, J.K., Di Rienzi, S.C., Poole, A.C., Koren, O., Walters, W.A., Caporaso, J.G., et al., 2014a. Conducting a microbiome study. Cell 158, 250–262.

Goodrich, J.K., Waters, J.L., Poole, A.C., Sutter, J.L., Koren, O., Blekhman, R., et al., 2014b. Human genetics shape the gut microbiome. Cell 159, 789–799.

Guarner, F., Malagelada, J.R., 2003. Role of bacteria in experimental colitis. Best Pract. Res. Clin. Gastroenterol. 17, 793–804.

Gueimonde, M., Collado, M.C., 2012. Metagenomics and probiotics. Clin. Microbiol. Infect. 18 (Suppl. 4), 32–34.

Han, Y., Li, L., Zhang, Y., Yuan, H., Ye, L., Zhao, J., et al., 2014. Phenomics of vascular disease: the systematic approach to the combination therapy. Curr. Vasc. Pharmacol. Available from: http://dx.doi.org/10.2174/1570161112666141014144829.

Handelsman, J., Rondon, M.R., Brady, S.F., Clardy, J., Goodman, R.M., 1998. Molecular biological access to the chemistry of unknown soil microbes: a new frontier for natural products. Chem. Biol. 5, R245–R249.

Hess, D.J., Henry-Stanley, M.J., Wells, C.L., 2011. Gentamicin promotes *Staphylococcus aureus* biofilms on silk suture. J. Surg. Res. 170, 302–308.

HMPC (The Human Microbiome Project Consortium), 2012a. Structure, function and diversity of the healthy human microbiome. Nature 486, 207–214.

HMPC (The Human Microbiome Project Consortium), 2012b. A framework for human microbiome research. Nature 486, 215–221.

Hongoh, Y., 2011. Toward the functional analysis of uncultivable, symbiotic microorganisms in the termite gut. Cell Mol. Life Sci. 68, 1311–1325.

Howden, B.P., Beaume, M., Harrison, P.F., Hernandez, D., Schrenzel, J., Seemann, T., et al., 2013. Analysis of the small RNA transcriptional response in multidrug-resistant *Staphylococcus aureus* after antimicrobial exposure. Antimicrob. Agents Chemother. 57, 3864–3874. Available from: http://dx.doi.org/10.1128/AAC.00263-13.

Hussein, H.S., Omaye, S.T., 2003. Introduction to the food safety concerns of verotoxin-producing *Escherichia coli*. Exp. Biol. Med. (Maywood) 228, 331–332.

Ideker, T., Thorsson, V., Ranish, J.A., Christmas, R., Buhler, J., Eng, J.K., et al., 2001. Integrated genomic and proteomic analyses of a systematically perturbed metabolic network. Science 292, 929–934.

Iguchi, A., Iyoda, S., Kikuchi, T., Ogura, Y., Katsura, K., Ohnishi, M., et al., 2014. A complete view of the genetic diversity of the *Escherichia coli* O-antigen biosynthesis gene cluster. DNA Res.

Joeres-Nguyen-Xuan, T.H., Boehm, S.K., Joeres, L., Schulze, J., Kruis, W., 2010. Survival of the probiotic *Escherichia coli* Nissle 1917 (EcN) in the gastrointestinal tract given in combination with oral mesalamine to healthy volunteers. Inflamm. Bowel Dis. 16, 256–262.

Kim, H.B., Borewicz, K., White, B.A., Singer, R.S., Sreevatsan, S., Tu, Z.J., et al., 2012. Microbial shifts in the swine distal gut in response to the treatment with antimicrobial growth promoter, tylosin. Proc. Natl. Acad. Sci. USA. 109, 15485–15490.

Kittelmann, S., Seedorf, H., Walters, W.A., Clemente, J.C., Knight, R., Gordon, J.I., et al., 2013. Simultaneous amplicon sequencing to explore co-occurrence patterns of bacterial, archaeal and eukaryotic microorganisms in rumen microbial communities. PLoS One 8, e47879.

Koch, G., Yepes, A., Förstner, K.U., Wermser, C., Stengel, S.T., Modamio, J., et al., 2014. Evolution of resistance to a last-resort antibiotic in *Staphylococcus aureus* via bacterial competition. Cell 158, 1060–1071.

Kong, Y., Teather, R., Forster, R., 2010. Composition, spatial distribution, and diversity of the bacterial communities in the rumen of cows fed different forages. FEMS Microbiol. Ecol. 74, 612–622.

Laehnemann, D., Pena-Miller, R., Rosenstiel, P., Beardmore, R., Jansen, G., Schulenburg, H., 2014. Genomics of rapid adaptation to antibiotics: convergent evolution and scalable sequence amplification. Genome Biol. Evol. 6, 1287–1301.

Lamendella, R., Domingo, J.W., Ghosh, S., Martinson, J., Oerther, D.B., 2011. Comparative fecal metagenomics unveils unique functional capacity of the swine gut. BMC Microbiol. 11, 103.

Ley, R.E., Backhed, F., Turnbaugh, P., Lozupone, C.A., Knight, R.D., Gordon, J.I., 2005. Obesity alters gut microbial ecology. Proc. Natl. Acad. Sci. USA. 102, 11070–11075.

Ley, R.E., Turnbaugh, P.J., Klein, S., Gordon, J.I., 2006. Microbial ecology: human gut microbes associated with obesity. Nature 444, 1022–1023.

Lievin-Le Moal, V., Servin, A.L., 2006. The front line of enteric host defense against unwelcome intrusion of harmful microorganisms: mucins, antimicrobial peptides, and microbiota. Clin. Microbiol. Rev. 19, 315–337.

Lin, J., 2014. Antibiotic growth promoters enhance animal production by targeting intestinal bile salt hydrolase and its producers. Front. Microbiol. 5, 33.

Lodemann, U., Lorenz, B.M., Weyrauch, K.D., Martens, H., 2008. Effects of *Bacillus cereus* var. toyoi as probiotic feed supplement on intestinal transport and barrier function in piglets. Arch. Anim. Nutr. 62, 87–106.

Marcinakova, M., Klingberg, T.D., Laukova, A., Budde, B.B., 2010. The effect of pH, bile and calcium on the adhesion ability of probiotic enterococci of animal origin to the porcine jejunal epithelial cell line IPEC-J2. Anaerobe 16, 120–124.

Marco, D., 2008. Metagenomics and the niche concept. Theory Biosci. 127, 241–247.

Martin, M., Biver, S., Steels, S., Barbeyron, T., Jam, M., Portetelle, D., et al., 2014. Identification and characterization of a halotolerant, cold-active marine endo-beta-1,4-glucanase by using functional metagenomics of seaweed-associated microbiota. Appl. Environ. Microbiol. 80, 4958–4967.

Mashego, M.R., Rumbold, K., De Mey, M., Vandamme, E., Soetaert, W., Heijnen, J.J., 2007. Microbial metabolomics: past, present and future methodologies. Biotechnol. Lett. 29, 1–16.

Mathijs, E., Stals, A., Baert, L., Botteldoorn, N., Denayer, S., Mauroy, A., et al., 2012. A review of known and hypothetical transmission routes for noroviruses. Food Environ. Virol. 4, 131–152.

McFarland, L.V., 2014. Use of probiotics to correct dysbiosis of normal microbiota following disease or disruptive events: a systematic review. BMJ Open 4, e005047.

McHardy, I.H., Goudarzi, M., Tong, M., Ruegger, P.M., Schwager, E., Weger, J.R., et al., 2013. Integrative analysis of the microbiome and metabolome of the human intestinal mucosal surface reveals exquisite inter-relationships. Microbiome 1, 17.

Messelhausser, U., Kampf, P., Colditz, J., Bauer, H., Schreiner, H., Holler, C., et al., 2011. Qualitative and quantitative detection of human pathogenic *Yersinia enterocolitica* in different food matrices at retail level in Bavaria. Foodborne Pathog. Dis. 8, 39–44.

Miller, R.R., Montoya, V., Gardy, J.L., Patrick, D.M., Tang, P., 2013. Metagenomics for pathogen detection in public health. Genome Med. 5, 81.

Morelli, G., Song, Y., Mazzoni, C.J., Eppinger, M., Roumagnac, P., Wagner, D.M., et al., 2010. *Yersinia pestis* genome sequencing identifies patterns of global phylogenetic diversity. Nat. Genet. 42, 1140–1143.

Naccache, S.N., Federman, S., Veeraraghavan, N., Zaharia, M., Lee, D., Samayoa, E., et al., 2014. A cloud-compatible bioinformatics pipeline for ultrarapid pathogen identification from next-generation sequencing of clinical samples. Genome Res. 24, 1180–1192.

Nicholson, J.K., Holmes, E., Elliott, P., 2008. The metabolome-wide association study: a new look at human disease risk factors. J. Proteome Res. 7, 3637–3638.

Oh, S., Caro-Quintero, A., Tsementzi, D., DeLeon-Rodriguez, N., Luo, C., Poretsky, R., et al., 2011. Metagenomic insights into the evolution, function, and complexity of the planktonic microbial community of Lake Lanier, a temperate freshwater ecosystem. Appl. Environ. Microbiol. 77, 6000–6011.

Ohlrich, E.J., Cullinan, M.P., Leichter, J.W., 2010. Diabetes, periodontitis, and the subgingival microbiota. J. Oral. Microbiol. 2.

Overbeek, R., Begley, T., Butler, R.M., Choudhuri, J.V., Chuang, H.Y., Cohoon, M., et al., 2005. The subsystems approach to genome annotation and its use in the project to annotate 1000 genomes. Nucleic Acids Res. 33, 5691–5702.

Park, H.J., Lee, S.E., Kim, H.B., Isaacson, R.E., Seo, K.W., Song, K.H., 2014a. Association of obesity with serum leptin, adiponectin, and serotonin and gut microflora in beagle dogs. J. Vet. Intern. Med. Available from: http://dx.doi.org/10.1111/jvim.12455 (e-Pub ahead of print).

Park, H.J., Lee, S.E., Oh, J.H., Seo, K.W., Song, K.H., 2014b. Leptin, adiponectin and serotonin levels in lean and obese dogs. BMC Vet. Res. 10, 113.

Parvaneh, K., Jamaluddin, R., Karimi, G., Erfani, R., 2014. Effect of probiotics supplementation on bone mineral content and bone mass density. Sci. World J. 2014, 595962.

Paster, B.J., Dewhirst, F.E., Weisburg, W.G., Tordoff, L.A., Fraser, G.J., Hespell, R.B., et al., 1991. Phylogenetic analysis of the spirochetes. J. Bacteriol. 173, 6101–6109.

Petrosino, J.F., Highlander, S., Luna, R.A., Gibbs, R.A., Versalovic, J., 2009. Metagenomic pyrosequencing and microbial identification. Clin. Chem. 55, 856–866.

Price, L.B., Liu, C.M., Melendez, J.H., Frankel, Y.M., Engelthaler, D., Aziz, M., et al., 2009. Community analysis of chronic wound bacteria using 16S rRNA gene-based pyrosequencing: impact of diabetes and antibiotics on chronic wound microbiota. PLoS One 4, e6462.

Raes, J., Bork, P., 2008. Molecular eco-systems biology: towards an understanding of community function. Nat. Rev. Microbiol. 6, 693–699.

Reilly, K., Attwood, G.T., 1998. Detection of *Clostridium proteoclasticum* and closely related strains in the rumen by competitive PCR. Appl. Environ. Microbiol. 64, 907–913.

Russell, J.B., Cotta, M.A., Dombrowski, D.B., 1981. Rumen bacterial competition in continuous culture: *Streptococcus bovis* versus *Megasphaera elsdenii*. Appl. Environ. Microbiol. 41, 1394–1399.

Salim, H.M., Kang, H.K., Akter, N., Kim, D.W., Kim, J.H., Kim, M.J., et al., 2013. Supplementation of direct-fed microbials as an alternative to antibiotic on growth performance, immune response, cecal microbial population, and ileal morphology of broiler chickens. Poult. Sci. 92, 2084–2090.

Saulnier, D.M., Spinler, J.K., Gibson, G.R., Versalovic, J., 2009. Mechanisms of probiosis and prebiosis: considerations for enhanced functional foods. Curr. Opin. Biotechnol. 20, 135–141.

Saunders, N.J., Trivedi, U.H., Thomson, M.L., Doig, C., Laurenson, I.F., Blaxter, M.L., 2011. Deep resequencing of serial sputum isolates of *Mycobacterium tuberculosis* during therapeutic failure due to poor compliance reveals stepwise mutation of key resistance genes on an otherwise stable genetic background. J. Infect. 62, 212–217. Available from: http://dx.doi.org/10.1016/j.jinf.2011.01.003.

Schloss, P.D., Handelsman, J., 2005. Metagenomics for studying unculturable microorganisms: cutting the Gordian knot. Genome Biol. 6, 229.

Singh, B., Bhat, T.K., Kurade, N.P., Sharma, O.P., 2008. Metagenomics in animal gastrointestinal ecosystem: a microbiological and biotechnological perspective. Indian J. Microbiol. 48, 216–227.

Siqueira Jr., J.F., Rocas, I.N., 2010. The oral microbiota: general overview, taxonomy, and nucleic acid techniques. Methods Mol. Biol. 666, 55–69.

Skogerson, K., Harrigan, G.G., Reynolds, T.L., Halls, S.C., Ruebelt, M., Iandolino, A., et al., 2010. Impact of genetics and environment on the metabolite composition of maize grain. J. Agric. Food Chem. 58, 3600–3610.

Stefka, A.T., Feehley, T., Tripathi, P., Qiu, J., McCoy, K., Mazmanian, S.K., et al., 2014. Commensal bacteria protect against food allergen sensitization. Proc. Natl. Acad. Sci. USA. 111, 13145–13150.

Sun, Z., Liu, W., Bao, Q., Zhang, J., Hou, Q., Kwok, L., et al., 2014. Investigation of bacterial and fungal diversity in tarag using high-throughput sequencing. J. Dairy Sci. 97, 6085–6096.

Turnbaugh, P.J., Gordon, J.I., 2009. The core gut microbiome, energy balance and obesity. J. Physiol. 587, 4153–4158.

Ubeda, C., Taur, Y., Jenq, R.R., Equinda, M.J., Son, T., Samstein, M., et al., 2010. Vancomycin-resistant *Enterococcus* domination of intestinal microbiota is enabled by antibiotic treatment in mice and precedes bloodstream invasion in humans. J. Clin. Invest. 120, 4332–4341. Available from: http://dx.doi.org/10.1172/JCI43918.

Walker, R., Buckley, M., 2006. Probiotic microbes: the scientific basis. A report from The American Academy of Microbiology.

Wang, Y., Qian, P.Y., 2009. Conservative fragments in bacterial 16S rRNA genes and primer design for 16S ribosomal DNA amplicons in metagenomic studies. PLoS One 4, e7401.

Warnecke, F., Luginbuhl, P., Ivanova, N., Ghassemian, M., Richardson, T.H., Stege, J.T., et al., 2007. Metagenomic and functional analysis of hindgut microbiota of a wood-feeding higher termite. Nature 450, 560–565.

World Health Organization, 2014. Antimicrobial resistance: global report on surveillance. <www.who.int.drugresistance/documents/surveillancereport/en/>.

Yang, X., Xie, L., Li, Y., Wei, C., 2009. More than 9,000,000 unique genes in human gut bacterial community: estimating gene numbers inside a human body. PLoS One 4, e6074.

Yelton, A.P., Comolli, L.R., Justice, N.B., Castelle, C., Denef, V.J., Thomas, B.C., et al., 2013. Comparative genomics in acid mine drainage biofilm communities reveals metabolic and structural differentiation of co-occurring archaea. BMC Genomics 14, 485.

Zhu, W., Lomsadze, A., Borodovsky, M., 2010. Ab initio gene identification in metagenomic sequences. Nucleic Acids Res. 38, e132.

Zoetendal, E.G., Akkermans, A.D., De Vos, W.M., 1998. Temperature gradient gel electrophoresis analysis of 16S rRNA from human fecal samples reveals stable and host-specific communities of active bacteria. Appl. Environ. Microbiol. 64, 3854–3859.

Index

Note: Page numbers followed by *"f " and "t"* refer to figures and tables respectively.

Printed in the United States
by Bookmasters

Printed in the United States
By Bookmasters